INTRODUCTION TO MOLECULAR ENERGIES AND SPECTRA

MARLIN D. HARMONY

The University of Kansas

HOLT, RINEHART AND WINSTON, INC.
New York   Chicago   San Francisco   Atlanta
Dallas   Montreal   Toronto   London   Sydney

PREFACE

In writing this book I have attempted to put into one volume the material that I have taught at various times in courses on molecular spectroscopy. The impetus has been the feeling that no single text has adequately covered the major areas of molecular spectroscopy in a uniform and systematic fashion. I have perhaps erred in not discussing certain fields, such as NQR or Mössbauer spectroscopy, but the intent has been to treat those fields which, in my opinion, have made the greatest impact upon chemistry. During the task of writing this text I have been appreciative of the many fine books that have been written in the field of molecular spectroscopy, particularly the three-volume set by Herzberg, and the monographs by Townes and Schawlow, and Pople, Schneider, and Bernstein. The specialist will find these books to be of foremost importance, but it is hoped that the unified view of molecular spectroscopy presented in this text will be of value to the nonspecialist and the specialist wishing information about other areas.

The book is intended to be used as an introductory text by students who have had the equivalent of an elementary course in quantum mechanics, which now-a-days may be one semester of a typical two or three semester course in physical chemistry at the undergraduate level. Basically, it is assumed that the reader is acquainted with standard wave mechanics including perturbation theory. At the University of Kansas a course on molecular spectroscopy follows a senior–graduate level course on quantum chemistry, and is populated primarily by first- and second-year graduate students.

The first six chapters of the book are introductory in one way or another. Chapter 2 is meant to key the reader to the quantum mechanical approaches and symbolism to be followed in the text, particularly the matrix mechanics method. Chapter 3 sets out the general nature of the molecular energy problems to be dealt with in later chapters, and chapter 4 provides the solutions to two quantum mechanical problems which find extensive use in later chapters. In chapter 5, a brief treatment of selection rule theory has been given, since my experience has been that this topic is usually covered only very cursorily in beginning quantum theory courses. Chapter 6 on group theory is provided because of the belief that no student should study molecular spectroscopy without the benefits of its powerful and easily applied logic. Indeed group theory is used extensively in chapters 7–11. A student with a good background in quantum theory may find that only chapters 3 and 6 need to be studied before beginning chapters 7–11.

The remaining five chapters contain the pure spectroscopy material, covering microwave, infrared, NMR, ESR and electronic spectroscopy in that order. Each chapter takes its beginning principles from chapter 3 and makes use of results from other of the introductory chapters, but is otherwise nearly complete in itself. In each spectroscopic chapter except the last, a brief discussion of the instrumental and experimental methods is given, but the emphasis is placed upon the elucidation of the energy levels and selection rules, and the interpretation of spectra via the principles of quantum mechanics. In all cases the aim has been to show clearly the physical basis of the spectra, and the ways in which the spectra lead to information about molecular properties.

In contrast to most general spectroscopy books, no treatment of atomic spectra has been given, although some of the principles appear when needed. It is hoped that the general principles of atomic spectra will be known to the student using this book, but the lack of such background should not be particularly detrimental.

I have one principal apology to the reader, namely the omission of a discussion of Raman spectroscopy. The field is important and expanding, so I hope the interested reader will search out Herzberg's Volume 2. Of course, much of the material presented here in Chapter 8 is useful for this field.

My appreciation goes to many people, particularly the Berkeley Chemistry Department staff, to whom I owe my spectroscopic spirit; to my first six graduate students, LMW, MSD, GL, YSL, KWC and DKH, who struggled along with me during the first several years at Kansas; and to those later students who continue to stimulate me. I have received helpful advice from numerous people including Professor Christoffersen, Professor Klemperer and Professor Flygare, and have been constantly encouraged by Professor Kleinberg. The professional work of the H. R. & W. staff is gratefully acknowledged.

Finally, I thank my parents and my wife, JAH, without whose cheerful encouragement and great forebearance this book might not have reached the press.

Marlin D. Harmony

Lawrence, Kansas
April 1971

CONTENTS

Preface   *iii*
List of Tables   *ix*

1   INTRODUCTION   *1*
   1.1   Scope of Molecular Spectroscopy   *1*
   1.2   Emission and Absorption Spectroscopy   *2*
   1.3   Resonance Spectroscopy   *3*
   1.4   Electromagnetic Radiation   *3*
   1.5   Applications of Molecular Spectroscopy to Chemistry and Physics   *6*

2   QUANTUM MECHANICS AND SPECTROSCOPY   *9*
   2.1   Relevance of Quantum Mechanics to Spectroscopy   *9*
   2.2   Wave and Matrix Mechanics   *10*
   2.3   Relationship of Matrix and Wave Mechanics   *15*
   2.4   Intrinsic Spin   *17*
   2.5   Symmetry Properties of Wave Functions; Indistinguishability of Particles   *20*

3   NATURE AND TREATMENT OF THE MOLECULAR ENERGY TERMS   *23*
   3.1   General Molecular Hamiltonian; Separation of Translational Energy   *23*
   3.2   Molecular Born–Oppenheimer Approximation; Separation of Electronic Energies   *28*
   3.3   Separation of Vibrational and Rotational Energies   *33*
   3.4   Other Molecular Energy Terms, Including External-Field Terms; Perturbations   *46*

4   ANGULAR MOMENTUM AND HARMONIC OSCILLATION IN MATRIX MECHANICS   *53*
   4.1   Harmonic Oscillator in One Dimension   *53*
   4.2   Angular Momentum Commutation Relations   *65*

4.3   Matrix Elements of Angular Momentum   *69*
4.4   Properties of the Direction Cosines   *78*

5   ABSORPTION OF ELECTROMAGNETIC RADIATION   *85*
5.1   Calculation of Transition Probabilities by Time-Dependent Perturbation Theory   *86*
5.2   Transition Probability in a Periodic Perturbation   *90*
5.3   Electric Dipole Transition Probability   *95*
5.4   Some More General Considerations   *99*
5.5   Relationship of Transition Probability to Observable Quantities   *103*

6   GROUP THEORY AND ITS RELATIONSHIP TO QUANTUM MECHANICS AND SPECTROSCOPY   *109*
6.1   Symmetry Operations and Transformations about a Point   *111*
6.2   Some Properties of Groups   *116*
6.3   Theory of Matrix Representations of Groups   *122*
6.4   Application of Group Theory to Quantum Mechanics   *136*

7   MICROWAVE SPECTROSCOPY   *147*
7.1   General Characteristics and Experimental Aspects of Microwave Spectroscopy   *147*
7.2   Principal Axes   *153*
7.3   Microwave Spectrum of the Linear Molecule   *158*
7.4   Microwave Spectrum of the Symmetric Rotor Molecule   *177*
7.5   Microwave Spectrum of the Asymmetric Rotor Molecule   *184*
7.6   Stark Effect of Linear and Symmetric-Rotor Molecules   *204*
7.7   Quadrupole Hyperfine Interaction   *209*
7.8   Hindered Internal Rotation   *215*
7.9   Structural Determination   *218*

8   INFRARED SPECTROSCOPY   *225*
8.1   General Characteristics and Experimental Aspects of Infrared Spectroscopy   *225*
8.2   Infrared Spectra of Diatomic Molecules   *230*
8.3   Infrared Spectra of Polyatomic Molecules   *247*
8.4   Group-Theoretical Aspects of Molecular Vibrations   *263*
8.5   Normal Coordinate Analysis   *282*
8.6   Some Qualitative Aspects of Vibrational Analyses   *305*

9   NUCLEAR MAGNETIC RESONANCE SPECTROSCOPY   *313*
9.1   General Characteristics and Experimental Aspects   *313*
9.2   Behavior of Nuclei in Magnetic Fields   *318*
9.3   General Features of Liquid and Gas Phase NMR Spectra   *327*

9.4   Time-Dependent Phenomena *344*
9.5   Quantum Mechanical Treatment of Complex NMR Spectra in Liquids *360*
9.6   Theory of the Chemical Shift and Spin–Spin Interaction *379*
9.7   Other Nuclear Magnetic Resonance Techniques *387*

10   ELECTRON PARAMAGNETIC RESONANCE *393*
10.1   General Considerations *393*
10.2   Energy Levels and Spectrum of the Hydrogen Atom *397*
10.3   Spectra of Organic Radicals in Liquid Solutions *403*
10.4   Organic Radicals With Triplet States *410*
10.5   Electron Paramagnetic Resonance of Transition Metal Complexes *420*
10.6   Time-Dependent Phenomena *435*
10.7   Theoretical Interpretation of EPR Parameters *437*
10.8   Other EPR Studies *444*

11   ELECTRONIC SPECTROSCOPY *447*
11.1   Introduction *447*
11.2   Electronic Spectra of Diatomic Molecules *450*
11.3   Electronic States of Diatomic Molecules *464*
11.4   Theory of the Electronic Energies of Diatomic Molecules *478*
11.5   Electronic Spectra of Polyatomic Molecules *485*
11.6   Molecular Orbitals and Spectra of Polyatomic Molecules *492*
11.7   Spectra of Transition Metal Complexes *506*

APPENDICES
Appendix A   Classical Mechanics *513*
Appendix B   Some Frequently Used Quantum Mechanical Theorems *519*
Appendix C   Properties of Matrices *523*
Appendix D   Sign Change in Angular Momentum Commutation Relations *531*
Appendix E   Character Tables *535*
Appendix F   Atomic Masses and Natural Abundances of the Stable Isotopes of Elements with Atomic Number $\leq$ 54 *545*
Appendix G   Physical Constants and Conversion Factors *551*
Appendix H   Derivation of Eq. (8.128) *553*

INDEX *559*

LIST OF TABLES

Table 3.1   "Small" energy terms observable in molecular spectra. *49*

Table 4.1   Matrix elements of angular momentum. *76*

Table 4.2   Direction cosine matrix elements. *82*

Table 6.1   Symmetry elements and point groups for some representative molecules. *113*

Table 6.2   Group multiplication table for C_{2h} group. *117*

Table 6.3   Multiplication table for C_{3v} group. *119*

Table 6.4   Rudimentary character tables for C_{2h} and C_{3v} groups. *128*

Table 6.5   Development of character table for D_4 group. *130*

Table 6.6   Results of applying operations of D_3 group to ϕ_a of Fig. 6.5. *144*

Table 7.1   Comparison of measured B_v values of $Cs^{133}Cl^{35}$ with those calculated from Eq. (7.25). *161*

Table 7.2   Parameters of diatomic and triatomic molecules. *162*

Table 7.3   Experimental microwave spectrum of cesium bromide. *169*

Table 7.4   Some observed transition frequencies and computed intensities of OCS. *171*

Table 7.5   Rotational transitions of fluoroacetylene in ground vibrational state. *173*

Table 7.6   Rotational constants obtained by analysis of data in Table 7.5. *174*

Table 7.7   Experimental atomic coordinates and bond distances in fluoroacetylene. *176*

Table 7.8   Rotational parameters for some symmetric rotor molecules. *180*

Table 7.9   Microwave spectrum of trifluoromethyl acetylene. *182*

Table 7.10 Asymmetric rotor matrix elements for $J = 0$, 1, and 2 states. *191*

Table 7.11 Asymmetric rotor energy levels which arise from linear or quadratic secular equations. *192*

Table 7.12 Character table for D_2 or V in asymmetric rotor notation. *196*

Table 7.13 Symmetry selection rules for asymmetric rotor transitions. *199*

Table 7.14 Low-J asymmetric-rotor transitions. *199*

Table 7.15 Rotational transitions of bicyclobutane (C_4H_6) in MHz. *200*

Table 7.16 Observed transitions and rotational constants of SF_4. *201*

Table 7.17 Representative quadrupole coupling constants. *213*

Table 7.18 Barriers to internal rotation from microwave spectroscopy. *219*

Table 7.19 r_0 and r_s structures for carbonyl sulfide (OCS). *220*

Table 8.1 Energy levels of HCl^{35}. *232*

Table 8.2 Vibrational parameters for diatomic molecules. *234*

Table 8.3 Vibrational fundamentals of diatomic molecules observed in the infrared spectrum. *239*

Table 8.4 A portion of the infrared rotation-vibration band for the $v = 0 \rightarrow v = 1$ transition of HCl^{35}. *246*

Table 8.5 Some observed infrared bands of N_2O. *260*

Table 8.6 Some observed infrared bands of CS_2. *261*

Table 8.7 Frequencies of the Q branch subbands of ν_1 of NF_2. *263*

Table 8.8 Contribution to the character of the $3N$-dimensional representation per unshifted atom. *269*

Table 8.9 Some observed infrared bands of NH_3 in the gas phase. *283*

Table 8.10 Some observed infrared bands of gaseous acetylene (C_2H_2). *283*

Table 8.11 Typical bond stretching and bending force constants. *306*

Table 8.12 Typical group frequencies in cm^{-1}. *308*

Table 9.1 Properties of some common nuclei. *319*

Table 9.2 Representative coupling constants involving proton, fluorine and C^{13} nuclei. *340*

Table 9.3 Frequencies and relative intensities of transitions for an AB system of spin-$\frac{1}{2}$ nuclei. *370*

Table 9.4 Symmetry functions and their $M_{F,T}$ values for an AB_2 system of protons. *376*

Table 10.1 Representative parameters for triplet radicals. *419*

Table 10.2 Representative spin-orbit coupling constants and crystal-field splitting parameters. *426*

Table 10.3 Representative g values for transition-metal complexes. *433*

Table 10.4 Experimental values of the isotropic Fermi contact coupling constant. *441*

Table 10.5 g values of organic radicals in liquid solution. *444*

Table 11.1 Allowed transitions for diatomic molecules. *453*

Table 11.2 Electronic spectral data for O_2^{16}. *455*

Table 11.3 Measured bandheads in the $v'' = 0$ progression in the $X \rightarrow B$ electronic transition of molecular iodine. *462*

Table 11.4   Configurations and terms of some homonuclear diatomic molecules. *476*

Table 11.5   LCAO–MO–SCF results for the fluorine molecule. *480*

Table 11.6   Observed electronic states and transitions of ammonia. *488*

Table 11.7   Ground-state configurations of some non-linear XH_2 molecules. *491*

Table 11.8   Selection rules for some low-lying states of formaldehyde. *496*

Table 11.9   Carbonyl absorption frequencies and intensities. *497*

Table 11.10   Characteristic electronic transitions of simple structural units. *498*

Table 11.11   Characters of representations of S, P, D, F states in an octahedral field. *508*

Table 11.12   Characters of representations of S, P, D, F states in a tetragonal (D_{4h}) field. *509*

Table 11.13   Tables for reducing representations of O under the group D_4. *510*

CHAPTER 1

INTRODUCTION

1.1 SCOPE OF MOLECULAR SPECTROSCOPY

Few developments have had such a revolutionary effect on chemistry as have those in the area of spectroscopy. Whereas the chemist of the early 20th century was often content with the spectroscopic knowledge that sodium produced a yellow flame and potassium a violet one, the chemist of the present day has become beseiged with an often bewildering array of spectroscopic methods scanning an extremely broad range of wavelength and frequency. These diverse spectroscopic techniques owe their development and present widespread use to several factors. The first of these occurred in the late 1920s with the advent of quantum mechanics, a new theory which produced a firm basis for describing the absorption and emission of electromagnetic radiation. The second important factor was the improvement of technology, particularly in the area of electronics, during and after the Second World War. And finally must be listed the chemist's recently developed inclination to pursue his discipline on the *microscopic*, rather than on the *macroscopic*, level.

For the purposes of this book, we define spectroscopy as the science dealing with the absorption or emission of electromagnetic radiation by matter. More explicitly, the absorption or emission is governed by the Bohr frequency relationship

$$\nu = \frac{E_2 - E_1}{h} \tag{1.1}$$

where E_2 and E_1 are upper and lower stationary energy states, ν is the frequency of the radiation, and h is Planck's constant. By means of this traditional definition, it is clear that mass spectroscopy (or spectrometry) is not included in the subject matter of this book; nor are most of the various light scattering experiments, although the inelastic Raman scattering technique does come within our definition.

It follows that molecular spectroscopy is that branch of spectroscopy which deals with molecular energy states as opposed to atomic or nuclear ones. But the definition is rather broadly interpreted to include spectra whose origins are largely nuclear (such as nuclear magnetic resonance) or atomic (such as transition metal complex spectra). To be more specific, we treat primarily the following areas of molecular spectroscopy:

(1) Rotational
(2) Vibrational
(3) Electronic
(4) Electron spin resonance
(5) Nuclear magnetic resonance

1.2 EMISSION AND ABSORPTION SPECTROSCOPY

Two distinct methods can be utilized for spectroscopic studies. In one, atoms or molecules are produced in excited states, and the characteristic *emission* spectrum is observed as the species return to the ground state. In the second method, electromagnetic radiation is passed through a sample and the absorption of radiation as a function of frequency is noted. These two methods are illustrated very schematically in Figs. 1.1(a) and (b).

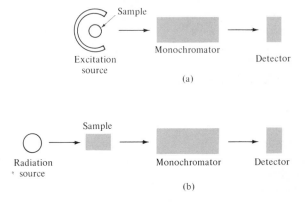

FIG. 1.1.   Schematic representation of (a) emission spectroscopy and (b) absorption spectroscopy.

The theory defining the absorption and emission processes is covered in some detail in chapter 5, but it may be mentioned here that standard emission spectroscopy refers to the *spontaneous* emission of radiation as opposed to *stimulated* emission. Stimulated emission processes have recently become important as the source of *maser* and *laser* action.

It should also be mentioned here that the emission technique is of relatively little importance for all molecular studies except those involving electronic spectra. This is primarily because the lifetimes of all but excited electronic states are usually so long that nonradiative collisional deexcitation is more probable than spontaneous emission. In the case of electronic spectroscopy, most emission studies are currently aimed at investigations of *fluorescence* and *phosphorescence*, two phenomena which are discussed briefly in later chapters. For these reasons, our future discussions deal almost exclusively with absorption spectroscopy.

1.3 RESONANCE SPECTROSCOPY

All absorption-spectroscopy experiments result, in a certain sense, in *resonance* spectra. As Eq. (1.1) shows, a transition occurs only when the frequency corresponds to the spacing between an appropriate pair of energy levels; that is, when Eq. 1.1 is satisfied, the system is in resonance. The term is usually reserved, however, for an experiment in which the spacing between energy levels can be varied (by means of an external electric or magnetic field) so that it is possible to achieve a resonance condition by varying either the frequency of the electromagnetic radiation or the energy-level spacing, or both simultaneously. For example, in chapter 9 it is shown that the energy difference between the upper and lower levels of a proton in an external magnetic field is given by

$$E_2 - E_1 = g_n\beta_nH_0 \tag{1.2}$$

where g_n and β_n are constants and H_0 is the static magnetic field applied to the sample. Combining this with Eq. 1.1, we find the resonance condition to be

$$\nu = g_n\beta_nH_0/h \tag{1.3}$$

1.4 ELECTROMAGNETIC RADIATION

The description of the nature and behavior of electromagnetic radiation is well known from the earliest theories of optics, and the later theories as given by Maxwell's equations.[1] It is sufficient here to point out that in

---

[1] See, for example, R. P. Winch, *Electricity and Magnetism* (Prentice-Hall, Englewood Cliffs, N.J., 1963), 2nd ed.

a vacuum, the frequency ν and wavelength λ of the radiation are related by

$$\nu = c/\lambda \tag{1.4}$$

where c is the velocity of light in a vacuum. While the cgs unit of wavelength is cm, other units are often more appropriate; for example, the Angstrom Å; micron μ; millimicron $m\mu$. These are related as follows: 1 cm = 10^8 Å = 10^4 μ = 10^7 $m\mu$. More recently, the nanometer (1 nm = 1 $m\mu$ has come into use. In a like manner, frequencies occur naturally in cgs units of sec^{-1} or cycles per second. It is becoming common practice to call the basic frequency unit the Hertz (1 Hz = 1 cycle/sec). In other cases, megacycles per second (Mc/s) or kilocycles per second (Kc/s) are more useful; or, in terms of the Hertz, the latter would be MHz and KHz, respectively.

In the infrared (and to a lesser extent, the visible) spectral region, the unit of cm^{-1} or *wave number* is commonly used. This is simply $1/\lambda$ with λ in cm. Many authors have used the symbol $\bar{\nu}$ for this quantity,

$$\bar{\nu} = 1/\lambda = \nu/c \tag{1.5}$$

From the identity in (1.5), it is clear that the wave number is a frequency unit related to sec^{-1} by the conversion factor c. For this reason we normally do not endow it with a special symbol, but consider it to be merely another unit of frequency measurement.

Although early man had little cognizance of radiation other than the fraction of a wavelength decade falling between about 4000 and 7000 Å, modern man's experience extends well over 15 decades of wavelength. Indeed, the growth of science and technology has, to a great extent, paralleled the extension of the range of useful wavelengths. Figure 1.2 summarizes a great deal of information concerning the *electromagnetic spectrum*. Included in the figure are a variety of wavelength and frequency calibrations and the common names for the various regions. Typical sources of radiation are indicated in addition to some of the communication and spectroscopic applications.

By virtue of Eq. (1.1), it is apparent that the energy emitted or absorbed per molecule is equal to $h\nu$. In accord with the wave-particle dualism of nature, the quantity of radiation of frequency ν and energy $h\nu$ is given the particlelike name, *photon*. By using de Broglie's relation between wavelength and linear momentum, p,

$$\lambda = h/p \tag{1.6}$$

with the previously stated energy of a photon

$$E = h\nu \tag{1.7}$$

we obtain the linear momentum of a photon

$$p = h\nu/c \tag{1.8}$$

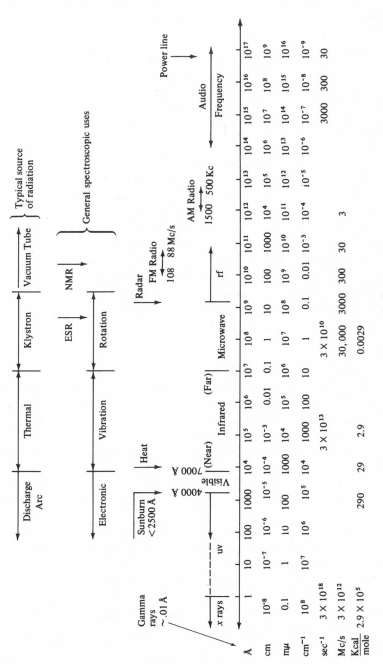

FIG. 1.2. The electromagnetic spectrum.

5

Figure 1.2 gives one scale calibrated in photon energy, namely, kcal/mole of photons. At this point, it is interesting to observe that photon energies of ordinary chemical interest lie in or near to the visible region of the spectrum.

1.5 APPLICATIONS OF MOLECULAR SPECTROSCOPY TO CHEMISTRY AND PHYSICS

Because of the great versatility of molecular spectroscopy it is impossible to present, within the content of any single text, an inclusive exposition of the variety of applications of molecular spectroscopy in the physical and biological sciences. Still it seems useful to attempt a rather general summary of these applications before going further into the specific details of the spectroscopic theory and methodology. If one accepts the general premise of quantized molecular energy it follows that an analysis of a spectrum leads, by virtue of Eq. (1.1), to a knowledge of the energy states of the molecular species. If it is possible to separate the energy effects of nuclei from those of electrons (which is possible to a good approximation), we can identify two general spectroscopic applications: the determination of molecular structure (geometric placement of nuclei) and the determination of electronic structure or distribution. Since the spectra of different molecular species are more or less dissimilar in general, it is possible to perform chemical analysis, which may be listed as the third general application.

A. Qualitative and Quantitative Analysis

It becomes apparent in later chapters that each of the five types of spectroscopy enumerated in Sec. 1.1 has some application for qualitative identification of molecular species. The most useful techniques for this type of application are those that involve transitions that may be related to certain atomic arrangements or functional groups. For example, NMR, infrared and, to a lesser extent, electronic spectroscopy owe much of their importance to this kind of behavior. A microwave spectrum, on the other hand, is specific for the particular molecule as a whole, but contains no gross features related to any of the component atoms or groups. In succeeding chapters we attempt to show why these types of spectral behavior occur, but do not treat extensively the numerous practical applications of the results to qualitative analysis. These applications are treated in a number of excellent books which have been listed as references at the end of each chapter.

The majority of the quantitative analytical applications result by application of the Lambert–Beer expression, which for solutions may be written

$$\log I_0/I = \epsilon(\nu)cl \qquad (1.9)$$

in which I_0 is the radiation intensity incident upon the sample, I is the intensity of the radiation transmitted by the sample, $\epsilon(\nu)$ is the extinction (or absorption) coefficient at frequency ν and c is the concentration of the sample of path length l. If $\epsilon(\nu)$ is known at some frequency ν_0 for a particular species of interest, it is obvious that a measurement of the transmission of a solution at ν_0 results in the determination of the concentration of the desired species. While quantitative analysis of this type is not treated, chapter 5 does describe the theoretical connection between the absorption coefficient and certain other molecular properties.

B.   Determination of Molecular Structure

Along with electron and x-ray diffraction techniques, spectroscopy provides nearly all the quantitative knowledge that we have about molecular structure. While some qualitative-structural information may be obtained from nearly all the spectroscopic methods, only those which provide information about rotational energy states may be used to obtain quantitative structural results,[2] that is, internuclear distances and angles. In later chapters it is seen that electronic and vibrational spectra, in addition to rotational spectra, can provide the required data, although the practical applications are rather limited because of experimental difficulties. It is apparent from all this that microwave spectroscopy (see Fig. 1.2) is the source for most quantitative-structural results, since this frequency region is the principal one for observing rotational spectra.

Whereas microwave spectroscopy serves as a quantitative structural tool, infrared and NMR spectroscopy find extremely widespread qualitative use. Infrared spectroscopy, for example, may permit one to infer the symmetry of a molecule or to infer the structural arrangements of atoms and groups. NMR studies provide similar results, again relying heavily upon symmetry-type arguments.

The appropriate technique depends, of course, on the type of information desired and on the applicability of the technique to the particular molecular system of interest. For example, microwave spectroscopy can be applied only to gaseous systems, while infrared and NMR spectroscopy are applicable to studies of both gaseous and condensed systems. The applicability and limitations of the various methods are stressed in future chapters, so it should become clear how any given structural problem might be attacked.

---

[2] This is not strictly true. There are other types of energy terms which depend upon nuclear coordinates in rather direct ways. The dipolar interaction of two nuclear spins, for example, makes possible the determination of the separation of the two nuclei by nuclear magnetic resonance studies of solids (see chapter 9). These methods are of less general utility and are not treated in any great detail.

C. Determination of Electronic Energies and Properties

Each of the spectroscopic methods may provide some information about electronic energy or electronic distribution and density in a molecule. Electronic (visible and ultraviolet) spectroscopy provides directly the energies of the electronic states of molecules and thereby produces some knowledge about their electron configurations. From vibrational spectroscopy it is possible to obtain information concerning the variation of electronic energy as a function of nuclear coordinates. More specifically, one can determine force constants,

$$k_i = \left(\frac{\partial^2 V}{\partial Q_i{}^2} \right) \tag{1.10}$$

where the derivative with respect to normal coordinate is evaluated at the equilibrium nuclear configuration. Microwave spectroscopy provides electric dipole moments and quadrupole coupling constants, each of which are related to electron distribution. The resonance techniques yield a wealth of detailed information concerning electron density. For example, NMR chemical shifts give a measure of the diamagnetic shielding produced by electrons, while ESR coupling constants may be used to infer unpaired spin densities in organic radicals.

This enumeration could be extended to much greater length. It has been our intention, however, to merely indicate a few of the results related to electron density. In succeeding chapters these and other measurements are discussed in detail and in such a way as to indicate their relationship to concepts of interest to chemists.

It should not be imagined that the above discussion has exhausted the applications of molecular spectroscopy to chemistry and other sciences; the truth is far from this. Still, it is hoped that the enumeration has been of sufficient breadth to provide the uninitiated reader with some idea of what to expect from molecular spectroscopy. It may have been observed that what we have spoken of have been essentially *intramolecular* properties. There are, in addition, numerous spectroscopic applications providing information about *intermolecular* properties, including intermolecular forces and energy transfer. Some of these topics are treated since they have important roles in the theoretical and experimental development of spectroscopy. In particular, we discuss the important aspects of absorption linewidths and relaxation phenomena as they relate to spectroscopy.

CHAPTER 2

QUANTUM MECHANICS AND SPECTROSCOPY

2.1 RELEVANCE OF QUANTUM MECHANICS TO SPECTROSCOPY

Although a considerable amount of spectroscopy was performed before the advent of quantum mechanics, most of the developments in the field have relied heavily upon the quantum theories that were developed just prior to 1930. In this regard the two major applications of quantum mechanics to spectroscopy are:

(a) Prediction of stationary-state energy levels and the relationship of the permitted energies to the physical properties of the molecular species.

(b) Prediction of transition selection rules which lead to satisfaction of the Bohr frequency relation, Eq. (1.1), and of the intensities of the permitted transitions.

While much spectroscopic work can be carried on by adherence to empirical rules and standard formulas, the proper interpretation, insight, and ability to handle new problems requires a sound quantum-mechanical basis. In Secs. 2.2–2.5 of this chapter we discuss some quantum-mechanical aspects which are of direct concern to us in the treatment of spectroscopic problems. The emphasis is placed upon the matrix formulation of quantum mechanics, since we use this approach almost exclusively in later chapters. The reader is urged to consult his favorite quantum-mechanics text for complete descriptions of the theory.

Our use of quantum mechanics is quite elementary in general, except perhaps in chapter 4, where two important problems are solved by the matrix technique, and in chapter 5, where some aspects of the interaction of matter with radiation are discussed. A number of textbooks and monographs which treat quantum mechanics have been listed as general references at the end of this chapter.

2.2   WAVE AND MATRIX MECHANICS

Two rather distinct, but closely related, formulations of nonrelativistic quantum mechanics are available for our use, namely, wave mechanics[1] and matrix mechanics.[2] The former method was developed with an emphasis upon the wave-particle dualism of microscopic matter and leads to answers via solutions (eigenvalues and eigenfunctions) of differential equations. Rather strong emphasis is placed upon both the eigenvalues and eigenfunctions in this method even though only the former are observable quantities. The matrix-mechanics approach was developed with most emphasis upon observable quantities and leads to answers via solutions of algebraic equations. We will find both formalisms useful at various times, but the evaluation of stationary-state energies proceeds more neatly in the latter mode.

In wave mechanics the stationary-state energies are solutions of Schrödinger's time-independent equation

$$\widetilde{\mathcal{H}}(p, q)\psi(q) = E\psi(q) \tag{2.1}$$

where $\widetilde{\mathcal{H}}(p, q)$ is the Hamiltonian operator for the atomic or molecular system expressed in terms of momenta p, and conjugate coordinates q,[3] $\psi(q)$ is the wave function, and E the energy. Similar expressions may be written for observables other than the energy by substitution of the proper operator.[4] Often experimental measurements of other dynamical quantities can be made in the stationary-energy states given by Eq. (2.1). If $\psi(q)$ of Eq. (2.1) is an eigenfunction of one of these other dynamical variables, the stationary-state value of the variable is given by an equation like Eq. (2.1) using the *same* wave function and the appropriate operator. If the eigenfunctions of the energy are not eigenfunctions of the variable of interest, the *average* value (or *expectation* value) of the variable in the

---

[1] E. Schrödinger, *Ann. Physik* **79**, 361 (1926); **79**, 478 (1926); **80**, 437 (1926); **81**, 109 (1926).

[2] W. Heisenberg, *Z. Physik* **33**, 879 (1925).

[3] See Appendix A for a brief discussion of classical mechanics.

[4] All proper quantum mechanical operators must be Hermitian. See Appendix B for some important quantum-mechanical theorems.

energy state $\psi_i(q)$ is given by

$$\bar{M} = \langle M \rangle = \frac{\int \psi_i^* \tilde{M} \psi_i d\tau}{\int \psi_i^* \psi_i d\tau} \tag{2.2}$$

where $\tilde{M}$ is the operator form of the variable.

In matrix mechanics every variable is represented by a matrix, for example, the Hamiltonian matrix $\mathfrak{K}$ is that for the energy. The values of the elements of $\mathfrak{K}$ are obtained by solving matrix equations arising from the commutation relations of the various dynamical variables of the system. When $\mathfrak{K}$ is diagonal[5] the values are the same stationary-state energies given by Eq. (2.1). When, as is often the case, the Hamiltonian matrix is not diagonal, the stationary-state energies are obtained by diagonalizing the matrix. This diagonalization is performed formally by the similarity transformation[6] with the unitary matrix $\mathbf{S}$

$$\mathbf{S}^{-1}\mathfrak{K}\mathbf{S} = \mathbf{E} \tag{2.3}$$

which yields the diagonal eigenvalue matrix $\mathbf{E}$. Equation (2.3) may be written also as

$$\mathfrak{K}\mathbf{S} = \mathbf{S}\mathbf{E} \tag{2.4}$$

which is formally similar to the wave-mechanical result in Eq. (2.1).

While the solution of Eq. (2.1) involves finding the eigenvalues of a partial differential equation of second order, the solution of (2.3) or (2.4) requires the solution of a system of linear homogeneous equations. A typical (ik) element of matrix equation (2.4) is

$$\sum_j^h (\mathfrak{K}_{ij}S_{jk} - S_{ij}E_{jk}) = 0 \tag{2.5}$$

but since $\mathbf{E}$ is diagonal,

$$\sum_j^h (\mathfrak{K}_{ij}S_{jk} - S_{ik}E_{kk}) = 0 \tag{2.6}$$

Introducing the Kronecker δ, Eq. (2.6) becomes

$$\sum_j^h (\mathfrak{K}_{ij} - \delta_{ij}E_{kk})S_{jk} = 0 \tag{2.7}$$

More explicitly Eq. (2.7) is a member of the set of homogeneous linear

---

[5] See Appendix C for some properties of matrices.
[6] See footnote 5.

equations,

$$(\mathcal{K}_{11} - E)\,S_{1k} + \mathcal{K}_{12}S_{2k} + \cdots + \mathcal{K}_{1n}S_{nk} = 0$$

$$\mathcal{K}_{21}S_{1k} + (\mathcal{K}_{22} - E)\,S_{2k} + \cdots + \mathcal{K}_{2n}S_{nk} = 0 \qquad (2.8)$$

$$\mathcal{K}_{n1}S_{1k} + \mathcal{K}_{n2}S_{2k} + \cdots + (\mathcal{K}_{nn} - E)\,S_{nk} = 0$$

$$k = 1, 2, 3 \cdots n$$

where $E = E_{kk}$ for simplicity. The set of equations (2.8) has a nontrivial solution (that is, all $S_{ik} \neq 0$) only if the determinant of the coefficients of the S_{ik} equals zero.[7] This leads to Eq. (2.9)

$$\begin{vmatrix} \mathcal{K}_{11} - E & \mathcal{K}_{12} & \mathcal{K}_{13} \cdots \mathcal{K}_{1n} \\ \mathcal{K}_{21} & \mathcal{K}_{22} - E & \mathcal{K}_{23} \cdots \mathcal{K}_{2n} \\ \mathcal{K}_{31} & \mathcal{K}_{32} & \mathcal{K}_{33} - E \cdots \\ \vdots & \vdots & \\ \mathcal{K}_{n1} & \mathcal{K}_{n2} \cdots \mathcal{K}_{nn} - E \end{vmatrix} = 0 \qquad (2.9)$$

or in matrix form

$$|\, \mathcal{K} - E\mathbf{I}\,| = 0 \qquad (2.10)$$

where $\mathbf{I}$ is the unit matrix. The n roots (not necessarily all distinct) of the polynomial in E^n given by (2.9) or (2.10) are clearly the elements of the diagonal matrix $\mathbf{E}$ of Eqs. (2.3) or (2.4), and by our postulates are the energies of the stationary states of the system.

After the n roots E_{kk} have been determined in this fashion, the S_{jk} of Eqs. (2.7) or (2.8) may be determined for each value of E_{kk}. Each value of E_{kk} leads to a *column* of the matrix $\mathbf{S}$ of Eq. (2.3), so that by this procedure the entire matrix $\mathbf{S}$, and consequently $\mathbf{S}^{-1}$ may be found. Application of $\mathbf{S}$ as indicated by Eq. (2.3) then leads directly to the energy matrix $\mathbf{E}$.

We find both quantum-mechanical formulations to be useful, but the actual problem of obtaining energy-level expressions is performed in the matrix formalism. Since, as shown in Sec. 2.3, the two methods have a very close relationship, no harm occurs by interspersing them when it is convenient.

In addition to the determination of stationary-state energies, we need to determine selection rules as mentioned in Sec. 2.1. This requires the use of the time-dependent wave equation,

$$\tilde{\mathcal{K}}(p, q, t)\Psi(q, t) = -\frac{\hbar}{i}\frac{\partial}{\partial t}\Psi(q, t) \qquad (2.11)$$

---

[7] H. Margenau and G. M. Murphy, *The Mathematics of Physics and Chemistry* (Van Nostrand, Princeton, N.J., 1943), p. 313.

The application of Eq. (2.11) to the absorption and emission of electromagnetic radiation is postponed until chapter 5.

We occasionally have need to treat molecular energy terms which are very small compared to other dominant terms. For this reason it is profitable to use the well-known methods of perturbation theory which are described in any quantum-mechanics textbook.

EXAMPLE 2-1

While most students are aware of the means of solving the differential equations of wave mechanics, namely Eq. (2.1), they may be less aware of the practical significance of the similarity transformation expressed by Eq. (2.3). This diagonalization is easily performed for any two-dimensional, real Hermitian matrix. Suppose

$$\mathcal{3C} = \begin{pmatrix} a & b \\ b & c \end{pmatrix} \tag{2.12}$$

then the diagonalization is performed by using

$$\mathbf{S} = \begin{pmatrix} \cos\theta & -\sin\theta \\ \sin\theta & \cos\theta \end{pmatrix} \tag{2.13}$$

where

$$\tan 2\theta = 2b/(a - c) \tag{2.14}$$

Now $\mathbf{S}^{-1} = \mathbf{S}'$, the transpose of $\mathbf{S}$, since $\mathbf{S}$ is real orthogonal (see Appendix C); therefore

$$\mathbf{S}^{-1} = \begin{pmatrix} \cos\theta & \sin\theta \\ -\sin\theta & \cos\theta \end{pmatrix} \tag{2.15}$$

Multiplying the matrices as in Eq. (2.3) gives

$$\mathbf{S}^{-1}\mathcal{3C}\mathbf{S} = \begin{pmatrix} a\cos^2\theta + c\sin^2\theta \\ + 2b\sin\theta\cos\theta & 0 \\ & a\sin^2\theta + c\cos^2\theta \\ 0 & - 2b\sin\theta\cos\theta \end{pmatrix} \tag{2.16}$$

We have, then, an analytical method for diagonalizing any real Hermitian, two-dimensional matrix. The proper value of θ for use in Eq. (2.16) is gotten from Eq. (2.14) for any set of (a, b, c) values.

There is an interesting and valuable geometrical interpretation of the similarity transformation illustrated by this example. The matrix $\mathbf{S}$ is the rotational transformation which takes the vector r_1 into the vector r_2 as shown in Fig. 2.1(a). Mathematically, if r_1 and r_2 are written as column vectors,

$$r_1 = \begin{pmatrix} X_1 \\ Y_1 \end{pmatrix}, \qquad r_2 = \begin{pmatrix} X_2 \\ Y_2 \end{pmatrix} \tag{2.17}$$

the transformation is

$$Sr_1 = \begin{pmatrix} \cos\theta & -\sin\theta \\ \sin\theta & \cos\theta \end{pmatrix}\begin{pmatrix} X_1 \\ Y_1 \end{pmatrix} = r_2 = \begin{pmatrix} X_2 \\ Y_2 \end{pmatrix} \tag{2.18}$$

By matrix multiplication,

$$X_2 = X_1 \cos\theta - Y_1 \sin\theta$$

$$Y_2 = X_1 \sin\theta + Y_1 \cos\theta \tag{2.19}$$

It is easily verified that S^{-1} is the matrix which transforms r_2 into r_1, that is, the reverse of (2.18). On the other hand, while S may be considered to rotate r_1 by an angle $+\theta$, it may be considered also to be the matrix which rotates the X, Y coordinate system by an angle $-\theta$ into the X', Y' coordinate system of Fig. 2.1(b). In Fig. 2.1(b), the vector r has the coordinates

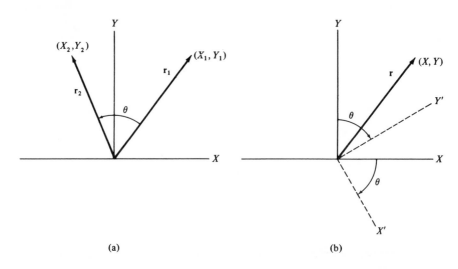

(a)   (b)

FIG. 2.1.   Rotational transformation in two dimensions. In (a), a vector r_1 is rotated by an angle θ to produce r_2. In (b), the vector r is observed in the X, Y-axis system and also the axis system resulting from a rotation of the axes by $-\theta$.

(X_1, Y_1) in the X, Y axis system, while it has the coordinates (X_2, Y_2) in the X', Y' system.

Either of these interpretations is correct, and they both are useful. Further considerations show that the *similarity* transformation is itself an axis system transformation; when the similarity transformation is a real orthogonal one leading to a diagonal matrix as in Eqs. (2.3) or (2.16), it is known as a *principal-axis* transformation.[8]

2.3 RELATIONSHIP OF MATRIX AND WAVE MECHANICS

In later chapters several applications of the matrix-mechanics formulation are described. At this point we wish to show the close connection between the matrix- and wave-mechanical formulations. In this way, the student who is primarily familiar with wave mechanics should obtain a clearer understanding of the meaning and usefulness of the matrix approach.

Let us suppose that we desired to solve Schrödinger's equation, which was

$$\tilde{\mathcal{H}}\psi = E\psi \tag{2.1}$$

that is, we desire the eigenvalues E_i and eigenfunctions ψ_i which satisfy

$$\tilde{\mathcal{H}}\psi_i = E_i\psi_i \tag{2.20}$$

We imagine, as is often the case, that an analytical solution of (2.20) is not possible for the problem of interest. It is possible, in general, to expand the (unknown) functions ψ_i in terms of a complete, *orthonormal* set of (known) functions ϕ_j.[9]

$$\psi_i = \sum_j^n a_{ji}\phi_j \tag{2.21}$$

where the ϕ_j are orthonormal, that is

$$\int \phi_j^*\phi_k d\tau = \delta_{jk} \tag{2.22}$$

As shown in Appendix B, the a_{ji} would be given by the integrals

$$a_{ji} = \int \psi_j^*\phi_i d\tau = (\psi_j \mid \phi_i) = (j \mid i) \tag{2.23}$$

---

[8] See, for example, Ref. 7, p. 326.

[9] See Appendix B for some related results. We do not delve into the mathematical requirements for the expansion to be valid. We simply remark that "completeness" implies the use of functions spanning the same space. For more detailed discussion, see L. Landau and E. Lifshitz, *Quantum Mechanics, Non-Relativistic Theory* (Addison-Wesley, Reading, Mass., 1958).

where the latter two equalities are useful, common shorthand notations. Note that the a_{ji} may be considered to be elements of the matrix $\mathbf{A}$. This matrix is unitary (see Appendix C), that is,

$$\mathbf{A}^{-1} = \mathbf{A}^{*\prime}$$

or

$$(\mathbf{A}^{-1})_{ij} = a_{ji}^{*} \tag{2.24}$$

From Eq. (2.20), multiplication by ψ_l^* and integration over all space leads to

$$\int \psi_l^* \widetilde{\mathcal{H}} \psi_i d\tau = E_i \int \psi_l^* \psi_i d\tau = E_i \, \delta_{li} \tag{2.25}$$

assuming the ψ_i to be orthonormal. Substitution of Eq. (2.21) into (2.25) leads to Eq. (2.26).

$$\sum_k^n \sum_j^n a_{kl}^* a_{ji} \int \phi_k^* \widetilde{\mathcal{H}} \phi_j d\tau = E_i \delta_{li} \tag{2.26}$$

The integrals which appear in Eq. (2.26) may now be considered to be elements of the matrix $\mathcal{H}^\phi$, the superscript indicating the functions used to evaluate the integrals. These *matrix elements* may be written in the various forms,

$$\int \phi_k^* \widetilde{\mathcal{H}} \phi_j d\tau = \mathcal{H}_{kj}^\phi = (\phi_k \mid \widetilde{\mathcal{H}} \mid \phi_j)$$

$$= (k \mid \widetilde{\mathcal{H}} \mid j) \tag{2.27}$$

Using (2.27) and taking only the nonzero elements of Eq. (2.26), we obtain

$$\sum_{k,j}^n a_{ki}^* a_{ji} \mathcal{H}_{kj}^\phi = E_i = E_{ii} \tag{2.28}$$

where the last definition $E_i = E_{ii}$ is made so that it is clear that the E_i may be considered to be elements of a diagonal matrix. In fact, the E_{ii} are, from Eq. (2.20), the elements of the diagonal matrix $\mathcal{H}^\psi$, that is

$$\int \psi_i^* \widetilde{\mathcal{H}} \psi_i d\tau = \mathcal{H}_{ii}^\psi = (\psi_i \mid \widetilde{\mathcal{H}} \mid \psi_i)$$

$$= E_{ii} \tag{2.29}$$

Rearranging the left-hand side of Eq. (2.28) and noting that $a_{ki}^* = (\mathbf{A}^{-1})_{ik}$ from Eq. (2.24), we arrive at

$$\sum_{k,j}^n (\mathbf{A}^{-1})_{ik} \mathcal{H}_{kj}^\phi a_{ji} = E_{ii} = \mathcal{H}_{ii}^\psi \tag{2.30}$$

But this is just the iith element of the matrix equation

$$\mathbf{A}^{-1}\mathfrak{3C}^{\phi}\mathbf{A} = \mathfrak{3C}^{\psi} = \mathbf{E} \tag{2.31}$$

Note now that Eq. (2.31) is in precisely the same form as Eq. (2.3), but its derivation provides a new insight into the matrix technique. For Eq. (2.31) says that to obtain the energy states for any system, we simply calculate $\mathfrak{3C}^{\phi}$ using a complete set of known functions, and then diagonalize this matrix. The desired energies are then the values of the matrix $\mathfrak{3C}^{\psi} = \mathbf{E}$ which results.

Although the matrix-mechanics solutions of many physically interesting problems can be performed without any specific reference to wave functions, in most cases it is useful to consider the solutions in the spirit of the preceding derivation. It should be mentioned that any problem which is exactly solvable by wave mechanics should also be exactly solvable by the matrix formulation.[10] The great value of the matrix method lies, however, in its suitability for numerical solution; namely, numerical diagonalization of matrices of large order. In many cases, the proper expansion functions ϕ_i are evident from the problem which must be solved. Usually, they are functions which result from exact solutions of similar but simpler quantum-mechanical problems. For example, the well-known harmonic oscillator wave functions would be an appropriate set for a solution of the quantum-mechanical problem involving a particle moving with a potential energy

$$V = \tfrac{1}{2}kx^2 + k'x^3 \tag{2.32}$$

where the second term on the right-hand side is a *small* anharmonic contribution. In any case, the functions used (implicitly or explicitly) in the evaluation of the matrix elements of $\mathfrak{3C}^{\phi}$ are known as *basis* functions; and we say that $\mathfrak{3C}$ is evaluated in the ϕ *representation*.

2.4 INTRINSIC SPIN

What we have been considering (at least implicitly) up to this point have been quantum-mechanical problems involving three-dimensional space coordinates (and in some cases, the time variable). For example, Schrödinger's equation (2.1) or (2.4) has been treated as having a Hamiltonian whose variables are simply the spatial coordinates (and conjugate momenta) of the particles being described, so that the wave function also has been a function of these same variables. Although the molecular energy terms of largest magnitude can be treated satisfactorily by Hamiltonians

---

[10] The matrix solution of the well-known nonrelativistic hydrogen-atom problem has been given recently by M. Bander and C. Itzykson, Rev. Mod. Phys. **38**, 330 (1966).

of this type, there are a number of exceedingly interesting small contributions which arise from properties which cannot be described by spatial variables of the classical type.

The properties we speak of here are the *intrinsic* electron and nuclear angular momenta and concommitant magnetic moments. Electron spin angular momentum was first proposed by Goudsmit and Uhlenbeck[11] to explain the multiplet structure of atomic spectra and the Stern-Gerlach[12] experiment, wherein a beam of silver atoms was divided into two distinct beams by passage through an inhomogeneous magnetic field. If the total electron spin angular momentum is $\tilde{S}$ and the components are $\tilde{S}_X$, $\tilde{S}_Y$, and $\tilde{S}_Z$, we may define the properties of electron spin by the commutation relations

$$\tilde{S}_X \tilde{S}_Y - \tilde{S}_Y \tilde{S}_X = -(\hbar/i)\,\tilde{S}_Z$$

$$\tilde{S}_Y \tilde{S}_Z - \tilde{S}_Z \tilde{S}_Y = -(\hbar/i)\,\tilde{S}_X \qquad (2.33)$$

$$\tilde{S}_Z \tilde{S}_X - \tilde{S}_X \tilde{S}_Z = -(\hbar/i)\,\tilde{S}_Y$$

which are of identical form (except possibly the sign) to those for orbital or rotational angular momentum (see chap. 4). We do not delve into the matrix elements of spin angular momentum here, but simply point out the well-known result that the electron has a spin of $S = \frac{1}{2}$ and may have projections of angular momentum in the Z direction of $m_s = \pm\frac{1}{2}\hbar$. The latter statement implies

$$\tilde{S}_Z \alpha(\sigma) = \tfrac{1}{2}\hbar\alpha(\sigma)$$

and   (2.34)

$$\tilde{S}_Z \beta(\sigma) = -\tfrac{1}{2}\hbar\beta(\sigma)$$

where $\alpha(\sigma)$ and $\beta(\sigma)$ are the eigenfunctions for $m_s = \frac{1}{2}$ and $m_s = -\frac{1}{2}$, respectively, and σ is the *spin-space* variable. The eigenfunctions are normalized and orthogonal, that is,

$$\int |\alpha(\sigma)|^2 d\sigma = \int |\beta(\sigma)|^2 d\sigma = 1$$

and   (2.35)

$$\int \alpha^*(\sigma)\beta(\sigma) d\sigma = 0$$

and it should be clear that the volume element $d\sigma$ is entirely unrelated to ordinary three-dimensional space. The magnetic moment of the electron is proportional to its angular momentum and may be written

$$\tilde{\mu}_e = -g_e \beta_e \tilde{S}/\hbar \qquad (2.36)$$

[11] G. Goudsmit and S. Uhlenbeck, *Naturwissen.* **13**, 953 (1925).

[12] O. Stern and W. Gerlach, *Z. Physik.* **8**, 110; **9**, 349 (1922).

where $g_e = 2.0002$ for a free electron and $\beta_e = e\hbar/2mc$ is the Bohr magneton and has the value 0.927×10^{-20} erg/G. The expression (2.36) may be considered to apply to any system with intrinsic spin S, the factor g_e then being considered to be an experimentally determined factor.

Since protons and neutrons are also found to have intrinsic spin of $\frac{1}{2}$, nuclei may have, in general, a nuclear spin I, and a magnetic moment

$$\tilde{\mu}_n = g_n \beta_n \tilde{I}/\hbar \qquad (2.37)$$

g_n is again an experimentally determined quantity while $\beta_n = \beta_e(m/M)$ where m = mass of electron and M = mass of proton. Nuclear spin values may be half-integral, integral or zero (see chapter 9 for a more detailed discussion of nuclear properties). All the commutation relations for electron spin, Eqs. (2.33), and the Eqs. (2.34) and (2.35) apply to nuclear spin properties also with appropriate change in symbolism.

It should be clear that electron spin-space (σ) and nuclear spin-space $(\sigma_n$, say) are totally unrelated and do not span the same variables. We illustrate this point further by the following example. Suppose $\tilde{\mathcal{K}}(q, \sigma, \sigma_n) = \tilde{\mathcal{K}}(q) + \tilde{\mathcal{K}}(\sigma) + \tilde{\mathcal{K}}(\sigma_n)$ where the Hamiltonian contains energy terms described by ordinary classical coordinates q and electron and nuclear spin coordinates σ and σ_n, respectively. Then Schrödinger's equation becomes

$$[\tilde{\mathcal{K}}(q) + \tilde{\mathcal{K}}(\sigma) + \tilde{\mathcal{K}}(\sigma_n)]\psi(q, \sigma, \sigma_n) = E\psi(q, \sigma, \sigma_n) \qquad (2.38)$$

A separation of variables is obtained promptly by assuming

$$\psi(q, \sigma, \sigma_n) = \phi(q) \cdot \phi(\sigma) \cdot \phi(\sigma_n) \qquad (2.39)$$

Since $\tilde{\mathcal{K}}(q)$ can operate only on $\phi(q)$, $\tilde{\mathcal{K}}(\sigma)$ on $\phi(\sigma)$, and $\tilde{\mathcal{K}}(\sigma_n)$ on $\phi(\sigma_n)$, because of the mutual exclusiveness of the three coordinate spaces, we find

$$\phi(\sigma)\phi(\sigma_n)\tilde{\mathcal{K}}(q)\phi(q) + \phi(q)\phi(\sigma_n)\tilde{\mathcal{K}}(\sigma)\phi(\sigma)$$
$$+\phi(q)\phi(\sigma)\tilde{\mathcal{K}}(\sigma_n)\phi(\sigma_n) = E\phi(q)\phi(\sigma)\phi(\sigma_n) \qquad (2.40)$$

or

$$\frac{\tilde{\mathcal{K}}(q)\phi(q)}{\phi(q)} + \frac{\tilde{\mathcal{K}}(\sigma)\phi(\sigma)}{\phi(\sigma)} + \frac{\tilde{\mathcal{K}}(\sigma_n)\phi(\sigma_n)}{\phi(\sigma_n)} = E \qquad (2.41)$$

But by the usual arguments,[13] each term on the left must be equal to a constant, so that the solution to (2.38) is given by the solutions of

$$\tilde{\mathcal{K}}(q)\phi(q) = E_q\phi(q)$$
$$\tilde{\mathcal{K}}(\sigma)\phi(\sigma) = E_\sigma\phi(\sigma) \qquad (2.42)$$
$$\tilde{\mathcal{K}}(\sigma_n)\phi(\sigma_n) = E_{\sigma_n}\phi(\sigma_n)$$

---

[13] See, for example, L. Pauling and E. B. Wilson, Jr., *Introduction to Quantum Mechanics* (Wiley, New York, 1938), pp. 90–91.

with E being given by

$$E = E_q + E_\sigma + E_{\sigma_n} \tag{2.43}$$

Equation (2.39) is always a valid solution of the wave equation as long as the Hamiltonian is of the form of that in (2.38). In ordinary, nonrelativistic quantum problems this is always the case. The most well-known example in which the spin variables cannot be exactly separated is that of spin-orbit coupling $(\tilde{S} \cdot \tilde{L})$. This coupling of spin and space variables makes an exact separation impossible, although the separation is often a useful first approximation.

2.5 SYMMETRY PROPERTIES OF WAVE FUNCTIONS; INDISTINGUISHABILITY OF PARTICLES

Although a more complete description of the symmetry properties of wavefunctions is given in later chapters, it is appropriate to mention here those simple properties arising from the inherent *indistinguishability* of similar particles. This indistinguishability is closely related to Heisenberg's *uncertainty principle*, since this concept shows that it is impossible to follow precisely the path of a particle in quantum mechanics.[14] Hence, even though we may label and locate a particular particle at some particular time, we cannot know at some later time whether this *same* particle has arrived at some other point in space. According to wave mechanics we can be certain only that $|\psi|^2 d\tau$ predicts the probability distribution for the collection of indistinguishable identical particles.

This indistinguishability has the following important result. If $\tilde{R}$ is the operator corresponding to permutation or interchange of two identical particles,

$$\tilde{R}^2\psi = \psi \tag{2.44}$$

must hold, since the application of $\tilde{R}$ twice must leave ψ unchanged. We see then that the wave function must either change sign or remain unchanged when two identical particles are interchanged, that is

$$\tilde{R}\psi = \pm\psi \tag{2.45}$$

We say that any correct wave function must be either symmetrical ($+$ sign) or antisymmetrical ($-$ sign) with respect to interchange of identical particles. If the former case applies, the particles are said to obey *Bose–Einstein* statistics, while in the latter case the particles obey *Fermi–Dirac* statistics. It is an experimental fact that electrons, protons, and neutrons are Fermi–Dirac particles while photons and alpha particles are Bose–Einstein in behavior. From this it is clear that the symmetry or parity of any nucleus

---

[14] See footnote 9, Landau and Lifshitz, Chapter IX.

is determined simply by the number of protons and neutrons. Thus nuclei of odd atomic number obey Fermi statistics, while those of even atomic number obey Bose statistics.

The further consequence of these symmetry properties may be seen as follows. If $\psi = \psi(q_i)\psi_s$, where $\psi(q_i)$ is the ordinary spatial or *orbital* wave function, and ψ_s is the electron spin wave function, the above discussion shows that ψ must be antisymmetric with respect to interchange of pairs of electrons. This means that if $\tilde{R}\psi(q_i) = +\psi(q_i)$, $\tilde{R}\psi_s = -\psi_s$; while if $R\psi(q_i) = -\psi(q_i)$, $\tilde{R}\psi_s = +\psi_s$ is required. This emphasizes the fact that, although the orbital wave functions resulting from Schrödinger's equation may in general be either symmetric or antisymmetric, only those which give the proper over-all symmetry when combined with the spin functions are permitted. Therefore, we must always consider these symmetry aspects in any problem, since otherwise we may predict energy states which cannot exist.

SUPPLEMENTARY REFERENCES

H. Eyring, J. Walter, and G. E. Kimball, *Quantum Chemistry* (Wiley, New York, 1944).

W. Kauzmann, *Quantum Chemistry* (Academic, New York, 1957).

L. D. Landau and E. M. Lifshitz, *Quantum Mechanics, Non-Relativistic Theory* (Addison-Wesley, Reading, Mass., 1958).

I. N. Levine, *Quantum Chemistry* (Allyn and Bacon, Boston, 1970), Vol. I.

H. Margenau and G. M. Murphy, *The Mathematics of Physics and Chemistry* (Van Nostrand, Princeton, 1956), Vol. 1.

A. Messiah, *Quantum Mechanics* (Wiley, New York, 1961).

L. C. Pauling and E. B. Wilson, *Introduction to Quantum Mechanics* (McGraw-Hill, New York, 1935).

L. S. Schiff, *Quantum Mechanics* (McGraw-Hill, New York, 1955).

CHAPTER 3

NATURE AND TREATMENT OF THE MOLECULAR ENERGY TERMS

3.1 GENERAL MOLECULAR HAMILTONIAN; SEPARATION OF TRANSLATIONAL ENERGY

As indicated in chapter 1, the chief task of this book is the description and analysis of molecular spectra of several types. Our problem is twofold: (1) we must solve the quantum-mechanical Schrödinger's equation (2.1) or its matrix mechanical analog (2.4) in order to obtain the permissible stationary-state energies for any molecule; and (2) we must determine which of these stationary-state energies can be involved in the absorption or emission of radiation in the manner prescribed by the Bohr relationship (1.1); that is, we must determine the *selection rules*. In this chapter, we deal with the problem of finding the Hamiltonian for the molecular energy, and undertake a very preliminary solution of the quantum-mechanical problem. The more complete solutions are undertaken in the later chapters as each type of spectroscopy is encountered. Chapter 5 describes the general selection-rule theory, and chapter 6 presents further selection-rule results. Again, the selection rules are described in detail for each type of spectra in the later chapters.

While it would be possible to obtain the appropriate energy levels (to at least a first approximation) for several of the energy-types of interest by the use of simple approximate models, the approach here is to begin the treatment in a more or less rigorous fashion and then to proceed to suitable approximations when necessary. For example, it becomes evident that the spectra occurring in the microwave region may be described to a good approximation by considering the three-dimensional rotation of a system of point masses whose separations are fixed, that is, by considering the *rigid-rotor* model. Similarly, it will be shown that the spectra occurring in the infrared region are adequately described by a multidimensional *harmonic oscillator* model. Furthermore, although it is possible to adequately describe much of ESR and NMR spectroscopy in an *ad hoc* fashion which is independent of the other larger molecular energy terms, these spectra are also treated within the same general framework, at least initially.

It should be admitted at the outset that, although we begin the problem in a correct fashion, we rapidly lose rigor as we proceed. This is primarily because the quantum-mechanical problems are very difficult but also because the intent of the book is to provide an introduction to a very broad field. Our approach has the following advantages.

(i) The reader is not confronted with a series of simplified Hamiltonian models which are difficult to relate conceptually to the real, physical problem. For example, the novice may reasonably ask, "Why should the rotational problem be treated by a rigid-rotor model when the molecule is known to be vibrating simultaneously?"

(ii) Since the correct Hamiltonian is described, the approximations and limitations involved in the use of simplified models can be made clear. Then, although the exact quantum-mechanical problem may not be solvable, the correct formulation is useful as a guide for making small corrections to the simplified models.

(iii) Finally, but not of least importance, our approach should give spectroscopy a unified appearance. All too often the beginning student, and in many cases even the experienced research scientist, fails to recognize that quantum mechanics provides the link which unifies all of spectroscopy. Instead he views each field of spectroscopy as a distinct creature whose being has nothing in common with other types of spectroscopy. The format of this chapter should counteract this by providing the basis for a broad understanding of molecular spectroscopy.

Let us begin by considering a molecule having N nuclei and n electrons, situated in field-free space. The kinetic energy of the system of nuclei and electrons can then be specified classically in the laboratory space-fixed axis system (X, Y, Z) of Fig. 3.1. The result is

$$T = \frac{1}{2} \sum_{\alpha}^{N} \frac{p_{X\alpha}^2 + p_{Y\alpha}^2 + p_{Z\alpha}^2}{m_\alpha} + \frac{1}{2m_e} \sum_{i}^{n} (p_{Xi}^2 + p_{Yi}^2 + p_{Zi}^2) \quad (3.1)$$

where the m_α are the nuclear masses, m_e is the electron mass, and the $p_{X\alpha}$, etc., are the linear-momentum components. Writing the momenta in operator form the kinetic-energy operator becomes

$$\tilde{T} = -\frac{\hbar^2}{2}\sum_{\alpha}^{N}\frac{\nabla_{\alpha}^2}{m_{\alpha}} - \frac{\hbar^2}{2m_e}\sum_{i}^{n}\nabla_i^2 \tag{3.2}$$

with
$$\nabla_{\alpha}^2 = \frac{\partial^2}{\partial X_{\alpha}^2} + \frac{\partial^2}{\partial Y_{\alpha}^2} + \frac{\partial^2}{\partial Z_{\alpha}^2}$$

$$\tag{3.3}$$

and
$$\nabla_i^2 = \frac{\partial^2}{\partial X_i^2} + \frac{\partial^2}{\partial Y_i^2} + \frac{\partial^2}{\partial Z_i^2}$$

The major portion of the molecular potential energy can be obtained by recognizing that the electron-nuclei system undergoes Coulombic inter-

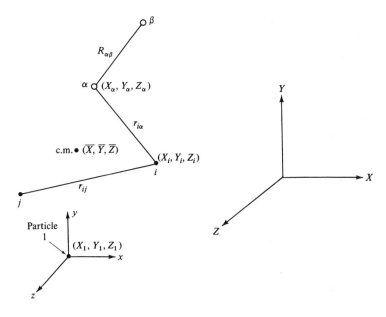

FIG. 3.1.   Coordinate systems used in locating electrons and nuclei in space. The X, Y, Z-axis system is imagined to be fixed in the laboratory (space fixed). Points 1, i, and j are representative electrons and points α and β are representative nuclei. The center of mass of the system is located at $\bar{X}$, $\bar{Y}$, $\bar{Z}$ relative to the space-fixed axes. A second (molecule-fixed) axis system x, y, z is located at electron 1. A *few* representative interparticle distances are labeled explicitly.

actions, so that by reference to Fig. 3.1 we see that the Coulombic potential energy is

$$V_c = e^2 \sum_{\alpha > \beta} \frac{Q_\alpha Q_\beta}{R_{\alpha\beta}} + e^2 \sum_{i>j} \frac{1}{r_{ij}} - e^2 \sum_{i,\alpha} \frac{Q_\alpha}{r_{i\alpha}} \qquad (3.4)$$

where the first term sums up nuclear-nuclear repulsions, the second electron-electron repulsions and the third nuclear-electron attractions. e is the magnitude of the electronic charge and $Q_\alpha e$ is the nuclear charge. Note that V_c is the correct form for the Coulombic potential energy in both classical and quantum mechanics. Several other types of molecular potential energy terms may still exist in field-free space, and additional terms arise when external electric and magnetic fields are present. For the time being, let us indicate by the symbol V' all other potential energy terms. These are considered in more detail later.

The complete Hamiltonian is then given in the X, Y, Z axis system as

$$\widetilde{\mathcal{H}} = \widetilde{T} + \widetilde{V}_c + \widetilde{V}' \qquad (3.5)$$

and, as shown by Eq. (3.2) and (3.3), is a function of $3(n + N)$ variables. As is well known from elementary quantum mechanics, the translational energy of a particle in a field-free space is not quantized.[1] It should be suspected, then, that the parts of (3.1) and (3.5) which correspond to translation of the entire molecule are uninteresting spectroscopically, and might just as well be dispensed with immediately. This can be done by introducing the following coordinates in place of the $3(n + N)$ coordinates previously described: three coordinates $\bar{X}$, $\bar{Y}$, $\bar{Z}$ which locate the center of mass of the molecule, and $3(n + N) - 3$ Cartesian coordinates which locate all the particles *relative* to some particular particle, say particle $i = 1$. Explicitly, these new variables are

$$\bar{X} = \frac{1}{M} \sum_{i=2}^{(n+N)} (m_i X_i) + m_1 X_1$$

$$\bar{Y} = \frac{1}{M} \sum_{i=2}^{(n+N)} (m_i Y_i) + m_1 Y_1 \qquad (3.6)$$

$$\bar{Z} = \frac{1}{M} \sum_{i=2}^{(n+N)} (m_i Z_i) + m_1 Z_1$$

---

[1] Even in an infinite-potential "box" of modest laboratory size, say 1 cm^3, the energy levels for a proton or electron are nearly continuous due to the very small spacing of levels. See, for example, K. S. Pitzer, *Quantum Chemistry* (Prentice-Hall, Englewood Cliffs, N.J., 1953).

and

$$X_i - X_1 = x_i$$
$$Y_i - Y_1 = y_i \qquad i = 2, 3 \ldots (n + N) \qquad (3.7)$$
$$Z_i - Z_1 = z_i$$

In these expressions, the index i runs over both electrons and nuclei with the range $i = 2 \ldots (n + N)$. $i = 1$ has been excluded since the coordinates of 3–7 are relative to particle 1. M is the total mass of the molecule ($M = m_1 + \sum_i m_i$) and m_i represents either electron or nuclear masses. Then it is easily shown that slightly tedious calculus that the substitution of the variables of 3–6 and 3–7 into 3–2 and 3–3 leads to

$$\tilde{T} = -\frac{\hbar^2}{2} \left\{ \sum_{i=2}^{(n+N)} \left(\frac{1}{m_1} + \frac{1}{m_i} \right) \left(\frac{\partial^2}{\partial x_i^2} + \frac{\partial^2}{\partial y_i^2} + \frac{\partial^2}{\partial z_i^2} \right) \right.$$
$$\left. + \frac{2}{m_1} \sum_{k>i}^{n+N} \sum_{i=2}^{n+N} \left(\frac{\partial^2}{\partial x_k \, \partial x_i} + \frac{\partial^2}{\partial y_k \, \partial y_i} + \frac{\partial^2}{\partial z_k \, \partial z_i} \right) + \frac{1}{M} \left(\frac{\partial^2}{\partial \bar{X}^2} + \frac{\partial^2}{\partial \bar{Y}^2} + \frac{\partial^2}{\partial \bar{Z}^2} \right) \right\}$$

$$(3.8)$$

or

$$\tilde{T} = -\frac{\hbar^2}{2} \left\{ f(x_i, y_i, z_i) + \frac{1}{M} \nabla_{\text{c.m.}}^2 \right\} \qquad (3.9)$$

with obvious identifications of terms. The first term represents the internal, relative kinetic energy and the $\nabla_{\text{c.m.}}^2$ part represents the kinetic energy due to translation of the c.m. of the molecule. If $\tilde{T}_R$ is the relative kinetic energy and $\tilde{T}_t$ the c.m. translational energy, we may write Eq. (3.9) in the simpler form

$$\tilde{T} = \tilde{T}_R[x_2 \ldots z_{(n+N)}] + \tilde{T}_t(\bar{X}, \bar{Y}, \bar{Z}) \qquad (3.10)$$

Noting now that $V_c = V_c(x_2 \ldots z_{(n+N)})$, the wave equation may be written

$$(\tilde{T}_R + \tilde{T}_t + V_c)\psi(x_2 \ldots z_{(n+N)}, \bar{X}, \bar{Y}, \bar{Z}) = E\psi(x_2 \ldots z_{(n+N)}, \bar{X}, \bar{Y}, \bar{Z})$$

$$(3.11)$$

when V' is zero. The variables may now be separated by choosing

$$\psi = \psi_R(x_2 \ldots z_{(n+N)}) \cdot \psi_t(\bar{X}, \bar{Y}, \bar{Z}) \qquad (3.12)$$

Inserting Eq. (3.12) into Eq. (3.11) we obtain, after division by $\psi_R \psi_t$,

$$(\tilde{T}_R + \tilde{V}_c)\psi_R/\psi_R + \tilde{T}_t \psi_t/\psi_t = E \qquad (3.13)$$

The variables have therefore been separated, so we have one equation for translation of the c.m. of the molecule,

$$\tilde{T}_t \psi_t = E_t \psi_t \qquad (3.14)$$

and one for the internal relative energies,

$$(\tilde{T}_R + \tilde{V}_c)\psi_R = E_R\psi_R \tag{3.15}$$

where

$$E = E_t + E_R \tag{3.16}$$

When the potential energy terms V' of Eq. (3.5) are nonzero and are not explicitly dependent upon the external environment, that is, when $V' \neq V'(X, Y, Z)$, the internal energy equation, (3.15), can be written

$$(\tilde{T}_R + \tilde{V}_c + \tilde{V}')\psi_R = E_R\psi_R \tag{3.17}$$

and the translational energy still separates. It occurs for most of what is of interest to molecular spectroscopy that the translational energy *is* separable; so most of what follows involves the internal energies as given by Eq. (3.15) or (3.17).

One may ask under what experimental conditions this separation of translation is possible. Certainly the separation is an excellent approximation when the molecule exists in a low-density medium with no externally applied electromagnetic fields. Hence, under normal laboratory conditions a low-pressure gas fulfills the requirements. For under these conditions the intermolecular forces are so weak as to perturb the free-translational motions only slightly. When the molecule exists in the more dense liquid phase, Eq. (3.17) becomes less rigorously applicable, but remains as a valid first approximation. This may be thought to arise qualitatively because the intermolecular forces (which are now quite large) produce an average potential energy V', which is constant over the range of the internal motions. In the solid state, the separation of translational energy is not possible, since the strong crystalline restoring forces prohibit free translation from occurring.

Although the electrons and nuclei of a molecule may experience forces from externally applied electric and/or magnetic fields, relatively little effect will be produced upon the translational motions in practical applications, that is, the translational motion can be separated approximately from the internal motions. The net result of this brief discussion is that the internal (nontranslational) energies of molecules in nondense media can be treated by an equation of the form of (3.17).

3.2   MOLECULAR BORN–OPPENHEIMER APPROXIMATION; SEPARATION OF ELECTRONIC ENERGIES

One of the first features of atomic structure learned by a student of physics or chemistry is the great disparity of the electron and nuclear masses. This same feature is involved in the Born–Oppenheimer approximation,[2] a principle which is of extraordinary importance in the field of molecular spectroscopy. Simply stated, the principle says that the motions of the electrons may be treated by assuming the nuclei to be fixed, since the lighter electrons undergo many periods of their motion during the time of one nuclear motion cycle. The nuclear motion problem, on the other hand, is treated by considering the nuclei to move in an effective potential produced by all the electrons. Another way of rationalizing this separability is to observe (as is evident later) that the electronic wave function is a very slowly varying function of internuclear separation R_α while the nuclear wave function varies rapidly with R_α (see Fig. 3.2).

The separation can be shown and understood qualitatively in the following way. We have written the molecular Hamiltonian previously as

$$\tilde{\mathcal{3C}} = -\frac{\hbar^2}{2} \sum_\alpha^N \frac{\nabla_\alpha^2}{m_\alpha} - \frac{\hbar^2}{2m_e} \sum_i^n \nabla_i^2 + V_n + V_e + V_{ne} \qquad (3.18)$$

in the absence of other than Coulombic potential energy terms. In (3.18) the potential energy terms V_n, V_e, and V_{ne} represent nuclear repulsion, electron repulsion and nuclear-electron attraction, respectively. It may be mentioned that the Hamiltonian which results after elimination of the translational motion [see Eq. (3.15)] could be used here as well as that involving all $3(n + N)$ coordinate variables. The essential features of what follows are independent of the translational separation. It is useful now

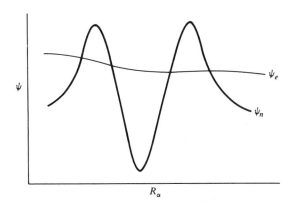

FIG. 3.2. Qualitative dependence of electronic and nuclear wave functions upon nuclear coordinate.

[2] M. Born and J. R. Oppenheimer, *Ann. Physik.* **84**, 457 (1927).

to rewrite $\tilde{\mathcal{3C}}$ as

$$\tilde{\mathcal{3C}} = \tilde{\mathcal{3C}}_n + \tilde{\mathcal{3C}}_e \tag{3.19}$$

where

$$\tilde{\mathcal{3C}}_n = -\tfrac{1}{2}\hbar^2 \sum_{\alpha}^{N} (1/m_\alpha) \nabla_\alpha^2 \tag{3.20}$$

and

$$\tilde{\mathcal{3C}}_e = -(\hbar^2/2m_e) \sum_{i}^{n} \nabla_i^2 + V_n + V_e + V_{ne} \tag{3.21}$$

Note that $\tilde{\mathcal{3C}}_e$ would be the proper Hamiltonian for the electronic energy if the nuclei were infinitely heavy, or if the nuclei were clamped rigidly (fixed internuclear separations). The wave equation for the electronic energy is then

$$\tilde{\mathcal{3C}}_e \psi_e(r_i, R_\alpha) = E_e(R_\alpha)\psi_e(r_i, R_\alpha) \tag{3.22}$$

in which the electronic eigenfunctions are explicit functions of the electron coordinates r_i, and are dependent parametrically upon the nuclear coordinates R_α. The eigenvalues of Eq. (3.22) clearly depend upon the nuclear coordinates also, since the terms V_n and V_{ne} involve the R_α's. The validity of Eq. (3.22) for treating electronic energies can be seen now if we return to the total Hamiltonian (3.19) and the complete wave equation

$$\tilde{\mathcal{3C}}\psi = (\tilde{\mathcal{3C}}_n + \tilde{\mathcal{3C}}_e)\psi = E\psi \tag{3.23}$$

Let us look for a solution in the form

$$\psi = \psi_e(r_i, R_\alpha) \cdot \psi_n(R_\alpha) \tag{3.24}$$

Upon substitution of (3.24) into (3.23) we obtain

$$\left\{ -\frac{\hbar^2}{2} \sum_{\alpha} \frac{1}{m_\alpha} \psi_n \nabla_\alpha^2 \psi_e - \hbar^2 \sum_{\alpha} \frac{1}{m_\alpha} \nabla_\alpha \psi_e \cdot \nabla_\alpha \psi_n \right\}$$

$$-\frac{\hbar^2}{2m_e} \sum_{i} \psi_n \nabla_i^2 \psi_e - \frac{\hbar^2}{2} \sum_{\alpha} \frac{1}{m_\alpha} \psi_e \nabla_\alpha^2 \psi_n \tag{3.25}$$

$$+ \psi_e \psi_n (V_n + V_e + V_{ne}) = E\psi_e \psi_n$$

where use has been made of the relations

$$\nabla_\alpha^2 \psi_e \psi_n = \nabla_\alpha(\nabla_\alpha \psi_e \psi_n)$$

$$= \nabla_\alpha(\psi_e \nabla_\alpha \psi_n + \psi_n \nabla_\alpha \psi_e) \tag{3.26}$$

$$= \psi_e \nabla_\alpha^2 \psi_n + \psi_n \nabla_\alpha^2 \psi_e + 2\nabla_\alpha \psi_e \cdot \nabla_\alpha \psi_n$$

and

$$\nabla_i^2 \psi_e \psi_n = \psi_n \nabla_i^2 \psi_e \tag{3.27}$$

It is interesting now to neglect temporarily the terms in brackets in Eq. (3.25). Then Eq. (3.25) may be written as

$$\psi_e \widetilde{\mathcal{3C}}_n \psi_n + \psi_n \widetilde{\mathcal{3C}}_e \psi_e = E \psi_e \psi_n \tag{3.28}$$

if use is made of Eqs. (3.20) and (3.21). Finally, after division of both sides of Eq. (3.28) by $\psi_e \psi_n$, we obtain

$$\widetilde{\mathcal{3C}}_n \psi_n / \psi_n + \widetilde{\mathcal{3C}}_e \psi_e / \psi_e = E \tag{3.29}$$

Equation (3.29) is, at first glance, in the form of a partial differential equation with separated variables, similar to (3.15). But, it must be remembered that the second term on the left-hand side of (3.29) depends parametrically upon the R_α's, which are the variables of the first term. Hence, the second term leads to Eq. (3.22), with $E_e(R_\alpha)$ being the value of $\widetilde{\mathcal{3C}}_e \psi_e / \psi_e$. We can then write

$$\widetilde{\mathcal{3C}}_n \psi_n / \psi_n + E_e = E \tag{3.30}$$

or finally, we obtain the Schrödinger equation for nuclear motions,

$$(\widetilde{\mathcal{3C}}_n + E_e) \psi_n = E \psi_n \tag{3.31}$$

Subject to the neglect of the terms in brackets in Eq. (3.25), it is seen that the wave equation can be separated into a part dependent upon electron coordinates [Eq. (3.22)] and a part dependent upon nuclear coordinates [Eq. (3.31)]. The electronic wave equation depends, however, upon the values of the R_α's, so that the $E_e(R_\alpha)$ acts as the potential energy for the nuclear motions, as is clearly shown in (3.31). In general, (3.22) leads to a manifold of states $\psi_e^{(i)}$, and energies $E_e^{(i)}$; and for each of these states, (3.31) lead to a set of nuclear states $\psi_n^{(i,j)}$ with total energies $E_{i,j}$. Figure 3.3 illustrates the situation for a hypothetical diatomic molecule. Four electronic states are illustrated, viz., $E_e^{(0)}$, $E_e^{(1)}$, $E_e^{(2)}$, and $E_e^{(3)}$. As is seen in chapter 11, the lower three electronic states are stable with respect to the free atoms, while the fourth, $E_e^{(3)}$, is unstable for all internuclear separations. For each of the three stable molecular states, Eq. (3.31) leads to the energies E_{ij} as shown in the figure. Although the eigenvalues of (3.31) appear to be the total energy (electronic plus nuclear), the calculation really involves only the determination of nuclear framework energy. This is most easily seen by choosing the zero of energy to be at the minimum of each (stable) electronic state. Then (3.31) is properly written as

$$(\widetilde{\mathcal{3C}}_n + E_e) \psi_n = E_n \psi_n \tag{3.32}$$

which form indicates clearly that the energies are for the nuclear motion.

Let us return now to a consideration of the neglected terms in Eq. (3.25). The approximation is satisfactory if these terms are much smaller than the smallest of the remaining terms, namely, the nuclear kinetic energy. Mathematically, we require

$$\psi_n \nabla_\alpha^2 \psi_e / \psi_e \nabla_\alpha^2 \psi_n \ll 1, \qquad \nabla_\alpha \psi_e \cdot \nabla_\alpha \psi_n / \psi_e \nabla_\alpha^2 \psi_n \ll 1 \qquad (3.33)$$

This holds as long as ψ_e varies much more slowly with respect to R than does ψ_n, since in this case $\nabla_\alpha \psi_e \ll \nabla_\alpha \psi_n$ and $\nabla_\alpha^2 \psi_e \ll \nabla_\alpha^2 \psi_n$ (see Fig. 3.2). We do not go into this further, but simply note that the situation is qualitatively as shown in Fig. 3.2 for most molecules, so the approximation is

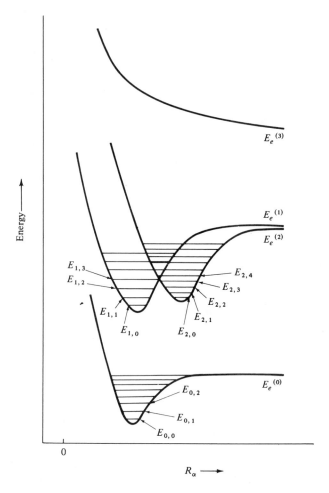

FIG. 3.3. Qualitative sketch of electronic energies, $E_e^{(i)}(R_\alpha)$ arising from solution of Eq. (3.22); and nuclear framework energies $E_{i,j}$ arising from Eq. (3.31) [or (3.32)].

excellent in most cases of interest.[3] The reader is urged to consult the literature for the quantitative justification of the Born–Oppenheimer approximation.[4]

The wave equation for electronic motions, (3.22), is considered in detail in chapter 11. Some of the results of chapter 11 [particularly the functional form of $E_e(R_\alpha)$] are utilized in the preceding chapters, but little detailed knowledge is necessary. Equation (3.32) is analyzed further in the Sec. 3.3, where one final, approximate separation of variables is performed.

3.3 SEPARATION OF VIBRATIONAL AND ROTATIONAL ENERGIES

It has been shown in Secs. 3.1 and 3.2 that it is possible to treat separately the over-all molecular translational energy, the electronic energy for a rigid nuclear framework, and the nuclear framework internal (nontranslational) energies. It should be apparent that the orientation of the nuclear framework in space has no direct dependence upon $E_e(R_\alpha)$ to the extent that previously described approximations are valid. For this reason, the rotational energy of the field-free nuclear framework might be expected to be separable from the remaining internal motions. This separation must depend, of course, upon the possibility of writing the wave equation in such a form that there is no coupling between rotational and nonrotational variables. It is not intuitively obvious that this is possible, and indeed, it can be done only approximately.

Before proceeding let us note that the only masses involved in Eq. (3.32) are the nuclear masses m_α. Since the nuclear mass differs from the atomic mass by only about 0.05%, it is common practice to write this expression in terms of the atomic rather than the nuclear masses. For most spectroscopic applications this transcription produces results with negligibly small errors. In fact, better experimental agreement is usually obtained when atomic masses are utilized. This is because the inner shell electrons in molecules remain essentially spherically distributed about their parent nuclei, and therefore act as though they were concentrated at the nuclei. It is only the "valence" electrons in nonspherical molecular orbitals which do not move as though their mass were concentrated at the nuclei. This phenomenon is known as electron "slip," and is really a breakdown of the Born–Oppenheimer approximation. Although there are several

---

[3] As a very practical example of the breakdown of the approximation, the Λ doubling of $^2\Pi$ diatomic molecules may be mentioned. It arises from the coupling of electronic and rotational motions and was treated quite early by E. Hill and J. van Vleck, *Phys. Rev.* **32**, 250 (1928).

[4] See footnote 2 and also W. Kolos and L. Wolniewicz, *J. Chem. Phys.* **41**, 3663 (1964).

spectroscopic measurements with sufficient sensitivity to detect this property,[5] we continue our treatment with atomic masses substituted for nuclear masses.

The kinetic energy of a system of point masses that are free to translate, rotate and vibrate in three dimensions has been described in a number of places. The treatment we follow is essentially that of Wilson, Decius and Cross.[6] Most students of physical science are probably familiar with the qualitative aspects. Our molecular system is considered to consist of N point masses, the atoms, with $3N$ degrees of freedom; that is, the positions in space of the N masses can be completely specified by a suitable set of $3N$ coordinates. Three coordinates suffice to define the translational motion of the center of mass of the molecule, as shown in Sec. 3.1. In a similar fashion, the angular orientation of the molecule in space can be specified, in general, by three (two for a linear molecule) angular coordinates. This leaves $3N - 6$ ($3N - 5$ for linear molecules) coordinates to describe the atomic positions relative to some internal axis system which *rotates* and *translates* with the molecule. Since strong restoring forces prevent the molecule from flying apart, but instead, cause the atoms to undergo an oscillatory motion about an equilibrium position, these $3N - 6$ (or $3N - 5$) coordinates are correctly considered to represent the vibrational degrees of freedom.

The situation is represented in Fig. 3.4 for a linear triatomic molecule, such as N_2O. In the laboratory (space-fixed) axis system, the c.m. is located by X_c, Y_c, Z_c. If the axes fixed in the molecule are labeled x, y, and z, and have their origin at the c.m., the orientation of the molecule is given by the angular variables ϑ and φ. The remaining $3N - 5 = 4$ coordinates have been chosen to be θ_1, θ_2, $(z_3 - z_2) - R_{23}$ and $(z_1 - z_2) - R_{12}$. θ_1 represents bending of the two bonds in the xz plane, while θ_2 is the similar bend in the yz plane. $(z_3 - z_2) - R_{23}$ is the extension of the 2–3 bond whose equilibrium length is R_{23}. $(z_1 - z_2) - R_{12}$ is the similar extension coordinate for the 1–2 bond. It becomes clear later in this chapter and in chapter 8 that these are suitable, but not unique, vibrational coordinates.

To proceed to the more general problem, we now make use of the coordinate systems outlined in Fig. 3.5. Although we have proved that translational kinetic energy is separable, we again put in the full set of $3N$ coordinates for completeness. The c.m. of the nuclear (atomic) system is located in the laboratory space-fixed axis system (X, Y, Z) by the vector $\mathbf{R}$. The (x, y, z) axis system is fixed in, and moves with the molecule, and its

---

[5] See C. H. Townes and A. L. Schawlow, *Microwave Spectroscopy* (McGraw-Hill Book Co., New York, 1955), for a detailed discussion. Michael Tinkham, *Group Theory and Quantum Mechanics* (McGraw-Hill Book Co., New York, 1964), gives a brief, clear discussion also.

[6] E. B. Wilson, Jr., J. C. Decius, P. C. Cross, *Molecular Vibrations* (McGraw-Hill Book Co., New York, 1955).

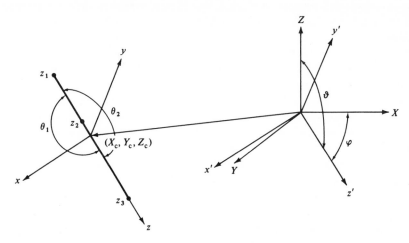

FIG. 3.4. Possible set of coordinates used to specify the positions of the atoms of a linear triatomic molecule undergoing translational, rotational and vibrational motions. X, Y, Z are spaced-fixed axes and x, y, z are molecule-fixed axes located at center of mass (X_c, Y_c, Z_c). Angular position of molecule in space is determined by ϑ and φ. θ_1 and θ_2 are bending coordinates in the xz and yz planes, respectively.

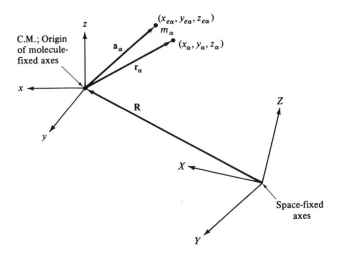

FIG. 3.5. Axis systems used to specify positions of N atoms (nuclei) of any molecule in space. Molecule-fixed axes have their origin at center of mass (CM). $\mathbf{r}_\alpha$ is position vector of αth atom at any time; $\mathbf{a}_\alpha$ is the equilibrium position vector.

orientation with respect to the (X, Y, Z) axis system may be specified by three angular coordinates not shown in the figure. These would be the same two used in Fig. 3.4 plus one more which we need not specify any further.[7] Any atom of mass m_α can now be located in the molecule-fixed axis system by the vector $\mathbf{r}_\alpha$ with components $x_\alpha, y_\alpha, z_\alpha$. These coordinates, plus the previously mentioned ones, are sufficient to define the position of every atom in space. Note that we have $3N + 3 + 3 = 3N + 6$ coordinates in all, so that six coordinates are *redundant*; that is, six of the coordinates are not independent. We indicate later how these redundant coordinates may be removed.

It becomes convenient later if the equilibrium position of each atom in the molecule-fixed axes is specified by the vector $\mathbf{a}_\alpha$, with components $x_{\alpha e}, y_{\alpha e}, z_{\alpha e}$. At this point, the equilibrium position may be considered to be arbitrary, but we see later that it is the position about which the atom oscillates. The displacement from equilibrium may be written

$$\varrho_\alpha = \mathbf{r}_\alpha - \mathbf{a}_\alpha \tag{3.34}$$

A. Kinetic Energy of the Nuclear (Atomic) System

If $\mathbf{u}_\alpha$ is the velocity of the αth particle in the space-fixed axis system, the kinetic energy may be written

$$2T = \sum_\alpha^N m_\alpha \mathbf{u}_\alpha \cdot \mathbf{u}_\alpha = \sum_\alpha^N m_\alpha u_\alpha^2 \tag{3.35}$$

Our problem is to express $\mathbf{u}_\alpha$ in terms of the coordinate systems and position vectors of Fig. 3.5. From the nature of the problem, it is clear that $\mathbf{u}_\alpha$ contains contributions from the translation of the c.m., the rotation of the system of atoms, and the vibrational motion of the atoms in the molecule-fixed axis system:

$$\mathbf{u}_\alpha = \mathbf{u}_{\text{tran}} + \mathbf{u}_{\text{rot}} + \mathbf{u}_{\text{vib}} \tag{3.36}$$

The translational velocity is simply the velocity at which the c.m. moves with respect to the space-fixed axes, that is, $\mathbf{u}_{\text{tran}} = \dot{\mathbf{R}}$. If the x, y, z axes are rotating at an angular velocity $\boldsymbol{\omega}$, or equivalently, if the vector $\mathbf{r}_\alpha$ is rotating with angular velocity $\boldsymbol{\omega}$, the linear rotational velocity of mass m_α is $\boldsymbol{\omega} \times \mathbf{r}_\alpha$.[8] The motion of m_α with respect to the molecule-fixed axes (vibrational motion) may be written simply as $\mathbf{v}_\alpha$, where $\mathbf{v}_\alpha$ has the components $\dot{x}_\alpha, \dot{y}_\alpha,$

---

[7] These are the Eulerian angles. See footnote 6 for a definition of the complete set.

[8] See for example, J. L. Synge and B. A. Griffith, *Principles of Mechanics*, 3rd ed. (McGraw-Hill Book Co., New York, 1959). Everyone is certainly aware from very elementary physics that the magnitude of the linear velocity of a point moving about the circumference of a circle is ωr. This results from $|\boldsymbol{\omega} \times \mathbf{r}| = \omega r \sin\theta$, since $\theta = 90°$ in this case.

and $\dot{z}_\alpha$. The final result for $\mathbf{u}_\alpha$ is then

$$\mathbf{u}_\alpha = \dot{\mathbf{R}} + \boldsymbol{\omega} \times \mathbf{r}_\alpha + \mathbf{v}_\alpha \tag{3.37}$$

The kinetic energy of the system of point masses is now obtained by substitution of (3.37) into (3.35).

$$2T = \dot{R}^2 \sum_\alpha m_\alpha + \sum_\alpha m_\alpha (\boldsymbol{\omega} \times \mathbf{r}_\alpha) \cdot (\boldsymbol{\omega} \times \mathbf{r}_\alpha) + \sum_\alpha m_\alpha v_\alpha{}^2$$
$$+ 2\boldsymbol{\omega} \cdot \sum_\alpha (m_\alpha \mathbf{r}_\alpha \times \mathbf{v}_\alpha) + 2\dot{\mathbf{R}} \cdot \boldsymbol{\omega} \times \sum_\alpha m_\alpha \mathbf{r}_\alpha + 2\dot{\mathbf{R}} \cdot \sum_\alpha m_\alpha \mathbf{v}_\alpha \tag{3.38}$$

This rather complicated expression can be simplified immediately if we recall that there must exist six relations which remove the six redundant coordinates. The first three of these are quite straightforward, being merely the relations which locate the c.m.;

$$\sum_\alpha m_\alpha x_\alpha = 0 \tag{3.39a}$$

$$\sum_\alpha m_\alpha y_\alpha = 0 \tag{3.39b}$$

$$\sum_\alpha m_\alpha z_\alpha = 0 \tag{3.39c}$$

or in vector form,

$$\sum_\alpha m_\alpha \mathbf{r}_\alpha = 0 \tag{3.40}$$

From this, it follows that also[9]

$$\sum_\alpha m_\alpha \mathbf{v}_\alpha = 0 \tag{3.41}$$

By use of (3.40) and (3.41) we obtain Eq. (3.38) in the form

$$2T = \dot{R}^2 \sum_\alpha m_\alpha + \sum_\alpha m_\alpha (\boldsymbol{\omega} \times \mathbf{r}_\alpha) \cdot (\boldsymbol{\omega} \times \mathbf{r}_\alpha) + \sum_\alpha m_\alpha v_\alpha{}^2$$
$$+ 2\boldsymbol{\omega} \cdot \sum_\alpha (m_\alpha \mathbf{r}_\alpha \times \mathbf{v}_\alpha) \tag{3.42}$$

The three conditions necessary in addition to those of Eqs. (3.39) specify the manner in which the vibrational motions occur. Specifically, they must restrict the internal motions in such a way that the atomic system does not rotate with respect to the molecule-fixed axes. This is a rather subtle point, but it should be clear that up to now we have not defined exactly what is meant by "molecule-fixed" axes; since evidently we are permitting the atoms to undergo some kind of oscillatory motion. The necessary conditions have been discussed by Eckart,[10] and we merely reproduce them, since they are not critical to our treatment. In vector form,

[9] By differentiation (3.40) becomes $\sum_\alpha m_\alpha \dot{\mathbf{r}}_\alpha = 0$. But $\dot{\mathbf{r}}_\alpha$ is the velocity of a point α as observed in the space-fixed axis system. Therefore, $\dot{\mathbf{r}}_\alpha = \boldsymbol{\omega} \times \mathbf{r}_\alpha + \mathbf{v}_\alpha$. We get then, $\boldsymbol{\omega} \times \sum_\alpha m_\alpha \mathbf{r}_\alpha + \sum_\alpha m_\alpha \mathbf{v}_\alpha = 0$, or $\sum_\alpha m_\alpha \mathbf{v}_\alpha = 0$.

[10] C. Eckart, *Phys. Rev.* **47**, 552 (1935).

they may be written as

$$\sum_{\alpha} m_{\alpha} \mathbf{a}_{\alpha} \times \mathbf{r}_{\alpha} = 0 \qquad (3.43a)$$

or

$$\sum_{\alpha} m_{\alpha} \mathbf{a}_{\alpha} \times \mathbf{v}_{\alpha} = 0 \qquad (3.43b)$$

Essentially, (3.43b) states that there shall be no angular momentum with respect to the molecule-fixed axes. In this fashion, we are guaranteed that the molecule cannot "get away" from its axis system. When (3.43b) is substituted into (3.42) and use is made of (3.34), we obtain the slightly modified equation

$$2T = \dot{R}^2 \sum_{\alpha} m_{\alpha} + \sum_{\alpha} m_{\alpha} (\boldsymbol{\omega} \times \mathbf{r}_{\alpha}) \cdot (\boldsymbol{\omega} \times \mathbf{r}_{\alpha}) + \sum_{\alpha} m_{\alpha} v_{\alpha}^2$$

$$+ 2\boldsymbol{\omega} \cdot \sum_{\alpha} m_{\alpha} (\mathbf{p}_{\alpha} \times \mathbf{v}_{\alpha}) \qquad (3.44)$$

It can be noted that only the first term contains the variable R. Therefore, in any wave equation containing a potential energy term that is independent of R, this translational energy term separates, a result which was derived previously in Sec. 3.1. We neglect this term from this point on.

The remaining three terms now have the following significance. If $\mathbf{v}_{\alpha} = 0$ the third and fourth terms in Eq. (3.44) would go to zero, leaving only the second term which would clearly be the rotational kinetic energy of a rigid, nonvibrating molecule with constant $\mathbf{r}_{\alpha}$'s. On the other hand, if $\boldsymbol{\omega} = 0$, the second and fourth terms are zero, leaving only the third term, which would correspond to the kinetic energy of the atoms relative to the molecular framework or molecule-fixed axis system. Or as we presently see, this term would be the kinetic energy of the oscillating nuclei in the field of the electrons. When neither $\boldsymbol{\omega}$ nor $\mathbf{v}_{\alpha}$ are zero, all of the terms are present, and it is evident that the fourth term produces a *coupling* between vibration and rotation, such that neither has a strictly distinct meaning.

The kinetic energy given by Eq. (3.44) could be put into quantum mechanical operator form as shown by Wilson, Decius, and Cross.[11] We do not go into this rather lengthy procedure, but simplify the classical kinetic energy by noting that the coupling term can usually be neglected except for those cases when the vibrational amplitudes (which are related to ϱ_{α}) are large, or when certain degeneracies occur among the vibrational states.[12] It may be mentioned that large vibrational amplitudes are usually associated with low frequency vibrations such as torsional motions of CH_3 groups, or out-of-plane ring-bending vibrations in cyclic species. As a good first approximation, therefore, the coupling term of Eq. (3.44) can be neglected, leaving only the rotational and vibrational kinetic energy terms.

[11] See footnote 6, pp. 279–283.
[12] See footnote 6, p. 367.

To the approximation discussed in Sec. 3.2, the classical Hamiltonian for nuclear motions (exclusive of translation) may be written

$$\mathcal{H}_{vr} = \tfrac{1}{2}\sum_\alpha m_\alpha(\boldsymbol{\omega} \times \mathbf{a}_\alpha) \cdot (\boldsymbol{\omega} \times \mathbf{a}_\alpha) + \tfrac{1}{2}\sum_\alpha m_\alpha v_\alpha^2 + V(r_\alpha) \quad (3.45)$$

where $V(r_\alpha)$ is the potential energy term that was labeled $E_e(R_\alpha)$ in Eq. (3.31), and for the limit of small vibrations we have replaced $\mathbf{r}_\alpha$ by $\mathbf{a}_\alpha$ in the rotational term.

It should be recalled now that the rotational term is a function of only the three angles relating X, Y, Z to x, y, z, while the second and third terms are functions of x_α, y_α, and z_α. If we write

$$\tilde{\mathcal{H}}_{vr} = \tilde{\mathcal{H}}_r + \tilde{\mathcal{H}}_v \quad (3.46)$$

where $\tilde{\mathcal{H}}_r$ is the operator form of the rotational kinetic energy and $\tilde{\mathcal{H}}_v$ is the operator form for the vibrational kinetic and potential energies. Writing

$$\psi_{vr} = \psi_v \cdot \psi_r \quad (3.47)$$

it is easily shown by procedures similar to those of Secs. 3.1 and 3.2 that the variables of the wave equation

$$\tilde{\mathcal{H}}_{vr}\psi_{vr} = E_{vr}\psi_{vr} \quad (3.48)$$

separate, yielding

$$\tilde{\mathcal{H}}_r\psi_r = E_r\psi_r \quad (3.49)$$

and

$$\tilde{\mathcal{H}}_v\psi_v = (\tilde{T}_v + \tilde{V})\psi_v = E_v\psi_v \quad (3.50)$$

with

$$E_r + E_v = E_{vr} \quad (3.51)$$

The result, as we suspected earlier, is that the vibrational and rotational energies are separable, and may be treated as separate problems as long as the vibration-rotation coupling term is small. In practice, the separation is never perfect, since $\tilde{\mathcal{H}}_r$ contains some vibrational effects because of the dependence of the $\mathbf{a}_\alpha$'s upon the vibrational states. The approximation is, nevertheless, an excellent one for the analysis of most of the observed molecular spectra.

B. Rotational Hamiltonian

From Eq. (3.45) the classical rotational Hamiltonian is

$$\mathcal{H}_r = \tfrac{1}{2}\sum_\alpha m_\alpha(\boldsymbol{\omega} \times \mathbf{a}_\alpha) \cdot (\boldsymbol{\omega} \times \mathbf{a}_\alpha) \quad (3.52)$$

When this is expanded by standard methods of vector algebra, Eq. (3.53) is obtained

$$\mathcal{H}_r = \tfrac{1}{2}\sum_\alpha m_\alpha\{ (\omega_x^2 + \omega_y^2 + \omega_z^2)(x_{ae}^2 + y_{ae}^2 + z_{ae}^2)$$

$$- (\omega_x x_{ae} + \omega_y y_{ae} + \omega_z z_{ae})^2\} \quad (3.53)$$

where x_{ae}, etc., are the components of the equilibrium position vectors $\mathbf{a}_\alpha$. By defining in the usual way the moments of inertia

$$I_{xx} = \sum_\alpha m_\alpha(y_{ae}^2 + z_{ae}^2) \text{ etc.}, \tag{3.54}$$

and the products of inertia

$$I_{xy} = I_{yx} = \sum_\alpha m_\alpha x_{ae} y_{ae}, \text{ etc.}, \tag{3.55}$$

the Hamiltonian may be written

$$\mathcal{H}_r = \tfrac{1}{2}\{I_{xx}\omega_x^2 + I_{yy}\omega_y^2 + I_{zz}\omega_z^2 - 2I_{xy}\omega_x\omega_y - 2I_{xz}\omega_x\omega_z - 2I_{yz}\omega_y\omega_z \tag{3.56}$$

It is interesting to note, also, that (3.56) may be written in the form (see Appendix C)

$$2\mathcal{H}_r = \boldsymbol{\omega}\mathbf{I}\boldsymbol{\omega}$$

$$= (\omega_x \omega_y \omega_z) \begin{pmatrix} I_{xx} & -I_{xy} & -I_{xz} \\ -I_{yx} & I_{yy} & -I_{yz} \\ -I_{zx} & -I_{zy} & I_{zz} \end{pmatrix} \begin{pmatrix} \omega_x \\ \omega_y \\ \omega_z \end{pmatrix} \tag{3.57}$$

$\mathbf{I}$ is known as the *inertial* tensor and is obviously in the form of a symmetric matrix. Since it is always possible to find an axis system for which the inertial tensor is diagonal,[13] we may write for this *principal axis system*,

$$\mathcal{H}_r = \tfrac{1}{2}\{I_{x'x'}\omega_{x'}^2 + I_{y'y'}\omega_{y'}^2 + I_{z'z'}\omega_{z'}^2\} \tag{3.58}$$

where the primes indicate the new molecule-fixed axis system.

In order to put the Hamiltonian into the proper quantum mechanical form the following substitutions of angular momentum components are made:

$$P_x = I_{xx}\omega_x$$

$$P_y = I_{yy}\omega_y \tag{3.59}$$

$$P_z = I_{zz}\omega_z$$

in which the primes have been suppressed. The rotational Hamiltonian becomes

$$\mathcal{H}_r = \frac{P_x^2}{2I_{xx}} + \frac{P_y^2}{2I_{yy}} + \frac{P_z^2}{2I_{zz}} \tag{3.60}$$

in the inertial principal axis system. The moments of inertia are evaluated for the equilibrium atomic positions. As such, Eq. (3.60) represents the

---

[13] See footnote 8, pp. 282–288. Also see chapter 7 for a practical discussion of principal axes in molecules.

rigid-rotor approximation. Since the equilibrium positions depend upon vibrational amplitudes as previously mentioned, the moments of inertia differ, in general, in different vibrational states.

It is possible to write the quantum-mechanical operators for P_x, P_y, and P_z in terms of the angular variables previously described. A general analytical solution of the rotational wave equation (3.49) is not, however, possible. For this reason, the matrix mechanics approach is most useful, and is the method that will be utilized in chapter 7. The necessary matrix elements of angular momentum are derived in chapter 4.

EXAMPLE 3-1

The preceding theory can be illustrated for the simple case of the diatomic molecule of Fig. 3.6. Since the atoms are taken as point masses, I_{zz} is zero, and there is no angular momentum about the z axis. The origin of the axis system of Fig. 3.6 is at the c.m., so $I_{xx} = I_{yy} = I$. Noting that $P^2 = P_x{}^2 + P_y{}^2 + P_z{}^2$, Equation (3.60) becomes for the diatomic molecule,

$$\mathcal{K}_r = P^2/2I \tag{3.61}$$

By use of Eq. (3.54), the moment of inertia can be written

$$I = m_2 z_2{}^2 + m_1 z_1{}^2 \tag{3.62}$$

or, by noting that

$$m_1 z_1 + m_2 z_2 = 0 \tag{3.63}$$

and

$$R = z_2 - z_1 \tag{3.64}$$

the moment of inertia becomes

$$I = \frac{m_1 m_2}{m_1 + m_2} R^2 \tag{3.65}$$

By introducing the reduced mass

$$\mu = \frac{m_1 m_2}{m_1 + m_2} \tag{3.66}$$

the Hamiltonian becomes

$$\mathcal{K}_r = \frac{P^2}{2\mu R^2} \tag{3.67}$$

The quantum mechanical solution of Eq. (3.49) with the Hamiltonian (3.67) is easily obtained as is illustrated in chapter 7.

C. Vibrational Hamiltonian

The remaining part (terms two and three) of Eq. (3.45) represents the vibrational energy of the nuclei in the field of the electrons. Inserting the linear momentum $\mathbf{p}_\alpha = m_\alpha \mathbf{v}_\alpha$ into (3.45), the Hamiltonian becomes

$$\mathcal{H}_v = \sum_\alpha \frac{p_\alpha^2}{2m_\alpha} + V(r_\alpha) \tag{3.68}$$

$V(r_\alpha)$, it is recalled, is the electronic energy as a function of the r_α's. In general it has no simple functional form, so an exact solution of Eq. (3.50), with the operator form of (3.68) substituted, is not possible. It is observed experimentally, though, that the behavior of the potential function is often very nearly parabolic near the minimum (see Fig. 3.3), that is, $V(r_\alpha) \sim \frac{1}{2}k_\alpha(r_\alpha - a_\alpha)^2$. In such a case, the problem reduces to the well-known classical *harmonic oscillator* and the quantum-mechanical wave equation (3.50) is solvable in the proper system of coordinates. More exact potential functions have been utilized successfully for the diatomic molecule in an attempt to reproduce the correct behavior of $V(r_\alpha)$ at $r_\alpha = 0$ and $r_\alpha = \infty$. Because the wave equation becomes many dimensional for polyatomic molecules, the solution of (3.50) is normally restricted to the simple, but still very useful, harmonic approximation.

It is most convenient to rewrite the vibrational kinetic energy as

$$T = \frac{1}{2} \sum_{\alpha=1}^{N} m_\alpha(\dot{x}_\alpha^2 + \dot{y}_\alpha^2 + \dot{z}_\alpha^2) \tag{3.69}$$

by introducing the components of $\mathbf{v}_\alpha$. Recalling that $\mathbf{r}_\alpha = \boldsymbol{\varrho}_\alpha + \mathbf{a}_\alpha$ from Eq. (3.34), it follows that $x_\alpha = \Delta x_\alpha + x_{e\alpha}$ and $\dot{x}_\alpha = (\Delta \dot{x}_\alpha)$ since the equilibrium coordinate $x_{e\alpha}$ is constant. If mass weighted displacement coordinates are defined by

$$q_1 = \sqrt{m_1}\,\Delta x_1$$
$$q_2 = \sqrt{m_1}\,\Delta y_1$$
$$\vdots$$
$$q_4 = \sqrt{m_2}\,\Delta x_2 \tag{3.70}$$
$$q_5 = \sqrt{m_2}\,\Delta y_2$$
$$\vdots$$
$$q_{3N} = \sqrt{m_N}\,\Delta z_N$$

the kinetic energy becomes

$$T = \frac{1}{2} \sum_{i=1}^{3N} \dot{q}_i^2 \tag{3.71}$$

Since the precise form of $V(r_\alpha)$ is seldom known, it is usual to expand $V(r_\alpha) \propto V(q_i)$ in a power series in q_i.

$$V = V_0 + \sum_i^{3N} \left(\frac{\partial V}{\partial q_i}\right)_0 q_i + \frac{1}{2} \sum_{i,j}^{3N} \left(\frac{\partial^2 V}{\partial q_i \partial q_j}\right)_0 q_i q_j + \cdots. \tag{3.72}$$

By adjusting the energy scale, the constant term V_0 can be eliminated, and since $(\partial V/\partial q_i)_0$ must be zero when all $q_i = 0$,[14]

$$V = \frac{1}{2}\sum_{i,j}^{3N} f_{ij}q_iq_j \tag{3.73}$$

where the f_{ij} are the second derivatives of Eq. (3.72) and are known as force constants. The Hamiltonian then becomes

$$\mathfrak{IC}_v = \frac{1}{2}\sum_{i=1}^{3N} \dot{q}_i{}^2 + \frac{1}{2}\sum_{i,j}^{3N} f_{ij}q_iq_j \tag{3.74}$$

Although this equation could now be put into quantum-mechanical operator form, we do not, since the variables of Schrödinger's equation (3.50) would not be separable in this coordinate system. Equation (3.74) is, however, a convenient form for a classical treatment of the vibrational problem. Both the classical and quantum-mechanical treatment are presented in chapter 8 with Eq. (3.74) as a starting point.

It should be pointed out that we have written (3.74) in terms of $3N$ coordinate variables, which means that $3N - 6$ (or $3N - 5$) of these variables are not involved in the vibrational motion, but instead refer to translation and rotation. These variables could be removed by use of the redundancy conditions (3.40) and (3.43), but it is often easier to start the problem over with a set of $3N - 6$ (or $3N - 5$) *internal* coordinates which satisfy the redundancy conditions immediately. The use of internal coordinates is illustrated in chapter 8. It turns out that no theoretical problems occur, however, if we carry along the redundant coordinates. When the classical or quantum-mechanical problem is treated, the redundant variables lead to zero vibrational frequencies and energies, a perfectly reasonable result since the rotational and translational motions are subject to no restoring forces (that is, the force constants for the translational and rotational coordinates are zero).

EXAMPLE 3-2

Equation (3.74) and the redundancy conditions can be nicely illustrated for the diatomic molecule of Fig. 3.6. Writing the kinetic energy (3.71) in terms of the displacement coordinates (3.70), we find

$$2T = m_1[(\Delta\dot{x}_1)^2 + (\Delta\dot{y}_1)^2 + (\Delta\dot{z}_1)^2]$$
$$+ m_2[(\Delta\dot{x}_2)^2 + (\Delta\dot{y}_2)^2 + (\Delta\dot{z}_2)^2] \tag{3.75}$$

[14] This is true since at equilibrium (all $q_i = 0$), the force on every particle must be zero. Therefore, $(\partial V/\partial q_i)_0 = f_i = 0$, and the equilibrium position corresponds to the minimum of the potential curve.

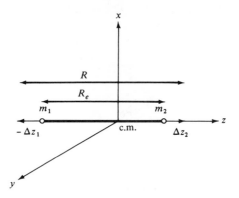

FIG. 3.6.   Diatomic molecule with axis system at center of mass. R is bond distance at any time and R_e is equilibrium bond distance. Δz_1 and Δz_2 are displacement coordinates.

The redundancy relations lead to expressions of the form[15]

$$m_1 \Delta x_1 + m_2 \Delta x_2 = 0$$

or

$$\Delta x_1 = -\Delta x_2 \left(\frac{m_2}{m_1}\right) \tag{3.76}$$

By differentiation,

$$(\Delta \dot{x}_1) = -(\Delta \dot{x}_2) \left(\frac{m_2}{m_1}\right) \tag{3.77}$$

so we may eliminate $(\Delta \dot{x}_1)$, $(\Delta \dot{y}_1)$, and $(\Delta \dot{z}_1)$ from (3.75). Furthermore, the redundancy conditions (3.43a) lead to an expression of the form

$$m_1(z_{e1}\Delta x_1 - x_{e1}\Delta z_1) + m_2(z_{e2}\Delta x_2 - x_{e2}\Delta z_2) = 0 \tag{3.78}$$

Since the coordinate system is situated at the c.m., it is clear that $y_{e1} = y_{e2} = 0 = x_{e1} = x_{e2}$. With these results, the substitution of (3.78) into (3.76) leads to

$$(\Delta x_2) = -\Delta x_1 \left(\frac{m_1}{m_2}\right)\left(\frac{z_{e1}}{z_{e2}}\right) \tag{3.79}$$

$$= \Delta x_2 \left(\frac{z_{e1}}{z_{e2}}\right)$$

[15] Note that Eqs. (3.39) may be written $\sum_\alpha m_\alpha \Delta x_\alpha = 0$, etc., since (3.39) must apply to the equilibrium position, that is $\sum_\alpha m_\alpha x_{e\alpha} = 0$.

But (3.79) can be valid only if $\Delta x_2 = 0$, since z_{e1}/z_{e2} cannot be zero. We conclude, then, that Δx_2 and Δy_2 are zero, which leads to

$$(\Delta \dot{x}_2) = (\Delta \dot{x}_1) = (\Delta \dot{y}_1) = (\Delta \dot{y}_2) = 0 \tag{3.80}$$

The kinetic energy may now be written

$$2T = \frac{m_2}{m_1}(m_1 + m_2)(\Delta \dot{z}_2)^2$$

$$\tag{3.81}$$

$$= \frac{m_2}{m_1}(m_1 + m_2)\dot{z}_2^2$$

which shows that the redundancy relations have reduced the kinetic energy to a function of one variable.

It is now convenient to introduce a new variable ξ defined by

$$\xi = R - R_e = (z_2 - z_1) - (z_{e2} - z_{e1})$$

$$\tag{3.82}$$

$$= \Delta z_2 - \Delta z_1$$

Introducing Eq. (3.82) into Eq. (3.81) and using the z-axis redundancy relation Eq. (3.39) once more, we get

$$2T = \frac{m_1 m_2}{m_1 + m_2}\dot{\xi}^2 \tag{3.83a}$$

or

$$2T = \mu\dot{\xi}^2 \tag{3.83b}$$

using the *reduced mass*

$$\mu = \frac{m_1 m_2}{m_1 + m_2} \tag{3.84}$$

By defining the linear momentum

$$p = \mu\dot{\xi} \tag{3.85}$$

Eq. (3.83b) becomes

$$T = \frac{p^2}{2\mu} \tag{3.83c}$$

Note that Eqs. (3.83) show that the vibrational kinetic energy of the diatomic molecule reduces to that of a particle of mass μ moving in the ξ direction.

Turning now to the potential energy, we note that $f_{36} = f_{63} = f$ of Eq. (3.73) is the only independent nonzero force constant. We get, therefore,

$$V = fq_3q_6 \tag{3.86}$$

which becomes

$$V = \left(\frac{\partial^2 V}{\partial \Delta z_1 \partial \Delta z_2}\right)_0 \Delta z_1 \Delta z_2 \tag{3.87}$$

by introduction of the nonmass weighted coordinates. By using Eq. (3.82) and the redundancy relation between Δz_1 and Δz_2, we can transform Eq. (3.87) to the variable ξ, giving

$$V = \left(\frac{\partial^2 V}{\partial \xi^2}\right)_0 \xi^2 \tag{3.88a}$$

Using the definition $(\partial^2 V/\partial \xi^2)_0 = \frac{1}{2}k$ we obtain

$$V = \tfrac{1}{2}k\xi^2 \tag{3.88b}$$

where k is the conventional Hooke's Law force constant. The Hamiltonian is obtained by summing Eqs. (3.83c) and (3.88b), which gives

$$\mathcal{H}_v = (p^2/2\mu) + \tfrac{1}{2}k\xi^2 \tag{3.89}$$

This is the well-known form for a one-dimensional harmonic oscillator. As will be shown in chapters 4 and 8, the Hamiltonian matrix of Eq. (3.89) is easily obtained in diagonal form.

3.4 OTHER MOLECULAR ENERGY TERMS, INCLUDING EXTERNAL-FIELD TERMS; PERTURBATIONS

We have seen that to a good approximation the largest energy terms may be described by the molecular Hamiltonian

$$\widetilde{\mathcal{H}} = \widetilde{\mathcal{H}}_e + \widetilde{\mathcal{H}}_v + \widetilde{\mathcal{H}}_r \tag{3.90}$$

Writing the wave function as a product

$$\psi = \psi_e \cdot \psi_v \cdot \psi_r \tag{3.91}$$

the molecular energies were found to be

$$E = E_e + E_v + E_r \tag{3.92}$$

The magnitudes of these terms become clearer later, but we may mention here that $E_e \sim 10^2 \, E_v \sim 10^4 \, E_r$. The other energy terms of interest arise from the Hamiltonian potential energy terms which were indicated by V' in Eq. (3.5). These terms are in many cases quite small (on the order of $10^{-3} \, E_r$, say) so they can be treated by the well-known methods of perturbation theory using the functions given by (3.91) as the basis functions for the perturbation treatment.

To show the method more clearly, consider the effect of applying a static electric field on a polar diatomic molecule. The perturbation term in the Hamiltonian is now of the form $\tilde{\mathcal{H}}' = -\mu \cdot \mathcal{E} = -\mu \mathcal{E} \cos \theta$, where μ and $\mathcal{E}$ are the electric dipole moment and electric field, respectively, and θ is the angle between the field direction and μ. The dipole moment μ depends on the electron and nuclear coordinates, and therefore depends upon ψ_e and ψ_v; it is independent of ψ_r to the extent that vibration-rotation coupling is absent. The $\cos\theta$ term involves the orientation of the molecule with respect to space-fixed axes, and therefore has rotational dependence, but not electronic or vibrational. Since $\mu\mathcal{E} \sim 10^{-3} \tilde{\mathcal{H}}_r$, the Stark effect appears as a fine structure on the rotational levels E_r. The first-order perturbation correction to the $\psi_e \psi_v \psi_r$ state is then

$$E^{(1)} = -\mathcal{E}(ev \mid \tilde{\mu} \mid ev)(r \mid \cos \theta \mid r) \tag{3.93}$$

where e, v, and r are quantum labels for the states. It turns out, in fact, (see chapter 7) that this first-order term goes to zero, so we obtain the second-order term which is

$$E^{(2)} = -\sum_{r'} \frac{\mathcal{E}^2 \mid (ev \mid \tilde{\mu} \mid ev) \mid^2 \mid (r \mid \cos \theta \mid r') \mid^2}{E_{evr} - E_{evr'}} \tag{3.94}$$

Note that we do not sum over e' and v' since the electronic and vibrational energy levels are widely spaced and the perturbation operator is small. $(ev \mid \tilde{\mu} \mid ev)$ is just the average value of the dipole moment ($\bar{\mu}$ or $\langle \mu \rangle$) in the state $\psi_e \psi_v$. This illustrates the important fact that low-energy perturbations normally turn out to be averages over the electronic and vibrational wave functions, although this result is often suppressed in formulating the problem. It may be further mentioned that the term $E^{(2)}$ breaks a degeneracy of the rotational levels, so that each rotational state is in general split into a fine structure by the Stark effect.

Many of the small Hamiltonian terms of interest involve the nuclear and electron spin angular momenta. In these cases it is necessary to include the electron spin and nuclear spin wave functions[16] in the total wave function. The nuclear wave function may be included satisfactorily simply by multiplying (3.91) by ψ_n. This is proper (in practical cases) since the spatial variables of ψ_v and ψ_r are entirely independent of the spin variables of ψ_n, and there are no Hamiltonian terms which couple the space and spin vari-

---

[16] It was noted in chapter 2 that the electron and nuclear spin functions were needed to insure the proper permutation symmetry of the total molecular wave function. Here the need is for explicit matrix element evaluation involving spin variables.

ables (see chapter 2).[17] Essentially the same thing is true for the electron spin wave function, although it is common to write ψ_e as a Slater determinant with spin included, in which case the spin functions generally are not factorable from the spatial part. Nevertheless, for the treatment of Hamiltonian terms containing the electron spin variables, we add the electron spin function ψ_s as a factor to (3.91) in the same way as for ψ_n. As for the nuclear spin case, this approach is suitable as long as there are no large coupling terms between electron spin and electronic spatial variables. The resulting *total* wave function is

$$\psi_T = \psi_e \cdot \psi_v \cdot \psi_r \cdot \psi_n \cdot \psi_s \tag{3.95}$$

As an illustration, consider a molecule having one nucleus with non-zero spin I in the presence of a large external magnetic field of strength $H = H_Z$. The Hamiltonian perturbation term is now

$$\mathfrak{H}' = -\mu_n \cdot \mathbf{H} = g_n \beta_n \mathbf{I} \cdot \mathbf{H} = g_n \beta_n I_Z H_Z,$$

as shown in detail in chapter 9.[18] $\mathfrak{H}'$ is typically on the order of $10^{-3} E_R$ when $H \sim 10\ 000$ G, so it produces a hyperfine structure on the rotational energy levels. g_n and β_n are nuclear constants in the preceding expression, so only the matrix elements of the Z component of the nuclear spin angular momentum need be considered. Assuming no interaction of the nuclear spin with other angular momenta such as electron spin angular momentum or rotational angular momentum (that is, only $\mathfrak{H}'$ is considered), the first-order energy correction to the state ψ_{evrs} would be

$$E^{(1)} = g_n \beta_n H_Z (\psi_e \psi_v \psi_r \psi_s \mid \psi_e \psi_v \psi_r \psi_s)(\psi_n \mid \tilde{I}_Z \mid \psi_n)$$

$$= g_n \beta_n H_Z (n \mid \tilde{I}_Z \mid n) \tag{3.96}$$

There are always at least two functions ψ_n, for example, α and β when $I = \frac{1}{2}$, so the state ψ_{evrs} is split into a hyperfine structure. The energy $E^{(1)}$ is the dominant term when we consider the area of nuclear magnetic resonance spectroscopy, in which case we have no interest in the other larger energy terms, but indeed, are interested in even smaller terms in the Hamiltonian which produce further splittings of the nuclear spin states given by $E^{(1)}$ above.

From the preceding discussion, it should be clear how all molecular energy terms of interest may be treated. To the approximation described,

---

[17] This is not strictly true when relativistic effects are included. The simple product is, however, a satisfactory zero-order approximation, even when a phenomenological coupling of variables occurs, such as in the spin-rotation interaction in $^1\Sigma$ molecules.

[18] We neglect here the shielding of the nucleus by the electrons. See chapter 9.

the electronic, vibrational and rotational energy levels may be obtained from the appropriate wave equations or matrices. Other energy terms are treated by perturbation theory or by a complete matrix mechanical formulation if necessary. It should not be imagined that the interesting perturbations are always smaller than the rotational energy terms as in the two instances described above. For example, spin-orbit ($\mathbf{S} \cdot \mathbf{L}$ type) interaction may produce a fine structure on E_e which is larger than either the vibrational or rotational terms. What we have tried to indicate is the general scheme of the energies in which we will be most interested in the following chapters. Table 3.1 provides a summary of some of the "smaller" Hamiltonian terms (exclusive of $\mathfrak{IC}_e$, $\mathfrak{IC}_v$, and $\mathfrak{IC}_r$) which can be observed in molecular spectra. The typical magnitudes of the various terms have been indicated, and the most important spectroscopic methods for their study have been listed also.

TABLE 3.1

"Small" energy terms observable in molecular spectra.

Term	Magnitude	Spectroscopic method
Spin-orbit coupling ($\propto \mathbf{L} \cdot \mathbf{S}$)	10^2–10^4 cm^{-1}	Electronic, ESR
Vibrational potential barriers (inversion, hindered rotation)	1–100 cm^{-1}	Vibrational, microwave
Electron Zeeman effect ($-\mathbf{u}_s \cdot \mathbf{H}$)	~ 1 cm^{-1} for $H = 5000$ G	ESR
Stark effect (electric dipole, $-\mathbf{u} \cdot \mathcal{E}$)	$\sim 10^{-3}$ cm^{-1} for $\mathcal{E} = 1000$ V/cm	Microwave
Nuclear Zeeman effect ($-\mathbf{u}_n \cdot \mathbf{H}$)	$\sim 10^{-3}$ cm^{-1} for $H = 5000$ G	NMR
Nuclear electric quadrupole interaction	10^{-5}–10^{-2} cm^{-1}	Microwave, NQR
Coriolis coupling ($\boldsymbol{\omega} \times \mathbf{v}_\alpha$)	10–10^3 cm^{-1}	Vibrational, microwave
Electron spin-nuclear spin coupling ($\mathbf{S} \cdot \mathbf{I}$)	$\sim 10^{-3}$ cm^{-1}	ESR, microwave
Nuclear spin-spin coupling ($\mathbf{I}_i \cdot \mathbf{I}_j$)	$\sim 10^{-9}$ cm^{-1}	NMR
Spin-rotation interaction ($\propto$ to rotational magnetic moment)	10^{-4}–10^{-5} cm^{-1} for $H \sim 5000$ G	Microwave

Finally, we have shown in Fig. 3.7 the general characteristics of the energy levels for the water (H_2O) molecule as determined by molecular spectroscopy. Only a few of the lower electronic, vibrational and rotational states are indicated; the numbers above each state are the wavenumber energies measured relative to the ground vibrational state, and the symbols in parentheses are the standard spectroscopic labels which are described in later chapters. Also shown are the effects of electric and magnetic fields

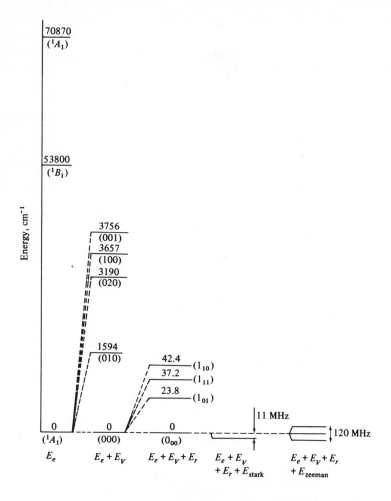

FIG. 3.7.   Some of the lower energy levels of the water molecule (H_2O). Numbers above lines are energies in wave numbers, numbers in parentheses are spectroscopic labels. Energy scale is only schematic.

on the lowest state ψ_{evr}. In an electric field of 5000 V/cm, the lowest rotational state shifts by 11 MHz, while a magnetic field of 14 100 G produces a triplet[19] with spacings of 60 MHz.

SUPPLEMENTARY REFERENCES

H. C. Allen and P. C. Cross, *Molecular Vib-Rotors* (Wiley, New York, 1963).

H. Eyring, J. Walter, and G. E. Kimball, *Quantum Chemistry* (Wiley, New York, 1944).

H. J. Goldstein, *Classical Mechanics* (Addison-Wesley, Reading, Mass., 1959).

H. Margenau and G. M. Murphy, *The Mathematics of Physics and Chemistry* (Van Nostrand, Princeton, 1956), Vol. 1.

L. C. Pauling and E. B. Wilson, *Introduction to Quantum Mechanics* (McGraw-Hill, New York, 1935).

C. H. Townes and A. L. Schawlow, *Microwave Spectroscopy* (McGraw-Hill Book Co., New York, 1955).

E. B. Wilson, J. C. Decius, and P. C. Cross, *Molecular Vibrations* (McGraw-Hill, New York, 1955).

[19] The central component is actually a very closely spaced doublet. See chapter 9.

CHAPTER 4

ANGULAR MOMENTUM AND HARMONIC OSCILLATION IN MATRIX MECHANICS

In the later chapters which treat the various kinds of spectroscopy, we find the quantum-mechanical solutions of two model problems to be extremely useful. The first of these is the harmonic oscillator which plays the key role in the theoretical treatment of infrared spectra. The second involves the properties of angular momenta in three dimensions which provide most of the necessary results for treating microwave, ESR, and NMR spectra. Some other properties closely related to angular momenta, namely the direction cosines, are discussed, also. While a complete understanding of the material in this chapter is not necessary to the study of later chapters, it is necessary that the important results be fully appreciated.

4.1 HARMONIC OSCILLATOR IN ONE DIMENSION

In Sec. 2.2 we mentioned that the matrix solutions of quantum-mechanical systems must proceed through the commutation relations appropriate to the problem. Our first job, then, is to define the dynamical system and then write the operator commutation relations. We consider a particle of mass m moving in the x direction under the influence of a potential energy $V(x)$ that is precisely specified later.

By the fundamental rules of quantum mechanics the linear momentum operator is

$$\tilde{p}_x = \frac{\hbar}{i} \frac{d}{dx} \tag{4.1}$$

If $f = f(x)$, we find easily that

$$(\tilde{p}_x x)f = \frac{\hbar}{i}\left\{ f + x\frac{df}{dx}\right\}$$

and   (4.2a)

$$(x\tilde{p}_x)f = \frac{\hbar}{i}\left\{ x\frac{df}{dx}\right\}$$

so we get the well-known result

$$(\tilde{p}_x x - x\tilde{p}_x)f = \frac{\hbar}{i}f \tag{4.2b}$$

This provides us with the *commutator* of $\tilde{p}_x$ and x which we write simply as

$$[\tilde{p}_x, x] = \tilde{p}_x x - x\tilde{p}_x = \frac{\hbar}{i} \tag{4.3}$$

It is obvious that

$$[x, x] = 0$$
$$[\tilde{p}_x, \tilde{p}_x] = 0 \tag{4.4}$$

The starting point in matrix mechanics involves the following assumption.

Any equation written in operator form is still true if matrices are substituted for operators. With this principle, Eq. (4.3) becomes

$$[\mathbf{p}_x, \mathbf{x}] = \mathbf{p}_x\mathbf{x} - \mathbf{x}\mathbf{p}_x = (\hbar/i)\mathbf{I} \tag{4.5}$$

where $\mathbf{I}$ is the unit matrix. For simplicity hereafter we suppress the subscript on $\mathbf{p}_x$.

The kinetic energy of the particle in operator form is

$$\tilde{T} = \frac{\tilde{p}^2}{2m} \tag{4.6a}$$

or in matrix form

$$\mathbf{T} = \frac{\mathbf{p}^2}{2m} \tag{4.6b}$$

If the particle moves in a potential field such that the force on the particle is proportional to its displacement, the potential energy is the well-known Hooke's-law type,

$$\mathbf{V}(x) = \tfrac{1}{2}k\mathbf{x}^2 \tag{4.7}$$

The Hamiltonian can then be written

$$\mathcal{3C} = T + V = \frac{p^2}{2m} + \tfrac{1}{2}kx^2 \tag{4.8}$$

In classical mechanics a particle having a Hamiltonian of this form is known as a linear *harmonic oscillator*.

In order to find the elements of $\mathcal{3C}$ we need to write down the commutation relations between $\mathcal{3C}$, p, and x. For example,

$$[\mathcal{3C}, x] = \mathcal{3C}x - x\mathcal{3C} = \frac{1}{2m}ppx + \frac{k}{2}x^3 - \frac{1}{2m}xpp - \frac{k}{2}x^3$$

$$= \frac{1}{2m}(ppx - xpp) \tag{4.9}$$

$$= \frac{1}{2m}\{p(px - xp) + (px - xp)p\}$$

But by Eq. (4.5) the terms in parentheses are simply $(\hbar/i)I$, so Eq. (4.9) becomes

$$[\mathcal{3C}, x] = \frac{1}{2m}\left(\frac{\hbar}{i}p + \frac{\hbar}{i}p\right) = \frac{\hbar}{im}p \tag{4.10a}$$

where we have used the fact that the product of the identity matrix with any other matrix simply gives the matrix. By an entirely analogous procedure we may show that

$$[\mathcal{3C}, p] = -\frac{k\hbar}{i}x \tag{4.11a}$$

Equations (4.10a) and (4.11a) may be solved for p and x, respectively, which gives

$$p = \frac{mi}{\hbar}[\mathcal{3C}, x] \tag{4.10b}$$

$$x = -\frac{i}{k\hbar}[\mathcal{3C}, p] \tag{4.11b}$$

These matrix equations can now be rewritten as ordinary algebraic equations by finding the general jkth elements.

$$p_{jk} = \frac{mi}{\hbar}\sum_n (H_{jn}x_{nk} - x_{jn}H_{nk}) \tag{4.10c}$$

$$x_{jk} = -\frac{i}{k\hbar}\sum_n (H_{jn}p_{nk} - p_{jn}H_{nk}) \tag{4.11c}$$

In (4.10c) and (4.11c) n spans all the rows and columns of the matrices which later we see to be infinite.

Let us now look for solutions of Eqs. (4.10c) and (4.11c) for which $\mathfrak{IC}$ is diagonal, since a diagonal Hamiltonian matrix provides the eigenvalues or energy levels of the harmonic oscillator. The requirement that $\mathfrak{IC}$ be diagonal means that H_{jn} and H_{nk} are zero except when $j = n$ and $n = k$. Therefore, the sums reduce to single terms as follows:

$$p_{jk} = \frac{mi}{\hbar}\,(H_{jj}x_{jk} - x_{jk}H_{kk}) = \frac{mi}{\hbar}\,x_{jk}(H_{jj} - H_{kk}) \tag{4.12}$$

$$x_{jk} = -\frac{i}{k\hbar}\,(H_{jj}p_{jk} - p_{jk}H_{kk}) = -\frac{i}{k\hbar}\,p_{jk}(H_{jj} - H_{kk}) \tag{4.13}$$

The right-hand equalities arise because the x_{jk} and p_{jk} are merely numbers so there are no noncommuting quantities. From (4.12) and (4.13) it is seen that $\mathbf{p}$ and $\mathbf{x}$ have no nonzero diagonal matrix elements, that is

$$p_{jj} = 0 \tag{4.14}$$

$$x_{jj} = 0 \tag{4.15}$$

since the right-hand side of the equations goes to zero when $j = k$.

When $j \neq k$ and $H_{jj} \neq H_{kk}$ (that is, states are nondegenerate) it is possible to find an expression relating p_{jk} and x_{jk} by eliminating $H_{jj} - H_{kk}$ from Eqs. (4.12) and (4.13). Direct division of (4.12) by (4.13) leads to

$$\frac{p_{jk}}{x_{jk}} = -mk\,\frac{x_{jk}}{p_{jk}} \tag{4.16a}$$

or

$$p_{jk}{}^2 = -mkx_{jk}{}^2 \tag{4.16b}$$

which gives finally

$$p_{jk} = \pm x_{jk}i(mk)^{1/2} \tag{4.16c}$$

This result is used eventually to calculate the p_{jk} from the x_{jk}. We note here that if one of the quantities is real the other is pure imaginary.

Substituting Eq. (4.16c) into Eq. (4.13) provides the first important quantization restriction upon the diagonal elements of $\mathfrak{IC}$. We find

$$H_{jj} - H_{kk} = \pm\hbar\left(\frac{k}{m}\right)^{1/2} \tag{4.17a}$$

or

$$H_{jj} - H_{kk} = \pm h\nu \tag{4.17b}$$

where

$$\nu = \frac{1}{2\pi} \left(\frac{k}{m}\right)^{1/2} \tag{4.18}$$

Implicit in the result (4.17) is that p_{jk} and x_{jk} are not zero, since otherwise the solution could not have been obtained. Equation (4.17b) states that for any eigenvalue H_{jj} there exists an eigenvalue H_{kk} which differs in energy by either $+h\nu$ or $-h\nu$. A further important observation is that when p_{jk} and x_{jk} are zero,

$$H_{jj} - H_{kk} \neq \pm h\nu \tag{4.19}$$

since otherwise the solutions would be contradictory.

We do not yet know the values of the H_{ii}, but since $\mathfrak{K}$ contains only squared Hermitian matrices we know that the H_{ii} are real and equal to or greater than zero.[1] For this reason it is clear that there exists an H_{ii} for which there is no lower element. If we call the lowest element H_{00} it is possible to rearrange the diagonal Hamiltonian matrix as in Fig. 4.1 so that

$$\mathfrak{K} = \begin{pmatrix} H_{00} & & & & 0 \\ & H_{00}+h\nu & & & \\ & & H_{00}+2h\nu & & \\ & & & H_{00}+3h\nu & \\ 0 & & & & \ddots \end{pmatrix}$$

FIG. 4.1. Hamiltonian matrix for harmonic oscillator arranged in ascending order of eigenvalues.

the elements are in ascending order with spacings of $h\nu$ as required. We are arbitrarily but conveniently numbering the rows and columns starting with zero. From Fig. 4.1 it is seen that once H_{00} is known, all the eigenvalues are given by the expression

$$H_{vv} = H_{00} + vh\nu \tag{4.20a}$$

with $v = 0, 1, 2, 3\ldots$ being simply the row and column matrix labels. Since the Hamiltonian elements of Fig. 4.1 are the stationary-state energies, we may write Eq. (4.20a) as

$$E_v = E_0 + vh\nu \tag{4.20b}$$

[1] See Appendix C.

$$\mathbf{p} = \begin{pmatrix} 0 & p_{01} & 0 & 0 & \cdots \\ p_{10} & 0 & p_{12} & 0 & \\ 0 & p_{21} & 0 & p_{23} & \\ 0 & 0 & p_{32} & 0 & \ddots \\ \vdots & & & & \end{pmatrix}$$

$$\mathbf{x} = \begin{pmatrix} 0 & x_{01} & 0 & 0 & \cdots \\ x_{10} & 0 & x_{12} & 0 & \\ 0 & x_{21} & 0 & x_{23} & \\ 0 & 0 & x_{32} & 0 & \ddots \\ \vdots & & & & \end{pmatrix}$$

FIG. 4.2. Matrices of momentum and coordinate arranged to correspond with ordering of $\mathcal{H}$ in Fig. 4.1.

Since the elements of $\mathcal{H}$ have been arranged in a particular fashion it is necessary that the matrices of **p** and **x** be similarly ordered. When this is done the matrices are as shown in Fig. 4.2. All diagonal elements of **p** and **x** are zero as previously found. Also, all off-diagonal elements of **p** and **x** are zero except those which satisfy Eqs. (4.17), that is, those elements whose row-column labels correspond to Hamiltonian elements which are adjacent in Fig. 4.1. With this new ordering scheme the nonzero elements of **p** and **x** are

$$p_{jj\pm1} \quad \text{and} \quad x_{jj\pm1}$$

The lowest (unknown) eigenvalue of $\mathcal{H}$, H_{00}, can be found easily if $|p_{01}|^2$ and $|x_{01}|^2$ are known. This follows from the expanded form of Eq. (4.8),

$$H_{ij} = \frac{1}{2m} \sum_n p_{in}p_{nj} + \frac{k}{2} \sum_n x_{in}x_{nj} \tag{4.21}$$

which becomes for the *lowest* diagonal element

$$\begin{aligned} H_{00} &= \frac{1}{2m} p_{01}p_{10} + \frac{k}{2} x_{01}x_{10} \\ &= \frac{1}{2m} |p_{01}|^2 + \frac{k}{2} |x_{01}|^2 \end{aligned} \tag{4.22}$$

by using the results of Fig. 4.2 and the Hermitian properties of p and x,

$$\begin{aligned} p_{01}p_{10} &= p_{01}p_{01}^* = |p_{01}|^2 \\ x_{01}x_{10} &= x_{01}x_{01}^* = |x_{01}|^2 \end{aligned} \tag{4.23}$$

These elements are found by expanding Eq. (4.5) which gives

$$p_{jj+1}x_{j+1,j} + p_{jj-1}x_{j-1,j} - x_{jj+1}p_{j+1,j} - x_{jj-1}p_{j-1,j} = \frac{\hbar}{i} \quad (4.24a)$$

Writing Eq. (4.24a) for the lowest possible index j, namely $j = 0$, gives

$$p_{01}x_{10} - x_{01}p_{10} = \frac{\hbar}{i} \quad (4.24b)$$

When $j = 0$ (and $k = 1$) Eq. (4.16c) gives

$$p_{01} = -ix_{01}(mk)^{1/2} \quad (4.25a)$$

where the minus sign was chosen since it must correspond to the result in (4.17b) which must have a minus sign when $j = 0$ and $k = 1$. From Eq. (4.25a),

$$p_{10} = p_{01}{}^* = ix_{01}{}^*(mk)^{1/2} \quad (4.25b)$$

Substituting Eqs. (4.25) into (4.24b) gives

$$-i(mk)^{1/2}x_{01}x_{10} - i(mk)^{1/2}x_{01}x_{01}{}^* = \frac{\hbar}{i} \quad (4.26)$$

or

$$|x_{01}|^2 = \frac{\hbar}{2(mk)^{1/2}} \quad (4.27)$$

Equation (4.25a) then leads to

$$|p_{01}|^2 = |x_{01}|^2 mk = \frac{\hbar(mk)^{1/2}}{2} \quad (4.28)$$

The unknown eigenvalue H_{00} is finally obtained by substitution of Eqs. (4.27) and (4.28) into (4.22). We obtain

$$H_{00} = \frac{\hbar}{2}\left(\frac{k}{m}\right)^{1/2} = \frac{1}{2}h\nu \quad (4.29)$$

Equations (4.20) then become

$$H_{vv} = E_v = h\nu(v + \tfrac{1}{2}) = \hbar\left(\frac{k}{m}\right)^{1/2}(v + \tfrac{1}{2}) \quad (4.30)$$

which gives the quantum-mechanical expression for the linear harmonic oscillator energy levels. These are shown in Fig. 4.3 along with the po-

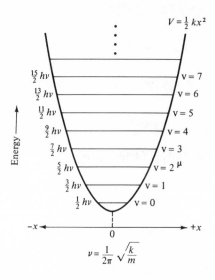

FIG. 4.3.   Potential energy curve and energy levels of linear harmonic oscillator.

tential energy curve $V = \frac{1}{2}kx^2$. The results are of course identical to those obtained by the wave-mechanical solution of Schrödinger's differential equation.[2] Note that the quantization of energy arose naturally from the algebraic equations derived from commutation relations.

It is possible now to complete the determination of the elements of $\mathbf{p}$ and $\mathbf{x}$. We note first that Eq. (4.16c) can now be written

$$p_{jj\pm1} = \mp x_{jj\pm1}(mk)^{1/2} i \tag{4.31}$$

for the possible nonzero elements. Here the top signs all go together and the bottom signs do likewise. The minus sign goes with the $j, j+1$ element because in this case we know that the sign in (4.17a) is negative. Substituting Eq. (4.31) into Eq. (4.24a) leads after some algebra to

$$| x_{jj+1} |^2 - | x_{j-1,j} |^2 = \frac{\hbar}{2(km)^{1/2}} \tag{4.32}$$

We see that the square of the magnitude of the elements of $\mathbf{x}$ form a progres-

[2] See for example, L. Pauling and E. B. Wilson, Jr., *Introduction to Quantum Mechanics* (McGraw-Hill, New York, 1935).

sion with spacing $\hbar/2(km)^{1/2}$. For example, Eq. (4.32) states that

$$| x_{12} |^2 - | x_{01} |^2 = \frac{\hbar}{2(km)^{1/2}}$$

$$| x_{23} |^2 - | x_{12} |^2 = \frac{\hbar}{2(km)^{1/2}} \text{ etc.}$$

(4.33)

In terms of the index v, we can write

$$| x_{v,v+1} |^2 = | x_{01} |^2 + v \left(\frac{\hbar}{2(km)^{1/2}} \right) \tag{4.34}$$

or using (4.27)

$$| x_{v,v+1} |^2 = (v + 1) \frac{\hbar}{2(km)^{1/2}} \tag{4.35}$$

Extracting the square root gives

$$| x_{v,v+1} | = \left[(v + 1) \frac{\hbar}{2(km)^{1/2}} \right]^{1/2} \tag{4.36}$$

At this point we have an option as to whether $x_{v,v+1}$ is real or imaginary. Formally, $x_{v,v+1}$ may be written

$$x_{v,v+1} = | x_{v,v+1} | \exp(i\phi) \tag{4.37}$$

We make the choice of phase angle so that the elements of **x** are real, that is, $\phi = 0$. This gives

$$x_{v,v+1} = \left[(v + 1) \frac{\hbar}{2(km)^{1/2}} \right]^{1/2} = x_{v+1,v} \tag{4.38}$$

where the second equality recognizes the Hermitian nature of **x**. With no difficulty, Eqs. (4.31) and (4.38) give

$$p_{v,v+1} = -i \left[(v + 1) \frac{\hbar(km)^{1/2}}{2} \right]^{1/2} = -p_{v+1,v} \tag{4.39}$$

which completes the evaluation of the matrix elements of $\mathcal{K}$, **p**, and **x**. Any more complicated matrix elements of quantities involving the basic matrices can be determined by matrix multiplication.

Of particular interest is the evaluation of expectation values (average values) of various dynamical variables of interest. For example, the expectation value of the potential energy in any state v is easily determined from the elements of $\mathbf{x}$.

$$
\begin{aligned}
\langle V \rangle &= \tfrac{1}{2}k\langle x^2 \rangle = \tfrac{1}{2}k\langle v \mid \tilde{x}^2 \mid v \rangle \\
&= \tfrac{1}{2}k \sum_{v'} \langle v \mid \tilde{x} \mid v' \rangle \langle v' \mid \tilde{x} \mid v \rangle \\
&= \tfrac{1}{2}k\{x_{v,v+1}x_{v+1,v} + x_{v,v-1}x_{v-1,v}\}
\end{aligned}
\tag{4.40}
$$

In Eq. (4.40) the last equality arises because $\mathbf{x}$ has only $v, v \pm 1$ elements. From Eq. (4.38) we obtain

$$
\begin{aligned}
\langle V \rangle &= \tfrac{1}{2}k\left\{\left[(v+1)\,\frac{\hbar}{2(km)^{1/2}}\right]^{1/2}\left[(v+1)\,\frac{\hbar}{2(km)^{1/2}}\right]^{1/2}\right. \\
&\qquad\qquad \left. + \left[v\,\frac{\hbar}{2(km)^{1/2}}\right]^{1/2}\left[v\,\frac{\hbar}{2(km)^{1/2}}\right]^{1/2}\right\} \\
&= k\left(\frac{\hbar}{2\sqrt{km}}\right)(v+\tfrac{1}{2}) \\
&= \frac{\hbar}{2}\sqrt{\frac{k}{m}}\,(v+\tfrac{1}{2})
\end{aligned}
\tag{4.41}
$$

Note now that the expectation value of the kinetic energy is gotten easily by difference,

$$
\begin{aligned}
\langle T \rangle &= \langle E \rangle - \langle V \rangle \\
&= E_v - \langle V \rangle
\end{aligned}
\tag{4.42}
$$

which by Eqs. (4.30) and (4.41) gives

$$
\begin{aligned}
\langle T \rangle &= \hbar\sqrt{\frac{k}{m}}\,(v+\tfrac{1}{2}) - \frac{\hbar}{2}\sqrt{\frac{k}{m}}\,(v+\tfrac{1}{2}) \\
&= \frac{\hbar}{2}\sqrt{\frac{k}{m}}\,(v+\tfrac{1}{2})
\end{aligned}
\tag{4.43}
$$

We see in chapter 8 that the results of this section are directly applicable to the vibrations of molecules as a first (good) approximation. This has already been shown for the diatomic molecule which has the Hamiltonian shown in Eq. (3.89). The energy levels in this case will be given by Eq. (4.30) with μ substituted for m. Although the vibrational energy levels are considered in detail later, it may be well to show the order of magnitude of the terms involved. A typical force constant is $\sim 4 \times 10^5$ dyne cm^{-1}. For a reduced mass of 10^{-23} grams, the energy is

$$E_v \sim \hbar \sqrt{4 \times 10^{28}} \, (v + \tfrac{1}{2})$$

$$\sim 2 \times 10^{-13}(v + \tfrac{1}{2}) \text{ erg} \qquad (4.44)$$

and the separation between energy levels is 2×10^{-13} erg, which corresponds to a frequency of

$$\nu = \frac{2 \times 10^{-13}}{6.62 \times 10^{-27}} \text{ sec}^{-1} \sim 3 \times 10^{13} \text{ sec}^{-1}$$

$$\sim 1000 \text{ cm}^{-1} \qquad (4.45)$$

EXAMPLE 4-1

It is perhaps helpful to show that the matrices $\mathbf{p}$ and $\mathbf{x}$ do indeed satisfy the commutation relation of Eq. (4.5).

Method (A): $\mathbf{px}$ and $\mathbf{xp}$ can have at most only v, v and $v, v \pm 2$ nonzero elements because $\mathbf{p}$ and $\mathbf{x}$ have only $v, v \pm 1$ elements. Since the matrix $(\hbar/i)\mathbf{I}$ has only diagonal elements we need to show that

$$(\mathbf{px})_{v,v\pm2} - (\mathbf{xp})_{v,v\pm2} = 0 \qquad (4.46a)$$

in addition to

$$(\mathbf{px})_{v,v} - (\mathbf{xp})_{v,v} = \frac{\hbar}{i} \qquad (4.47a)$$

Working with Eq. (4.46a) we find with the aid of Eqs. (4.38) and (4.39)

$$p_{v,v+1}x_{v+1,v+2} - x_{v,v+1}p_{v+1,v+2} = -i\frac{\hbar}{2}[(v+1)(v+2)]^{1/2}$$

$$+ i\frac{\hbar}{2}[(v+1)(v+2)]^{1/2} = 0 \qquad (4.46b)$$

and similarly for the $v, v - 2$ elements. The left-hand side of (4.47a) can now be evaluated as

$$p_{v,v-1}x_{v-1,v} + p_{v,v+1}x_{v+1,v} - x_{v,v-1}p_{v-1,v} - x_{v,v+1}p_{v+1,v}$$

$$= i\frac{\hbar}{2}v - i\frac{\hbar}{2}(v+1) + i\frac{\hbar}{2}v - i\frac{\hbar}{2}(v+1) = -i\hbar = \frac{\hbar}{i} \qquad (4.47b)$$

Method (B) : Here we calculate a finite portion of the $\mathbf{p}$ and $\mathbf{x}$ matrices and multiply the two matrices together. From Eqs. (4.38) and (4.39) we find for example,

$$x_{01} = x_{10} = \alpha, \qquad\qquad x_{12} = x_{21} = \sqrt{2}\,\alpha$$

$$x_{23} = x_{32} = \sqrt{3}\,\alpha, \qquad\qquad x_{34} = x_{43} = \sqrt{4}\,\alpha$$

$$p_{01} = -p_{10} = -i\beta, \qquad\qquad p_{12} = -p_{21} = -i\sqrt{2}\,\beta \qquad (4.48)$$

$$p_{23} = -p_{32} = -i\sqrt{3}\,\beta, \qquad p_{34} = -p_{43} = -i\sqrt{4}\,\beta$$

where

$$\alpha = \left(\frac{\hbar}{2(km)^{1/2}}\right)^{1/2}$$

$$(4.49)$$

$$\beta = \left(\frac{\hbar(km)^{1/2}}{2}\right)^{1/2}$$

The left-hand side of Eq. (4.5) can now be written

$$
\mathbf{px} - \mathbf{xp} = i\alpha\beta
\begin{pmatrix}
0 & -1 & 0 & 0 & 0 \\
1 & 0 & -\sqrt{2} & 0 & 0 \\
0 & \sqrt{2} & 0 & -\sqrt{3} & 0 \\
0 & 0 & \sqrt{3} & 0 & -\sqrt{4} \\
0 & 0 & 0 & \sqrt{4} & 0
\end{pmatrix}
\begin{pmatrix}
0 & 1 & 0 & 0 & 0 \\
1 & 0 & \sqrt{2} & 0 & 0 \\
0 & \sqrt{2} & 0 & \sqrt{3} & 0 \\
0 & 0 & \sqrt{3} & 0 & \sqrt{4} \\
0 & 0 & 0 & \sqrt{4} & 0
\end{pmatrix}
$$

$$
-i\alpha\beta
\begin{pmatrix}
0 & 1 & 0 & 0 & 0 \\
1 & 0 & \sqrt{2} & 0 & 0 \\
0 & \sqrt{2} & 0 & \sqrt{3} & 0 \\
0 & 0 & \sqrt{3} & 0 & \sqrt{4} \\
0 & 0 & 0 & \sqrt{4} & 0
\end{pmatrix}
\begin{pmatrix}
0 & -1 & 0 & 0 & 0 \\
1 & 0 & -\sqrt{2} & 0 & 0 \\
0 & \sqrt{2} & 0 & -\sqrt{3} & 0 \\
0 & 0 & \sqrt{3} & 0 & -\sqrt{4} \\
0 & 0 & 0 & \sqrt{4} & 0
\end{pmatrix}
$$

$$(4.50a)$$

by truncating the matrices at $v = 4$. Performing the matrix multiplication indicated in Eq. (4.50a) and noting that $\alpha\beta = \frac{1}{2}\hbar$ gives

$$\mathbf{px} - \mathbf{xp} = i\frac{\hbar}{2}\begin{pmatrix} -1 & 0 & -\sqrt{2} & 0 & 0 \\ 0 & -1 & 0 & -\sqrt{6} & 0 \\ \sqrt{2} & 0 & -1 & 0 & -\sqrt{12} \\ 0 & \sqrt{6} & 0 & -1 & 0 \\ 0 & 0 & \sqrt{12} & 0 & \star \end{pmatrix}$$

$$-i\frac{\hbar}{2}\begin{pmatrix} 1 & 0 & -\sqrt{2} & 0 & 0 \\ 0 & 1 & 0 & -\sqrt{6} & 0 \\ \sqrt{2} & 0 & 1 & 0 & -\sqrt{12} \\ 0 & \sqrt{6} & 0 & 1 & 0 \\ 0 & 0 & \sqrt{12} & 0 & \star \end{pmatrix} \quad (4.50b)$$

where the $\star$ indicates that the element obtained by multiplication of the truncated matrices of Eq. (4.50a) is incorrect for the infinite matrices. Finally the subtraction in Eq. (4.50b) leads to

$$\mathbf{px} - \mathbf{xp} = \frac{\hbar}{i}\begin{pmatrix} 1 & 0 & 0 & 0 \cdots \\ 0 & 1 & 0 & 0 \\ 0 & 0 & 1 & 0 \\ 0 & 0 & 0 & 1 \\ & \cdot & & \cdot \\ & \cdot & & \cdot \\ & \cdot & & \cdot \end{pmatrix} = \frac{\hbar}{i}\mathbf{I} \quad (4.50c)$$

which again verifies that Eqs. (4.38) and (4.39) do indeed satisfy the basic commutation relation.

4.2 ANGULAR MOMENTUM COMMUTATION RELATIONS

In order to describe the properties of systems with nuclear framework angular momentum, electronic orbital angular momentum, or intrinsic spin angular momentum, it is necessary to begin with appropriate commutation relations among the variables of interest. These can be derived by beginning with the classical expression for the angular momentum of a point mass with respect to a space-fixed (laboratory) axis system,

$$\mathbf{P} = \mathbf{R} \times \mathbf{p} \tag{4.51a}$$

where $\mathbf{R}$ is the position vector of the point and $\mathbf{p} = m\mathbf{v}$ is the linear momentum of the point mass of velocity $\mathbf{v}$ and mass m.[3] Recalling the alternative form of Eq. (4.51a)

$$\mathbf{P} = \begin{vmatrix} \mathbf{i} & \mathbf{j} & \mathbf{k} \\ X & Y & Z \\ p_X & p_Y & p_Z \end{vmatrix} \tag{4.51b}$$

the components of the total angular momentum along space-fixed axes are easily seen to be

$$P_X = Yp_Z - Zp_Y$$
$$P_Y = Zp_X - Xp_Z \tag{4.52}$$
$$P_Z = Xp_Y - Yp_X$$

It is clear also that the square of the total angular momentum (the scalar product of $\mathbf{P}$ with itself) is related to the components by

$$P^2 = P_X{}^2 + P_Y{}^2 + P_Z{}^2 \tag{4.53}$$

The quantum-mechanical commutation relations of interest can now be obtained by replacing the dynamical variables in (4.52) by operators and using the commutation relations for linear momentum which are

$$\tilde{p}_F F - F\tilde{p}_F = \frac{\hbar}{i} \qquad (F = X, Y, Z)$$

$$\tilde{p}_F \tilde{p}_{F'} - \tilde{p}_{F'} \tilde{p}_F = 0 \qquad (F, F' = X, Y, Z) \tag{4.54}$$

$$\tilde{p}_F F' - F' \tilde{p}_F = 0 \qquad (F, F' = X, Y, Z; F \neq F')$$

$$FF' - F'F = 0 \qquad (F, F' = X, Y, Z)$$

Note that Eqs. (4.54) include those of Eqs. (4.3) and (4.4) but extend them to three dimensions. Let us now evaluate $[\tilde{P}_X, \tilde{P}_Y] = \tilde{P}_X \tilde{P}_Y - \tilde{P}_Y \tilde{P}_X$. We find by substitution of Eqs. (4.52)

$$[\tilde{P}_X, \tilde{P}_Y] = (Y\tilde{p}_Z - Z\tilde{p}_Y)(Z\tilde{p}_X - X\tilde{p}_Z)$$
$$- (Z\tilde{p}_X - X\tilde{p}_Z)(Y\tilde{p}_Z - Z\tilde{p}_Y) \tag{4.55a}$$

---

[3] The angular momentum of a collection of point masses with no interactions is simply the sum of the individual angular momenta. Similarly, the angular momentum of a rigid continuous body is gotten by continuous summation, that is, by integration.

Performing the indicated operator multiplication leads to

$$
\begin{aligned}
[\tilde{P}_X, \tilde{P}_Y] &= Y\tilde{p}_Z(Z\tilde{p}_X - X\tilde{p}_Z) - Z\tilde{p}_Y(Z\tilde{p}_X - X\tilde{p}_Z) \\
&\quad - Z\tilde{p}_X(Y\tilde{p}_Z - Z\tilde{p}_Y) + X\tilde{p}_Z(Y\tilde{p}_Z - Z\tilde{p}_Y) \\
&= \tilde{p}_Z(YZ\tilde{p}_X - YX\tilde{p}_Z + XY\tilde{p}_Z - XZ\tilde{p}_Y) \\
&\quad - Z(\tilde{p}_YZ\tilde{p}_X - \tilde{p}_YX\tilde{p}_Z + \tilde{p}_XY\tilde{p}_Z - \tilde{p}_XZ\tilde{p}_Y) \\
&= \tilde{p}_ZZ(Y\tilde{p}_X - X\tilde{p}_Y) - Z\tilde{p}_Z(\tilde{p}_XY - \tilde{p}_YX) \\
&= (\tilde{p}_ZZ - Z\tilde{p}_Z)(Y\tilde{p}_X - X\tilde{p}_Y)
\end{aligned}
\tag{4.55b}
$$

We recognize now that the second factor in Eq. (4.55b) is $-P_Z$ and that the first factor has the value $\hbar/i$ from Eqs. (4.54). This gives

$$
[\tilde{P}_X, \tilde{P}_Y] = -\frac{\hbar}{i}\tilde{P}_Z
\tag{4.56a}
$$

Entirely similar considerations show that

$$
[\tilde{P}_Y, \tilde{P}_Z] = -\frac{\hbar}{i}\tilde{P}_X
$$

$$
\tag{4.56b}
$$

$$
[\tilde{P}_Z, \tilde{P}_X] = -\frac{\hbar}{i}\tilde{P}_Y
$$

These results make it possible to show also that

$$
\begin{aligned}
[\tilde{P}^2, \tilde{P}_X] &= 0 \\
[\tilde{P}^2, \tilde{P}_Y] &= 0 \\
[\tilde{P}^2, \tilde{P}_Z] &= 0
\end{aligned}
\tag{4.57}
$$

For example,

$$
\begin{aligned}
[\tilde{P}^2, \tilde{P}_X] &= (\tilde{P}_X^2 + \tilde{P}_Y^2 + \tilde{P}_Z^2)\tilde{P}_X - \tilde{P}_X(\tilde{P}_X^2 + \tilde{P}_Y^2 + \tilde{P}_Z^2) \\
&= \tilde{P}_X^3 + \tilde{P}_Y(\tilde{P}_Y\tilde{P}_X) + \tilde{P}_Z(\tilde{P}_Z\tilde{P}_X) \\
&\quad - \tilde{P}_X^3 - (\tilde{P}_X\tilde{P}_Y)\tilde{P}_Y - (\tilde{P}_X\tilde{P}_Z)\tilde{P}_Z \\
&= \tilde{P}_Y\left(\tilde{P}_X\tilde{P}_Y + \frac{\hbar}{i}\tilde{P}_Z\right) + \tilde{P}_Z\left(\tilde{P}_X\tilde{P}_Z - \frac{\hbar}{i}\tilde{P}_Y\right) \\
&\quad - \left(\tilde{P}_Y\tilde{P}_X - \frac{\hbar}{i}\tilde{P}_Z\right)\tilde{P}_Y - \left(\tilde{P}_Z\tilde{P}_X + \frac{\hbar}{i}\tilde{P}_Y\right)\tilde{P}_Z \\
&= 0
\end{aligned}
$$

This completes the derivation of the basic commutation relations for angular momentum measured with respect to space-fixed axes, although some further relations are derived later. It is necessary at this point to remember (see Sec. 3.3) that the treatment of the rotational energy terms in the Hamiltonian required the use of a rotating, molecule-fixed axis system. Indeed, the rotational Hamiltonian of Eq. (3.60) involves angular momenta referred to the molecule-fixed axis system, not the space-fixed axis system. One must naturally inquire whether the commutation relations just derived are appropriate for molecule-fixed axes also. The answer is that they are not quite correct; the sign of the right sides of Eq. (4.56) must be changed! This rather curious result has been discussed briefly by many authors. Van Vleck[4] gives a rather satisfying algebraic proof of this phenomenon which we have reproduced in Appendix IV. The result of changing the sign of the commutators has no clear physical interpretation but is akin to the sign change in velocity observed by a passenger when he gets off a moving train. We have then, for angular momenta referred to molecule-fixed axes,[5]

$$[\tilde{P}_x, \tilde{P}_y] = \frac{\hbar}{i} \tilde{P}_z$$

$$[\tilde{P}_y, \tilde{P}_z] = \frac{\hbar}{i} \tilde{P}_x \qquad (4.58)$$

$$[\tilde{P}_z, \tilde{P}_x] = \frac{\hbar}{i} \tilde{P}_y$$

Since the magnitude of the angular momentum vector must be independent of the axis system to which it is referred it follows that

$$\tilde{P}_X{}^2 + \tilde{P}_Y{}^2 + \tilde{P}_Z{}^2 = \tilde{P}_x{}^2 + \tilde{P}_y{}^2 + \tilde{P}_z{}^2 \qquad (4.59)$$

The commutation relations similar to Eqs. (4.57) apply for molecule-fixed axes also,

$$[\tilde{P}^2, \tilde{P}_x] = 0$$
$$[\tilde{P}^2, \tilde{P}_y] = 0 \qquad (4.60)$$
$$[\tilde{P}^2, \tilde{P}_z] = 0$$

---

[4] J. H. Van Vleck, *Rev. Mod. Phys.* **23**, 213 (1951).

[5] To be slightly more explicit we should point out that Eqs. (4.56) and (4.58) apply to the *total* (sum of rotational, orbital, spin, etc.) angular momentum of a molecule or to the rotational angular momentum of the heavy nuclear framework of a molecule, but that Eqs. (4.58) (with the anomalous signs) are not appropriate for electronic orbital or spin angular momentum. Instead, the commutation relations for these internal angular momenta have the form of Eqs. (4.56) for *either* the space- or molecule-fixed axes. The reason for this difference is that within the confines of the Born–Oppenheimer approximation the internal angular momenta of a molecule are treated with the assumption of fixed or clamped nuclei; therefore they can have nothing to do with the rigid framework rotation and consequently have the same commutation relations in either axis system. See Ref. 4 for a more thorough discussion.

and since the space- and molecule-fixed axes are independent it must be true that all the $\tilde{P}_g$ commute with all the $\tilde{P}_F$.

$$[\tilde{P}_F, \tilde{P}_g] = 0 \qquad (g = x, y, z; \ F = X, Y, Z) \qquad (4.61)$$

4.3 MATRIX ELEMENTS OF ANGULAR MOMENTUM

In Appendix B we have proved the well-known quantum-mechanical theorem which states that for each member of a set of simultaneously commuting operators or matrices there exists a set of simultaneously diagonal matrices. Or in wave-mechanical language there exists a set of wave functions which are simultaneously eigenfunctions of each operator. From the work of Sec. 4.2 we see that $\tilde{P}^2$, $\tilde{P}_Z$, and $\tilde{P}_z (\mathbf{P}^2, \mathbf{P}_Z, \mathbf{P}_z)$ form a set of commuting operators (matrices). Note that $\tilde{P}^2$, $\tilde{P}_X$ and $\tilde{P}_x$ or $\tilde{P}^2$, $\tilde{P}_Y$, $\tilde{P}_y$ would be acceptable sets also, but the conventional choice is the former one.

The previously mentioned theorem shows that it is possible to expect that $\mathbf{P}^2$, $\mathbf{P}_Z$ and $\mathbf{P}_z$ have only diagonal elements

$$\langle JKM \mid \tilde{P}^2 \mid JKM \rangle = k_J \qquad (4.62a)$$

$$\langle JKM \mid \tilde{P}_Z \mid JKM \rangle = k_M \qquad (4.62b)$$

$$\langle JKM \mid \tilde{P}_z \mid JKM \rangle = k_K \qquad (4.62c)$$

where J, K, and M are labels which are associated with $\tilde{P}^2$, $\tilde{P}_z$ and $\tilde{P}_Z$, respectively, and k_J, k_M, and k_K are the still to be determined diagonal values or eigenvalues of the matrices $\mathbf{P}^2$, $\mathbf{P}_Z$ and $\mathbf{P}_z$. In wave-mechanical parlance, Eqs. (4.62) are written

$$\tilde{P}^2 \phi_{JKM} = k_J \phi_{JKM} \qquad (4.62a')$$

$$\tilde{P}_Z \phi_{JKM} = k_M \phi_{JKM} \qquad (4.62b')$$

$$\tilde{P}_z \phi_{JKM} = k_K \phi_{JKM} \qquad (4.62c')$$

where the ϕ_{JK} are the eigenfunctions of the set of operators. We proceed now to the determination of the complete matrices of all the angular momentum operators ($\tilde{P}^2$, $\tilde{P}_X$, $\tilde{P}_x$, ..., etc.) in the JKM representation, that is, we use implicitly the ϕ_{JK} as the basis functions. The approach is somewhat more complicated than that for the harmonic oscillator (Sec. 4.1) for which matrix elements were derived by requiring only one matrix $\mathcal{H}$ to be diagonal.

We begin by simplifying our problem somewhat. Since each of the $\tilde{P}_g$ commute with each of the $\tilde{P}_F$, [see Eq. (4.61)] the parts of the matrix elements involving J and K can be treated separately from those involving J and M. And because of the similarity of the commutation relations the results for the $\tilde{P}_g$ can be written down by inspection once the $\tilde{P}_F$ matrices have been deduced. In the following derivation we therefore suppress the K labels temporarily.

Let us define now two useful operators

$$\tilde{P}^+ = \tilde{P}_X + i\tilde{P}_Y \qquad (4.63a)$$

and

$$\tilde{P}^- = \tilde{P}_X - i\tilde{P}_Y \qquad (4.63b)$$

from which it follows that

$$\tilde{P}^2\tilde{P}^+ - \tilde{P}^+\tilde{P}^2 = 0 \qquad (4.64a)$$

$$\tilde{P}^2\tilde{P}^- - \tilde{P}^-\tilde{P}^2 = 0 \qquad (4.64b)$$

and

$$\tilde{P}_Z\tilde{P}^+ - \tilde{P}^+\tilde{P}_Z = \hbar\tilde{P}^+ \qquad (4.65a)$$

$$\tilde{P}_Z\tilde{P}^- - \tilde{P}^-\tilde{P}_Z = -\hbar\tilde{P}^- \qquad (4.65b)$$

From Eq. (4.65a) it is possible to write the general matrix element for $\tilde{P}_Z\tilde{P}^+$ as

$$\langle JM \mid \tilde{P}_Z\tilde{P}^+ \mid J'M' \rangle = \sum_{J''M''} \langle JM \mid \tilde{P}_Z \mid J''M'' \rangle \langle J''M'' \mid \tilde{P}^+ \mid J'M' \rangle$$

$$= \sum_{J''M''} \langle JM \mid \tilde{P}^+ \mid J''M'' \rangle \langle J''M'' \mid \tilde{P}_Z \mid J'M' \rangle$$

$$+ \hbar\langle JM \mid \tilde{P}^+ \mid J'M' \rangle \qquad (4.66)$$

But $\langle J''M'' \mid \tilde{P}_Z \mid J'M' \rangle$ is zero unless $J'M' = J''M''$; and similarly $\langle JM \mid \tilde{P}_Z \mid J''M'' \rangle$ is zero unless $JM = J''M''$ because of our initial choice of representation. Furthermore, since $\tilde{P}^2$ commutes with $\tilde{P}^+$ by Eq. (4.64a), J must equal J' in the $\tilde{P}^+$ elements. Making these substitutions in Eq. (4.66) leads to

$$\langle JM \mid \tilde{P}_Z \mid JM \rangle \langle JM \mid \tilde{P}^+ \mid JM' \rangle = \langle JM \mid \tilde{P}^+ \mid JM' \rangle \langle J'M' \mid \tilde{P}_Z \mid J'M' \rangle$$

$$+ \hbar\langle JM \mid \tilde{P}^+ \mid JM' \rangle \qquad (4.67a)$$

or by rearranging

$$\langle JM \mid \tilde{P}^+ \mid JM' \rangle \{ \langle JM \mid \tilde{P}_Z \mid JM \rangle - \langle J'M' \mid \tilde{P}_Z \mid J'M' \rangle - \hbar \} = 0 \qquad (4.67b)$$

We see that when $\langle JM \mid \tilde{P}^+ \mid JM' \rangle$ is nonzero,

$$k_M - k_{M'} = \hbar \qquad (4.68)$$

which implies that the eigenvalues of $\tilde{P}_Z$ form a progression with spacings of $\hbar$; that is, the eigenvalues are spaced at integral values of $\hbar$. Although we do not yet know the actual values of k_M, it is useful to anticipate the result so as to define

$$k_M = M\hbar \qquad (4.69)$$

where M is a quantum number whose values are still to be found. It is useful to note also that $\langle JM \mid \tilde{P}^+ \mid JM' \rangle$ is nonzero only when $M - M' = 1$, so that the nonzero elements are of the form $\langle J, M + 1 \mid \tilde{P}^+ \mid JM \rangle$. A similar analysis shows the only nonzero values of $\tilde{P}^-$ to be $\langle JM \mid \tilde{P}^- \mid J, M + 1 \rangle$.

To obtain further information about the quantum number M, we find from Eqs. (4.63) that

$$\tilde{P}^- \tilde{P}^+ = \tilde{P}^2 - \tilde{P}_Z{}^2 - \hbar \tilde{P}_Z \tag{4.70a}$$

and

$$\tilde{P}^+ \tilde{P}^- = \tilde{P}^2 - \tilde{P}_Z{}^2 + \hbar \tilde{P}_Z \tag{4.70b}$$

Thus the matrices of $\tilde{P}^- \tilde{P}^+$ and $\tilde{P}^+ \tilde{P}^-$ are both diagonal with values

$$\langle JM \mid \tilde{P}^- \tilde{P}^+ \mid JM \rangle = k_J - k_M{}^2 - \hbar k_M$$

$$= k_J - M^2 \hbar^2 - M \hbar^2 \tag{4.71a}$$

$$\langle JM \mid \tilde{P}^+ \tilde{P}^- \mid JM \rangle = k_J - k_M{}^2 + \hbar k_M$$

$$= k_J - M^2 \hbar^2 + M \hbar^2 \tag{4.71b}$$

Equation (4.71a) shows that $k_J \geq M^2 \hbar^2 + M \hbar^2$ because of the Hermitian nature of $\tilde{P}^- \tilde{P}^+$, so we see that there is some upper algebraic limit to the permissible values of M. If negative values of M are permissible Eq. (4.71b) shows that there must also be a lower algebraic limit on M. Thus we conclude that the permissible values of M are bounded above by some value M_{max} and below by some value M_{min}.

Because $\tilde{P}^+$ has only $M + 1$, M elements, and $\tilde{P}^-$ has only M, $M + 1$ elements, (4.71a) becomes

$$\langle JM \mid \tilde{P}^- \mid JM + 1 \rangle \langle JM + 1 \mid \tilde{P}^+ \mid JM \rangle = k_J - M^2 \hbar^2 - M \hbar^2 \tag{4.72}$$

Suppose that $M = M_{max}$. In this case $M + 1$ does not exist so

$$k_J - M^2{}_{max} \hbar^2 - M_{max} \hbar^2 = 0 \tag{4.73a}$$

Similarly if $M = M_{min}$, Eq. (4.71b) gives

$$k_J - M^2{}_{min} \hbar^2 + M_{min} \hbar^2 = 0 \tag{4.73b}$$

Equating (4.73a) and (4.73b) leads to

$$M^2{}_{min} - M_{min} - M^2{}_{max} - M_{max} = 0 \tag{4.74}$$

which is algebraically equivalent to

$$(M_{max} + M_{min})(M_{min} - M_{max} - 1) = 0 \tag{4.75}$$

It is impossible for the second factor in Eq. (4.75) to be zero, thus we conclude

$$M_{max} = -M_{min} \tag{4.76}$$

We now define

$$J = M_{max} \tag{4.77}$$

and recall that the difference between M_{max} and M_{min} must be a positive integer, that is

$$M_{max} - M_{min} = I \qquad (4.78a)$$

or

$$J - (-J) = I \qquad (4.78b)$$

which gives

$$J = \frac{I}{2} \qquad (4.78c)$$

Thus we see that the commutation relations permit both *integral* and *half-integral* values of J (the maximum value of M) and consequently M. In addition it is clear that $M = 0$ must exist so that $J = 0$ is suitable also. While the complete significance of the quantum number J has not yet been established we should point out now that the half-integral values are not suitable for rotational or orbital angular momentum but are restricted to systems with electron or nuclear spin.

This is most easily seen by turning to a short wave-mechanical treatment. In spherical polar coordinates, $\tilde{P}_Z$ becomes[6]

$$\tilde{P}_Z = \frac{\hbar}{i} \frac{\partial}{\partial \varphi} \qquad (4.79)$$

where φ is the azimuthal angle. Then Eq. (4.62b′) becomes

$$\frac{\hbar}{i} \frac{\partial}{\partial \varphi} \phi_{JKM}(r, \vartheta, \varphi) = k_M \phi_{JKM}(r, \vartheta, \varphi) \qquad (4.80)$$

If ϕ_{JKM} is written as a product,

$$\phi_{JKM} = \phi_M(\varphi) f_{JK}(r, \vartheta) \qquad (4.81)$$

Eq. (4.80) gives

$$\frac{d\phi_M}{d\varphi} = \frac{i}{\hbar} k_M \phi_M \qquad (4.82)$$

which has normalized solutions

$$\phi_M = \frac{1}{(2\pi)^{1/2}} \exp(iM\varphi) \qquad (4.83)$$

Note that ϕ_M is single valued $[\phi_M(0) = \phi_M(2\pi)]$ only if $M = 0$ or if M is a positive or negative integer. This result does not apply to intrinsic

[6] H. Eyring, J. Walter and G. E. Kimball, *Quantum Chemistry* (Wiley, New York, 1944), p. 41.

spin systems, since they are not defined in real three-dimensional space and no operator like that in (4.79) exists.

Returning to Eq. (4.73a) we can now write down the eigenvalue of $\tilde{P}^2$,

$$k_J = J^2\hbar^2 - J\hbar^2 \tag{4.84}$$

so that Eq. (4.62a) becomes

$$\langle JKM \mid \tilde{P}^2 \mid JKM \rangle = \hbar^2 J(J+1) \tag{4.85a}$$

while Eq. (4.62b) is

$$\langle JKM \mid \tilde{P}_Z \mid JKM \rangle = \hbar M \tag{4.85b}$$

with

$$J = \begin{cases} 0, 1, 2, 3 \dots & \text{All angular momenta} \\ \frac{1}{2}, \frac{3}{2}, \frac{5}{2}, \frac{7}{2} \dots & \text{Spin systems only} \end{cases} \tag{4.86}$$

and

$$M = J, J-1, J-2, \dots, -J+2, -J+1, -J$$

In terms of the vector model it is clear that M represents the projection of the total angular momentum $\mathbf{P}$ upon the space-fixed Z axis. We commonly refer to J as being the magnitude of the angular momentum (in units of $\hbar$) even though we really mean $[J(J+1)]^{1/2}$.

The nonzero off-diagonal elements can be gotten by noting that Eq. (4.72) can be written

$$\langle JM \mid \tilde{P}^- \mid JM+1 \rangle \langle JM+1 \mid \tilde{P}^+ \mid JM \rangle$$
$$= \hbar^2 [J(J+1) - M(M+1)] \quad (4.87)$$

But by the Hermitian properties of $\tilde{P}_X$ and $\tilde{P}_Y$

$$\langle JM \mid \tilde{P}^- \mid JM+1 \rangle = \langle JM \mid \tilde{P}_X \mid JM+1 \rangle - i\langle JM \mid \tilde{P}_Y \mid JM+1 \rangle$$
$$= \langle JM+1 \mid \tilde{P}_X \mid JM \rangle^* + i\langle JM+1 \mid \tilde{P}_Y \mid JM \rangle^*$$
$$= \langle JM+1 \mid \tilde{P}^+ \mid JM \rangle^* \tag{4.88}$$

and Eq. (4.87) becomes

$$\mid \langle JM+1 \mid \tilde{P}^+ \mid JM \rangle \mid^2 = \mid \langle JM \mid \tilde{P}^- \mid JM+1 \rangle \mid^2$$
$$= \hbar^2 [J(J+1) - M(M+1)] \quad (4.89)$$

Choosing the phase angle such that the matrix elements are real [as was done in Eq. (4.37) for the harmonic oscillator]

$$\langle JM + 1 \mid \tilde{P}^+ \mid JM \rangle = \langle JM \mid \tilde{P}^- \mid JM + 1 \rangle$$
$$= \hbar[J(J + 1) - M(M + 1)]^{1/2} \qquad (4.90)$$

Since $\tilde{P}^+ + \tilde{P}^- = 2\tilde{P}_X$, we can write

$$\langle JM + 1 \mid \tilde{P}^+ \mid JM \rangle + \langle JM + 1 \mid \tilde{P}^- \mid JM \rangle = \langle JM + 1 \mid 2\tilde{P}_X \mid JM \rangle$$
$$(4.91)$$

which gives

$$\langle JKM + 1 \mid \tilde{P}_X \mid JKM \rangle = \tfrac{1}{2}\hbar[J(J + 1) - M(M + 1)]^{1/2} \qquad (4.92)$$

because the second term in Eq. (4.91) is zero. Note that the Hermitian property requires

$$\langle JKM \mid \tilde{P}_X \mid JKM + 1 \rangle = \langle JKM + 1 \mid \tilde{P}_X \mid JKM \rangle$$

In a similar manner we find

$$\langle JKM + 1 \mid \tilde{P}_Y \mid JKM \rangle = \tfrac{1}{2}i\hbar[J(J + 1) - M(M + 1)]^{1/2}$$
$$= -\langle JKM \mid \tilde{P}_Y \mid JKM + 1 \rangle \qquad (4.93)$$

If we solve the similar commutation relations involving $\tilde{P}_x$, $\tilde{P}_y$, and $\tilde{P}_z$ we find

$$\langle JKM \mid \tilde{P}_z \mid JKM \rangle = \hbar K \qquad (4.94)$$

$$\langle JKM \mid \tilde{P}_x \mid JK + 1M \rangle = \tfrac{1}{2}\hbar[J(J + 1) - K(K + 1)]^{1/2}$$
$$= i\langle JKM \mid \tilde{P}_y \mid JK + 1M \rangle \qquad (4.95)$$

where $K = J, J - 1, \ldots, -J + 1, -J$ is the projection of J upon the molecule-fixed z axis. Comparison of Eq. (4.95) with Eqs. (4.92) and (4.93) shows the only difference to be a change in the sign of i, which is not unexpected since that was the only difference in the commutation relations.

In Table 4.1 we have summarized the angular momentum matrix elements using the symbols that are conventional for the common cases. We emphasize here that since spin angular momentum is defined *only* by commutation relations with the normal sign as mentioned in Sec. 4.2, and not by an operator like $\tilde{P}_Z$ in Eq. (4.79), the appropriate matrix elements are those derived for J and M. The conventions that we use for labeling the angular momentum quantum numbers are as follows.

(1) For rotational (nuclear framework) angular momentum we use the symbols J, K, and M as described above.

(2) For orbital (electron) angular momentum we use the symbols L and M_L for the total angular momentum quantum number and Z-axis component, respectively; when the molecule-fixed z-axis component is needed (as for diatomic molecules) we use Λ.

(3) For electron spin angular momentum we use S, M_S, and Σ (when needed) for the total, Z-axis and z-axis angular momentum quantum numbers, respectively.

(4) For nuclear spin we use I and M_I as the total and Z-axis-component angular momentum quantum numbers, respectively.

(5) When it is necessary to sum several types of angular momenta other conventions are necessary. For the sum of rotational and nuclear we use F and M_F (Z axis), for example. For the sum of orbital and spin we use J, as we do also for rotational plus spin angular momentum. Thus there is some conflict in symbolism but these cases are clarified when they arise.

EXAMPLE 4-2

Systems with spin angular momentum of $S = \frac{1}{2}$ (or $I = \frac{1}{2}$) have particularly simple spin matrices. These are found by substituting $S = \frac{1}{2}$ and $M_S = \pm\frac{1}{2}$ into the expressions in Table 4.2, with the following results.

$$S_X = \tfrac{1}{2}\hbar \begin{pmatrix} 0 & 1 \\ 1 & 0 \end{pmatrix} \qquad S_Y = \tfrac{1}{2}\hbar \begin{pmatrix} 0 & -i \\ i & 0 \end{pmatrix}$$

$$S_Z = \tfrac{1}{2}\hbar \begin{pmatrix} 1 & 0 \\ 0 & -1 \end{pmatrix} \qquad S^2 = \hbar^2 \begin{pmatrix} \frac{3}{4} & 0 \\ 0 & \frac{3}{4} \end{pmatrix}$$

(4.96)

We can verify easily the basic commutation rules, for example,

$$S_X S_Y - S_Y S_X = \tfrac{1}{4}\hbar^2 \left\{ \begin{pmatrix} 0 & 1 \\ 1 & 0 \end{pmatrix}\begin{pmatrix} 0 & -i \\ i & 0 \end{pmatrix} - \begin{pmatrix} 0 & -i \\ i & 0 \end{pmatrix}\begin{pmatrix} 0 & 1 \\ 1 & 0 \end{pmatrix} \right\}$$

$$= \tfrac{1}{4}\hbar^2 \left\{ \begin{pmatrix} i & 0 \\ 0 & -i \end{pmatrix} - \begin{pmatrix} -i & 0 \\ 0 & i \end{pmatrix} \right\}$$

(4.97)

$$= \tfrac{1}{4}\hbar^2 \begin{pmatrix} 2i & 0 \\ 0 & -2i \end{pmatrix}$$

$$= i\hbar \left\{ \hbar \begin{pmatrix} \frac{1}{2} & 0 \\ 0 & -\frac{1}{2} \end{pmatrix} \right\}$$

$$= -(\hbar/i) S_Z$$

TABLE 4.1

Matrix elements of angular momentum in representation which diagonalizes the square of the angular momentum and its Z (or z) component. All values in units of $\hbar$ or $\hbar^2$.

A. Molecular (framework) rotational angular momentum

Element	Value
$\langle JKM \mid \tilde{P}^2 \mid JKM \rangle$	$J(J+1)$
$\langle JKM \mid \tilde{P}_Z \mid JKM \rangle$	M
$\langle JKM \mid \tilde{P}_z \mid JKM \rangle$	K
$\langle JKM \mid \tilde{P}_X \mid JK,\, M+1 \rangle =$ $\langle JK, M+1 \mid \tilde{P}_X \mid JKM \rangle$	$\frac{1}{2}[J(J+1) - M(M+1)]^{1/2}$
$\langle JKM \mid \tilde{P}_Y \mid JK, M+1 \rangle =$ $-\langle JK, M+1 \mid \tilde{P}_Y \mid JKM \rangle$	$\frac{i}{2}[J(J+1) - M(M+1)]^{1/2}$
$\langle JKM \mid \tilde{P}_x \mid J, K+1, M \rangle =$ $\langle J, K+1, M \mid \tilde{P}_x \mid JKM \rangle$	$\frac{1}{2}[J(J+1) - K(K+1)]^{1/2}$
$\langle JKM \mid \tilde{P}_y \mid J, K+1, M \rangle =$ $-\langle J, K+1, M \mid \tilde{P}_y \mid JKM \rangle$	$-\frac{i}{2}[J(J+1) - K(K+1)]^{1/2}$

B. Orbital angular momentum[a]

Element	Value
$\langle LM_L \mid \tilde{P}^2 \mid LM_L \rangle$	$L(L+1)$
$\langle LM_L \mid \tilde{P}_Z \mid LM_L \rangle$	M_L
$\langle LM_L \mid \tilde{P}_X \mid L, M_L+1 \rangle =$ $\langle L, M_L+1 \mid \tilde{P}_X \mid LM_L \rangle$	$\frac{1}{2}[L(L+1) - M_L(M_L+1)]^{1/2}$
$\langle LM_L \mid \tilde{P}_Y \mid L, M_L+1 \rangle =$ $-\langle L, M_L+1 \mid \tilde{P}_Y \mid LM_L \rangle$	$\frac{i}{2}[L(L+1) - M_L(M_L+1)]^{1/2}$

C. Electron spin angular momentum[b]

Element	Value
$\langle SM_S \mid \tilde{P}^2 \mid SM_S \rangle$	$S(S+1)$
$\langle SM_S \mid \tilde{P}_Z \mid SM_S \rangle$	M_S
$\langle SM_S \mid \tilde{P}_X \mid S, M_S+1 \rangle =$ $\langle S, M_S+1 \mid \tilde{P}_X \mid SM_S \rangle$	$\frac{1}{2}[S(S+1) - M_S(M_S+1)]^{1/2}$
$\langle SM_S \mid \tilde{P}_Y \mid S, M_S+1 \rangle =$ $-\langle S, M_S+1 \mid \tilde{P}_Y \mid SM_S \rangle$	$\frac{i}{2}[S(S+1) - M_S(M_S+1)]^{1/2}$

TABLE 4.1 (*continued*)

D. Nuclear spin angular momentum

Element	Value
$\langle IM_I \mid \tilde{P}^2 \mid IM_I \rangle$	$I(I+1)$
$\langle IM_I \mid \tilde{P}_z \mid IM_I \rangle$	M_I
$\langle IM_I \mid \tilde{P}_X \mid I, M_I + 1 \rangle =$ $\langle I, M_I + 1 \mid \tilde{P}_X \mid IM_I \rangle$	$\frac{1}{2}[I(I+1) - M_I(M_I+1)]^{1/2}$
$\langle IM_I \mid \tilde{P}_Y \mid I, M_I + 1 \rangle =$ $-\langle I, M_I + 1 \mid \tilde{P}_Y \mid IM_I \rangle$	$\frac{i}{2}[I(I+1) - M_I(M_I+1)]^{1/2}$

ᵃ The matrix elements of $\tilde{P}_x$, $\tilde{P}_y$, and $\tilde{P}_z$ are identical to those of Table 4.1 A if Λ is substituted for K.

ᵇ The matrix elements of $\tilde{P}_x$, $\tilde{P}_y$, and $\tilde{P}_z$ are identical to those of $\tilde{P}_X$, $\tilde{P}_Y$, and $\tilde{P}_Z$ if Σ is substituted for M_S.

The matrices of Eq. (4.96) are known as the Pauli matrices and are found to be of great importance for the discussion of ESR and NMR spectroscopy.

According to the wave mechanical format, if α is the eigenfunction of $\tilde{S}_z$ whose eigenvalue is $+\frac{1}{2}$ and β that for $-\frac{1}{2}$, Eqs. (4.96) show that

$$\tilde{S}_z\alpha = \tfrac{1}{2}\hbar\alpha \qquad \tilde{S}_z\beta = -\tfrac{1}{2}\hbar\beta$$

$$\tilde{S}^2\alpha = \tfrac{3}{4}\hbar^2\alpha \qquad \tilde{S}^2\beta = \tfrac{3}{4}\hbar^2\beta \tag{4.98}$$

EXAMPLE 4-3

In many problems of interest, several independent, or approximately independent, angular momenta may exist. While our need in later chapters for a complete treatment of the addition of angular momenta is slight, we illustrate briefly how the situation is handled. If we have two noninteracting angular momenta L_1 and L_2,[7] the total angular momentum is given by

$$\mathbf{L} = \mathbf{L}_1 + \mathbf{L}_2 \tag{4.99}$$

If $\tilde{L}_1$, $\tilde{L}_2$, $\tilde{L}^2$, and $\tilde{L}_z$ all commute (which is the case when L_1 and L_2 are independent) the matrix elements of $\tilde{L}^2$ and $\tilde{L}_z$ in the $L_1L_2LM_L$ representation are the diagonal ones of the form previously derived,

$$\langle L_1L_2LM_L \mid \tilde{L}^2 \mid L_1L_2LM_L \rangle = \hbar^2 L(L+1) \tag{4.100}$$

$$\langle L_1L_2LM_L \mid \tilde{L}_z \mid L_1L_2LM_L \rangle = \hbar M_L \tag{4.101}$$

---

[7] We use L_1 and L_2 but the treatment is not limited to orbital angular momentum.

where

$$L = L_1 + L_2, L_1 + L_2 - 1, \ldots, |L_1 - L_2| \qquad (4.102a)$$

and

$$M_L = L, L - 1, \ldots, -L + 1, -L \qquad (4.102b)$$

Note that Eq. (4.102a) is the well-known result of the quantum-mechanical vector model.

To illustrate the computation of other matrix elements we choose those of the scalar product $\mathbf{L}_1 \cdot \mathbf{L}_2$. From Eq. (4.99) we find

$$\tilde{L}^2 = \tilde{L}_1{}^2 + 2\tilde{L}_1\tilde{L}_2 + \tilde{L}_2{}^2$$

or

$$\tilde{L}_1\tilde{L}_2 = \tfrac{1}{2}(\tilde{L}^2 - \tilde{L}_1{}^2 - \tilde{L}_2{}^2) \qquad (4.103)$$

Thus the matrix elements of $\tilde{L}_1\tilde{L}_2$ are

$$\langle L_1L_2LM \mid \tilde{L}_1\tilde{L}_2 \mid L_1'L_2'L'M' \rangle = \tfrac{1}{2}\{\langle L_1L_2LM \mid \tilde{L}^2 \mid L_1'L_2'L'M' \rangle$$
$$- \langle L_1L_2LM \mid \tilde{L}_1{}^2 \mid L_1'L_2'L'M' \rangle - \langle L_1L_2LM \mid \tilde{L}_2{}^2 \mid L_1'L_2'L'M \rangle\}$$

$$(4.104a)$$

But each of the matrices $\mathbf{L}^2$, $\mathbf{L}_1{}^2$ and $\mathbf{L}_2{}^2$ is diagonal so that $\mathbf{L}_1\mathbf{L}_2$ must be diagonal also. Inserting the values from Table 4.2, we find

$$\langle L_1L_2LM \mid \tilde{L}_1\tilde{L}_2 \mid L_1L_2LM \rangle = \tfrac{1}{2}\{L(L+1) - L_1(L_1+1) - L_2(L_2+1)\}\hbar^2$$

$$(4.104b)$$

In a similar manner, since

$$\tilde{L}\tilde{L}_1 = \tilde{L}_1{}^2 + \tilde{L}_2\tilde{L}_1 = \tfrac{1}{2}(\tilde{L}^2 + \tilde{L}_1{}^2 - \tilde{L}_2{}^2)$$

we find

$$\langle L_1L_2LM \mid \tilde{L}\tilde{L}_1 \mid L_1L_2LM \rangle = \tfrac{1}{2}\{L(L+1) + L_1(L_1+1) - L_2(L_2+1)\}\hbar^2$$

$$(4.105)$$

The complete calculation of the matrices of $\tilde{L}_{1X}$, $\tilde{L}_{1Y}$, $\tilde{L}_{1Z}$, $\tilde{L}_{2X}$, $\tilde{L}_{2Y}$ and $\tilde{L}_{2Z}$ is beyond the scope of this treatment, and the results are not needed in later chapters, although some special cases and simplifications are encountered. The interested reader is urged to consult Condon and Shortley[8] for an excellent discussion of the complete problem.

4.4   PROPERTIES OF THE DIRECTION COSINES

In Sec. 4.2 we discussed the properties of angular momentum when referred to both space-fixed and molecule-fixed axes. The angular mo-

---

[8] E. U. Condon and G. H. Shortley, *The Theory of Atomic Spectra* (Cambridge U.P., Cambridge, 1959), pp. 56–66.

mentum components in the two axis systems are related by the direction cosines Φ_{gF} which are the cosines of the angles between the molecule- and space-fixed axes. For example,

$$P_x = \Phi_{xX}P_X + \Phi_{xY}P_Y + \Phi_{xZ}P_Z \tag{4.106a}$$

It is seen that Eq. (4.106a) merely expresses the fact that P_x is the sum of components of P_X, P_Y, and P_Z projected along the x direction. In a more general fashion we write this as

$$F = X, Y, Z$$
$$P_g = \sum_F \Phi_{gF}P_F \tag{4.107a}$$
$$g = x, y, z$$

or

$$\mathbf{P}^{(g)} = \mathbf{\Phi}\mathbf{P}^{(F)} \tag{4.107b}$$

where
$$\mathbf{P}^{(g)} = \begin{pmatrix} P_x \\ P_y \\ P_z \end{pmatrix} \quad \text{and} \quad \mathbf{P}^{(F)} = \begin{pmatrix} P_X \\ P_Y \\ P_Z \end{pmatrix}$$

and $\mathbf{\Phi}$ is the matrix of the direction cosines. It should be noted that $\mathbf{\Phi}$ is a real orthogonal matrix so that the reverse transformation would be

$$\mathbf{P}^{(F)} = \mathbf{\Phi}'\mathbf{P}^{(g)} \tag{4.107c}$$

In order that the form of ϕ is understood, we write it explicitly as

$$\mathbf{\Phi} = \begin{pmatrix} \Phi_{xX} & \Phi_{xY} & \Phi_{xZ} \\ \Phi_{yX} & \Phi_{yY} & \Phi_{yZ} \\ \Phi_{zX} & \Phi_{zY} & \Phi_{zZ} \end{pmatrix} \tag{4.108}$$

As we shall see in chapter 7, the direction cosines are involved anytime it becomes necessary to relate the rotating molecular-fixed axes (or angular momenta) to the stationary spaced-fixed axes, as for example when a rotating molecule is in the presence of an external static electric field. In classical mechanics the Φ_{gF} would naturally specify the time dependence of the rotating axes with respect to the space-fixed axis system.

The quantum-mechanical properties of the direction cosines, namely their matrix elements, can be obtained by solving their commutation relations with the angular momenta. Since the solution of the matrix equations is tedious (but not unlike that for the angular momenta) we shall be content here with merely a brief discussion of the commutators and the resulting matrices.

We observe first that the direction cosines behave like the components of a vector. This is seen from Eq. (4.106a) which can be rewritten

$$P_x = (\Phi_{xX} \Phi_{xY} \Phi_{xZ}) \begin{pmatrix} P_X \\ P_Y \\ P_Z \end{pmatrix} \qquad (4.106b)$$

or

$$P_x = \mathbf{\Phi}_x \cdot \mathbf{P}^{(F)} \qquad (4.106c)$$

where $\mathbf{\Phi}_x$ is the three-dimensional direction-cosine row vector. Moreover, $\mathbf{\Phi}_x$ has the same transformation properties as an ordinary Cartesian vector in the x direction. This is obvious because the components of $\mathbf{\Phi}_x$ are the cosines of the angles which project the Cartesian vector $\mathbf{x}$ onto the X, Y, and Z axes. For example,

$$\Phi_{xX} = \cos\theta_{xX} = \frac{x_X}{x}$$

$$\Phi_{xY} = \frac{x_Y}{x} \qquad (4.109)$$

$$\Phi_{xZ} = \frac{x_Z}{x}$$

where x_X, x_Y, and x_Z are the projections of the vector $\mathbf{x}$ of length x upon the X, Y, and Z axes, respectively. Thus to within a constant factor [x^{-1} in (4.109)], $\mathbf{\Phi}_x$ transforms like $\mathbf{x}$. We can therefore obtain the commutation relations of all the Φ_{gF} if those for X, Y, Z and x, y, and z are known.

As an example, if the commutator $[X, \tilde{P}_Y]$ is known, that for $[\tilde{\Phi}_{gX}, \tilde{P}_Y]$ is immediately known. For the former case we obtain

$$[X, \tilde{P}_Y] = X\tilde{P}_Y - \tilde{P}_Y X = XZ\tilde{p}_x - XX\tilde{p}_z - Z\tilde{p}_x X + X\tilde{p}_z X \qquad (4.110a)$$

by use of Eq. (4.52). Rearranging we find

$$[X, \tilde{P}_Y] = -Z(\tilde{p}_x X - X\tilde{p}_x) + X(\tilde{p}_z X - X\tilde{p}_z) \qquad (4.110b)$$

which becomes

$$[X, \tilde{P}_Y] = -\frac{\hbar}{i} Z \qquad (4.110c)$$

when Eqs. (4.54) are used. We conclude then, that

$$[\tilde{\Phi}_{gX}, \tilde{P}_Y] = \tilde{\Phi}_{gX}\tilde{P}_Y - \tilde{P}_Y\tilde{\Phi}_{gX} = -\frac{\hbar}{i} \tilde{\Phi}_{gZ} \qquad (4.111a)$$

In this manner we can arrive at the complete set of commutation relations which are summarized as follows:

$$[\tilde{\Phi}_{gF}, \tilde{\Phi}_{g'F'}] = 0 \qquad (F, F' = X, Y, Z; g, g' = x, y, z) \quad (4.111b)$$

$$[\tilde{P}_g, \tilde{\Phi}_{gF}] = [\tilde{P}_F, \tilde{\Phi}_{gF}] = 0 \quad (4.111c)$$

$$[\tilde{\Phi}_{gX}, \tilde{P}_Y] = -[\tilde{\Phi}_{gY}, \tilde{P}_X] = -\frac{\hbar}{i}\tilde{\Phi}_{gZ}, \text{ etc.} \quad (4.111d)$$

$$[\tilde{\Phi}_{xF}, \tilde{P}_y] = -[\tilde{\Phi}_{yF}, \tilde{P}_x] = \frac{\hbar}{i}\tilde{\Phi}_{zF}, \text{ etc.} \quad (4.111e)$$

Note the sign difference in (4.111d) and (4.111e) caused by the previously mentioned sign change in the angular momentum commutation relations. The resulting matrix equations can be solved by the general methods described by Condon and Shortley.[9] King, Hainer, and Cross[10] as well as others have given the results of the solution.

It has been shown that the matrix elements can be written as the product of three factors

$$\langle JKM \mid \tilde{\Phi}_{gF} \mid J'K'M' \rangle$$

$$= \langle J \parallel \tilde{\Phi}_{gF} \parallel J' \rangle \cdot \langle JK \parallel \tilde{\Phi}_{gF} \parallel J'K' \rangle \cdot \langle JM \parallel \tilde{\Phi}_{gF} \parallel J'M' \rangle \quad (4.112)$$

These factors have the form of matrix elements but are written with the double bars to indicate that they are not complete matrix elements. The first factor depends only on J, the second upon J and K, and the third upon J and M. All the permissible values of these factors have been tabulated in Table 4.2, where it is seen that nonzero elements occur only when $\Delta J = 0$, ± 1, $\Delta K = 0, \pm 1$, and $\Delta M = 0, \pm 1$.

To illustrate the use of the table we write down the diagonal elements of $\tilde{\Phi}_{xZ}$ and $\tilde{\Phi}_{zZ}$.

$$\langle JKM \mid \tilde{\Phi}_{xZ} \mid JKM \rangle = \langle J \parallel \tilde{\Phi}_{xZ} \parallel J \rangle \langle JK \parallel \tilde{\Phi}_{xZ} \parallel JK \rangle \langle JM \parallel \tilde{\Phi}_{xZ} \parallel JM \rangle$$

$$= \frac{1}{4J(J+1)} \cdot (0) \cdot 2M = 0 \quad (4.113)$$

$$\langle JKM \mid \tilde{\Phi}_{zZ} \mid JKM \rangle = \langle J \parallel \tilde{\Phi}_{zZ} \parallel J \rangle \langle JK \parallel \tilde{\Phi}_{zZ} \parallel JK \rangle \langle JM \parallel \tilde{\Phi}_{zZ} \parallel JM \rangle$$

$$= \frac{1}{4J(J+1)} \cdot 2K \cdot 2M = \frac{KM}{J(J+1)} \quad (4.114)$$

[9] See footnote 8, pp. 59–64.
[10] P. C. Cross, R. M. Hainer, and G. W. King, *J. Chem. Phys.* **12**, 210 (1944).

TABLE 4.2
Direction cosine matrix elements in JKM representation.

$$\langle JKM \mid \bar{\Phi}_{gF} \mid J'K'M' \rangle = \langle J \| \bar{\Phi}_{gF} \| J' \rangle \cdot \langle JK \| \bar{\Phi}_{gF} \| J'K' \rangle \cdot \langle JM \| \bar{\Phi}_{gF} \| J'M' \rangle$$

	$J' = J+1$	$J' = J$	$J' = J-1$
$\langle J \| \bar{\Phi}_{gF} \| J' \rangle$	$\{4(J+1)[(2J+1)(2J+3)]^{1/2}\}^{-1}$	$[4J(J+1)]^{-1}$	$[4J(4J^2-1)^{1/2}]^{-1}$
$\langle JK \| \bar{\Phi}_{zF} \| J'K \rangle$	$2[(J+1)^2 - K^2]^{1/2}$	$2K$	$2(J^2 - K^2)^{1/2}$
$\langle JM \| \bar{\Phi}_{gZ} \| J'M \rangle$	$2[(J+1)^2 - M^2]^{1/2}$	$2M$	$2(J^2 - M^2)^{1/2}$
$\langle JK \| \bar{\Phi}_{zF} \| J'K \pm 1 \rangle$ $= \mp i \langle JK \| \bar{\Phi}_{yF} \| J'K \pm 1 \rangle$	$\mp[(J \pm K+1)(J \pm K+2)]^{1/2}$	$[J(J+1) - K(K \pm 1)]^{1/2}$	$\pm[(J \mp K)(J \mp K - 1)]^{1/2}$
$\langle JM \| \bar{\Phi}_{gX} \| J'M \pm 1 \rangle$ $= \pm i \langle JM \| \bar{\Phi}_{gY} \| J'M \pm 1 \rangle$	$\mp[(J \pm M+1)(J \pm M+2)]^{1/2}$	$[J(J+1) - M(M \pm 1)]^{1/2}$	$\pm[(J \mp M)(J \mp M - 1)]^{1/2}$

EXAMPLE 4-4

We verify here the first of the commutators written in Eq. (4.111d). From the results of Sec. 4.3, we know that $\tilde{P}_X$ and $\tilde{P}_Y$ have only $M, M \pm 1$ elements and are diagonal in J and K. Therefore Eq. (4.111d) can be written for the diagonal element and $g = z$

$$\langle JKM \mid \Phi_{zX} \mid JKM - 1 \rangle \langle JKM - 1 \mid \tilde{P}_Y \mid JKM \rangle$$

$$+ \langle JKM \mid \Phi_{zX} \mid JKM + 1 \rangle \langle JKM + 1 \mid \tilde{P}_Y \mid JKM \rangle$$

$$- \langle JKM \mid \tilde{P}_Y \mid JKM - 1 \rangle \langle JKM - 1 \mid \Phi_{zX} \mid JKM \rangle$$

$$- \langle JKM \mid \tilde{P}_Y \mid JKM + 1 \rangle \langle JKM + 1 \mid \Phi_{zX} \mid JKM \rangle \qquad (4.115)$$

$$= -\frac{\hbar}{i} \langle JKM \mid \Phi_{zz} \mid JKM \rangle$$

Evaluating the left-hand side by means of Tables 4.1 and 4.2, we obtain

$$\frac{1}{4J(J+1)} \cdot 2K \cdot ([J(J+1) - M(M-1)]^{1/2})$$

$$\cdot \left(\frac{i\hbar}{2} [J(J+1) - (M-1)(M)]^{1/2} \right) + \frac{1}{4J(J+1)} \cdot 2K$$

$$\cdot ([J(J+1) - M(M+1)]^{1/2}) \cdot \left(-\frac{i\hbar}{2} [J(J+1) - M(M+1)]^{1/2} \right)$$

$$- \left(-\frac{i\hbar}{2} [J(J+1) - (M-1)(M)]^{1/2} \right) \cdot \frac{1}{4J(J+1)} \cdot 2K$$

$$\cdot ([J(J+1) - (M-1)(M)]^{1/2}) - \left(\frac{i\hbar}{2} [J(J+1) - M(M+1)]^{1/2} \right)$$

$$\cdot \frac{1}{4J(J+1)} \cdot 2K \cdot ([J(J+1) - M(M+1)]^{1/2}) = \frac{K}{2J(J+1)} \cdot \frac{i\hbar}{2}$$

$$\cdot \{ J(J+1) - M(M-1) - J(J+1) + M(M+1)$$

$$+ J(J+1) - M(M-1) - J(J+1) + M(M+1) \}$$

$$= \frac{MK\hbar i}{J(J+1)} = -\frac{\hbar}{i} \frac{MK}{J(J+1)} \qquad (4.116)$$

By comparison of this result to that of Eq. (4.114) we see clearly that the commutation relation is satisfied.

SUPPLEMENTARY REFERENCES

H. C. Allen and P. C. Cross, *Molecular Vib-Rotors* (Wiley, New York, 1963).

E. U. Condon and G. H. Shortley, *Theory of Atomic Spectra* (Cambridge U.P., Cambridge, 1935).

A. R. Edmonds, *Angular Momentum in Quantum Mechanics* (Princeton U.P., Princeton, 1960).

L. D. Landau and E. M. Lifshitz, *Quantum Mechanics, Non-Relativistic Theory* (Addison-Wesley, Reading, Mass., 1958).

M. E. Rose, *Elementary Theory of Angular Momentum* (Wiley, London, 1957).

CHAPTER 5

ABSORPTION OF ELECTROMAGNETIC RADIATION

In chapter 2 we indicated the general quantum-mechanical approaches for determining the energy levels of any atomic or molecular system. chapter 4 illustrated the matrix-mechanics solution for the eigenvalues and matrix elements of angular momenta and also the matrix elements appropriate for a harmonic oscillator. The basic molecular Hamiltonian was thoroughly discussed in chapter 3 and thus the groundwork has been laid for the determination of the various energy levels of molecular systems in later chapters. This is, however, only the first half of the spectroscopic problem since we need to know which molecular states can be involved in the absorption (or emission) of electromagnetic radiation, that is, we need to know which states can be observed to satisfy the Bohr frequency relation Eq. (1.1).

In this chapter we show in a rather simple formalism the criteria which must be satisfied for the absorption of electromagnetic radiation. At the outset we should mention that the approach is not the most rigorous or general treatment, since this would require consideration of the quantization of the radiation field itself,[1] a topic whose complexity is unnecessary for the results that we require. Indeed, our perturbation treatment via Schrödinger's time-dependent equation provides the correct results for the *stimulated* absorption and emission processes in which we are chiefly interested. The perturbation treatment says nothing about the *spontaneous*

---

[1] W. Heitler, *The Quantum Theory of Radiation* (Oxford U.P., London, 1954), 3rd ed.

emission of radiation, but in this case we can fall back upon the well-known phenomenological absorption and emission coefficients of Einstein.[2]

5.1 CALCULATION OF TRANSITION PROBABILITIES BY TIME-DEPENDENT PERTURBATION THEORY

The behavior of an atomic or molecular system under the influence of a time-dependent potential is described by Schrödinger's equation with time included,

$$\widetilde{\mathcal{3C}}\Psi(q, t) = -\frac{\hbar}{i}\frac{\partial}{\partial t}\Psi(q, t) \tag{5.1}$$

where Ψ is shown as a function of coordinates and time and $\widetilde{\mathcal{3C}}$ may be written

$$\widetilde{\mathcal{3C}} = \widetilde{\mathcal{3C}}^0 + \widetilde{\mathcal{3C}}'(t) \tag{5.2}$$

In Eq. (5.2) $\widetilde{\mathcal{3C}}^0$ is the time-independent Hamiltonian for the system in the absence of the time-dependent potential energy term $\widetilde{\mathcal{3C}}'$. In the cases of interest to us, $\widetilde{\mathcal{3C}}'$ is small so that its effect is found by treating it as a small perturbation on the stationary behavior of the system. It is easy to verify[3] that when $\widetilde{\mathcal{3C}}'(t) = 0$, the stationary-state eigenfunctions of Eq. (2.1) are related to the eigenfunctions of Eq. (5.1) by

$$\Psi_n^0 = \psi_n^0 \exp[-(i/\hbar)E_n^0 t] \tag{5.3}$$

where E_n^0 is the energy of the nth stationary state ψ_n^0. We use the superscript zero to signify the stationary states, since they serve as suitable zero-order basis functions for the following perturbation theory.

Since the solutions of Eq. (5.1) are of the form of Eq. (5.3) when the perturbation is absent, we should expect that a suitable approximation for Ψ when $\widetilde{\mathcal{3C}}'(t) \neq 0$ would be

$$\Psi = \sum_n c_n(t)\Psi_n^0 \tag{5.4}$$

where the expansion coefficients are functions of time and the summation is over the complete set of zero-order functions. By the usual statistical interpretation of quantum mechanics the probability that an atom or molecule exists in a certain state E_k^0 is simply

$$\left| \int \Psi^* \Psi_k^0 d\tau \right|^2 = c_k^* c_k = |c_k|^2 \tag{5.5}$$

where the right-hand side of Eq. (5.5) follows by substituting Eq. (5.4) and using the orthonormal properties of the Ψ_n^0. Thus, if at time $t = 0$ a

---

[2] A. Einstein, Z. Physik **18**, 121 (1917).

[3] Simply substitute Eq. (5.3) into Eq. (5.1) and (2.1) to show that it satisfies each equation when $\widetilde{\mathcal{3C}}'(t) = 0$.

molecule is known to exist in some state $\Psi_0{}^0$ (that is, $c_0 = 1$), the probability of finding the molecule in some other state at some later time after applying $\widetilde{\mathfrak{IC}}'(t)$ is given by $|c_n(t)|^2$. Our job therefore, is to find the time dependence of the $c_n(t)$ under the influence of the perturbation $\widetilde{\mathfrak{IC}}'(t)$, the particular case of interest being that when $\widetilde{\mathfrak{IC}}'(t)$ represents the interaction of an electromagnetic wave with the molecular system.

We now substitute Eq. (5.4) into Eq. (5.1) and use the generál Hamiltonian of Eq. (5.2) to give

$$\sum_n c_n \widetilde{\mathfrak{IC}}^0 \Psi_n{}^0 + \sum_n c_n \widetilde{\mathfrak{IC}}' \Psi_n{}^0 = -\frac{\hbar}{i} \sum_n \dot{c}_n \Psi_n{}^0 - \frac{\hbar}{i} \sum_n c_n \dot{\Psi}_n{}^0$$

or

$$\sum_n c_n \widetilde{\mathfrak{IC}}' \Psi_n{}^0 = i\hbar \sum_n \dot{c}_n \Psi_n{}^0 \tag{5.6}$$

where we have used

$$\Psi_n{}^0 = -\frac{i}{\hbar} E_n{}^0 \dot{\Psi}_n{}^0 \tag{5.7a}$$

and

$$\widetilde{\mathfrak{IC}}^0 \Psi_n{}^0 = E_n{}^0 \Psi_n{}^0 \tag{5.7b}$$

If we multiply both sides of Eq. (5.6) on the left-hand side by $\Psi_m{}^{0*}$ and integrate over all coordinate space we obtain

$$\sum_n c_n \int \Psi_m{}^{0*} \widetilde{\mathfrak{IC}}' \Psi_n{}^0 d\tau = i\hbar \dot{c}_m \tag{5.8a}$$

in which the right-hand side of Eq. (5.6) has been reduced to one term $(n = m)$ because of the orthogonality of the $\Psi_n{}^0$. Denoting the matrix element

$$H_{mn}{}'^t = \int \Psi_m{}^{0*} \widetilde{\mathfrak{IC}}' \Psi_n{}^0 d\tau \tag{5.9a}$$

where the superscript t denotes the use of time-dependent wave functions, we can rewrite Eq. (5.8a) as

$$\dot{c}_m = -\frac{i}{\hbar} \sum_n c_n H_{mn}{}'^t(t) \tag{5.8b}$$

For the usual case when $\widetilde{\mathfrak{IC}}'(t)$ is not a differential operator, the exponential time dependence of the $\Psi_n{}^0$ can be taken out of the integrand of Eq. (5.8a) so that we can write

$$\dot{c}_m = -\frac{i}{\hbar} \sum_n c_n \exp[-(i/\hbar)(E_n{}^0 - E_m{}^0)t] H_{mn}{}' \tag{5.8c}$$

where the matrix element is now

$$H_{mn}' = \int \psi_m^{0*}\widetilde{\mathfrak{IC}}'\psi_n^0 d\tau \qquad (5.9b)$$

Equations (5.8b and c) express the time dependence of the expansion coefficients for any system under the influence of a time-dependent perturbation. There are, in general, an infinite or very large number of equations of this form corresponding to all the stationary states of the molecular system. An exact solution of this set of simultaneous equations for the coefficients c_n is not possible in general so some approximate means of solution must be utilized. In any case it should be evident that if the system is defined at some time $t = 0$ by $c_0 = 1$, $c_{n\neq0} = 0$, the probability that a *transition* to some other state ψ_n^0 has occurred at some later time is given by $|c_n|^2$ as determined by appropriate solution of Eqs. (5.8).

EXAMPLE 5-1

As a very simple example of how Eqs. (5.8) may be solved, let us consider a molecular system under the influence of a perturbation whose time dependence is of the type of Fig. 5.1(a). We further assume that only two stationary states exist as shown in Fig. 5.1(b), so that

$$\Psi = c_1\Psi_1^0 + c_2\Psi_2^0 \qquad (5.10)$$

We imagine that the perturbation is small and that the perturbation operator is Hermitian, so that

$$|H_{12}'|^2 = |H_{21}'|^2 \ll (\Delta E)^2 \qquad (5.11)$$

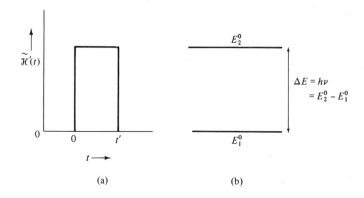

FIG. 5.1.   (a) A simple time dependent perturbation. (b) A simple two-level system in the presence of the perturbation of (a).

It is assumed implicitly that $\tilde{\mathcal{3C}}'(t)$ contains some functional dependence upon the coordinates q_i of the system, but we leave this unspecified. Then Eq. (5.8c) becomes for c_1 and c_2

$$\dot{c}_2 = - \frac{i}{\hbar} \left[H_{21}'c_1 \exp[(i/\hbar)\,\Delta Et] + c_2 H_{22}' \right] \qquad (5.12a)$$

$$\dot{c}_1 = - \frac{i}{\hbar} \left[H_{12}'c_2 \exp[-(i/\hbar)\,\Delta Et] + c_1 H_{11}' \right] \qquad (5.12b)$$

Suppose that at time $t = 0$ the system is specified by

$$c_1 = 1 \quad \text{and} \quad c_2 = 0$$

At some time $0 < t < t'$, if the perturbation is small, the time dependence of c_2 can be obtained by substituting $c_2 = 0$ and $c_1 = 1$ in the right-hand side of Eq. (5.12a), which gives

$$\dot{c}_2 = - \frac{i}{\hbar} H_{21}' \exp[(i/\hbar)\,\Delta Et] \qquad (5.13)$$

This expression can be integrated to yield

$$\int_0^{c_2} dc_2 = - \frac{i}{\hbar} H_{21}' \int_0^t \exp[(i/\hbar)\,\Delta Et]\, dt$$

$$c_2 = \frac{H_{21}'}{\Delta E} \{1 - \exp[(i/\hbar)\,\Delta Et]\} \qquad (5.14)$$

By Eq. (5.5) and c_2 as given by Eq. (5.14) the probability that the molecule exists in state ψ_2^0 at time t is

$$|c_2|^2 = \frac{4\,|H_{21}'|^2}{(\Delta E)^2} \left[\sin \frac{\Delta Et}{2\hbar} \right]^2 \qquad (5.15a)$$

where use has been made of the identity

$$e^x + e^{-x} = 2 \cos x$$

We can see that there is a nonzero but small probability that the molecule has undergone a transition to the excited state. If t is small the sin can be replaced by its argument so that

$$|c_2|^2 = \frac{|H_{21}'|^2\, t^2}{\hbar^2} \qquad (5.15b)$$

Because of the normalization condition of c_1 and c_2 the probability that the molecule has remained in the lowest state is

$$|c_1|^2 = 1 - |c_2|^2 = 1 - \frac{|H_{21}'|^2\, t^2}{\hbar^2} \qquad (5.16)$$

In a very simple way we can illustrate one further feature of the time-dependent perturbation. Let us calculate the energy of the states E_1 and E_2 correct to first order. We find

$$E_1 = E_1{}^0 + \int \Psi_1{}^{0*}\widetilde{\mathcal{JC}}'(t)\,\Psi_1{}^0 d\tau$$

$$= E_1{}^0 + H_{11}'(t) \qquad (5.17a)$$

and

$$E_2 = E_2{}^0 + H_{22}'(t) \qquad (5.17b)$$

where the usual first-order correction term is now time dependent. Thus the energy separation of the levels or the transition frequency between them varies with time, that is

$$\frac{E_2 - E_1}{h} = \frac{E_2{}^0 - E_1{}^0}{h} + \frac{H_{22}'(t) - H_{11}'(t)}{h} \qquad (5.18a)$$

or

$$\nu = \nu^0 + \frac{\Delta H'(t)}{h} \qquad (5.18b)$$

It is clear that the energy separation is not sharp in general, or stated differently, the absorption line is not infinitely narrow but has a width that is determined by the amplitude and time dependence of the perturbing term.

5.2   TRANSITION PROBABILITY IN A PERIODIC PERTURBATION

We consider here a rather general perturbation of the form

$$\widetilde{\mathcal{JC}}'(t) = \tilde{F}^0[\exp(2\pi i\nu t) + \exp(-2\pi i\nu t)] \qquad (5.19a)$$

which has the equivalent form

$$\widetilde{\mathcal{JC}}'(t) = 2\tilde{F}^0 \cos(2\pi\nu t) \qquad (5.19b)$$

In Eqs. (5.19), ν is the frequency of the periodic perturbation and $\tilde{F}^0$, the amplitude, is considered to be independent of t, but may depend upon ν and also upon the appropriate coordinates of the molecular system. It should be apparent that the form of Eq. (5.19) is such that it is suitable for describing the interaction of a molecule with electromagnetic radiation.

Substituting Eq. (5.19a) into Eq. (5.8a), and setting $c_o = 1$ and $c_{n \neq o} = 0$ we find the time dependence of the mth stationary state to be

$$
\begin{aligned}
\dot{c}_m &= -\frac{i}{\hbar} \int \exp[(i/\hbar)E_m^0 t] \psi_m^{0*} \tilde{F}^0 [\exp(2\pi i \nu t) \\
&\qquad\qquad + \exp(-2\pi i \nu t)] \exp[-(i/\hbar)E_o^0 t] \psi_o^0 d\tau \\
&= -\frac{i}{\hbar} F_{mo} \left\{ \exp\left[i\left(\frac{E_m^0 - E_o^0}{\hbar} + 2\pi \nu\right)t\right] \right. \\
&\qquad\qquad\qquad \left. + \exp\left[i\left(\frac{E_m^0 - E_o^0}{\hbar} - 2\pi \nu\right)t\right] \right\}
\end{aligned} \qquad (5.20)
$$

in which

$$
F_{mo} = \int \psi_m^{0*} \tilde{F}^0 \psi_o^0 d\tau \qquad (5.21)
$$

Using

$$
\frac{E_m^0 - E_o^0}{h} = \nu_{mo} \qquad (5.22)
$$

and integrating Eq. (5.20) from $t = 0$ to $t = t$, we find

$$
c_m = -\frac{1}{h} F_{mo} \left\{ \frac{\exp[2\pi i (\nu_{mo} + \nu)t] - 1}{\nu_{mo} + \nu} + \frac{\exp[2\pi i (\nu_{mo} - \nu)t] - 1}{\nu_{mo} - \nu} \right\} \qquad (5.23)
$$

Note that Eq. (5.23) has been derived in a manner similar to that used in Example 5-1, namely we assume that a suitable approximation for c_m can be obtained by setting $c_o = 1$ and $c_{n \neq o} = 0$ in Eq. (5.8a). The approximation should be valid if the perturbation is small, which is always true for sufficiently short time intervals after $t = 0$, or for sufficiently small values of F_{mo}.

Then, since the molecule was assumed to be present in state ψ_o^0 at $t = 0$, the square of the magnitude of Eq. (5.23) gives the probability that the molecule undergoes a transition into any state ψ_m^0 at some later time $t > 0$. Investigation of Eq. (5.23) shows two interesting cases. When $\nu_{mo} > 0$ (that is, $E_m^0 > E_o^0$), c_m is large only when $(\nu_{mo} - \nu) \sim 0$, and in this case the first term in brackets is small. This case would correspond to *absorption* of electromagnetic radiation if the perturbation term $\mathcal{H}'(t)$ were caused by the interaction of light with the molecule. On the other hand, if $\nu_{mo} < 0$ (that is $E_m^0 < E_o^0$), c_m is large only when $(\nu_{mo} + \nu) \sim 0$, in which case the second term in brackets is small. This case represents the induced *emission* of electromagnetic radiation since the molecule is undergoing a transition to a lower state. Thus the simple theory as presented here embraces both the absorption and induced emission of electromagnetic radiation.

Considering the former case, we neglect the first term in Eq. (5.23) and obtained for the transition probability

$$| c_m |^2 = \frac{4}{h^2} | F_{mo} |^2 \left\{ \frac{\sin^2(\pi t[\nu_{mo} - \nu])}{[\nu_{mo} - \nu]^2} \right\} \qquad (5.24)$$

The magnitude of the transition probability depends, of course, upon the explicit form of $\tilde{F}^0$ and the value of the matrix element F_{mo}, which we may call the *transition moment*. It is interesting to note that Eq. (5.24) provides a rationale for the Bohr frequency relation Eq. (1.1). This is because $| c_m |^2$ has its maximum value when $\nu_{mo} = \nu$, which by use of Eq. (5.22) leads directly to the Bohr frequency expression. Note, however, that the transition probability is not negligible for a small frequency range above and below ν_{mo}, which leads us to conclude that the absorption lines are not infinitely sharp but have instead a finite natural linewidth.[4]

We can gain some further insight into this problem if we plot the oscillating function in brackets in Eq. (5.24) versus $(\nu_{mo} - \nu) \equiv \delta\nu$. This is shown in Fig. 5.2 from which it is evident that the transition probability

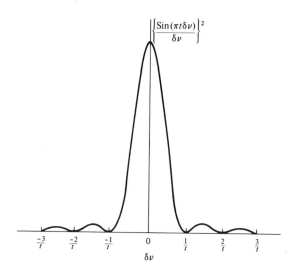

FIG. 5.2.   Plot of angular factor in Eq. (5.24).

[4] We see later that other factors usually broaden absorption lines beyond the natural linewidth.

is high only when

$$\delta\nu < \frac{1}{t} \qquad (5.25a)$$

or

$$t\delta\nu < 1 \qquad (5.25b)$$

Equation (5.25b) has an interesting interpretation by Heisenberg's uncertainty relation,

$$\delta E \delta t \leqq h \qquad (5.26a)$$

which may be considered to express the accuracy with which the energy of a particle can be measured if the measurement is carried out in the time interval δt. Suppose it were desired to measure spectroscopically the energy difference ΔE between states $E_m{}^0$ and $E_o{}^0$ by measuring the frequency ν_{mo}. Equation (5.26a) would then be written

$$\delta(\Delta E)\delta t \leqq h \qquad (5.26b)$$

or

$$\delta(h\nu_{mo})\delta t \leqq h$$

$$\delta\nu_{mo}\delta t \leqq 1 \qquad (5.26c)$$

Now the uncertainty in t, δt, can be no less than the length of time in which we are certain that the molecule exists in the excited state $\psi_m{}^0$, assuming that we knew definitely that the molecule was in $\psi_o{}^0$ immediately before the photon was absorbed. Einstein's theory of spontaneous emission (see Sec. 5.3) shows that each excited state molecule has some characteristic *lifetime* τ before it returns to the ground state. It is this lifetime which determines the uncertainty δt in Eq. (5.26c), since only for a length of time τ can we be certain that the molecule of interest has reached the state $\psi_m{}^0$. Consequently we set $\delta t = \tau$ in Eq. (5.26c) which gives

$$\delta\nu_{mo}\cdot\tau \leqq 1 \qquad (5.26d)$$

Expression (5.26d) provides then a means of estimating the maximum natural linewidth if the lifetime of the excited state in known.

Returning now to the point which brought about this slight digression, we see that Eqs. (5.25b) and (5.26d) are of a similar form. In fact, since $\delta\nu$ and $\delta\nu_{mo}$ are essentially the same quantity, we find that

$$t < \tau \qquad (5.27)$$

when we choose the equal sign in the uncertainty relation (5.26d). Equation (5.27) expresses the fact that the transition probability is large only for times which are small compared to the natural lifetime of the excited state. This usually places no severe restriction on the absorption process because the interaction of the molecule with the radiation field requires only a time t on the order of the period of one cycle of the electromagnetic radiation which is normally less than the lifetime of the excited state. For example, electronic states of atoms or molecules have typical lifetimes of 10^{-12} sec or greater while the period of one cycle of visible light is about 2×10^{-15} sec.

In many spectroscopic experiments the monochromaticity of the light is not great, that is, the light contains a range of frequencies of varying amplitudes. In this case it would be appropriate to sum the transition probability of Eq. (5.24) over all frequencies ν within the bandwidth $\Delta\nu$ of the irradiating light. We must recognize then that F_{mo}, which contains the amplitude of the electromagnetic radiation, is in general a function of ν. As shown by Fig. 5.2, however, the transition probability is large only when $\nu \sim \nu_{mo}$. Thus we may perform the summation over ν by integrating from ν_1 to ν_2 and setting $F_{mo}(\nu) = F_{mo}(\nu_{mo})$. Here we imagine ν_1 and ν_2 to be more or less symmetrically spaced about ν_{mo} and the range $\nu_2 - \nu_1$ to be the bandwidth of the radiation in which the amplitudes of the electromagnetic fields are appreciable. Equation (5.24) becomes then

$$| c_m |^2 \Delta\nu = \frac{4}{h^2} | F_{mo}(\nu_{mo}) |^2 \int_{\nu_1}^{\nu_2} \frac{\sin^2(\pi t[\nu_{mo} - \nu])}{[\nu_{mo} - \nu]^2} \, d\nu \qquad (5.28a)$$

or with a change of variables

$$| c_m |^2 \Delta\nu = \frac{4\pi t}{h^2} | F_{mo}(\nu_{mo}) |^2 \int_{-\infty}^{+\infty} \frac{\sin^2 X}{X^2} \, dX \qquad (5.28b)$$

where the limits of integration $X_1 = \pi t(\nu_{mo} - \nu_1)$ and $X_2 = \pi t(\nu_{mo} - \nu_2)$ have been extended from $-\infty$ to $+\infty$. No harm comes from this physically meaningless range since the integral has important contributions only when ν_1 and ν_2 are close to ν_{mo}. Substituting the value of the well-known integral we have

$$| c_m |^2 \Delta\nu = \frac{4\pi^2 t}{h^2} | F_{mo}(\nu_{mo}) |^2$$

$$= \frac{t}{\hbar^2} | F_{mo}(\nu_{mo}) |^2 \qquad (5.28c)$$

According to our earlier discussion we expect this equation to apply for very short periods of time, during which the transition probability increases linearly with time. Defining the transition probability per unit time P_{mo},

$$P_{mo} = \frac{| c_m |^2 \Delta\nu}{t} = \frac{1}{\hbar^2} | F_{mo}(\nu_{mo}) |^2 \qquad (5.29)$$

We have arrived finally at a very simple expression for the absorption transition probability for the relatively general interaction Hamiltonian of Eq. (5.19). As already noted the factor of most importance is the square of the transition moment matrix element. Note that Eq. (5.29) applies to the stimulated emission process as well as the absorption case.

5.3 ELECTRIC DIPOLE TRANSITION PROBABILITY

Although a complete treatment of the interaction of electromagnetic radiation with a system of charged particles requires the use of the vector and scalar potentials of Maxwell's equations, a satisfactory treatment of the electric dipole interaction can be obtained in a much simpler fashion.

If the electric part of the electromagnetic radiation is written as

$$\boldsymbol{\varepsilon} = 2\boldsymbol{\varepsilon}^o(\nu)\ \cos 2\pi\nu t \tag{5.30}$$

the time-dependent potential energy of the system of charged particles is

$$\widetilde{\mathcal{K}}'(t) = \boldsymbol{\varepsilon} \cdot \sum_i e_i \mathbf{r}_i \tag{5.31a}$$

in which e_i is the charge and $\mathbf{r}_i$ the position vector of the particle i. Defining the electric dipole moment,

$$\boldsymbol{\mu} = \sum_i e_i \mathbf{r}_i \tag{5.32}$$

we obtain

$$\widetilde{\mathcal{K}}'(t) = 2\boldsymbol{\varepsilon}^o(\nu) \cdot \boldsymbol{\mu} \cos 2\pi\nu t \tag{5.31b}$$

This perturbation term is of the general form of Eq. (5.19b), with

$$\widetilde{F}^0 = \boldsymbol{\varepsilon}^o(\nu) \cdot \boldsymbol{\mu} \tag{5.33}$$

We have assumed implicitly that $\boldsymbol{\varepsilon}^o(\nu)$ is constant over the dimensions of the molecule since otherwise the factor $\boldsymbol{\varepsilon}^o(\nu)$ would need to depend upon the coordinates of the particles. Our *quasistatic* approximation is the desired one for the electric dipole interaction and is physically realistic for long wavelength radiation (> 2000 Å say) since reasonably sized molecules have dimensions of less than 50 Å.

Substituting Eq. (5.33) into our previous general result Eq. (5.29), with the aid of Eq. (5.21) we find

$$P_{mo} = \frac{1}{\hbar^2} \mid \langle \boldsymbol{\varepsilon}^o(\nu_{mo}) \cdot \boldsymbol{\mu} \rangle_{mo} \mid^2 \tag{5.34}$$

where

$$\langle \boldsymbol{\varepsilon}^0(\nu_{mo}) \cdot \boldsymbol{\mu} \rangle_{mo} = \int \psi_m{}^{0*} \boldsymbol{\varepsilon}^o(\nu_{mo}) \cdot \boldsymbol{\mu}\psi_o{}^{0*} d\tau \tag{5.35}$$

If the radiation is isotropic we can write $\mathcal{E}^o{}_X = \mathcal{E}^o{}_Y = \mathcal{E}^o{}_Z$ and

$$\mathcal{E}^{o2} = \mathcal{E}^o{}_X{}^2 + \mathcal{E}^o{}_Y{}^2 + \mathcal{E}^o{}_Z{}^2$$

$$= 3\mathcal{E}^o{}_X{}^2 \tag{5.36}$$

The electric-field amplitude can then be factored out of the expanded matrix element of Eq. (5.35). Furthermore, if the radiation has random phases in the three directions, we can average over all phase angles, in which case the cross-product terms of the form $\mathcal{E}^o{}_X{}^2 \langle \mu_X \rangle_{mo}{}^* \langle \mu_Y \rangle_{mo}$ go to zero.[5] We can therefore rewrite Eq. (5.34) as

$$P_{mo} = \frac{\mathcal{E}^{o2}}{3\hbar^2} \{ |\langle \mu_X \rangle_{mo}|^2 + |\langle \mu_Y \rangle_{mo}|^2 + |\langle \mu_Z \rangle_{mo}|^2 \}$$

$$= \frac{\mathcal{E}^{o2}}{3\hbar^2} |\langle \boldsymbol{\mu} \rangle_{mo}|^2 \tag{5.37}$$

While this result is useful as written, it is common to express the transition probability in terms of the radiation density which is given by electromagnetic theory as

$$\rho = \frac{1}{2\pi} \mathcal{E}^{o2} \tag{5.38}$$

This leads finally to

$$P_{mo} = \frac{2\pi}{3\hbar^2} |\langle \boldsymbol{\mu} \rangle_{mo}|^2 \rho(\nu_{mo}) \tag{5.39}$$

which gives the probability of a transition from $\psi_o{}^0$ to $\psi_m{}^0$ per sec.[6] Note that the probability is nonzero only when the electric dipole transition moment matrix element is nonzero. It is this factor which is of primary importance in determining the allowed transitions for rotational, vibra-

---

[5] If the phases are φ_X, φ_Y, and φ_Z, we can write $\mathcal{E}^o \cdot \boldsymbol{\mu} = \mathcal{E}^o{}_X[\exp(-i\varphi_X)\mu_X + \exp(-i\varphi_Y)\mu_Y + \exp(-i\varphi_Z)\mu_Z]$ by assuming the radiation to be isotropic. Then $|\langle \mathcal{E}^o \cdot \boldsymbol{\mu} \rangle_{mo}|^2$ contains terms of the form $\mathcal{E}^o{}_X{}^2 \langle \mu_X \rangle_{mo}{}^* \langle \mu_Y \rangle_{mo} \exp[i(\varphi_X - \varphi_Y)]$. Since all phase differences $\Delta\varphi$ occur we can integrate over $\Delta\varphi$ which gives

$$\int_0^{2\pi} e^{i\Delta\varphi} d\Delta\varphi = 0$$

Thus only the squared terms remain.

[6] Using cgs units, that is, $\mu = $ esu cm, $\rho = $ erg cm^{-3}, P_{mo} in Eq. (5.39) has units of sec^{-2}. Hence P_{mo} is the transition probability per second integrated over the bandwidth $\Delta\nu$ of the absorbed radiation. $P_{mo}/\Delta\nu$ would be the average transition probability per second at the given frequency ν_{mo}. Later, when the lineshape function is introduced, the transition probability has units of sec^{-1}.

tional and electronic spectroscopy. Any set of rules which summarizes the allowed transitions shall henceforth be termed *selection rules*.

The coefficient of $\rho(\nu_{mo})$ in Eq. (5.39) is known as the Einstein[7] coefficient of absorption B_{mo} for the transition $o \to m$.

$$B_{mo} = \frac{2\pi}{3\hbar^2} \mid \langle \mathbf{\mu} \rangle_{mo} \mid^2 \tag{5.40}$$

From the treatment carried on thus far in this chapter it is clear that the probability of the reverse transition in an electromagnetic field is given by an expression identical to Eq. (5.39). Einstein's coefficient of induced emission is

$$B_{om} = \frac{2\pi}{3\hbar^2} \mid \langle \mathbf{\mu} \rangle_{mo} \mid^2 \tag{5.41}$$

The remaining link needed to complete the general treatment is that of *spontaneous emission* of radiation. While we cannot obtain the spontaneous emission rate by our method, we can use Einstein's result, namely,

$$A_{om} = \frac{16\pi^2 \nu_{om}{}^3 \hbar}{c^3} B_{om} \; \sec^{-1} \tag{5.42}$$

Thus, the rate of spontaneous emission per second is

$$P_{om}{}^{\text{spont}} = A_{om} = \frac{32\pi^3 \nu_{om}{}^3}{3c^3 \hbar} \mid \langle \mathbf{\mu} \rangle_{mo} \mid^2 \sec^{-1} \tag{5.43}$$

The reciprocal of A_{om} is termed the natural lifetime of the excited state (with respect to decay to the lower state)

$$\tau = \frac{1}{A_{om}} \tag{5.44}$$

Equation (5.43) therefore permits calculation of the natural lifetime of an excited state assuming the transition moment is of the electric dipole type. Assuming the dipole matrix element to be similar in all cases, we can compare the relative lifetimes of electronic (5000 Å), vibrational (1000 cm^{-1}) and rotational (30 GHz) energy states:

$$\tau(\text{electronic}) : \tau(\text{vibrational}) : \tau(\text{rotational}) = 1 : 8300 : 8.3 \times 10^{12}$$

An estimate of $\tau(\text{electronic})$ can be gotten from Eqs. (5.43) and (5.44) by assuming $\mid \langle \mathbf{\mu} \rangle_{mo} \mid^2 \sim 5 \times 10^{-36}$ esu^2 cm^2 [i.e., $(2.2)^2$ Debye^2] and $\lambda_{om} = 5000$ Å, which leads to $\tau(\text{electronic}) \sim 10^{-7}$ sec. Many electronic states

[7] See footnote 2.

are considerably shorter lived than this and many are much longer depending on the particular values of $\nu_{om}{}^3$ and $|\langle \mu \rangle_{mo}|^2$.

EXAMPLE 5-2

A simple illustration of the previous results can be obtained by considering an electron constrained to move in a one-dimensional harmonic potential, in which case the zero-field Hamiltonian is that of Eq. (4.8). The dipole matrix elements are of the form

$$\mu_{mn} = e\langle m \mid X \mid n \rangle = eX_{mn} \qquad (5.45)$$

As shown by Eq. (4.38) the only nonzero matrix elements of X are

$$X_{v+1,v} = \left[(v+1) \frac{\hbar}{2(km)^{1/2}} \right]^{1/2} \qquad (4.38)$$

so that the selection rules are $v \rightarrow v+1$ in absorption. The transition probability of Eq. (5.39) becomes

$$P_{v+1,v} = \frac{2\pi e^2}{3\hbar^2} \left[(v+1) \frac{\hbar}{2(km)^{1/2}} \right] \rho \qquad (5.46)$$

For the lowest transition, $v = 0 \rightarrow v = 1$, by introducing $\nu = (1/2\pi)(k/m)^{1/2}$ we find

$$P_{10} = \frac{e^2}{6\hbar\nu m} \rho \qquad (5.47a)$$

If the problem were treated as an isotropic three-dimensional harmonic oscillator we would obtain three identical terms $\mu_X{}^2$, $\mu_Y{}^2$ and $\mu_Z{}^2$ so the result would be

$$P_{10} = \frac{e^2}{2\hbar\nu m} \rho \qquad (5.47b)$$

This result is often used as a model for comparing experimentally measured transition probabilities. We define the *oscillator strength f* of an absorption or induced emission line as the ratio of the measured coefficient of induced absorption to the theoretical result of Eq. (5.47b).

$$f = \frac{B_{mn}(\text{measured})}{(e^2/2\hbar\nu m)} \qquad (5.48)$$

In this expression ν is taken as the frequency of the (electronic) transition $m \rightarrow n$ and m is the mass of the electron. Oscillator strengths of unity or greater correspond to strongly allowed transitions while smaller values indicate smaller values of the dipole matrix element.

5.4 SOME MORE GENERAL CONSIDERATIONS

A. Magnetic Dipole and Electric Quadrupole Transitions

The case treated in Sec. 5.3 is the most common situation arising in molecular spectroscopy. There are, however, numerous occasions when the electric dipole matrix elements go to zero, in which case the strong electric dipole transitions are not permitted. If transitions are still to be permitted there must be weaker interactions of the electromagnetic field with the molecule, producing small but nonzero transition probabilities. These interactions do exist and it is only because of their presence that electron spin resonance and nuclear magnetic resonance spectroscopy exist, for the electric dipole matrix elements are vanishing for these cases.

A complete expansion of the interaction of the electromagnetic field with a system of charged particles shows[8] that the terms next in importance to the electric dipole one are those due to the magnetic dipole and the electric quadrupole. The former term arises by interaction of the magnetic field with the magnetic moment produced by the moving electrons. For cases involving the *intrinsic* magnetic moments of electrons or nuclei, the classical electromagnetic calculation cannot give the correct results, but using the quantum mechanically deduced magnetic moments (see chapter 9) the appropriate results are easily obtained. For this case, Eq. (5.40) and the other similar ones are correct if we merely substitute μ_{mag} for the electric moment μ. Later we shall see that the electron and nuclear spin states of molecules have nonzero magnetic dipole matrix elements, whereas usually (to a good approximation at least) there are no nonzero electric dipole matrix elements. Because of the much smaller size of the spin magnetic dipole moments compared to electric dipole moments, magnetic dipole transition probabilities between electron spin states are roughly 10^5 times lower, while they are approximately 10^{11} times lower between nuclear spin states.

Electric quadrupole terms may arise in the general expression for the transition probability because of the interaction of the electric field with the nonspherical electron charge distribution present in many molecules. The quadrupole matrix elements are of the form

$$| \langle e\mathbf{Q} \rangle_{mo} |^2 \tag{5.49}$$

where

$$\mathbf{Q} = \sum_i \{ (3x_i{}^2 - R_i{}^2) + (3Y_i{}^2 - R_i{}^2) + (3Z_i{}^2 - R_i{}^2)$$
$$+ 6(X_iY_i + X_iZ_i + Y_iZ_i) \} \tag{5.50}$$

---

[8] E. U. Condon and G. H. Shortley, *The Theory of Atomic Spectra* (Cambridge U.P., Cambridge, 1959), pp. 83–87.

Whereas the electric dipole operator has components X, Y, and Z, we see that the electric quadrupole operator has components X^2, Y^2, Z^2, XY, XZ, YZ. Condon and Shortley[9] have shown that the Einstein coefficient for spontaneous emission by the quadrupole mechanism is

$$A_{om} = \frac{32\pi^6\nu^5}{10\hbar c^5} \mid \langle e\mathbf{Q}\rangle_{om} \mid^2 \qquad (5.51)$$

Because $Q \sim R^2$, Q^2 is about 10^{16} times smaller than μ^2 (electric dipole), so for radiation in the visible region the transition probability for quadrupole radiation given by Eq. (5.51) is 10^7–10^8 times smaller than that for electric dipole radiation. Quadrupole selection rules are of little concern to us because of their low probability and their relative rareness in practical spectroscopic applications.

B. Line Shapes

Equation (5.39) was derived by considering merely a single elementary event, the absorption of a photon of radiation of frequency ν_{mo} such that the molecule underwent a transition from $\psi_o{}^0$ to $\psi_m{}^0$. In fact we do not deal with single molecules and therefore the problem must really be more complex, for we must sometime worry about *nonradiative* energy-transfer processes caused by molecular collisions or interactions. These processes usually compete with and modify the elementary radiative process, the most obvious result of these processes being a broadening of the absorption lines far beyond their natural linewidth in most cases.

In the case of single absorption lines of gas-phase molecules in the visible, infrared or microwave regions the intermolecular interactions are predominantly due to electric interactions between pairs of molecules and are of the dipole-dipole, dipole-quadrupole, etc., type. In liquids and solids the interactions are, of course, much stronger and dispersed through many neighboring molecules, so consequently linewidths in the condensed phases are often very broad.

For magnetic spin systems the broadening effects are of a different origin, involving the mutual interactions of the spin centers and their interactions with the immediate surroundings. The effects are the same here as before, namely the absorption lines are considerably broadened.

In addition to line broadening by intermolecular interactions a number of other phenomena lead to the same effect. Many of these are instrumental in nature, having to do with radiation monochromaticity, instrument resolution and methods of detection or observation of the absorption lines. Line broadening also may occur in spectroscopic experiments on systems under-

---

[9] See footnote 8, p. 96.

going chemical reaction, if the reaction causes the radiatively excited species to disappear in a time scale equal to or less than the lifetime of the excited states as determined by all other causes. The explanation here proceeds along the same lines as that used in deriving Eq. (5.26d). We discuss some of these phenomena at appropriate times in later chapters, particularly when they are of critical importance to the interpretation of spectra or if they produce information of physical and chemical significance.

One final broadening phenomenon that is important in some spectroscopic applications with gases is that of Doppler broadening. The Doppler effect produces a shift in the absorption frequency

$$\delta\nu = \pm\nu_o\left(\frac{v}{c}\right) \tag{5.52}$$

where v is the velocity of the molecule and c is the velocity of the electromagnetic radiation. The effect is the same as the well-known one involving change in frequency of sound from a moving object, the shift being to high frequencies when the sound wave and the object are moving in the same direction and to lower frequencies when traveling in opposite directions. The distribution of velocities of molecules in a gas at temperature T is given by the well-known Maxwell velocity distribution

$$P(v) = N\exp(-mv^2/2kT) \tag{5.53}$$

which becomes a distribution of frequency shifts $\delta\nu$ by use of Eq. (5.52).

$$P(\delta\nu) = N\exp\left(-\frac{mc^2(\delta\nu)^2}{2kT\nu_o^2}\right) \tag{5.54}$$

Equation (5.54) gives the probability that a given frequency shift $\delta\nu = \nu - \nu_o$ occurs. The Gaussian curve extends over all frequencies of course, but is peaked at $\delta\nu = \nu - \nu_o = 0$. The half-width of the absorption line shape given by Eq. (5.54) at half-maximum intensity occurs at a frequency obtained from

$$N\exp\left(-\frac{mc^2(\delta\nu)^2}{2kT\nu_o^2}\right) = \frac{1}{2}N \tag{5.55}$$

which leads to

$$\delta\nu = \frac{\nu_o}{c}\sqrt{\frac{2kT}{m}\ln 2} \tag{5.56a}$$

In practical units the half-width of the absorption line is

$$\delta\nu = 3.58 \times 10^{-7}\left(\frac{T}{M}\right)^{1/2}\nu \tag{5.56b}$$

with M the molecular weight and ν in sec^{-1} as usual. We see that light molecules, high temperatures and high frequencies tend to increase the line width. Doppler broadening has been effectively reduced in some experiments by using molecular beams in which the molecular velocities, although still distributed over a wide range, are nearly unidirectional. Thus the electromagnetic radiation, if brought in at right angles to the molecular beam, produces very little Doppler shift and consequently very narrow linewidths are obtained. For the majority of spectroscopic applications the Doppler breadth given by Eq. (5.56b) is so much smaller than that from other broadening phenomena that it need not be considered.

From what has been said it should be clear that absorption (and emission) lines are not infinitely sharp and indeed are usually much broader than even the natural linewidth. While the calculation of the transition probability in the presence of the various broadening phenomena is difficult the results are generally of the form

$$P_{mo}' = \frac{1}{\hbar^2} \mid F_{mo} \mid^2 S(\nu) \text{ sec}^{-1} \qquad (5.57a)$$

which is simply the result given earlier in Eq. (5.29) multiplied by a line-shape factor $S(\nu)$. If $S(\nu)$ is normalized

$$\int_0^\infty S(\nu) d\nu = 1 \qquad (5.58)$$

the integral of P'_{mo} over all frequencies gives the total transition probability as we had it earlier. On the other hand Eq. (5.57a) itself gives the transition probability as a function of frequency and therefore gives the complete absorption line profile. For reference to other works we might point out that Eq. (5.57a) may be written in two alternative forms. In terms of angular frequency $\omega = 2\pi\nu$ we have

$$P_{mo}' = \frac{2\pi}{\hbar^2} \mid F_{mo} \mid^2 S(\omega) \text{ sec}^{-1} \qquad (5.57b)$$

while in terms of energy, $E = h\nu = \hbar\omega$,

$$P_{mo}' = \frac{2\pi}{\hbar} \mid F_{mo} \mid^2 S(E) \text{ sec}^{-1} \qquad (5.57c)$$

The two most common line-shape functions are the Lorentzian and Gaussian types. The Lorentzian is of the form

$$S(\nu) = \frac{2}{\pi} \left(\frac{\delta\nu}{[\nu - \nu_o]^2 + (\delta\nu)^2} \right) \qquad (5.59)$$

while the Gaussian form is

$$S(\nu) = \frac{2(\ln 2)^{1/2}}{\sqrt{\pi}\ \delta\nu} \exp[-(\nu - \nu_o)^2 \ln 2/(\delta\nu)^2] \qquad (5.60)$$

In each of these functions $\delta\nu$ is the half-width of the line shape at half-maximum intensity. In Fig. 5.3 these two shape functions have been plotted to show their general character. It is beyond the scope of our treatment to delve into the theoretical basis for the choice of the Lorentzian, Gaussian or some other line shape, but we point out the correct choice when it is appropriate to our spectroscopic treatment.

5.5   RELATIONSHIP OF TRANSITION PROBABILITY TO OBSERVABLE QUANTITIES

Transition probabilities are not directly measurable quantities. Instead, the common experimentally measurable quantities are the electromagnetic (light) intensity[10] absorbed or the fraction of electromagnetic intensity absorbed. In terms of transition probability, the power absorbed per cubic centimeter is

$$N_o P_{mo} h \nu_{mo} \qquad (5.61)$$

where N_o is the number of molecules per cubic centimeter that are available to undergo the transition $o \rightarrow m$, P_{mo} is our previous transition prob-

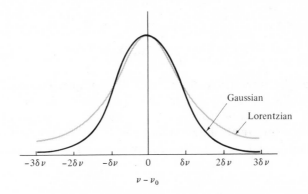

FIG. 5.3.   Lorentzian and Gaussian lineshape functions. Curves are normalized to same peak intensity rather than unit area as in Eqs. (5.59) and (5.60).

[10] Intensity has units of erg cm^{-2} sec^{-1} and therefore represents the power flowing through a 1 cm^2 cross-section area.

ability per second and ν_{mo} is the transition frequency. Note that in general we expect the power absorption to be a function of temperature, since the number of molecules in state $\psi_o{}^0$ is given by the Boltzmann distribution

$$\frac{N_o}{N_{\text{total}}} = \frac{g_o \exp(-E_o/kT)}{Z} \tag{5.62}$$

where g_o is the statistical weight and Z the partition function. In the study of electronic spectra the energy levels are widely spaced so $N_o/N_{\text{total}} = 1$, but in other types of spectroscopy more careful account of population distribution needs to be made.

To relate Eq. (5.61) to the light intensity we note that the decrease in the intensity of radiation passing through a homogeneous molecular sample is proportional to the light intensity and also to the path length, thus

$$-dI = \gamma I dl \tag{5.63}$$

where γ is the proportionality constant and is known as the absorption coefficient. Often Eq. (5.63) is written

$$-dI = \alpha I C dl \tag{5.64}$$

where C is the sample concentration in moles/liter and α is the molar absorption coefficient. The absorption coefficients are related to measurable quantities by writing Eqs. (5.63) and (5.64) in the integrated forms

$$\gamma(\nu) = \frac{1}{l} \ln \frac{I_0}{I} \text{ cm}^{-1} \tag{5.65a}$$

$$\alpha(\nu) = \frac{1}{Cl} \ln \frac{I_0}{I} \text{ liter mole}^{-1} \text{ cm}^{-1} \tag{5.65b}$$

where I_0 is the intensity incident upon the sample, I is transmitted intensity, and l is the path length of the homogeneous sample. We have indicated explicitly in Eqs. (5.65) that the absorption coefficients are functions of frequency. In the infrared and visible (or uv) spectral regions it is common to integrate over the absorption band, which gives the integrated coefficient

$$\gamma_{\text{int}} = \int_{\text{band}} \gamma(\nu) d\nu \tag{5.66}$$

We note now that Eq. (5.61) gives the power absorbed per cubic centimeter or the energy absorbed per sec per cubic centimeter. Thus multiplication by dl provides an alternative expression for dI,

$$-dI = N_o P_{mo} h\nu_{mo} dl \tag{5.67}$$

and by comparison to Eq. (5.63) we find

$$\gamma = N_o P_{mo} h \nu_{mo} / I \tag{5.68a}$$

When the radiation is isotropic and the transitions are of the electric dipole type, we may substitute Eq. (5.39) in (5.68a) which gives

$$\gamma = \frac{2\pi N_o h \nu_{mo} |\langle \boldsymbol{\mu} \rangle_{mo}|^2}{3\hbar^2 I} \rho_{mo}(\nu) \tag{5.68b}$$

But energy density and intensity are related by

$$I = c\rho$$

so we get finally

$$\gamma = \frac{8\pi^3 N_o \nu_{mo} |\langle \boldsymbol{\mu} \rangle_{mo}|^2}{3hc} \tag{5.68c}$$

Now actually there are transitions induced from state m to state o with the same probability as for $o \rightarrow m$. Therefore we should write $N_o - N_m$ in place of N_o in the preceding expressions since this gives the net absorption or absorption coefficient. Also, recall that the probability given by Eq. (5.39) was summed over the absorption line (and indeed had units of $\sec^{-2}$). Therefore we should identify the right-hand sides of Eqs. (5.68) with the integrated absorption coefficient, which makes everything dimensionally correct again. We get

$$\int \gamma(\nu) d\nu = \frac{8\pi^3 (N_o - N_m) \nu_{mo} |\langle \boldsymbol{\mu} \rangle_{mo}|^2}{3hc} \tag{5.69}$$

We see therefore that experimental measurements of γ can be used to determine the dipole matrix element which is the quantity of theoretical interest. In following chapters we give careful attention to the magnitudes and specific forms of $|\langle \boldsymbol{\mu} \rangle_{mo}|^2$ but are seldom interested in the exact values of γ since our primary interest is merely to determine selection rules.

EXAMPLE 5-3

Van Vleck and Weisskopf[11] and Karplus and Schwinger[12] have derived an expression for the absorption coefficient of a molecular transition broadened by "strong" intermolecular collisions in the gas phase. The meaning of strong in the derivation is that after each collision a Boltzmann energy distribution exists, which implies that the collisions are effective in transferring energy from excited molecules. While a correct derivation of this equation is beyond the scope of this book we can obtain the result

[11] J. H. Van Vleck and V. F. Weisskopf, *Rev. Mod. Phys.* **17**, 227 (1945).
[12] R. Karplus and J. Schwinger, *Phys. Rev.* **73**, 1020 (1948).

rather simply if we begin with the correct line-shape function. It has been shown that the correct line-shape function[13] is

$$S(\nu) = N \left\{ \frac{\delta\nu}{(\nu + \nu_0)^2 + (\delta\nu)^2} + \frac{\delta\nu}{(\nu - \nu_0)^2 + (\delta\nu)^2} \right\} \quad (5.70)$$

which is the sum of two Lorentzians of the form of Eq. (5.59), one resonant at $-\nu_0$ and the other at $+\nu_0$. Since the integral $\int S(\nu)d\nu$ from $\nu = 0$ to $\nu = \infty$ gives $\frac{1}{2}\pi$ for each term in Eq. (5.70), the normalizing factor is $N = 1/\pi$.

Using Eq. (5.70) and proceeding as described in this section we obtain, instead of Eq. (5.69),

$$\gamma(\nu) = \frac{8\pi^2(N_o - N_m)\nu_0 \, | \, \langle \mathbf{\mu} \rangle_{mo} \, |^2 \, S'(\nu)}{3hc} \quad (5.71)$$

where $S'(\nu) = (1/\pi)S(\nu)$. Note that the right-hand side of Eq. (5.71) differs from Eq. (5.69) by only the factor $1/\pi$ and the shape function $S'(\nu)$. Also we have identified the right-hand side with $\gamma(\nu)$ instead of $\int \gamma(\nu)d\nu$ since we have not integrated the right-hand side over all frequencies.

In the microwave or rf region two useful simplifications can be made. First, at reasonably low pressures (less than 1 Torr) $\delta\nu \ll \nu_0$ so only the second term in Eq. (5.70) is important in the region $\nu \sim \nu_0$. Secondly, $h\nu_0/kT \ll 1$ so that the population difference can be obtained from the Boltzmann distribution, which is

$$\frac{N_o}{N_m} = \exp(h\nu_0/kT) \sim 1 + h\nu_0/kT \quad (5.72)$$

for nondegenerate states $\psi_o{}^0$ and $\psi_m{}^0$. Thus,

$$N_o - N_m = N_o \left(\frac{h\nu_0}{kT} \right) \quad (5.73a)$$

where N_o is the number of molecules per cc in the state $\psi_o{}^0$. If f is the fraction of all molecules in the state $\psi_o{}^0$ and N is the total number of molecules per cc,

$$N_o - N_m = fNh\nu_0/kT \quad (5.73b)$$

Introducing the two approximations into Eq. (5.71) gives the absorption coefficient

$$\gamma(\nu) = \frac{8\pi^2 fN\nu_0{}^2 \, | \, \langle \mathbf{\mu} \rangle_{mo} \, |^2}{3ckT} \left\{ \frac{\delta\nu}{(\nu - \nu_0)^2 + (\delta\nu)^2} \right\} \quad (5.74)$$

[13] See footnotes 11 and 12.

which has units of cm^{-1}. As pointed out, the approximations involved make this expression particularly appropriate for microwave (rotational) spectra. The more exact treatments[14] show that ν_0^2 in the numerator of Eq. (5.74) should actually be ν^2, but since the linewidth $\delta\nu$ is very small, $\nu = \nu_0$ is an excellent approximation for this term. An expression that is of use later is γ_{max}, the value of $\gamma(\nu)$ when $\nu = \nu_0$,

$$\gamma_{max} = \frac{8\pi^2 f N \nu_0^2 \mid \langle \mathbf{\mu} \rangle_{mo} \mid^2}{3ckT\delta\nu} \tag{5.75}$$

The exact theory[15] shows that the linewidth is related to the collision time τ by

$$\delta\nu = \frac{1}{2\pi\tau} \tag{5.76}$$

and experimental values of linewidths are found to be predicted to at least an order of magnitude by use of kinetic theory collision times. Since the kinetic theory collision time is inversely proportional to pressure, and N is proportional to pressure

$$\frac{N}{\delta\nu} = f(T \text{ only}) \tag{5.77}$$

Thus we have the rather surprising prediction that the absorption coefficient is independent of pressure for a pure gas, a result that is experimentally borne out at low pressures in the absence of power saturation[16] effects.

Finally we note that $\gamma \ll 1$ for rotational microwave spectra so that the integrated form of Eq. (5.63) becomes

$$I = I_0 \exp(\gamma l) \sim I_0 + I_0 \gamma l \tag{5.78}$$

Thus the fraction of power absorbed is

$$\frac{I - I_0}{I} = \gamma l \tag{5.79}$$

γ is typically 10^{-7} cm^{-1} and $l \sim 300$ cm so the power absorption is $\sim 0.003\%$.

---

[14] See footnotes 11 and 12.
[15] See footnotes 11 and 12.
[16] Power saturation occurs when the transitions $o \to m$ occur so fast that a normal Boltzmann distribution cannot be obtained. The result is that as the population of the upper state m increases, γ begins to fall off and $\Delta\nu$ grows. See footnote 12 for a treatment of this problem.

SUPPLEMENTARY REFERENCES

E. U. Condon and G. H. Shortley, *Theory of Atomic Spectra* (Cambridge U.P., Cambridge, 1935).

D. W. Davies, *The Theory of the Electric and Magnetic Properties of Molecules* (Wiley, London, 1967).

W. Heitler, *The Quantum Theory of Radiation* (Oxford U.P., Cambridge, 1954).

L. D. Landau and E. M. Lifshitz, *Quantum Mechanics, Non-Relativistic Theory* (Addison-Wesley, Reading, Mass., 1958).

L. I. Schiff, *Quantum Mechanics* (McGraw-Hill, New York, 1955).

J. H. Van Vleck, *The Theory of Electric and Magnetic Susceptibilities* (Oxford U.P., Cambridge, England, 1932).

CHAPTER 6

GROUP THEORY AND ITS RELATIONSHIP TO QUANTUM MECHANICS AND SPECTROSCOPY

The appearance of symmetry in all realms of nature, and particularly in those areas described by the sciences of physics and chemistry, is familiar to everyone. The beauty of the symmetrical arrangement of the atoms (or ions) in a sodium chloride crystal or benzene molecule can hardly escape notice. That this symmetry manifests itself in many physical properties is also obvious, even to the novice. For example, the symmetrical arrangement of atoms in a benzene molecule clearly prohibits the existence of a permanent electric dipole moment, while, on the other hand, the less symmetrical chlorobenzene structure permits a dipole moment along the direction of the carbon-chlorine bond. And, as another example, the student of chemistry learns very early that the symmetrical arrangement of atoms in methane demands a chemical-bonding theory which produces equivalent carbon-hydrogen bonds.

Yet, the overriding importance of symmetry to physics and chemistry is not readily apparent from these strictly visual observations. In order to bring the picture into complete focus we must look rather carefully at some mathematical theories related to the subject. In this chapter we explore the field of group theory, particularly the theory of group representations. Our treatment is not extensive nor do we include many of the mathematically important proofs. Rather our efforts are aimed largely at presenting the mathematical machinery and showing its relationship and application to quantum mechanics and spectroscopy so that we have the use of this very valuable technique in later chapters. The reader who is interested in a deeper background in the subject is urged to consult the reference books listed at the end of this chapter.

Before proceeding it is perhaps worthwhile to show in a very simple way how symmetry and group theory enter into quantum-mechanical theory. Suppose R represents the operation of inversion of coordinates, that is $(x, y, z) \rightarrow (-x, -y, -z)$. If this operation, when performed on the Hamiltonian operator for some system, leaves the Hamiltonian operator unchanged or *invariant*, we can say that $\tilde{R}$ commutes with $\tilde{\mathfrak{IC}}$, that is

$$\tilde{R}\tilde{\mathfrak{IC}} = \tilde{\mathfrak{IC}}\tilde{R} \tag{6.1}$$

That this situation can occur is easily verified for a simple case such as the Hamiltonian for a linear harmonic oscillator [see Eq. (4.8) for example]. Since the kinetic-energy term contains only second derivatives ($\partial^2/\partial x^2$, etc.) it remains invariant to an inversion of coordinates, and the squared term in the potential energy guarantees that it too is invariant to the inversion operation. But a very well-known theorem of quantum mechanics[1] states that the matrices of commuting operators are simultaneously diagonal, or stated somewhat differently, an eigenfunction of any one of the commuting operators is simultaneously an eigenfunction of each of the commuting operators.

Thus in searching for eigenfunctions we can limit our search to those functions which have the proper symmetry with respect to the operators that leave $\tilde{\mathfrak{IC}}$ invariant. In the case described above, we know that the eigenfunctions of the harmonic oscillator must be either symmetric or antisymmetric with respect to inversion. This can be shown by applying $\tilde{R}$ to both sides of Schrödinger's equation and using Eq. (6.1), thus

$$\tilde{R}\tilde{\mathfrak{IC}}\psi_i = \tilde{\mathfrak{IC}}\tilde{R}\psi_i = \tilde{R}E_i\psi_i = E_i\tilde{R}\psi_i \tag{6.2}$$

The second and fourth terms show that $\tilde{R}\psi_i$ must be an eigenfunction of $\tilde{\mathfrak{IC}}$. Now if $\tilde{R}$ is applied a second time we get

$$\tilde{\mathfrak{IC}}\tilde{R}^2\psi_i = E_i\tilde{R}^2\psi_i \tag{6.3}$$

But $\tilde{R}^2 = \tilde{R}\tilde{R}$ must be no operation at all since it represents two consecutive inversions, whence we conclude that

$$\tilde{R}^2\psi_i = +\psi_i \tag{6.4}$$

so that

$$\tilde{R}\psi_i = \pm\psi_i \tag{6.5}$$

are the only possibilities. Thus ψ_i must be either symmetric $(+)$ or anti-

[1] See Appendix B.

symmetric $(-)$ to inversion. Indeed, everyone knows that the harmonic oscillator eigenfunctions have this property.

In the more general case the situation is similar but more complicated. We find that group theory provides the methods to find the eigenfunctions having the proper symmetry, thereby reducing our efforts considerably. In addition some more general results involving the existence or non-existence of matrix elements follow from the theory.

6.1   SYMMETRY OPERATIONS AND TRANSFORMATIONS ABOUT A POINT

A.   Symmetry Operators and Elements

The symmetry that exists in molecules can be systematically classified if we understand the meaning of a *symmetry operation* or *symmetry operator*. With respect to a molecule, a symmetry operation is a movement of atoms which leaves the molecule indistinguishable from its appearance before the operation. In particular, this requires that atoms either remain un-shifted or become shifted to equivalent positions. During any symmetry operation there must be no motion of the c.m. of the molecule, that is, we consider only symmetry about a point.

Our definition becomes clearer if we delineate the kinds of symmetry operations that are possible, namely:

(1) Rotation about an axis of symmetry (C_n).
(2) Reflection in or across a plane of symmetry (σ).
(3) Improper rotation, which is a rotation about a symmetry axis followed by reflection in a plane perpendicular to the axis (S_n).
(4) Inversion (i).
(5) Identity (E).

The last operation is trivial and means simply that the operation is "leave all atoms alone." The reason for identifying such an operation becomes clear very shortly.

All of the remaining operations are conveniently illustrated by the staggered form of the ethane molecule in Fig. 6.1. The z axis is an axis of three-fold symmetry, that is, rotations by $\pm\frac{2}{3}\pi$ rad cause equivalent atoms to be exchanged, for example $5 \to 6, 6 \to 4, 4 \to 5, 3 \to 1, 1 \to 2, 2 \to 3$. This rotational operation is known as a C_3 rotation, where the notation is a particular case of the general C_n or n-fold rotation. The staggered ethane molecule also contains three C_2 axes perpendicular to the planes 1, 7, 8, 4; 3, 7, 8, 6; and 2, 7, 8, 5.

The reflection operation is illustrated by reflection in the plane formed by atoms 1, 7, 8, 4, in which case $1 \to 1, 4 \to 4, 7 \to 7, 8 \to 8, 2 \to 3, 3 \to 2$,

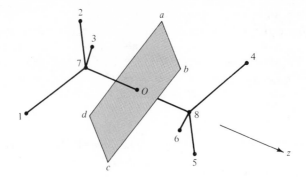

FIG. 6.1. Staggered conformation of the ethane molecule. z axis lies
          along C–C bond.

$5 \rightarrow 6$, $6 \rightarrow 5$. The reader should be able to find two more similar planes
of symmetry. These particular reflection operations are designated as σ_d,
the subscript signifying *di*agonal planes which are those that bisect the
angle between two C_2 axes. Symmetry planes that are perpendicular to the
major symmetry axis are designated σ_h, while those which contain the major
symmetry axis (but are not σ_d) are designated σ_v.

The improper rotation S_6 is obtained by first performing C_6 followed
by reflecting in the plane *abcd*. In this case the atom shifts are $8 \rightarrow 7$, $7 \rightarrow 8$,
$4 \rightarrow 3$, $1 \rightarrow 6$, $5 \rightarrow 1$, $6 \rightarrow 2$, $3 \rightarrow 5$, $2 \rightarrow 4$. It should be apparent that neither
C_6 nor σ_{abcd} are symmetry elements themselves.

Inversion i means to invert all atoms through the origin O so that
$7 \rightarrow 8$, $8 \rightarrow 7$, $1 \rightarrow 4$, $4 \rightarrow 1$, $2 \rightarrow 5$, $5 \rightarrow 2$, $3 \rightarrow 6$, $6 \rightarrow 3$. It should be noted
that a $+\frac{4}{3}\pi$ rotation, $C_3 \times C_3 = C_3{}^2$, is a symmetry operation also but
need not be specially noted since it is equivalent to a C_3 rotation in the
opposite sense. Some further insight into the identity operator can be
gotten now, since it is the result of two consecutive σ_d's, i's, or C_2's or three
consecutive C_3's, that is $\sigma_d\sigma_d = \sigma_d{}^2 = E$, $i^2 = E$, $C_3{}^3 = E$.

The symmetry elements of the staggered ethane molecule are then

$$E, \ 2C_3, \ 3C_2, \ 3\sigma_d, \ 2S_6, \ i$$

where the integers in front of the operations are the number of distinct
operations of each type. These 12 symmetry operators make up the elements
of the D_{3d} *point group*. The further properties and meaning of the point
group become apparent shortly, but we simply note here that all molecules
can be similarly classified with respect to their symmetry elements and
over-all point group. Table 6.1 lists the symmetry elements or operators
for several other molecules and gives the symbol used for the point group.

TABLE 6.1

Symmetry elements and point groups for some representative molecules.

Molecules	Symmetry elements	Point group
CO_2, O_2	E, $2S_\phi$, $\infty\,\sigma_v$, i, $2C_\phi$, $\infty\,C_2$	$D_{\infty h}$
H_2O, $H_2C{=}O$	E, C_2, σ_v, $\sigma_v{}'$	C_{2v}
NH_3, CH_3Cl	E, $2C_3$, $3\sigma_v$	C_{3v}
Staggered ethane	E, $2C_3$, $3C_2$, i, $2S_6$, $3\sigma_d$	D_{3d}
Eclipsed ethane	E, $2C_3$, $3C_2$, σ_h, $2S_3$, $3\sigma_v$	D_{3h}
$AuCl_4^-$	E, $2C_4$, C_2, $2C_2'$, $2C_2''$, i, $2S_4$, σ_h, $2\sigma_v$, $2\sigma_d$	D_{4h}
C_6H_6	E, $2C_6$, $2C_3$, C_2, $3C_2'$, $3C_2''$, i, $2S_3$, $2S_6$, σ_h, $3\sigma_d$, $3\sigma_v$	D_{6h}
CH_4	E, $8C_3$, $3C_2$, $6S_4$, $6\sigma_d$	T_d
SF_6	E, $8C_3$, $6C_2$, $6C_4$, $3C_2$, i, $6S_4$, $8S_6$, $3\sigma_h$, $6\sigma_d$	O_h

B. Symmetry Operators as Transformation Operators

Up to now we have considered the symmetry operators to be operations which move atoms about so as to leave a molecule invariant. These same operators can be considered to be coordinate transformation operators in the sense that

$$\hat{\imath}\mathbf{r} = -\mathbf{r} = \mathbf{r}' \tag{6.6a}$$

or

$$\hat{\imath}\begin{pmatrix} x \\ y \\ z \end{pmatrix} = \begin{pmatrix} -1 & 0 & 0 \\ 0 & -1 & 0 \\ 0 & 0 & -1 \end{pmatrix}\begin{pmatrix} x \\ y \\ z \end{pmatrix} = \begin{pmatrix} x' \\ y' \\ z' \end{pmatrix} \tag{6.6b}$$

That is, the inversion operator transforms $x \to -x$, $y \to -y$, and $z \to -z$ or it transforms the vector $\mathbf{r}$ into the vector $\mathbf{r}' = -\mathbf{r}$. We might say that the inversion operator is *represented* by the matrix

$$\begin{pmatrix} -1 & 0 & 0 \\ 0 & -1 & 0 \\ 0 & 0 & -1 \end{pmatrix}$$

Consider as a further example a system having the following symmetry elements: E, C_2, i, and σ_h (point group C_{2h}). The z axis is chosen to lie along the C_2 axis and x and y lie in the σ_h plane. Then we can easily see that

$$\tilde{E} \begin{pmatrix} x \\ y \\ z \end{pmatrix} = \begin{pmatrix} 1 & 0 & 0 \\ 0 & 1 & 0 \\ 0 & 0 & 1 \end{pmatrix} \begin{pmatrix} x \\ y \\ z \end{pmatrix} \tag{6.7a}$$

$$\tilde{C}_2 \begin{pmatrix} x \\ y \\ z \end{pmatrix} = \begin{pmatrix} -1 & 0 & 0 \\ 0 & -1 & 0 \\ 0 & 0 & 1 \end{pmatrix} \begin{pmatrix} x \\ y \\ z \end{pmatrix} \tag{6.7b}$$

$$\tilde{\imath} \begin{pmatrix} x \\ y \\ z \end{pmatrix} = \begin{pmatrix} -1 & 0 & 0 \\ 0 & -1 & 0 \\ 0 & 0 & -1 \end{pmatrix} \begin{pmatrix} x \\ y \\ z \end{pmatrix} \tag{6.7c}$$

$$\tilde{\sigma}_h \begin{pmatrix} x \\ y \\ z \end{pmatrix} = \begin{pmatrix} 1 & 0 & 0 \\ 0 & 1 & 0 \\ 0 & 0 & -1 \end{pmatrix} \begin{pmatrix} x \\ y \\ z \end{pmatrix} \tag{6.7d}$$

Later in this chapter we shall see that it is appropriate to say that x, y, and z form a basis for a representation of the group. For now we simply observe that the coordinates x, y, z, serve to generate a matrix representing the operators $\tilde{R}$.

More generally the rotational transformation matrices are not completely diagonal, for example for a case where the z axis is a C_3 axis. In this case,[2] if $\tilde{C}_3$ is the operator which rotates a point in the clockwise direction

$$\tilde{C}_3 \begin{pmatrix} x \\ y \\ z \end{pmatrix} = \begin{pmatrix} -\tfrac{1}{2} & -\sqrt{3}/2 & 0 \\ \sqrt{3}/2 & -\tfrac{1}{2} & 0 \\ 0 & 0 & 1 \end{pmatrix} \begin{pmatrix} x \\ y \\ z \end{pmatrix} \tag{6.8}$$

---

[2] Our convention concerning the transformation is as follows. If all *points* in the molecule are rotated clockwise (the sense determined by looking in the positive z direction), with the right-handed axis system fixed in space, the sign of θ in Eq. (2.13) is positive; while if the rotation is counterclockwise the sign of θ is negative. Note that if all the points in the molecule were considered fixed in space, and the coordinate system were rotated, the signs of θ would be reversed.

where the matrix representing $\tilde{C}_3$ contains a special case of the two-dimensional rotational transformation matrix given in Eq. (2.13).

The transformation properties of functions can be determined in a similar manner. For example, the transformation of some $f(x, y, z)$ by the operator $\tilde{\imath}$ proceeds as

$$\tilde{\imath}f(x, y, z) = f(\tilde{\imath}x, \tilde{\imath}y, \tilde{\imath}z) \tag{6.9}$$

If $f(x, y, z)$ were given for example by

$$f(x) = A\left(\frac{dx}{dt}\right)^2 + Bx^4 \tag{6.10}$$

where A and B are constants,

$$\tilde{\imath}f(x) = A\left(\frac{d(-x)}{dt}\right)^2 + B(-x)^4 \tag{6.11a}$$

or

$$\tilde{\imath}f(x) = A\left(\frac{dx}{dt}\right)^2 + Bx^4 \tag{6.11b}$$

Functions can also be used in the same way as coordinates for generating matrix representations of operators. The simple functions $x^2 - y^2$ and xy serve, for example, as a basis for a matrix representation of $\tilde{C}_3$ as can be seen in Eq. (6.12).

$$\tilde{C}_3\begin{pmatrix} x^2 - y^2 \\ xy \end{pmatrix} = \begin{pmatrix} \left(-\frac{1}{2}x - \frac{\sqrt{3}}{2}y\right)^2 - \left(+\frac{\sqrt{3}}{2}x - \frac{1}{2}y\right)^2 \\ \left(-\frac{1}{2}x - \frac{\sqrt{3}}{2}y\right)\left(+\frac{\sqrt{3}}{2}x - \frac{1}{2}y\right) \end{pmatrix}$$

$$= \begin{pmatrix} -\frac{1}{2}(x^2 - y^2) + \sqrt{3}\,xy \\ -\frac{\sqrt{3}}{4}(x^2 - y^2) - \frac{1}{2}xy \end{pmatrix} = \begin{pmatrix} -\frac{1}{2} & \sqrt{3} \\ -\frac{\sqrt{3}}{4} & -\frac{1}{2} \end{pmatrix}\begin{pmatrix} x^2 - y^2 \\ xy \end{pmatrix} \tag{6.12}$$

Note that the matrix representation generated here is not unitary, as were those of Eqs. (6.7) and (6.8).

When symmetry operators are applied successively, the meaning is as follows.

$$\tilde{R}\tilde{R}'f(x, y, z) = \tilde{R}f(\tilde{R}'x, \tilde{R}'y, \tilde{R}'z)$$

$$= f(\tilde{R}[\tilde{R}'x], \tilde{R}[\tilde{R}'y], \tilde{R}[\tilde{R}'z]) \tag{6.13}$$

That is, the operator on the right-hand side is applied first, followed by the operator(s) on the left-hand side. For the C_{2h} point group we can see that

$$\tilde{C}_2\tilde{\imath}\begin{pmatrix} x \\ y \\ z \end{pmatrix} = \tilde{C}_2\begin{pmatrix} -x \\ -y \\ -z \end{pmatrix} = \begin{pmatrix} 1 & 0 & 0 \\ 0 & 1 & 0 \\ 0 & 0 & -1 \end{pmatrix}\begin{pmatrix} x \\ y \\ z \end{pmatrix} \tag{6.14}$$

Note that $\tilde{C}_2\tilde{\imath}$ is equivalent to the operation $\tilde{\sigma}_h$, that is

$$\tilde{C}_2\tilde{\imath} = \tilde{\sigma}_h \tag{6.15}$$

We find shortly that this kind of equivalence is a general property of point-group symmetry operators. It should also be pointed out that symmetry operators, like quantum-mechanical operators, do not necessarily commute; that is, it is not always true that

$$\tilde{R}\tilde{R}' = \tilde{R}'\tilde{R} \tag{6.16}$$

where $\tilde{R}$ and $\tilde{R}'$ are symmetry operators of the same group.

6.2   SOME PROPERTIES OF GROUPS

A.   Group Multiplication

A group is made up of a set of different elements $\tilde{A}$, $\tilde{B}$, $\tilde{C}$... which have the property that "multiplication" of any two elements in the group yields an element in the group. The number of elements in the group is known as the order h and no element appears more than once. Group elements are typically represented by symmetry operators, algebraic numbers, or matrices, and "multiplication" is defined correspondingly. Group multiplication is always associative, that is $\tilde{A}(\tilde{B}\tilde{C}) = (\tilde{A}\tilde{B})\tilde{C}$, but not necessarily commutative. In addition every group has a *unit* or *identity* element $\tilde{E}$ such that

$$\tilde{E}\tilde{R} = \tilde{R}\tilde{E} = \tilde{R} \tag{6.17}$$

where $\tilde{R}$ is in the group. Furthermore, every group element has one *inverse* element, $\tilde{R}^{-1}$, such that

$$\tilde{R}\tilde{R}^{-1} = \tilde{R}^{-1}\tilde{R} = \tilde{E} \tag{6.18}$$

An element may be its own inverse.

EXAMPLE 6-1

In Sec. 6.1 A we described the C_{2h} group consisting of the elements E, C_2, i, and σ_h. An example of a molecule having this symmetry would be trans-H_2O_2 shown in Fig. 6.2. We can verify trivially that the elements listed do form a group as described above. If the original molecule as shown

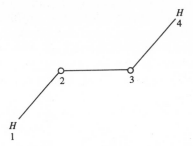

FIG. 6.2. Schematic drawing of trans-H_2O_2. Numbers are labels for the atoms. z axis passes through midpoint of O–O bond and is perpendicular to the plane of the paper.

in Fig. 6.2 is symbolized by $(1^+2^+3^+4^+)$, where the $+$ signs represent the side of the atoms above the plane, then $\tilde{C}_2(1^+2^+3^+4^+)$ yields $(4^+3^+2^+1^+)$. Application of $\tilde{C}_2$ twice must then yield $\tilde{C}_2\tilde{C}_2(1^+2^+3^+4^+) = \tilde{C}_2(4^+3^+2^+1^+) = (1^+2^+3^+4^+)$. Thus we have

$$\tilde{C}_2\tilde{C}_2 = \tilde{E} \qquad (6.19a)$$

Similarly we find that $\tilde{C}_2\tilde{\imath}(1^+2^+3^+4^+) = \tilde{C}_2(4^-3^-2^-1^-) = (1^-2^-3^-4^-)$. But this latter configuration is obtained directly from $\tilde{\sigma}_h(1^+2^+3^+4^+) = (1^-2^-3^-4^-)$, so we conclude that

$$\tilde{C}_2\tilde{\imath} = \tilde{\sigma}_h \qquad (6.19b)$$

In an identical fashion we find

$$\tilde{C}_2\tilde{\sigma}_h(1^+2^+3^+4^+) = \tilde{C}_2(1^-2^-3^-4^-) = (4^-3^-2^-1^-) = \tilde{\imath}(1^+2^+3^+4^+)$$

which gives

$$\tilde{C}_2\tilde{\sigma}_h = \tilde{\imath} \qquad (6.19c)$$

Obviously we have

$$\tilde{C}_2\tilde{E} = \tilde{C}_2 \qquad (6.19d)$$

so this completes all products of the form $\tilde{C}_2\tilde{R}$. Doing this for all the elements we obtain the "multiplication" table of Table 6.2. The meaning of

TABLE 6.2

Group multiplication table for C_{2h} group.

	E	C_2	i	σ_h
E	E	C_2	i	σ_h
C_2	C_2	E	σ_h	i
i	i	σ_h	E	C_2
σ_h	σ_h	i	C_2	E

the table is that multiplying the elements on the left-hand side onto those at the top yields the tabular entry. Note that for this group, each element is its own inverse, and all elements commute. Observe also that each element appears only *once* in each row or column. This must be true for any group multiplication table as can be easily proved. If, for example, the entries for $\tilde{C}_2\tilde{R}$ and $\tilde{C}_2\tilde{S}$ were identical, premultiplication by $\tilde{C}_2^{-1}$ would give

$$\tilde{C}_2^{-1}\tilde{C}_2\tilde{R} = \tilde{C}_2^{-1}\tilde{C}_2\tilde{S} \tag{6.20a}$$

or

$$\tilde{E}\tilde{R} = \tilde{E}\tilde{S} \tag{6.20b}$$

which means

$$\tilde{R} = \tilde{S} \tag{6.20c}$$

Thus two elements would be identical which contradicts our requirement that a group contains each element only once.

The reader should also observe that the multiplication table is satisfied if in place of operations we use any of the following four sets of numbers:

$$E = 1, \quad C_2 = 1, \quad i = 1, \quad \sigma_h = 1 \tag{6.21a}$$

$$E = 1, \quad C_2 = -1, \quad i = 1, \quad \sigma_h = -1 \tag{6.21b}$$

$$E = 1, \quad C_2 = 1, \quad i = -1, \quad \sigma_h = -1 \tag{6.21c}$$

$$E = 1, \quad C_2 = -1, \quad i = -1, \quad \sigma_h = 1 \tag{6.21d}$$

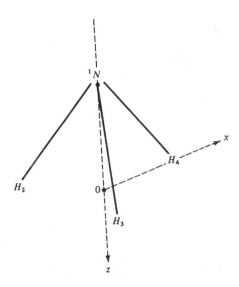

FIG. 6.3.   Symmetry elements of the ammonia molecule. Point O lies in the 234 plane and on the z axis (C_3 axis).

Moreover, the transformation matrices of Eqs. (6.7a–6.7d) also satisfy the multiplication table.

In order to see some further group properties let us consider a less trivial example than that just described. The ammonia molecule of Fig. 6.3 provides a suitable example. The group is the C_{3v} group of order 6. With the operator identifications given in Fig. 6.3 it can be verified that the multiplication table is as shown in Table 6.3. It can be verified also that the following sets of substitutions for operators also satisfy the multiplication table:

$$A = 1 \quad B = 1 \quad C = 1 \quad D = 1 \quad E = 1 \quad F = 1 \qquad \text{(6.22a)}$$

$$A = 1 \quad B = 1 \quad C = -1 \quad D = -1 \quad E = 1 \quad F = -1 \quad \text{(6.22b)}$$

$$A = \begin{pmatrix} -\frac{1}{2} & \frac{\sqrt{3}}{2} \\ -\frac{\sqrt{3}}{2} & -\frac{1}{2} \end{pmatrix} \qquad B = \begin{pmatrix} -\frac{1}{2} & -\frac{\sqrt{3}}{2} \\ \frac{\sqrt{3}}{2} & -\frac{1}{2} \end{pmatrix} \qquad C = \begin{pmatrix} -\frac{1}{2} & \frac{\sqrt{3}}{2} \\ \frac{\sqrt{3}}{2} & \frac{1}{2} \end{pmatrix}$$

$$D = \begin{pmatrix} -\frac{1}{2} & -\frac{\sqrt{3}}{2} \\ -\frac{\sqrt{3}}{2} & \frac{1}{2} \end{pmatrix} \qquad E = \begin{pmatrix} 1 & 0 \\ 0 & 1 \end{pmatrix} \qquad F = \begin{pmatrix} 1 & 0 \\ 0 & -1 \end{pmatrix} \qquad \text{(6.22c)}$$

TABLE 6.3

Multiplication table for C_{3v} group.

	E	A	B	C	D	F
E	E	A	B	C	D	F
A	A	B	E	F	C	D
B	B	E	A	D	F	C
C	C	D	F	E	A	B
D	D	F	C	B	E	A
F	F	C	D	A	B	E

Later we shall see the great significance of these three representations of the group elements.

B. Subgroups

If within a group there exists a smaller set of elements which also form a group, this subset of elements is known as a *subgroup*. For the group C_{2h} we can see that $E, C_2; E, i; E, \sigma_h$ are all subgroups of order 2. Of course, E is always a subgroup of order 1. For the C_{3v} group just described, the multiplication table shows that the possible subgroups are: $E; E, C; E, D; E, F; E, A, B$. That is, there are subgroups of order 1, 2, and 3. Note in both these example groups that

$$h/g_i = k_i \qquad (6.23)$$

where h is the order of the group, g_i the order of the ith subgroup and k_i is an integer. It can be shown formally that Eq. (6.23) is true for any (finite) group. We merely observe here that it is valid. For example, no subgroups of order 4 or 5 are possible for our C_{3v} group of order 6 which is the result predicted by Eq. (6.23). Physically, subgroups are simply groups of lower symmetry than the parent group.

C. Classes

Another important concept is that of *class* structure. A class of elements is a collection of all the elements within a group which are mutually *conjugate*. Two elements $\tilde{R}$ and $\tilde{S}$ are said to be conjugate if

$$\tilde{R} = \tilde{X}^{-1}\tilde{S}\tilde{X} \quad \text{or} \quad \tilde{S} = \tilde{X}\tilde{R}\tilde{X}^{-1} \qquad (6.24)$$

where $\tilde{X}$ is also a group element. It is easily proved that if $\tilde{R}$ and $\tilde{T}$ are each conjugate to a third element $\tilde{S}$ they are conjugate to each other. To show this assume that

$$\tilde{R} = \tilde{X}^{-1}\tilde{S}\tilde{X} \qquad (6.25a)$$

$$\tilde{T} = \tilde{Y}^{-1}\tilde{S}\tilde{Y} \qquad (6.25b)$$

Then we have from Eq. (6.25b)

$$\tilde{S} = \tilde{Y}\tilde{T}\tilde{Y}^{-1} \qquad (6.25c)$$

which upon substitution into Eq. (6.25a) gives

$$\tilde{R} = \tilde{X}^{-1}\tilde{Y}\tilde{T}\tilde{Y}^{-1}\tilde{X} = (\tilde{X}^{-1}\tilde{Y})\tilde{T}(\tilde{Y}^{-1}\tilde{X}) \qquad (6.25d)$$

But it is always true that $(\tilde{Y}^{-1}\tilde{X}) = (\tilde{X}^{-1}\tilde{Y})^{-1}$, since

$$(\tilde{X}\tilde{Y})(\tilde{Y}^{-1}\tilde{X}^{-1}) = \tilde{E} \qquad (6.26a)$$

or

$$(\tilde{X}\tilde{Y}) = \tilde{E}(\tilde{Y}^{-1}\tilde{X}^{-1})^{-1} = (\tilde{Y}^{-1}\tilde{X}^{-1})^{-1} \qquad (6.26b)$$

That is, *the inverse of a product is equal to the product of the inverses in reverse order.* Equation (6.25d) becomes therefore

$$\tilde{R} = (\tilde{X}^{-1}\tilde{Y})\,\tilde{T}\,(\tilde{X}^{-1}\tilde{Y})^{-1} = \tilde{Z}\tilde{T}\tilde{Z}^{-1} \tag{6.25e}$$

which shows that $\tilde{R}$ and $\tilde{T}$ are also conjugate.

It is easily shown that $\tilde{E}$ is always conjugate to itself only. For if $\tilde{R}$ is any other group element

$$\tilde{R}^{-1}\tilde{E}\tilde{R} = \tilde{E}\tilde{R}^{-1}\tilde{R} = \tilde{E} \tag{6.27}$$

We say that E forms a class in itself.

Once the multiplication table for a group is known it is a simple, but often tedious, matter to find the classes. For the C_{2h} group the class containing $\tilde{C}_2$ is found by forming all the products like Eq. (6.24). We find

$$\tilde{E}\tilde{C}_2\tilde{E}^{-1} = \tilde{C}_2, \quad \tilde{C}_2\tilde{C}_2\tilde{C}_2^{-1} = \tilde{C}_2, \quad \tilde{\imath}\tilde{C}_2\tilde{\imath}^{-1} = \tilde{C}_2, \quad \text{and} \quad \tilde{\sigma}_h\tilde{C}_2\tilde{\sigma}_h^{-1} = \tilde{C}_2.$$

Thus $\tilde{C}_2$ forms a class in itself also. In a similar manner we find that $\tilde{\imath}$ and $\tilde{\sigma}_h$ form classes in themselves, so the C_{2h} group contains only four complete classes.

The C_{3v} group provides a less trivial example of class structure. By use of the multiplication table the conjugate elements of A are found to be

$$EAE^{-1} = A$$
$$AAA^{-1} = BA^{-1} = A$$
$$BAB^{-1} = EB^{-1} = A$$
$$CAC^{-1} = DC^{-1} = B$$
$$DAD^{-1} = FD^{-1} = B$$
$$FAF^{-1} = CF^{-1} = B$$

Therefore A and B are conjugate elements and form a class. In a similar manner the elements conjugate to C are

$$ECE^{-1} = C$$
$$ACA^{-1} = FA^{-1} = D$$
$$BCB^{-1} = DB^{-1} = F$$
$$CCC^{-1} = EC^{-1} = C$$
$$DCD^{-1} = BD^{-1} = F$$
$$FCF^{-1} = AF^{-1} = D$$

Therefore C, D, and F are mutually conjugate elements and form a class. We see then that the C_{3v} group consists of three classes: E; A, B; C, D, F. As illustrated by the C_{2h} and the C_{3v} groups, the order of a class must, in general, be an integral divisor of the order of the group. Hence Eq. (6.23) applies to classes also, but now g_i is the number of elements in the class.

It can be noted now that the class structure of a group is closely related to the types of symmetry operations represented by the elements. In the last case considered we found that the two C_3 operations formed a class as did the three σ_v operations. Thus we see that a class is made up of operations of the same type, for example C_n types about the same axis, or reflections of the same nature. For the C_{2h} group we found that all four elements were classes in themselves. This is because E, C_2, i, and σ_h are all different types of operations.

Earlier in this chapter we described the symmetry operators of the D_{3d} group which describes the staggered ethane molecule. Without having the benefit of a group multiplication table we can still specify the classes with confidence by our knowledge of the types of operations. They are E; C_3, $-C_3$; $C_2{}^x$, $C_2{}^y$, $C_2{}^z$; $\sigma_d{}^{(1784)}$, $\sigma_d{}^{(3786)}$, $\sigma_d{}^{(2785)}$; S_6, $-S_6$; i, or more simply E, $2C_3$, $3C_2$, $3\sigma_d$, $2S_6$, and i. Thus this group of order 12 contains six classes, and we see again that the number of elements in a class is an integral divisor of the order of the group.

6.3   THEORY OF MATRIX REPRESENTATIONS OF GROUPS

A.   Reducible and Irreducible Representations

We have seen in earlier sections of this chapter that matrices form suitable representations of the elements of point groups. For example the matrices of Eqs. (6.7) form a representation of the C_{2h} group while the matrices of Eq. (6.22c) form a representation of the C_{3v} group. The number representations of Eqs. (6.22a) and (6.22b) may also be considered to be (one-dimensional) matrix representations of the C_{3v} group.

Representations of this type can always be generated by application of the group symmetry operators to the coordinates or vector components x, y, and z. This was illustrated by the Eqs. (6.7). It can be shown similarly that the matrix representation of Eq. (6.22c) can be generated by operating upon the two-dimensional vector

$$\begin{pmatrix} x \\ y \end{pmatrix}$$

We say that x and y form a *basis* for a representation of the C_{3v} group. Clearly, the dimension of a matrix representation is the same as the number of basis functions used in forming the representation. It is possible for many kinds of functions to serve as bases for group representations, and indeed our quantum-mechanical applications are associated with this fact. One restriction that we make is that our matrix representations be unitary (see Appendix C).

It should not be assumed from what has just been said that the number of matrix representations of a given group is limited. This is certainly not so since it is always possible to build other valid matrix representations from the simple ones. If $\mathbf{\Gamma}^{(1)}(R)$ and $\mathbf{\Gamma}^{(2)}(R)$ are the Rth elements of two different matrix representations, another can be formed by combining these in a supermatrix, namely

$$\mathbf{\Gamma}(R) = \begin{pmatrix} \mathbf{\Gamma}^{(1)}(R) & 0 \\ 0 & \mathbf{\Gamma}^{(2)}(R) \end{pmatrix} \tag{6.28}$$

If the matrices of all the group elements are formed in this fashion, the new set of matrices still obey the group multiplication table since the multiplication of block diagonal matrices proceeds simply by the multiplication of the individual blocks (Appendix C). Thus by combining Eq. (6.22b) and Eq. (6.22c) we get

$$A = \begin{pmatrix} 1 & 0 & 0 \\ 0 & -\frac{1}{2} & \frac{\sqrt{3}}{2} \\ 0 & -\frac{\sqrt{3}}{2} & -\frac{1}{2} \end{pmatrix}, \quad B = \begin{pmatrix} 1 & 0 & 0 \\ 0 & -\frac{1}{2} & -\frac{\sqrt{3}}{2} \\ 0 & \frac{\sqrt{3}}{2} & -\frac{1}{2} \end{pmatrix},$$

$$C = \begin{pmatrix} -1 & 0 & 0 \\ 0 & -\frac{1}{2} & \frac{\sqrt{3}}{2} \\ 0 & \frac{\sqrt{3}}{2} & \frac{1}{2} \end{pmatrix}, \quad D = \begin{pmatrix} -1 & 0 & 0 \\ 0 & -\frac{1}{2} & -\frac{\sqrt{3}}{2} \\ 0 & -\frac{\sqrt{3}}{2} & \frac{1}{2} \end{pmatrix},$$

$$E = \begin{pmatrix} 1 & 0 & 0 \\ 0 & 1 & 0 \\ 0 & 0 & 1 \end{pmatrix}, \quad F = \begin{pmatrix} -1 & 0 & 0 \\ 0 & 1 & 0 \\ 0 & 0 & -1 \end{pmatrix}$$

$$\tag{6.29}$$

which satisfy the C_{3v} multiplication table. In a similar manner we could form an unlimited number of matrix representations merely by building super-matrices with the simple matrices of Eqs. (6.22).

It is possible to form an unlimited number of additional 3-dimensional representations by performing unitary similarity transformations upon each of the matrices of Eq. (6.29), that is

$$\mathbf{U^{-1}AU} = \mathbf{\mathcal{a}}$$

$$\mathbf{U^{-1}BU} = \mathbf{\mathcal{B}} \qquad\qquad (6.30)$$

etc.

These new matrices still satisfy the multiplication table since all matrix equations retain their validity if each matrix is subjected to the same trans-formation. But note now that the transformed matrices would not show the block diagonal structure in general so that it would not be obvious that they contained smaller representations each of which satisfied the group multiplication table.

For our purposes we consider *equivalent* representations, that is, those related by similarity transformations, to be not essentially different or distinct. Thus if we are counting up distinct representations we will count those related by a similarity transformation only once. For example, the representation $\mathbf{\mathcal{a}}$, $\mathbf{\mathcal{B}}$, $\mathbf{\mathcal{C}}$... of Eq. (6.30) would be considered to be equiv-alent to the original representation $\mathbf{A}$, $\mathbf{B}$, $\mathbf{C}$....

From what has been said it is now possible to recognize two distinct types of matrix representations. A representation is said to be *reducible* if each of the group element matrices has the same block-diagonal form, that is, if they are of the form of Eq. (6.28), or if it is possible by some simi-larity transformation to reduce each of the matrices to a block-diagonal form. Thus the representation of C_{3v} given by Eqs. (6.29) is a reducible one. Further, the matrices $\mathbf{\mathcal{a}}$, $\mathbf{\mathcal{B}}$, $\mathbf{\mathcal{C}}$... of Eq. (6.30) are reducible also, since they can be brought to block-diagonal form by a similarity trans-formation.

If a representation is not block diagonal and if it is not possible to bring it to such a form by a similarity transformation upon all the group-element matrices, the representation is said to be *irreducible*. The three representations of Eqs. (6.22) are of this type. We see shortly that the number of irreducible representations that can exist for a given group is limited. For the C_{3v} group, it turns out that those of Eqs. (6.22) are the only ones.

One of our important tasks is to discover the means of reducing a given representation into its component irreducible representations. For simplicity of notation we can express any reducible representation Γ as

$$\Gamma = \sum_i a_i \Gamma^{(i)} \qquad\qquad (6.31a)$$

where $\Gamma^{(i)}$ are irreducible representations and the a_i specify how many times each occurs. For example, the reducible representation of Eq. (6.29) is

$$\Gamma = \Gamma^{(2)} + \Gamma^{(3)} \tag{6.31b}$$

where $\Gamma^{(2)}$ is the representation of Eq. (6.22b) and $\Gamma^{(3)}$ is that of Eq. (6.22c). Note that the symbolism of Eq. (6.31a) does not mean matrix addition.

Before proceeding to the basic theorems of irreducible representations, we wish to define the *character* χ of a matrix as

$$\chi(R) = \mathrm{Tr}\, \mathbf{\Gamma}(R) = \sum_i \Gamma(R)_{ii} \tag{6.32}$$

where Tr means the trace or sum of the diagonal elements. It is easy to prove that *the characters of all the matrices in a class are equal*. If R and S are in the same class

$$\mathbf{R} = \mathbf{X}^{-1}\mathbf{S}\mathbf{X} \tag{6.33a}$$

The character of $\mathbf{R}$ is then

$$\begin{aligned} \chi(R) &= \sum_i (\mathbf{X}^{-1}(R)\,\mathbf{S}(R)\mathbf{X}(R))_{ii} \\ &= \sum_i \sum_k \sum_j X^{-1}(R)_{ik} S(R)_{kj} X(R)_{ji} \tag{6.33b} \\ &= \sum_j \sum_k \sum_i X(R)_{ji} X^{-1}(R)_{ik} S(R)_{kj} \\ &= \sum_j [\mathbf{X}(R)\mathbf{X}^{-1}(R)\mathbf{S}(R)]_{jj} \end{aligned}$$

But $\mathbf{XX}^{-1} = \mathbf{E}$ and $\mathbf{ES} = \mathbf{S}$ so that

$$\chi(R) = \sum_j S(R)_{jj} = \chi(S) \tag{6.33c}$$

The C_{3v} representation of Eq. (6.29) provides a good example of this. The characters of $\mathbf{A}$ and $\mathbf{B}$ are 0, of $\mathbf{C}$, $\mathbf{D}$, and $\mathbf{F}$ are -1, and of $\mathbf{E}$ is 3. Thus although the matrices are not necessarily identical within a class, the characters are. We find that most of our quantum-mechanical problems can be solved by working only with the characters of the matrix representations, so the concept is extremely useful.

B. Theory of Irreducible Representations

The physical applications of group theory require invariably a knowledge of the properties of the irreducible representations of the groups.

The principal theorem of group-representation theory can be written[3]

$$\sum_R \Gamma^{(i)}(R)_{mn}{}^*\Gamma^{(j)}(R)_{m'n'} = \frac{h}{\sqrt{l_i l_j}} \delta_{ij}\delta_{mm'}\delta_{nn'} \tag{6.34}$$

where $\Gamma^{(i)}$ and $\Gamma^{(j)}$ are nonequivalent, irreducible, unitary representations of the group, h is the order of the group, l_i and l_j are the dimensions of the two representations and the δ's are the usual Kroneker deltas. Equation (6.34) is often called the "great orthogonality theorem" since it expresses

(a) The orthogonality of different irreducible representations, that is $\delta_{ij} = 0$ if $i \neq j$.

(b) The orthogonality of the elements in different rows of the same irreducible representations, that is, $\delta_{mm'} = 0$ unless $m = m'$.

(c) The orthogonality of the elements in different columns of the same irreducible representations, that is, $\delta_{nn'} = 0$ unless $n = n'$.

The reader should verify that Eq. (6.34) is satisfied for the representations of the C_{3v} group given by Eqs. (6.22).

The preceding three orthogonality statements indicate that the elements of the irreducible representations may be thought of as mutually independent, orthogonal vectors in h-dimensional group-element vector space. The total number of orthogonal vectors is $\sum_i l_i^2$, since each representation contains l^2 elements. But the total number of independent orthogonal vectors cannot exceed the dimension of the vector space, that is,

$$\sum_i l_i^2 \leqq h \tag{6.35a}$$

In fact it can be shown that the equality holds,[4]

$$\sum_i l_i^2 = h \tag{6.35b}$$

The importance of Eq. (6.35b) can be seen now as follows. For the C_{3v} group that we have considered, the order $h = 6$. Therefore the representations of Eq. (6.22) must be *all* the irreducible representations of this group, since $(1)^2 + (1)^2 + (2)^2 = 6$. Similarly, for the C_{2h} group of order $h = 4$, the four representations of Eq. (6.21) are all the irreducible representations. We can predict that a group of order 8 must have either: (a) eight one-dimensional irreducible representations, or (b) four one-dimensional and one two-dimensional representation. It is not possible to have two two-dimensional representations only, since there must always exist an

---

[3] See, for example: H. Eyring, J. Walter, and G. E. Kimball, *Quantum Chemistry* (Wiley, New York, 1944), p. 371; Michael Tinkham, *Group Theory and Quantum Mechanics* (McGraw-Hill, New York, 1964), p. 23.

[4] See footnote 3, Tinkham, p. 31.

identical or *totally symmetric* representation which consists of simply the one-dimensional unit matrix for each element. Indeed, both (a)- and (b)-type groups of order $h = 8$ exist, but no others.

Further useful information concerning the irreducible group representations can be obtained if the orthogonality theorem Eq. (6.34) is rewritten in terms of characters. To do this we write Eq. (6.34) for the case $m = n$ and $m' = n'$, and sum over m and m',

$$\sum_{m,m'} \sum_R \Gamma^{(i)}(R)_{mm}{}^* \Gamma^{(j)}(R)_{m'm'} = \frac{h}{(l_i l_j)^{1/2}} \delta_{ij} \sum_{m,m'} \delta_{mm'} \delta_{mm'} \quad (6.36a)$$

Using the expression for the character in Eq. (6.33), we get now

$$\sum_R \chi^{(i)}(R)^* \chi^{(j)}(R) = \frac{h}{(l_i l_j)^{1/2}} \delta_{ij} l_< \quad (6.36b)$$

since the sum over $\delta_{mm'}{}^2 = \delta_{mm'}$ gives the dimension $l_<$ of the smaller representation. For the case when $i = j$, we get

$$\sum_R \chi^{(i)}(R)^* \chi^{(i)}(R) = h \quad (6.36c)$$

Or if there are $N_{\mathcal{K}}$ elements of the class $\mathcal{K}$, Eq. (6.36c) becomes

$$\sum_{\mathcal{K}} \chi^{(i)}(\mathcal{K})^* \chi^{(i)}(\mathcal{K}) N_{\mathcal{K}} = h \quad (6.36d)$$

and Eq. (6.36b) becomes

$$\sum_{\mathcal{K}} \chi^{(i)}(\mathcal{K})^* \chi^{(j)}(\mathcal{K}) N_{\mathcal{K}} = \frac{h}{(l_i l_j)^{1/2}} \delta_{ij} l_< \quad (6.36e)$$

Now by Eq. (6.36e) the characters are seen to form an orthogonal set of vectors in the space of group classes. But, as we argued earlier, the number of orthogonal vectors can not exceed the dimension of the vector space, which in the present case can be written

number of irreducible representations $\leqq$ number of classes   (6.37)

As for Eq. (6.35a), we find that the equality always holds.

We can see that Eq. (6.37) is satisfied for the cases we have dealt with. For example, the C_{3v} group was found to have only three irreducible representations, those of Eq. (6.22), and only three classes of elements: E; A, B; C, D, F. Similarly for the C_{2h} group there were four classes and four irreducible representations.

C. Character Tables

It is very convenient to collect all the characters for the various classes of the irreducible representations of a particular group in a compact "character table" as shown in Table 6.4 for the C_{2h} and C_{3v} groups that we have been considering. The numbers in the tables are simply the characters of the matrices of Eqs. (6.21) and (6.22). We have labeled the irreducible representations with the symbols $\Gamma^{(i)}$ although we use a somewhat different labeling very shortly. The class structure is indicated by the types and numbers of symmetry operations that we have found previously.

From Eqs. (6.36d) and (6.36e) and variations of these equations it is seen that:

(1) The sum of the squares[5] of the characters of each element in a row is equal to h. For example, for $\Gamma^{(3)}$ of C_{3v}, we find $(2)^2 + 3(0)^2 + 2(-1)^2 = 6$.

(2) The sum of the squares of the characters in each column is equal to $h/N_\mathcal{K}$. Thus for the $3\sigma_v$ class of C_{3v} we find $(1)^2 + (1)^2 + (0)^2 = \frac{6}{3}$.

(3) Rows (irreducible representations) are orthogonal in the group elements. For C_{3v} again, $\Gamma^{(2)}$ and $\Gamma^{(3)}$ are orthogonal since $2(1) + 3(0)(-1) + 2(-1)(1) = 0$.

(4) Columns of the character tables are orthogonal also as shown by E and $2C_3$; $(1)(1) + (1)(1) + (2)(-1) = 0$.

By making use of the preceding four rules and Eqs. (6.35b) and (6.37) it is often possible to construct a character table like those of Table 6.4

TABLE 6.4

Rudimentary character tables for (A) C_{2h} and (B) C_{3v} groups.

A

C_{2h}	E	C_2	i	σ_h
$\Gamma^{(1)}$	1	1	1	1
$\Gamma^{(2)}$	1	-1	1	-1
$\Gamma^{(3)}$	1	1	-1	-1
$\Gamma^{(4)}$	1	-1	-1	1

B

C_{3v}	E	$2C_3$	$3\sigma_v$
$\Gamma^{(1)}$	1	1	1
$\Gamma^{(2)}$	1	1	-1
$\Gamma^{(3)}$	2	-1	0

[5] The conclusions stated here must be modified if complex characters are involved. That is, instead of "squares" we must state "squares of the absolute values." Similar modifications apply to the following discussion.

for a particular group without resorting to the task of determining the representation matrices.

EXAMPLE 6-2

Suppose we desire the character table for the group of order 8 which contains the following five classes:

$$E$$

$$2C_4$$

$$C_2$$

$$2C_2'$$

$$2C_2''$$

This group is known as D_4 and would be appropriate for a molecule having a square planar arrangement of atoms, in which case the various operations can be easily identified. Since there are five classes we know by Eq. (6.37) that there are five irreducible representations; we call them Γ_1, Γ_2, Γ_3, Γ_4, and Γ_5.

The first row Γ_1 of the character table can, as usual, be taken as the identical representation with all ones. By the normalization rules, the first column (class of E) must be $1,1,1,1,2$ since the class of E can be made up of only positive integers equal to the dimension of each representation. The partial character table may then be written as in Table 6.5A.

Now, except for the possible occurrence of imaginary characters, all the one-dimensional representations must have characters of $+1$ or -1. The signs are chosen so that the rows are orthogonal. Thus for Γ_2 we could write $1,1,1,-1,-1$. Similarly, Γ_3 and Γ_4 may be written $1,-1,1,1,-1$ and $1-1,1,-1,1$, respectively. Note that Γ_2, Γ_3, and Γ_4 are all orthogonal to Γ_1 and also are mutually orthogonal. Thus the nearly completed table is that of Table 6.5B.

The rule for normalization of columns (rule 2 above) can now be put to good use. The sum of squares should be $h/N_{\mathcal{K}}$, which is $\frac{8}{2} = 4$ for the $2C_4$, $2C_2'$, and $2C_2''$ classes. But this condition is already fulfilled in Table 6.5B so we conclude that the characters must be zero for these classes of Γ_5. The same rule applied to the C_2 class gives $h/N_{\mathcal{K}} = 8$, from which we see that $\chi(C_2) = \pm 2$ is required. The minus sign must be chosen in order that Γ_5 be orthogonal to the other rows. We have then the complete character table shown in Table 6.5 C.

In Appendix E we have compiled the character tables of many of the more important finite crystallographic point groups using the symbols conventional to spectroscopic and quantum-mechanical applications. Also included are two important infinite groups $C_{\infty v}$ and $D_{\infty h}$, which are appropriate for unsymmetrical and symmetrical linear molecules, respectively. Although these groups are infinite, all the results of interest to us may be obtained merely by considering the classes which have been explicitly indicated. C_ϕ and S_ϕ in these character tables mean rotation and improper rotation, respectively, by *any* angle ϕ about the major symmetry axis of the molecule.

Also listed in each character table are the simple coordinate functions which form bases for the irreducible representations. For example, x and y form a basis for the representation of the E irreducible representation of the C_{3v} group. We might also say that x and y have the transformation properties of the E representation. R_x, R_y, and R_z in the tables refer to rotations about the x, y, and z axes. The transformation properties of the polarizability components (z^2, xy, etc) are also given.

TABLE 6.5

Development of character table for D_4 group. See text for description.

A

	E	$2C_4$	C_2	$2C_2'$	$2C_2''$
Γ_1	1	1	1	1	1
Γ_2	1				
Γ_3	1				
Γ_4	1				
Γ_5	2				

B

	E	$2C_4$	C_2	$2C_2'$	$2C_2''$
Γ_1	1	1	1	1	1
Γ_2	1	1	1	-1	-1
Γ_3	1	-1	1	1	-1
Γ_4	1	-1	1	-1	1
Γ_5	2				

C

	E	$2C_4$	C_2	$2C_2'$	$2C_2''$
Γ_1	1	1	1	1	1
Γ_2	1	1	1	-1	-1
Γ_3	1	-1	1	1	-1
Γ_4	1	-1	1	-1	1
Γ_5	2	0	-2	0	0

EXAMPLE 6-3

We can illustrate that x and y form a basis for the irreducible representation of the $C_{\infty v}$ group. Consider the z axis (as usual) to lie along the C_ϕ axis. Then

$$\tilde{E}\begin{pmatrix} x \\ y \end{pmatrix} = \begin{pmatrix} 1 & 0 \\ 0 & 1 \end{pmatrix}\begin{pmatrix} x \\ y \end{pmatrix}$$

$$\tilde{C}_\phi\begin{pmatrix} x \\ y \end{pmatrix} = \begin{pmatrix} \cos\phi & -\sin\phi \\ \sin\phi & \cos\phi \end{pmatrix}\begin{pmatrix} x \\ y \end{pmatrix} \qquad (6.38)$$

$$\tilde{\sigma}_v\begin{pmatrix} x \\ y \end{pmatrix} = \begin{pmatrix} 1 & 0 \\ 0 & -1 \end{pmatrix}\begin{pmatrix} x \\ y \end{pmatrix}$$

where the reflection σ_v is taken in the xz plane. The characters are then

$$\chi(E) = 2$$

$$\chi(C_\phi) = 2\cos\phi \qquad (6.39)$$

$$\chi(\sigma_v) = 0$$

which correspond to the Π irreducible representation of $C_{\infty v}$ as shown in Appendix E.

D. Reduction of Reducible Representations

We have mentioned previously that a reducible representation can, in general, be written as a large block-diagonal matrix whose smaller submatrices are the irreducible representations. Thus the character of the Rth element of the reducible representation is simply

$$\chi(R) = \sum_i a_i \cdot \chi^{(i)}(R) \qquad (6.40)$$

where the a_i are the number of times the ith irreducible representation appears [see Eq. (6.31a)]. It is now an easy task to derive an expression which permits calculation of the a_i for any reducible representation.

To do this multiply $\chi^{(j)}(R)^*$ by $\chi(R)$ and sum over all R. We get

$$\sum_R \chi(R)\chi^{(j)}(R)^* = \sum_R \sum_i a_i \chi^{(i)}(R)\chi^{(j)}(R)^* \qquad (6.41a)$$

where we have used Eq. (6.40). But by Eq. (6.36c) the sum over R has only the nonzero value h when $i = j$, thus

$$\sum_R \chi(R)\chi^{(i)}(R)^* = a_i h \tag{6.41b}$$

or

$$a_i = \frac{1}{h}\sum_R \chi(R)\chi^{(i)}(R)^* \tag{6.41c}$$

or in terms of classes

$$a_i = \frac{1}{h}\sum_{\mathcal{K}} N_{\mathcal{K}}\chi(\mathcal{K})\chi^{(i)}(\mathcal{K})^* \tag{6.41d}$$

Consider our example group C_{3v}. Suppose a certain set of basis functions led to the generation of the reducible representation having the characters $\chi(E) = 7$, $\chi(C_3) = 1$, $\chi(\sigma_v) = -1$. By means of Eqs. (6.41) and the C_{3v} character table we find

$$a_{A_1} = \tfrac{1}{6}(7 + 2 - 3) = 1$$

$$a_{A_2} = \tfrac{1}{6}(7 + 2 + 3) = 2 \tag{6.42}$$

$$a_E = \tfrac{1}{6}(14 - 2 + 0) = 2$$

Thus

$$\Gamma = \Gamma^{(A_1)} + 2\Gamma^{(A_2)} + 2\Gamma^{(E)} \tag{6.43a}$$

or more simply

$$\Gamma = A_1 + 2A_2 + 2E \tag{6.43b}$$

Note that in many cases it may be possible to decompose or reduce a representation by inspection. For example, the C_{3v} representation having characters $\chi(E) = 3$, $\chi(C_3) = 0$, $\chi(\sigma_v) = 1$ leads easily to

$$\Gamma = \Gamma^{(A_1)} + \Gamma^{(E)} \tag{6.43c}$$

by simply looking at the C_{3v} character table and adding up a few rows mentally until the correct result is obtained. We find later that the majority of our physical applications require the reduction of representations so the Eqs. (6.41c) and (6.41d) are found extremely useful.

E. Direct Product Representations Within a Group

Later we find that it is useful to consider representations which are formed by the *direct product* of two matrix representations of the same group. If $\Gamma^{(1)}(R)$ and $\Gamma^{(2)}(R)$ are two such representations of the group, the direct product is

$$\mathbf{\Gamma}(R) = \mathbf{\Gamma}^{(1)}(R) \times \mathbf{\Gamma}^{(2)}(R) \tag{6.44a}$$

where the $\times$ means that *each* element of $\boldsymbol{\Gamma}^{(1)}(R)$ is multiplied onto *each* element of $\boldsymbol{\Gamma}^{(2)}(R)$. If $\boldsymbol{\Gamma}^{(1)}$ is n dimensional and $\boldsymbol{\Gamma}^{(2)}$ is m dimensional, $\boldsymbol{\Gamma}$ is nm dimensional. We illustrate this in Fig. 6.4, where the (R) have been suppressed for simplicity. We leave as an exercise for the reader the proof of

$$\chi(R) = \chi^{(1)}(R)\chi^{(2)}(R) \tag{6.44b}$$

which gives the character of a direct product representation. Note in Fig. 6.4 that $\boldsymbol{\Gamma}^{(1)}{}_{ij}\boldsymbol{\Gamma}^{(2)}$ is a matrix of order m. Equations (6.44) may be extended to triple and higher direct products.

To see that the direct product is a representation of the group, we proceed as follows. Suppose that $f_1, f_2 \ldots f_n$ and $g_1, g_2 \ldots g_m$ are bases for representations of $\boldsymbol{\Gamma}^{(1)}$ and $\boldsymbol{\Gamma}^{(2)}$, respectively, of a particular group. As we will show more clearly later, but as is readily apparent from Equation

$$
\begin{pmatrix}
\Gamma^{(1)}_{11} & \Gamma^{(1)}_{12} & \cdots & \Gamma^{(1)}_{1n} \\
\Gamma^{(1)}_{21} & \Gamma^{(1)}_{22} & & \\
\vdots & \vdots & \ddots & \\
\Gamma^{(1)}_{n1} & & & \Gamma^{(1)}_{nn}
\end{pmatrix}
\times
\begin{pmatrix}
\Gamma^{(2)}_{11} & \Gamma^{(2)}_{12} & \cdots & \Gamma^{(2)}_{1m} \\
\Gamma^{(2)}_{21} & \Gamma^{(2)}_{22} & & \\
\vdots & & \ddots & \\
\Gamma^{(2)}_{m1} & & & \Gamma^{(2)}_{mm}
\end{pmatrix}
=
$$

$$
\begin{pmatrix}
\Gamma^{(1)}_{11}\Gamma^{(2)}_{11} & \Gamma^{(1)}_{11}\Gamma^{(2)}_{12} & \cdots & \Gamma^{(1)}_{11}\Gamma^{(2)}_{1m} & \cdots & \Gamma^{(1)}_{1n}\Gamma^{(2)}_{1m} \\
\Gamma^{(1)}_{11}\Gamma^{(2)}_{21} & \Gamma^{(1)}_{11}\Gamma^{(2)}_{22} & & \ddots & & \\
\vdots & & & & & \\
\Gamma^{(1)}_{11}\Gamma^{(2)}_{m1} & & & & \ddots & \\
\vdots & & & & & \\
\Gamma^{(1)}_{n1}\Gamma^{(2)}_{m1} & & & & & \Gamma^{(1)}_{nn}\Gamma^{(2)}_{mm}
\end{pmatrix}
$$

or

$$
\boldsymbol{\Gamma}^{(1)} \times \boldsymbol{\Gamma}^{(2)} =
\begin{pmatrix}
\Gamma^{(1)}_{11}\boldsymbol{\Gamma}^{(2)} & \Gamma^{(1)}_{12}\boldsymbol{\Gamma}^{(2)} & \cdots & \Gamma^{(1)}_{1n}\boldsymbol{\Gamma}^{(2)} \\
\Gamma^{(1)}_{21}\boldsymbol{\Gamma}^{(2)} & \Gamma^{(1)}_{22}\boldsymbol{\Gamma}^{(2)} & & \\
\vdots & & \ddots & \\
\Gamma^{(1)}_{n1}\boldsymbol{\Gamma}^{(2)} & & & \Gamma^{(1)}_{nn}\boldsymbol{\Gamma}^{(2)}
\end{pmatrix}
$$

FIG. 6.4. Direct product multiplication.

(6.38), an operation of the group $\tilde{R}$ performed upon f_i or g_j leads to

$$\tilde{R}f_i = \sum_k \Gamma^{(1)}(R)_{ik} f_k \qquad (6.45)$$

$$\tilde{R}g_j = \sum_l \Gamma^{(2)}(R)_{jl} g_l$$

If $\tilde{R}$ is applied to any product $f_i g_j$, Eq. (6.9) and Eq. (6.45) show the result to be

$$\tilde{R}(f_i g_j) = \sum_k \Gamma^{(1)}(R)_{ik} f_k \sum_l \Gamma^{(2)}(R)_{jl} g_l$$

$$\qquad (6.46a)$$

$$= \sum_{k,l} f_k g_l \Gamma^{(1)}(R)_{ik} \Gamma^{(2)}(R)_{jl}$$

But as shown by Fig. 6.4, $\Gamma^{(1)}(R)_{ik} \Gamma^{(2)}(R)_{jl}$ is just an element of the direct-product matrix, that is

$$\tilde{R}(f_i g_j) = \sum_{k,l} \Gamma(R)_{ik,jl} f_k g_l \qquad (6.46b)$$

where $\Gamma(R)_{ik,jl}$ is the ik, jl element of the direct product matrix. Thus Eq. (6.46b) shows that the complete set of product functions form a basis for a representation of the group.

Now since this new direct product matrix is larger than the irreducible representations of the group in general, it follows that it must in general be reducible. The composition of the direct product may be found by the usual reduction formulas (6.41c) or (6.41d) after the characters of the representation have been obtained by Eq. (6.44b). We can show the reduction symbolically in a manner similar to Eq. (6.31a), thus

$$\Gamma^{(i)} \times \Gamma^{(j)} = \sum_k a_{ijk} \Gamma^{(k)} \qquad (6.47)$$

where the a_{ijk} are the numbers of times that the kth irreducible representation appears in the direct-product representation. The reduction of triple or higher cross products proceeds similarly.

Consider the representation formed by $A_2 \times E \times E$ for our C_{3v} group. The characters of this representation are readily found to be

$$\chi(E) = 4, \qquad \chi(C_3) = 1, \qquad \chi(\sigma_v) = 0$$

which reduces by inspection or by use of Eqs. (6.41) to

$$\Gamma = A_2 \times E \times E = A_1 + A_2 + E$$

Note that the triple-product was of dimension four $(1 \times 2 \times 2)$ and that the decomposed result is, as usual, of the same total dimension.

Some special cases of cross products are worth mentioning. If $\Gamma^{(A)}$ is the totally symmetrical representation of any particular group, it follows that

$$\Gamma^{(A)} \times \Gamma^{(j)} = \Gamma^{(j)} \tag{6.48}$$

This is obviously true since by Eq. (6.44b) the characters of the cross product are $\chi(R) = (1)[\chi^{(j)}(R)] = \chi^{(j)}(R)$.

It may be shown that the cross product of any representation with itself always contains the totally symmetrical representation $\Gamma^{(A)}$. That is,

$$\Gamma^{(i)} \times \Gamma^{(i)} \rightarrow \Gamma^{(A)} + \cdots \tag{6.49}$$

where the arrow means "contains" and we list only $\Gamma^{(A)}$ on the right-hand side, not being interested here in the remaining representations which may appear. The proof of (6.49) is obvious if $\Gamma^{(i)}$ is one-dimensional in which case $\chi^{(i)}(R)\chi^{(i)}(R) = 1$. It can be proved for higher-dimensional representations but is perhaps best verified by the reader for a few of the groups of Appendix E.

It is useful later to inquire whether a direct product contains the totally symmetrical representation $\Gamma^{(A)}$, that is, for example

$$\Gamma^{(i)} \times \Gamma^{(j)} \times \Gamma^{(k)} \rightarrow \Gamma^{(A)} + \cdots \tag{6.50a}$$

Multiplying in direct product fashion on both sides of Eq. (6.50a) by $\Gamma^{(i)}$ gives

$$\Gamma^{(i)} \times \Gamma^{(i)} \times \Gamma^{(j)} \times \Gamma^{(k)} \rightarrow \Gamma^{(i)} \times \Gamma^{(A)} + \cdots \tag{6.50b}$$

But by use of Eqs. (6.48) and (6.49) we may rewrite (6.50b) as

$$\Gamma^{(A)} \times \Gamma^{(j)} \times \Gamma^{(k)} \rightarrow \Gamma^{(i)} + \cdots \tag{6.51a}$$

or

$$\Gamma^{(j)} \times \Gamma^{(k)} \rightarrow \Gamma^{(i)} + \cdots \tag{6.51b}$$

by using Eq. (6.48) once again. Thus a triple direct product contains $\Gamma^{(A)}$ if the direct product of any two of the representations contains the third. If, instead of the triple direct product of Eq. (6.50a), we had been interested in only the double direct product,

$$\Gamma^{(i)} \times \Gamma^{(j)} \rightarrow \Gamma^{(A)} + \cdots \tag{6.52}$$

a similar analysis shows that instead of Eq. (6.51b) we get

$$\Gamma^{(i)} = \Gamma^{(j)} \tag{6.53}$$

That is, a double cross product may contain the totally symmetrical representation only if $\Gamma^{(i)}$ and $\Gamma^{(j)}$ are identical. Note that this is in accord with Eq. (6.49).

6.4   APPLICATION OF GROUP THEORY TO QUANTUM MECHANICS

A.   Symmetry of the Eigenfunctions of the Hamiltonian

Earlier in this chapter we showed in a very simple way that if $\tilde{R}$ was an operator that commuted with $\tilde{\mathcal{H}}$, that is, left it invariant, $\tilde{R}\psi_i$ was an eigenfunction of $\tilde{\mathcal{H}}$ if ψ_i was. If we search out all the symmetry operators which commute with $\tilde{\mathcal{H}}$, the resulting collection is known as the "group of Schrödinger's equation." Usually this group is readily apparent without actually applying all the operations to $\tilde{\mathcal{H}}$. For example, if we wish to find the group of operators which commute with the Hamiltonian for the vibrational motions of the nuclei of NH_3 in the average field of the electrons, the C_{3v} group must be appropriate since this group specifies the equilibrium symmetry of the molecule. This is obviously true for any analytically correct form of $\tilde{\mathcal{H}} = \tilde{T} + \tilde{V}$ since it cannot change if we merely permute similar atoms by the symmetry operations of C_{3v}.

As we saw earlier, $\tilde{R}\psi_n$ must give either $+\psi_n$ or $-\psi_n$ if the state n is nondegenerate. That is, there can be only one distinct eigenfunction corresponding to the energy E_n. If the E_n state has an l-fold degeneracy, operation by $\tilde{R}$ upon ψ_{ni} in general produces a linear combination of the l degeneration functions, but as before does not produce any function belonging to a different energy state. It should be emphasized that we are speaking of *normal* degeneracy cases, not cases of accidental degeneracy in which unrelated energy levels become nearly equal or equal in energy because of particular choices of properties or parameters.

We can show that the l-degenerate functions of any state E_n do form a basis for a representation of the group of Schrödinger's equation. If $\tilde{R}$ and $\tilde{S}$ are two members of the group,

$$\tilde{R}\psi_k^{(n)} = \sum_i^l \Gamma^{(n)}(R)_{ki}\psi_i^{(n)} \qquad (6.54a)$$

$$\tilde{S}\psi_i^{(n)} = \sum_j^l \Gamma^{(n)}(S)_{ij}\psi_j^{(n)} \qquad (6.54b)$$

But applying $\tilde{S}$ to (6.54a) gives

$$\tilde{S}\tilde{R}\psi_k^{(n)} = \sum_i^l \Gamma^{(n)}(R)_{ki}(\tilde{S}\psi_i^{(n)})$$

$$= \sum_{i,j} \Gamma^{(n)}(R)_{ki}\Gamma^{(n)}(S)_{ij}\psi_j^{(n)} \qquad (6.55)$$

Taking the transpose of the product in Eq. (6.55) we get

$$\tilde{S}\tilde{R}\psi_k^{(n)} = \sum_j \left[\mathbf{\Gamma}^{(n)}(R)\,\mathbf{\Gamma}^{(n)}(S) \right]'_{jk}\psi_j^{(n)}$$

$$= \sum_j \left[\mathbf{\Gamma}^{(n)}(S)'\,\mathbf{\Gamma}^{(n)}(R)' \right]_{jk}\psi_j^{(n)} \tag{6.56}$$

Since $\tilde{S}\tilde{R}$ must be some element $\tilde{T}$ in the group

$$\tilde{S}\tilde{R}\psi_k^{(n)} = \tilde{T}\psi_k^{(n)} = \sum_j \Gamma^{(n)}(T)_{kj}\psi_j^{(n)}$$

$$= \sum_j \Gamma^{(n)}(T)_{jk}'\psi_j^{(n)} \tag{6.57}$$

Comparing Eqs. (6.56) and (6.57) we find

$$\Gamma^{(n)}(T)' = \Gamma^{(n)}(S)'\Gamma^{(n)}(R)' \tag{6.58}$$

We see that the l-degenerate wave functions form a basis for a transpose matrix representation of the group, that is, the *transposes* of the transformation matrices of Eqs. (6.54) and (6.56) satisfy the group-element multiplication $\tilde{S}\tilde{R} = \tilde{T}$ as shown in Eq. (6.58). The fact that the transpose matrices are involved here is not particularly significant to our purposes. It arises because we chose to sum over columns of the transformation matrices [see Eqs. (6.54) for example], which has been our convention previously in this chapter. Note that if the matrices are symmetric or diagonal there is no difference between a matrix and its transpose. For these reasons the one-dimensional matrices formed by the numbers in Eqs. (6.21) were suitable even though we did not at that time have available to us the more general result.

Of more significance to our applications is the fact that the character of any matrix is the same as the character of the transpose matrix. Since all our results proceed through the use of characters we can obtain our results in all cases by considering $\mathbf{\Gamma}(R)$ rather than $\mathbf{\Gamma}(R)'$ and therefore usually do not discuss the latter matrix.

The representation formed above by the l-degenerate wave functions is irreducible. If it were not, it would be possible to find a unitary transformation which would bring the matrices into block-diagonal form. But if the $\mathbf{\Gamma}(R)$ were block diagonal each of the blocks would serve as a matrix representation for a different subset of the l-degenerate functions. Neglecting accidental degeneracies this is impossible, since there must exist a group operation $\tilde{R}$ which transforms any function into any others which are degenerate with it.

What we have shown is of great practical importance. *All eigenfunctions of a given problem must form bases for representations of the group, and the dimensions of the representations specify the normal degeneracy of the states.*

If the appropriate group for the physical problem of interest were our C_{3v} example group, we would know immediately that only three types of states were possible; nondegenerate A_1 and A_2 states and doubly degenerate E states. The symmetry labels A_1, A_2, and E are thus suitable forms of quantum numbers or labels. Of course, there may be, and usually are, further nonsymmetry quantum numbers or labels which distinguish different states of the same symmetry.

Note that these symmetry specifications are independent of the actual mathematical solution of the problem. Indeed, they are true even if the correct quantum-mechanical problem cannot be exactly solved. In later chapters we have many opportunities for classifying the various molecular energy levels according to the appropriate group of the Schrödinger equation, and these results are true even if the *exact* energy levels are not calculated.

B.   Matrix Element Theorems

Much of the effort in solving quantum-mechanical problems lies in the evaluation of integrals or matrix elements. For example, the determination of stationary-state energies requires the calculation of the Hamiltonian matrix elements in some basis representation, while the determination of transition probabilities or selection rules necessitates evaluation of the transition moment matrix elements. The greatest utility of group theory lies in its ability to specify when certain of these integrals are zero. This can be shown easily as follows.

Suppose we wish to investigate the overlap or orthogonality integral

$$\int f_i^{(B)} f_k^{(C)} d\tau \tag{6.59}$$

arising in some quantum-mechanical problem. We imagine that the functions $f_i^{(B)}$ and $f_k^{(C)}$ have been classified according to the appropriate symmetry group for the problem with the result that $f_i^{(B)}$ belongs to $\Gamma^{(B)}$ and $f_k^{(C)}$ belongs to $\Gamma^{(C)}$. By Sec. 6.3 E we know that $f_i^{(B)} f_k^{(C)}$ forms a basis for a representation of the direct product group. Thus

$$\tilde{R} \int f_i^{(B)} f_j^{(C)} d\tau = \sum_{kl} \Gamma^{(B)}(R)_{ik} \Gamma^{(C)}(R)_{jl} \int f_k^{(B)} f_l^{(C)} d\tau \tag{6.60}$$

by referring to Eq. (6.46a). The *value* of the definite integral cannot be affected by the operation $\tilde{R}$, so the left-hand side can be simply written $\int f_i^{(B)} f_j^{(C)} d\tau$.

If we now sum both sides of Eq. (6.60) over all group operations $\tilde{R}$

we get

$$h \int f_i{}^{(B)} f_j{}^{(C)} d\tau = \sum_R \sum_{kl} \Gamma^{(B)}(R)_{ik} \Gamma^{(C)}(R)_{jl} \int f_k{}^{(B)} f_l{}^{(C)} d\tau \qquad (6.61a)$$

or

$$\int f_i{}^{(B)} f_j{}^{(C)} d\tau = \frac{1}{h} \sum_R \sum_{kl} \Gamma^{(B)}(R)_{ik} \Gamma^{(C)}(R)_{jl} \int f_k{}^{(B)} f_l{}^{(C)} d\tau \qquad (6.61b)$$

By the basic orthogonality theorem given by Eq. (6.34) we conclude that

$$\int f_i{}^{(B)} f_j{}^{(C)} d\tau = 0 \qquad \text{unless } B = C \quad \text{and} \quad i = j \qquad (6.62)$$

since otherwise the sums in the right-hand side of Eq. (6.61b) go to zero. Thus Eq. (6.62) states that the integral (6.59) is zero unless the two functions belong to the same representation, that is,

$$\Gamma^{(B)} = \Gamma^{(C)} \qquad (6.63)$$

Furthermore, if the functions belong to a degenerate state, that is, the representation $\Gamma^{(B)} = \Gamma^{(C)}$ is of dimension two or higher, the integral is zero unless each of the two functions has the transformation properties of the same row of the matrix, that is, $i = j$.

Hence we have a way of determining whether a certain type of integral is necessarily vanishing. Our arguments do not permit evaluation of nonzero integrals, nor do they tell us about integrals which become zero or essentially zero for accidental (nonsymmetry) reasons.

For the investigation of more general integrals we observe that Eq. (6.63) implies [by virtue of Eq. (6.49)]

$$\Gamma^{(B)} \times \Gamma^{(C)} = \Gamma^{(A)}$$

where $\Gamma^{(A)}$ is the totally symmetrical representation. This is the representation to which any integral must belong, since, as we noted earlier, the integral must be invariant to all operations $\tilde{R}$. Thus the quantum-mechanical integral

$$\int f^{(B)} \tilde{O} f^{(C)} d\tau \qquad (6.64)$$

has the value zero unless

$$\Gamma^{(B)} \times \Gamma^{(O)} \times \Gamma^{(C)} \to \Gamma^{(A)} + \cdots \qquad (6.65a)$$

where $\Gamma^{(O)}$ is the representation to which the operator $\tilde{O}$ belongs. Equation (6.51) showed that this is true if

$$\Gamma^{(O)} \times \Gamma^{(C)} \to \Gamma^{(B)} + \cdots \qquad (6.65b)$$

One special case of Eq. (6.64) is particularly valuable. If $\tilde{O} = \tilde{\mathfrak{K}}$, the Hamiltonian operator, $\Gamma^{(O)} = \Gamma^{(\mathfrak{K})} = \Gamma^{(A)}$, since $\tilde{\mathfrak{K}}$ is always invariant to the group operations. Thus Eq. (6.65b) shows that

$$\int f_i^{(B)} \tilde{\mathfrak{K}} f_j^{(C)} d\tau = 0 \qquad (6.66)$$

unless

$$\Gamma^{(A)} \times \Gamma^{(C)} \to \Gamma^{(B)} + \cdots \qquad (6.67a)$$

which implies

$$\Gamma^{(C)} = \Gamma^{(B)} \qquad (6.67b)$$

Thus there can be no Hamiltonian matrix elements between states of different symmetry. Furthermore, if $\Gamma^{(C)}$ and $\Gamma^{(B)}$ are of dimension greater than 1, the integral is zero unless $f_i^{(B)}$ and $f_j^{(C)}$ transform according to the same row of $\Gamma^{(C)} = \Gamma^{(B)}$.

These remarkable theorems bring about great simplifications in solving quantum-mechanical problems, for they provide a means of sorting out the nonzero matrix elements before computation is actually begun. For the determination of spectroscopic selection rules, these symmetry arguments alone are sufficient to predict whether or not a given transition can occur.

C.   Symmetry Functions

Because of the matrix element results presented by Eqs. (6.66) and (6.67), it is clear that great simplifications in the Hamiltonian matrix occur if basis functions having the symmetry of the group of Schrödinger's equation are used. The procedure for forming such functions is relatively easy. As shown in several places,[6] if ϕ_1, ϕ_2, $\ldots$, ϕ_j, $\ldots$, ϕ_n are an arbitrary set of functions spanning the space of the group of interest, a function $\psi^{(i)}$ belonging to the ith irreducible representation can be formed from

$$\psi^{(i)} = N \sum_R \chi^{(i)}(R)^* \tilde{R} \phi_j \qquad (6.68)$$

where all the symbols have been previously defined except N which is simply a normalizing factor. Any member ϕ_j of the set may be used, and if each of the functions ϕ_j is used in Eq. (6.68) all the possible symmetry functions belonging to $\Gamma^{(i)}$ are found; *except for $l_i \geq 2$, only one function of the degenerate set is produced.* Then if the procedure is repeated for all the irreducible representations all the symmetry functions are obtained. Note that $\tilde{R}\phi_j$ is merely one of the functions $\phi_1 \ldots \phi_j \ldots \phi_n$. If it were not, the set of functions would not be spanning the space of the group which is a requirement of the method.

---

[6] See footnote 3, Tinkham, pp. 39–43.

In most cases it is not necessary to go through all the possibilities [that is, use all the ϕ_j in Eq. (6.68)] since many of them simply repeat the same result several times. It is perhaps best to first determine the reducible representation formed by the functions $\phi_1 \ldots \phi_n$, followed by reduction into the irreducible representations. In this way the number of possible symmetry functions of each type can be found. Then Eq. (6.68) is applied for each representation $\Gamma^{(i)}$ until the requisite number of symmetry functions are obtained.

The question arises as to how the other members of a degenerate set of functions are obtained if Eq. (6.68) produces only one of the set for $l_i \geqq 2$. The answer is simple. Form the remaining functions as linear combinations of the same subset of degenerate functions $\phi_{ki} \ldots \phi_{li}$ involved in the first $\psi^{(i)}$ in such a way that they are orthogonal to the first one. Thus, suppose a certain set of functions led to

$$\psi_1^{(i)} = \frac{1}{\sqrt{3}} (\phi_a + \phi_b + \phi_c) \tag{6.69}$$

by application of Eq. (6.68) for a particular representation of dimension $l_i = 2$. In Eq. (6.69) the functions ϕ_a, ϕ_b, and ϕ_c are assumed to have been an orthonormal set. Then the second function of symmetry $\Gamma^{(i)}$ could be written

$$\psi_2^{(i)} = \frac{1}{\sqrt{2}} (\phi_a - \phi_b)$$

or

$$\psi_2^{(i)} = \frac{1}{\sqrt{2}} (\phi_a - \phi_c) \tag{6.70}$$

or

$$\psi_2^{(i)} = \frac{1}{\sqrt{2}} (\phi_b - \phi_c)$$

or

$$\psi_2^{(i)} = \frac{1}{\sqrt{6}} (\phi_a + \phi_b - 2\phi_c)$$

$$\vdots$$

or any number of other ways which are orthogonal to the function of Eq. (6.69). The third function is found in a similar manner. This procedure for finding the remaining functions works because the complete representation matrices $\Gamma^{(i)}(R)$ must be unitary, that is, the functions forming a basis for $\Gamma^{(i)}$ must be orthonormal. The fact that the choice is not unique merely reflects the fact that the representation matrices are not unique, since any matrices that are related by a similarity transformation would be just as satisfactory. The character of the representation formed by $\psi_1^{(i)}$, $\psi_2^{(i)}$, and $\psi_3^{(i)}$ is, however, independent of the choice and must be that given by the character table for the $\Gamma^{(i)}$ representation. Note that the great simplicity of our method is that we deal only with the characters [see Eq. (6.68)] and do not require knowledge of the matrices $\Gamma^{(i)}(R)$.

EXAMPLE 6-4

To illustrate some of the principles of Sec. 6.4, consider a single electron moving in the field of three protons that are situated at the corners of an equilateral triangle as shown in Fig. 6.5. For our purposes we need not write down the Hamiltonian but merely note that it must be invariant to the symmetry operations which take the identical protons into each other. These operations include as a minimum E, $3C_2$, and $2C_3$, which constitute the D_3 point group given in Appendix E. Of course, the Hamiltonian must also be invariant to σ_h, $2S_3$, and $3\sigma_v$ in addition to the previous six elements, that is, it must be invariant to the operations of D_{3h}. D_3 is merely a subgroup of D_{3h} and is satisfactory for our treatment if we use wave functions whose transformation or symmetry properties are independent of σ_h, that is, do not change sign under σ_h. No harm would accrue from using the larger group except for a slight loss in simplicity.

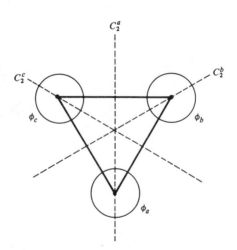

FIG. 6.5.   Protons situated at corner of an equilateral triangle. 1s hydrogen-atom wave functions are situated at each proton and are schematically indicated by the circles labeled ϕ_a, ϕ_b, and ϕ_c. Three C_2 axes of the molecule are labeled.

Let us choose basis functions which consist of a normalized hydrogen $1s$ orbital located at each proton. These functions must then form a basis for a representation of the group D_3. The character of this representation is found as follows. The operation E leaves everything unchanged, thus

$$\tilde{E} \begin{pmatrix} \phi_a \\ \phi_b \\ \phi_c \end{pmatrix} = \begin{pmatrix} 1 & 0 & 0 \\ 0 & 1 & 0 \\ 0 & 0 & 1 \end{pmatrix} \begin{pmatrix} \phi_a \\ \phi_b \\ \phi_c \end{pmatrix} \tag{6.71a}$$

$$\chi(E) = 3$$

The operation $\tilde{C}_3$ causes $\phi_a \to \phi_b$, $\phi_b \to \phi_c$, and $\phi_c \to \phi_a$ or

$$\tilde{C}_3 \begin{pmatrix} \phi_a \\ \phi_b \\ \phi_c \end{pmatrix} = \begin{pmatrix} 0 & 1 & 0 \\ 0 & 0 & 1 \\ 1 & 0 & 0 \end{pmatrix} \begin{pmatrix} \phi_a \\ \phi_b \\ \phi_c \end{pmatrix} \tag{6.71b}$$

$$\chi(C_3) = 0$$

Similarly, for one of the C_2 operations

$$\tilde{C}_2{}^a \begin{pmatrix} \phi_a \\ \phi_b \\ \phi_c \end{pmatrix} = \begin{pmatrix} 1 & 0 & 0 \\ 0 & 0 & 1 \\ 0 & 1 & 0 \end{pmatrix} \begin{pmatrix} \phi_a \\ \phi_b \\ \phi_c \end{pmatrix} \tag{6.71c}$$

$$\chi(C_2) = 1$$

By inspection or by use of Eq. (6.41), the reducible representation whose characters are given by Eqs. (6.71) decomposes as

$$\Gamma^\phi = A_1 + E \tag{6.72}$$

It must be possible, therefore, to generate symmetry functions which transform like A_1 and E of the D_3 point group.

In order to apply Eq. (6.68) we write down explicitly all the $\tilde{R}\phi_j$ for $\phi_j = \phi_a$. We find the results of Table 6.6 by reference to Fig. 6.5. Using Eq. 6.68, the symmetry function of type A_1 can now be obtained.

$$\psi^{(A_1)} = N\{(1)(\phi_a) + (1)(\phi_b + \phi_c) + (1)(\phi_a + \phi_c + \phi_b)\}$$

$$= 2N(\phi_a + \phi_b + \phi_c) \tag{6.73a}$$

$$= N'(\phi_a + \phi_b + \phi_c)$$

TABLE 6.6

Results of applying operations of D_3 group to ϕ_a of Fig. 6.5.

	E	$+C_3$	$-C_3$	$C_2{}^a$	$C_2{}^b$	$C_2{}^c$
$\tilde{R}\phi_a$	ϕ_a	ϕ_b	ϕ_c	ϕ_a	ϕ_c	ϕ_b

For computational purposes the normalization factor could be found so that the normalized function would be

$$\psi^{(A_1)} = \frac{1}{\sqrt{3}} (1 + 2S)^{-1/2}(\phi_a + \phi_b + \phi_c) \tag{6.73b}$$

where S is the overlap (nonorthogonality) integral between the $1s$ functions. The function of Eq. (6.73b) clearly transforms like A_1 and might have been guessed without even going through Eq. (6.68).

One function of E symmetry is similarly found by use of Table 6.6.

$$\psi_1{}^{(E)} = N\{(2)(\phi_a) + (-1)(\phi_b + \phi_c) + 0(\phi_a + \phi_c + \phi_b)\}$$

$$= N(2\phi_a - \phi_b - \phi_c) \tag{6.74}$$

$$= \frac{1}{\sqrt{6}} (1 - S)^{-1/2}(2\phi_a - \phi_b - \phi_c)$$

The second E-type function can be found as described earlier. We might choose

$$\psi_2{}^{(E)} = N(\phi_b - \phi_c) \tag{6.75a}$$

which is orthogonal to $\psi_1{}^{(E)}$ as is readily apparent by multiplying together the functional parts of Eqs. (6.74) and (6.75a) and then integrating over all space. If Eq. (6.75a) is normalized we find

$$\psi_2{}^{(E)} = \frac{1}{\sqrt{2}} (1 - S)^{-1/2}(\phi_b - \phi_c) \tag{6.75b}$$

The symmetry functions $\psi^{(A_1)}$, $\psi_1{}^{(E)}$, and $\psi_2{}^{(E)}$ provide the most satisfactory basis for proceeding with the quantum-mechanical calculation, specifically the evaluation of the Hamiltonian matrix elements. According to Eqs. (6.67) and the following discussion, the Hamiltonian matrix is diagonal when using the symmetry functions. That is

$$\mathfrak{IC} = \begin{pmatrix} H_{A_1,A_1} & 0 & 0 \\ 0 & H_{E(1),E(1)} & 0 \\ 0 & 0 & H_{E(2),E(2)} \end{pmatrix} \tag{6.76}$$

where the elements are labeled by the symmetry labels and $E(1)$ and $E(2)$ refer to the first and second E-type symmetry functions, respectively. Furthermore, since the E state is doubly degenerate the two $H_{E,E}$ matrix elements are identical. Hence only two matrix elements require numerical evaluation and these are immediately the eigenvalues of the Hamiltonian. According to the quantum-mechanical variational principle the lower of these two would be an upper bound to the true energy. Our group theory does not, however, provide us with any evidence concerning the correctness of this calculation.

Before leaving this example, we pause to verify that $\psi_1{}^{(E)}$ and $\psi_2{}^{(E)}$ do form a basis for the E irreducible representation of D_3. It is obvious that

$$\tilde{E}\begin{pmatrix}\psi_1{}^{(E)}\\ \psi_2{}^{(E)}\end{pmatrix} = \begin{pmatrix}1 & 0\\ 0 & 1\end{pmatrix}\begin{pmatrix}\psi_1{}^{(E)}\\ \psi_2{}^{(E)}\end{pmatrix} \tag{6.77a}$$

If we apply $+\tilde{C}_3$, we have found from Fig. 6.5 that $\phi_a \rightarrow \phi_b$, $\phi_b \rightarrow \phi_c$, and $\phi_c \rightarrow \phi_a$ which means that

$$\tilde{C}_3\psi_1{}^{(E)} = \frac{1}{\sqrt{6}}\,C(2\phi_b - \phi_c - \phi_a)$$

and $\tag{6.78a}$

$$\tilde{C}_3\psi_2{}^{(E)} = \frac{1}{\sqrt{2}}\,C(\phi_c - \phi_a)$$

where $C = (1 - S)^{-1/2}$. It is easily verified by a bit of algebra that Eqs. (6.78a) may be rewritten

$$\tilde{C}_3\psi_1{}^{(E)} = (-\tfrac{1}{2}\psi_1{}^{(E)} + \tfrac{1}{2}\sqrt{3}\,\psi_2{}^{(E)}) \tag{6.78b}$$

and

$$\tilde{C}_3\psi_2{}^{(E)} = (-\tfrac{1}{2}\sqrt{3}\,\psi_1{}^{(E)} - \tfrac{1}{2}\psi_2{}^{(E)})$$

The complete transformation is therefore

$$\tilde{C}_3\begin{pmatrix}\psi_1{}^{(E)}\\ \psi_2{}^{(E)}\end{pmatrix} = \begin{pmatrix}-\tfrac{1}{2} & \tfrac{1}{2}\sqrt{3}\\ -\tfrac{1}{2}\sqrt{3} & -\tfrac{1}{2}\end{pmatrix}\begin{pmatrix}\psi_1{}^{(E)}\\ \psi_2{}^{(E)}\end{pmatrix} \tag{6.77b}$$

In an entirely similar manner the transformation properties under $\tilde{C}_2{}^b$ are found to be

$$\tilde{C}_2{}^b\begin{pmatrix}\psi_1{}^{(E)}\\ \psi_2{}^{(E)}\end{pmatrix} = \begin{pmatrix}1 & 0\\ 0 & -1\end{pmatrix}\begin{pmatrix}\psi_1{}^{(E)}\\ \psi_2{}^{(E)}\end{pmatrix} \tag{6.77c}$$

Thus we see from Eqs. (6.77) that the characters of the representation are $\chi(E) = 2$, $\chi(C_3) = -1$, and $\chi(C_2) = 0$, which are the characters of the E irreducible representation of D_3 as expected. We have therefore verified that our method of generating symmetry basis functions for degenerate representations is correct.

SUPPLEMENTARY REFERENCES

F. A. Cotton, *Chemical Applications of Group Theory* (Interscience, New York, 1963).

V. Heine, *Group Theory in Quantum Mechanics* (Pergamon, New York, 1960).

D. S. Schonland, *Molecular Symmetry* (Van Nostrand, Princeton, N.J., 1965).

M. Tinkham, *Group Theory and Quantum Mechanics* (McGraw-Hill, New York, 1964).

E. P. Wigner, *Group Theory* (Academic, New York, 1959).

CHAPTER 7

MICROWAVE SPECTROSCOPY

7.1 GENERAL CHARACTERISTICS AND EXPERIMENTAL ASPECTS OF MICROWAVE SPECTROSCOPY

The field of spectroscopy covered in this chapter is that which developed rapidly after World War II as a consequence of the wartime advances in radar technology. While microwave spectroscopy covers a somewhat broader range of phenomena than just the study of pure rotational spectra, we concern ourselves with only this one area. Microwave spectroscopy is therefore considered to be synonymous with rotational spectroscopy.

The common frequency range covers from 5000 or 10 000 MHz to 40 000 or 50 000 MHz, although useful pure rotational studies are carried on at both higher and lower frequencies. In particular, the region above 100 GHz (1 gigahertz = 10^9 Hertz), which is known as the millimeter region, has been exploited increasingly in recent years, and, of course, pure rotational spectra may be observed in the far-infrared region (see Fig. 1.2). Several excellent books (including especially those of Townes and Schawlow, and Gordy, Smith, and Tramburulo) that have been on the market for several years, have been listed as general references at the end of this chapter. The more than casual reader can find in these sources very thorough experimental and theoretical descriptions of the field.

The molecular sample utilized in microwave spectroscopy is normally in the gaseous state, typically at pressures between 10^{-3} and 10^{-1} Torr, so the free rotational energy states are well defined. While there are reports of essentially free rotation of small molecules trapped in solid matrices,[1] no pure rotational microwave spectra have been observed in condensed phases. Relatively high pressures (≥ 1 Torr) are generally disadvantageous, since severe line broadening decreases the intensity and makes it difficult to distinguish microwave absorptions.

An important aspect of microwave spectroscopy is that the molecule must be polar,[2] that is, it must have a nonzero permanent electric dipole moment.[3] This is not the complete story, however, since it does not follow that *all* molecules with nonzero dipole moments have observable spectra, since other factors affect the absorption line intensities (see Sec. 7-3 B). In general, the microwave spectrum of an average-sized molecule (molecular weight of 100, say) are observable if the dipole moment is greater than 0.1 or 0.2 debye. On the other hand, Lide[4] has observed the microwave spectrum of propane ($\mu = 0.07$ debye) by conventional techniques, and Laurie, *et al.*,[5] have seen the spectrum of $HC \equiv CD$ ($\mu \sim 0.01$ debye) by using very high electric fields to induce a larger dipole moment.

In addition to the requirements that the molecules must be studied in the gas phase and must have nonzero dipole moments, there seems to be a practical upper limit to the size of the molecules which can be studied profitably by microwave spectroscopy. This limit cannot be stated precisely but is determined by intensity factors and factors relating to the complexity of the observed microwave spectra. The intensity problem is made clearer later, but qualitatively it involves two important factors: (i) When the size of the molecule increases, the moments of inertia usually increase also; this produces many energy states of relatively low rotational energy, so that a much greater number of states share in the molecular population than for small molecules. This effectively decreases the intensity of the individual spectral lines due to the decreased population of absorbing species in the state from which the transition originates. (ii) When the molecules of interest contain a large number of relatively heavy atoms a

[1] D. W. Robinson and W. G. Von Holle, *J. Chem. Phys.* **44**, 410 (1966).

[2] It must, however, be uncharged. Ionic species attenuate microwave radiation very strongly at all frequencies.

[3] In some rare instances this requirement has been broken down. For example, G. Birnbaum, A. Maryott and P. Wacker, *J. Chem. Phys.* **22**, 1782 (1954), have observed the pressure-induced microwave spectrum of CO_2. Very recently, A. Goertz, *J. Chem. Phys.* **48**, 523 (1968), has reported the microwave spectrum of allene ($CH_2=C=CH_2$) in the first and second excited states of the doubly degenerate bending vibration, even though the molecule has no permanent dipole moment. This result has been more recently disputed, however, by W. J. Lafferty, A. Maki, and W. C. Pringle, *J. Chem. Phys.* **50**, 564 (1969).

[4] D. R. Lide, *J. Chem. Phys.* **33**, 1514 (1960).

[5] J. S. Muenter and V. W. Laurie, *J. Am. Chem. Soc.* **86**, 3901 (1964).

greater number of low-energy vibrational states are available (see chapter 8). The molecules therefore become distributed in these low-energy states, thus diminishing the intensity of rotational lines originating from the ground vibrational state. The other major difficulty mentioned above, that is, spectrum complexity, arises simply because the large number of populated rotational states often produce more transitions for heavy molecules. This causes formidable, and often intractable, problems in the assignment of quantum numbers to the observed transitions.

With these points in mind we now point out that fluoronaphthalene is the largest (i.e., greatest number of atoms) hydrocarbon that has been investigated in any sort of critical fashion. Much heavier molecules have been studied, for example, methyl mercury bromide (CH_3HgBr), but they usually contain many fewer atoms. It had seemed until very recently that studies of molecules whose sizes were on the order of naphthalene would be useful only for qualitative or possibly quantitative analytical applications, an area which has been little investigated to the present time. However, recent studies have indicated that it may be possible to obtain useful conformational data from the microwave spectra of large molecules, so the future holds some hope for the larger hydrocarbons.

The most obvious visual characteristic of a microwave spectrum is the extraordinarily high resolution which is available. Assuming that a typical absorption line has a width at half-maximum intensity of 1 MHz, it is clear as shown in Fig. 7.1 that two such identical lines spaced at 1 MHz can be clearly but incompletely resolved. This easily achieved resolution is approximately 100 000 times greater than that of infrared spectroscopy. An interesting observation is that in the typical microwave range of 10 to 40 GHz, on the order of 30 000 absorption lines *could* be distinguished if they were continuously spaced at 1 MHz! Figure 7.2 shows a wide-band sweep of a portion of a microwave spectrum taken under relatively low-resolution conditions while the spectrum of Figure 7.3 was obtained under

FIG. 7.1.   Resolution of two microwave lines separated by the linewidth.

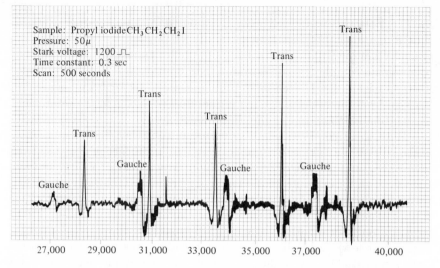

FIG. 7.2.   A portion of the low-resolution spectrum of propyl iodide.
Courtesy of Stuart Armstrong, Hewlett-Packard Co., Palo
Alto, California.

high-resolution conditions. It should be noted that the apparent linewidths
have been reduced to about 0.03 MHz in this latter spectrum, so the resolution has been significantly increased. Many of the important recent developments in microwave spectroscopy have involved the use of high-
resolution microwave techniques and the trend is likely to continue.

FIG. 7.3.   High-resolution spectrum of OCF_2 transition at 5872.4
MHz, showing splittings due to hyperfine interactions.
Markers on trace are spaced at 5 KHz (0.005 MHz). From
M. K. Lo, V. W. Weiss, W. H. Flygare, *J. Chem. Phys.*
45, 2442 (1966).

The principal features of a microwave spectrometer[6] can be briefly described with the aid of the block diagram of Figure 7.4. Until recent years the *reflex klystron* was used almost exclusively as the source of *monochromatic* microwave radiation, but the *backward-wave oscillator* has been finding increased use. Both of these sources have short-term free-running frequency stabilities of about 1 part in 10^5 or 10^6, and are continuously tuneable over a convenient frequency range. The stability can be increased to 1 part in 10^7 or 10^8 by means of electronic stabilization devices as indicated on the block diagram. The microwave radiation is transmitted conveniently by means of hollow, rectangular copper tubing, or *waveguide*, a portion of which (10 ft, say) is sealed off by means of mica windows for use as the sample cell. Detection of the microwave radiation is most commonly achieved by use of silicon crystal diodes that are specially constructed for use at microwave frequencies, although in some cases bolometers have found application.

Since typical experiments are conducted with incident radiation power levels of only a few tens of mW or less, and typical power absorptions of microwave transitions are on the order of 10^{-5} to 10^{-8} of the incident radiation, extremely sensitive detection and amplification schemes are necessary. The most successful method, and the heart of present-day

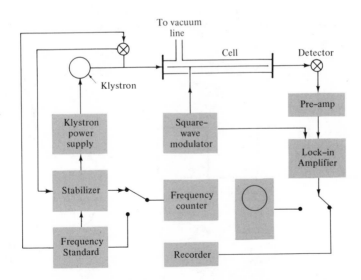

FIG. 7.4.   Schematic diagram of Stark-modulated microwave spectrometer.

[6] See, for example, C. H. Townes and A. L. Schawlow, *Microwave Spectroscopy* (McGraw-Hill, New York, 1955).

spectrometers, is the Hughes-Wilson Stark-modulation technique.[7] In this method, the sample is subjected to a square-wave electric field with an amplitude of several hundred to several thousand V/cm and a frequency of typically 1 to 100 KHz. This electric field is applied to a plane electrode placed in the center of the waveguide, running the length of cell as shown in Fig. 7.4. As the field alternately goes from zero to some value E (see Fig. 7.5) the positions of all pure rotational lines shift according to the Stark effect (see Sec. 7.6). Therefore, if the microwave source is at the zero-field frequency of an absorption line, the crystal detector senses an amplitude modulated signal at the frequency of the square-wave field. Following detection of this signal, amplification by a factor of $\sim 10^6$ is carried out by a narrow-banded pre-amp and amplifier system. The amplified signal is then null detected (also known as phase-sensitive or lock-in detection) and presented on an oscilloscope or strip-chart recorder. This latter detection process normally inverts the *nonzero* field lines (known as Stark components), so this accounts for the negative deflections in Fig. 7.5 and the spectra of Figs. 7.2 and 7.3.

Crude frequency measurements are obtained by means of cavity wavemeters, which are simply tuneable cavity resonators that absorb microwave power when the cavity length is typically some multiple of $\frac{1}{2}\lambda$, where λ is the wavelength of the radiation. Measurements of this type usually have an accuracy of about 0.05 to 0.1%. Precision frequency measurements are commonly performed in either of two ways. In the first, the frequency of the free-running microwave oscillator is compared to that

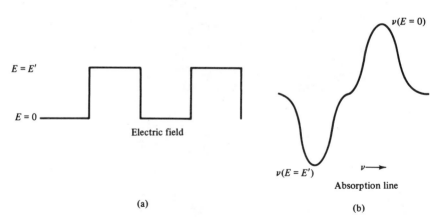

FIG. 7.5.   (a) Square-wave used for Stark-modulation and (b) zero-field and nonzero-field absorption line.

[7] R. H. Hughes and E. B. Wilson, Jr., *Phys. Rev.* **71**, 562 (1947).

of harmonics of a low-frequency standard whose stability is based upon a high-quality crystal oscillator with a typical stability of $1/10^7$. In practice, the standard frequency is mixed with the microwave frequency in a crystal diode so as to produce an intermediate frequency which can be measured with a communications receiver. After passing through the receiver, this frequency *marker* is superimposed on the absorption line of interest. In the best cases this procedure produces accuracies of about $1/10^6$.

The second method produces frequency measurements in a very direct fashion, but is considerably more sophisticated since it applies only to the stabilized microwave oscillator systems. Stabilization is obtained by electronically "locking" or synchronizing the microwave oscillator to harmonics of a low-frequency standard. By making the standard signal frequency variable, the microwave oscillator can be swept in a stabilized fashion by varying the standard frequency at a low rate. If the standard low frequency is measured with an electronic counter the klystron frequency is known at all points during the sweep. This method is capable of producing frequency precision as good as that of the crystal oscillator, which may be better than 1 part in 10^8. Of course, the accuracy to which an absorption line may be measured depends upon the width of the line, being greatest for narrow lines of the type illustrated in Fig. 7.3.

7.2   PRINCIPAL AXES

In chapter 3 it was seen that Cartesian *principal axes* produced the simplest form for the rotational kinetic energy. We can always locate these axes mathematically as shown below, but for molecules with symmetry one or more of the axes can be located by inspection. It should be emphasized first that the principal axes (PA) have their origin at the c.m. of the molecule.

For molecules with *axes* of symmetry, one of the three principal axes always lies along the axis of highest C_n symmetry. For example, H_2O, CH_3F, and SF_5Cl each have one of their principal axes along the C_2, C_3, and C_4 axes, respectively. When, in addition, the molecule has one or more planes of symmetry containing the axis of highest rotational symmetry, one of the remaining two principal axes must necessarily be perpendicular to one of these planes, while the other is in the plane. If a molecule has only a plane of symmetry, but no symmetry axes (C_s point group, say), one axis must be perpendicular to the plane, and the other two must lie in the plane. All of these results occur because of the vanishing of one or more of the products of inertia of Eq. (3.55). It should also be mentioned that when an axis of threefold or greater symmetry exists, only the principal axis along the symmetry axis is unique. The remaining two orthogonal axes

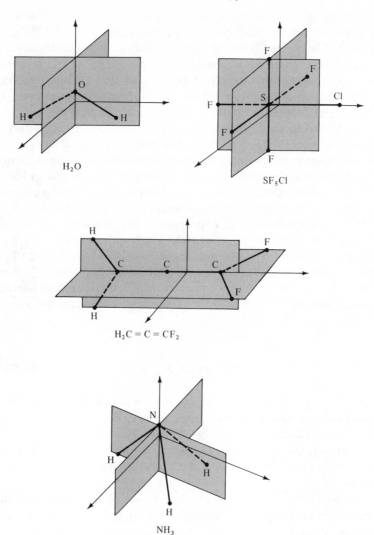

FIG. 7.6. Principal axes of some symmetrical molecules. Sketches are only qualitatively correct.

can be chosen arbitrarily. Figure 7.6 illustrates the principal axes for several molecules with symmetry.

In case there are no symmetry elements, the principal inertial axes cannot be located by inspection. The easiest general approach is to choose a Cartesian axis system arbitrarily in some convenient fashion. Then the moments and products of inertia can be calculated by using Eqs. (3.54) and (3.55), and the coordinates of the c.m. in the arbitrarily chosen axis system can be found from

$$\bar{X} = (1/M) \sum_i m_i x_i$$

$$\bar{Y} = (1/M) \sum_i m_i y_i \qquad (7.1)$$

$$\bar{Z} = (1/M) \sum_i m_i z_i$$

By means of the parallel-axis theorem of classical mechanics the moments and products of inertia may be computed in the c.m. axis system whose axes are parallel to the original ones. We obtain

$$I_{xx}{}^0 = I_{xx} - M(\bar{Y}^2 + \bar{Z}^2)$$

$$I_{yy}{}^0 = I_{yy} - M(\bar{X}^2 + \bar{Z}^2) \qquad (7.2)$$

$$I_{zz}{}^0 = I_{zz} - M(\bar{Y}^2 + \bar{X}^2)$$

and

$$I_{xy}{}^0 = I_{xy} - M\bar{X}\bar{Y}$$

$$I_{yz}{}^0 = I_{yz} - M\bar{Y}\bar{Z} \qquad (7.3)$$

$$I_{xz}{}^0 = I_{xz} - M\bar{X}\bar{Z}$$

where the superscript zero quantities refer to the c.m. axis system. The inertial tensor,

$$\mathbf{I}^0 = \begin{pmatrix} I_{xx}{}^0 & -I_{xy}{}^0 & -I_{xz}{}^0 \\ -I_{yx}{}^0 & I_{yy}{}^0 & -I_{yz}{}^0 \\ -I_{zy}{}^0 & -I_{zy}{}^0 & I_{zz}{}^0 \end{pmatrix} \qquad (7.4)$$

is not generally in diagonal form. It can always be brought to diagonal form, however, by means of a real orthogonal similarity transformation[8]

$$\mathbf{A}'\mathbf{I}^0\mathbf{A} = \mathbf{I}^{PA} \qquad (7.5)$$

where $\mathbf{I}^{PA}$ is the diagonal inertial tensor in the principal axis system. The columns of the matrix $\mathbf{A}'$ are the direction cosines of the principal axes with respect to the c.m. axes, that is,

$$\mathbf{A}' = \begin{pmatrix} \alpha_{x'x} & \alpha_{x'y} & \alpha_{x'z} \\ \alpha_{y'x} & \alpha_{y'y} & \alpha_{y'z} \\ \alpha_{z'x} & \alpha_{z'y} & \alpha_{z'z} \end{pmatrix} \qquad (7.6)$$

[8] See Example 2-1 and Appendix C for further discussions.

where $\alpha_{x'x} = \cos \theta_{x'x}$ and the primes refer to the principal axes. The first column of $\mathbf{A}'$ may be considered to be the components of the principal axis unit vector $\mathbf{i}$, plotted in the axis system parallel to the initial arbitrary one, but having its origin at the center of mass, that is,

$$\mathbf{i} = \begin{pmatrix} \alpha_{x'x} \\ \alpha_{y'x} \\ \alpha_{z'x} \end{pmatrix} \tag{7.7}$$

Similar considerations show that the second and third columns correspond to the j and k principal axis unit vectors.

The general methods of performing the transformation of Eq. (7.5) are identical to those described in chapter 2 for diagonalizing the Hamiltonian matrix. If possible $\mathbf{A}$ may be found analytically as in Example 2-1, in which case $\mathbf{I}^{PA}$ is obtained immediately by matrix multiplication. On the other hand, the determinantal equation

$$|\mathbf{I}^0 - \lambda \mathbf{1}| = 0 \tag{7.8}$$

may first be solved for the roots λ, which are the elements of $\mathbf{I}^{PA}$. The elements of the transformation matrix $\mathbf{A}$ are then obtained from equations of the form of (2.8). It should be stressed that the similarity transformation corresponds physically to a rotation of axes in three dimensions (see Example 2-1).

EXAMPLE 7-1

To illustrate some of the preceding principles, consider the nonlinear symmetrical AB_2 molecule of Fig. 7.7. We know that one of the principal

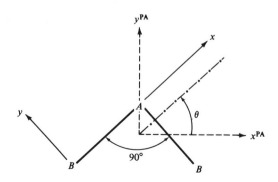

FIG. 7.7. AB_2 molecule with arbitrarily chosen axes x, y and computed principal axes x^{PA} and y^{PA}.

axes must lie along the C_2 axis while the other two lie in the two symmetry planes, as illustrated for H_2O in Fig. 7.6. Suppose, however, the axes are chosen arbitrarily as shown by the solid x, y axes in the figure. By means of Eqs. (7.1), (3.54), and (3.55), the following results are obtained:

$$I_{xx} = 2 \qquad I_{xy} = I_{yx} = -2$$

$$I_{yy} = 6 \qquad I_{xz} = I_{zx} = 0$$

$$I_{zz} = 8 \qquad I_{yz} = I_{zy} = 0$$

$$\bar{X} = \tfrac{3}{4} \qquad \bar{Y} = -\tfrac{1}{4} \qquad \bar{Z} = 0$$

Note that the zero values for four of the products of inertia result because of the plane of symmetry. The above values lead to

$$\mathbf{I^0} = \begin{pmatrix} \tfrac{3}{2} & \tfrac{1}{2} & 0 \\ \tfrac{1}{2} & \tfrac{3}{2} & 0 \\ 0 & 0 & 3 \end{pmatrix} \tag{7.9}$$

by use of Eqs. (7.2) and (7.3). The inertial tensor (7.9) may now be diagonalized by a transformation matrix similar to (2.13),

$$\mathbf{A} = \begin{pmatrix} \cos\theta & -\sin\theta & 0 \\ \sin\theta & \cos\theta & 0 \\ 0 & 0 & 1 \end{pmatrix} \tag{7.10}$$

with θ given by

$$\tan 2\theta = \frac{-2I_{xy}{}^0}{I_{xx}{}^0 - I_{yy}{}^0} \tag{7.11}$$

Substitution of values from (7.9) into (7.11) leads to $\theta = 45°$, and diagonalization of Eq. (7.9) by use of (7.10) and its transpose[9] leads to

$$\mathbf{I^{PA}} = \begin{pmatrix} 2 & 0 & 0 \\ 0 & 1 & 0 \\ 0 & 0 & 3 \end{pmatrix} \tag{7.12}$$

The columns of $\mathbf{A}^{-1}$ give the orientations of the principal axes as measured in the c.m. axis system that is parallel to the original x, y, z system. Since the transformation was effectively only two-dimensional (because z was

[9] Recall that $\mathbf{A}^{-1} = \mathbf{A}'$ for real orthogonal matrices.

chosen parallel to a principal axis) the angle $\theta = 45°$ gives the proper location of the principal axes x^{PA} and y^{PA}. The result we have obtained is as expected, that is, the axes are oriented as our earlier symmetry arguments predicted. It is verified easily that the same $\mathbf{I}^{PA}$ results if the moments of inertia are computed directly in the principal axis system.

7.3 MICROWAVE SPECTRUM OF THE LINEAR MOLECULE

A. Energy Levels

In the rigid-rotor approximation described in Sec. 3.3 B the classical Hamiltonian for the linear molecule was given by Eq. (3.61). The quantum-mechanical Hamiltonian is obtained by substitution of the operator for P^2 which gives

$$\widetilde{\mathfrak{IC}}_r = \frac{\tilde{P}^2}{2I} \tag{7.13}$$

where $\tilde{P}^2$ is the square of the total angular momentum and I is the principal moment of inertia of the molecule about the x or y axis. Since the problem involves only the operator of the total angular momentum the matrix elements of Eq. (7.13) may be obtained from Table 4.1. In the $\langle JKM \mid$ representation the desired matrix elements are

$$\langle JKM \mid \widetilde{\mathfrak{IC}}_r \mid J'K'M' \rangle = \frac{\langle JKM \mid \tilde{P}^2 \mid J'K'M' \rangle}{2I} \tag{7.14}$$

Table 4.1 shows that the only nonzero elements of $\tilde{P}^2$ are the diagonal ones so the Hamiltonian matrix elements are the stationary-state energy values. We find

$$E_r = \langle JKM \mid \widetilde{\mathfrak{IC}}_r \mid JKM \rangle = \frac{\hbar^2 J(J+1)}{2I} \tag{7.15}$$

with the rotational angular momentum quantum number J taking positive integral values or zero. Spectroscopists usually rewrite Eq. (7.15) in the form

$$E_r = hBJ(J+1) \tag{7.16}$$

with the *rotational constant* B defined by

$$B = \frac{h}{8\pi^2 I} \tag{7.17}$$

Figure 7.8 shows the extremely simple structure of the rigid-rotor levels predicted by Eq. (7.17).

Each of the J states of Eq. (7.16) has a $2J+1$ degeneracy in M, the value of the projection of $\tilde{P}$ on the space-fixed or Z axis. This "space-degeneracy" as it is called, always exists in the absence of external magnetic or electric fields, but may be more or less broken when external fields are applied. This problem is examined later in this chapter in the section on the Stark effect.

It might be imagined from our treatment that there exists also a $2J + 1$ degeneracy in K which is the projection of $\tilde{P}$ on the molecule-fixed axis. This is not true since, as observed in the derivation of (3.61), P_z is zero so $\langle JKM \mid \tilde{P}_z \mid JKM \rangle = K = 0$ for any value of J. This result is inferred in another way later in the chapter by recognizing that the linear molecule is a special case of the symmetric rotor.

It should be recalled that the internuclear distances are not constant, but rather vary as the vibrational excitation of the molecule changes. We expect then, that each vibrational state v has a slightly different rotational constant

$$B_v = \frac{h}{8\pi^2 I_v} \tag{7.18}$$

In view of the discussion of Sec. 3.4 it is correct to consider B_v to be the value of B averaged over the electronic and vibrational states of interest,[10]

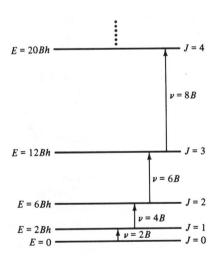

FIG. 7.8. Rotational energy levels and transitions of a rigid linear molecule.

[10] We carry no subscript to label the electronic state of interest. In order to avoid complications it is implicitly assumed in this chapter that all molecules are in $^1\Sigma$ electronic states (see chapter 11 also).

that is

$$B_v = \frac{h}{8\pi^2} \langle ev \mid I^{-1} \mid ev \rangle = \frac{h}{8\pi^2} \langle I^{-1} \rangle_v \qquad (7.19)$$

For the diatomic molecule Eq. (7.19) becomes

$$B_v = \frac{h}{8\pi^2 \mu} \langle ev \mid R^{-2} \mid ev \rangle = \frac{h}{8\pi^2 \mu} \langle R^{-2} \rangle_v \qquad (7.20)$$

by use of (3.65) and (3.66). The quantity which is defined as the internuclear distance or "bond length" in the vibrational state v is

$$R_v = \langle R^{-2} \rangle_v^{-1/2} \qquad (7.21)$$

for the diatomic molecule.

Since the moment of inertia of the general linear (or any polyatomic) molecule is a function of two or more internuclear parameters, the vibrational averaging cannot be undone as easily as for the diatomic molecule. For this reason Eq. (7.18) is used to define the *effective* moment of inertia I_v which is related to the effective z coordinates by

$$I_v = \sum_i m_i z_{vi}^2 \qquad (7.22a)$$

For the diatomic molecule, I_v may be expressed more simply as in Eq. (3.65),

$$I_v = \mu R_v^2 \qquad (7.22b)$$

where μ is the reduced mass. Note that according to Eq. (7.19) the effective moment of inertia should more properly be written

$$\frac{1}{I_v} = \langle I^{-1} \rangle_v = \left\langle \frac{1}{\sum_i m_i z_i^2} \right\rangle \qquad (7.23)$$

The effective coordinates of (7.22a) are not related in any simple way to averages of the z_i of (7.23), but as the vibrational motions go to zero the two expressions become identical since the z_i then become constants. Although such a vibrationless state is physically nonexistent, it is useful to define the behavior of this vibrationless *equilibrium* state by

$$B_e = \frac{h}{8\pi^2 I_e} = \frac{h}{8\pi^2 \sum_i m_i z_{ei}^2} \qquad (7.24)$$

This equilibrium state is that which would be appropriate for a molecule if it could exist at the minimum of the vibrational potential energy curve. For the diatomic molecule, it can be shown rather easily that

$$B_v = B_e - \alpha_e(v + \tfrac{1}{2}) + \cdots \qquad (7.25)$$

where v is the vibrational quantum number discussed in chapter 4 and α_e is a small positive constant that is characteristic of the molecule. For a polyatomic linear molecule having $3N - 5$ vibrational degrees of freedom, expression (7.25) becomes

$$B_v = B_e - \sum_i^{3N-5} \alpha_{ei}(v_i + \tfrac{1}{2}) \qquad (7.26)$$

The constants α_{ei} may be positive or negative in general, and are on the order of 1% or less of B. It should be pointed out that although a degenerate vibrational state has only a single, distinct α value, this value must be included in the sum d times if d is the degeneracy.

The validity of Eq. (7.25) can be seen from the data for CsCl in Table 7.1. Experimentally determined B_v values were used to evaluate α_e and B_e, and the best values of these equilibrium quantities were used to calculate the B_v for comparison purposes. It is seen that the two sets of numbers agree to within 0.1 MHz through the first seven vibrational states. Also shown in Table 7.1 are the r_v values for CsCl. From these it is clear that r_0 is a reasonably good approximation to the equilibrium bond distance r_e.

Unfortunately the determination of B_e and α_e values is very difficult for polyatomic molecules, even linear triatomic molecules, since measurements must be made on at least one excited vibrational state for each of the $3N - 5$ (or $3N - 6$ for nonlinear molecules) vibrational modes. Also, determination of r_e values requires the study of a number of isotopic mole-

TABLE 7.1

Comparison of measured B_v values of $Cs^{133}Cl^{35}$ with those calculated from Eq. (7.25) using $B_e = 2161.21$ MHz and $\alpha_e = 10.08$ MHz.[a]

v	B_v (measured)	B_v (calculated)	$r_v(\text{Å})$
0	2156.09	2156.17	2.910
1	2146.01	2146.09	2.916
2	2135.96	2136.05	2.923
3	2125.94	2125.93	2.930
4	2115.86	2115.85	2.937
5	2105.84	2105.77	2.944
6	2095.8	2095.7	2.951
7	2085.9	2085.6	2.958
8	2075.9	2075.5	2.965

[a] Data from A. Honig, M. L. Stitch, and M. Mandel, *Phys. Rev.* **92**, 901 (1953).

cules equal to the number of r_e parameters. For example, evaluation of $r_e(\text{NN})$ and $r_e(\text{NO})$ in N_2O could be performed by evaluating B_e for each of the molecules $N^{14}N^{14}O^{16}$ and $N^{15}N^{14}O^{16}$, and assuming that the internuclear distances are invariant to isotopic substitution (see Sec. 7.9). Table 7.2 lists equilibrium rotational parameters for several diatomic and triatomic molecules that have been carefully investigated.

There remains one other important contribution to the rotational energy that is caused by nonrigid-rotor behavior. This results from the stretching of the molecule as it rotates in space. For the diatomic molecule, the amount of stretching is clearly dependent upon the strength of the chemical bond, or more precisely, upon the force constant defined by Eqs. (3.88). While a correct treatment of this phenomenon requires a return to the complete Hamiltonian for a vibrating rotor, Eq. (3.44), an approximate result of the proper form may be obtained as follows.

TABLE 7.2
Parameters of diatomic and triatomic molecules.

Molecule	α_e (MHz)	B_e (MHz)	D_e (MHz)	r_e (Å)	Reference
$Na^{23}Cl^{35}$	42.28	6 537.07	0.0086	2.3606	a
$Br^{79}F^{19}$	156.3	10 706.9	...	1.759	a
$C^{12}S^{32}$	177.54	24 584.35	...	1.5349	a
$C^{12}O^{16}$	525.24	57 898.57	0.184	1.1282	a
$N^{14}N^{14}O^{16}$	$\alpha_1 = 51.6$ $\alpha_2 = -12.6$ $\alpha_3 = 104.0$	12 626.8	0.0057	$r_{NN} = 1.126$	b
$N^{15}N^{14}O^{16}$	$\alpha_1 = 46$ $\alpha_2 = -11$ $\alpha_3 = 101$	12 189		$r_{NO} = 1.186$	b
$HC^{12}N^{14}$	$\alpha_1 = 309$ $\alpha_2 = -111$ $\alpha_3 = 297$	44 508		$r_{CH} = 1.063$	b
$DC^{12}N^{14}$	$\alpha_1 = 198$ $\alpha_2 = -124$ $\alpha_3 = 312$	36 339		$r_{CN} = 1.154$	b

a *Microwave Spectral Tables*, NBS Monograph 70, Vol. 1 (U.S. Government Printing Office, Washington, D.C., 1964).
b See C. C. Costain, *J. Chem. Phys.* **29,** 864 (1958).

We observe first that the average separation of the atoms of, for example, a diatomic molecule is determined by the balance between the centrifugal force

$$\left(f_c = \mu r \omega^2 = \frac{P^2}{Ir} \right)$$

and the centripetal force produced by the potential field in which the nuclei move. If the molecule elongates by an amount $\Delta r = r - r_e$, this balance is expressed by

$$k \Delta r = \frac{P^2}{Ir} \tag{7.27}$$

where k is the force constant of Eq. (3.88), and we imagine that the molecule rotates about only one axis so that P^2 is the square of the total angular momentum. In Eq. (7.27) and all following ones we neglect differences between r and r_e and I and I_e except in the term Δr. The permitted elongation of the molecule is found from (7.27) to be

$$\Delta r = \frac{P^2}{kIr} \tag{7.28}$$

This elongation causes an increase in the moment of inertia which is calculated to be

$$I_{r+\Delta r} = \mu (r + \Delta r)^2$$
$$\sim \mu r^2 + 2\mu r \Delta r \tag{7.29}$$

Utilizing Eq. (7.28) and $I = \mu r^2$, we obtain for the moment of inertia of the elongated molecule

$$I_{r+\Delta r} = I + \frac{2\mu P^2}{kI} + \cdots$$
$$= I \left(1 + \frac{2\mu P^2}{kI^2} \right) \tag{7.30}$$

or

$$\frac{1}{I_{r+\Delta r}} = \frac{1}{I} \left(1 - \frac{2\mu P^2}{kI^2} + \cdots \right) \tag{7.31}$$

The change in the reciprocal moment of inertia is therefore

$$\Delta \left(\frac{1}{I} \right) = - \frac{2\mu P^2}{kI^2} \tag{7.32}$$

and the change in the rotational energy is

$$\Delta \mathcal{H} = \frac{P^2}{2} \Delta \left(\frac{1}{I} \right)$$

$$= - \frac{\mu P^4}{kI^3} \tag{7.33}$$

This shows clearly that the effect of centrifugal distortion is to decrease the rotational energy. The diagonal matrix element of $\tilde{P}^4$ from Table 4.1 is $\hbar^2 J^2 (J + 1)^2$ which produces finally the centrifugal distortion contribution to the energy,

$$E_{\text{c.d.}} \sim - \frac{\mu}{kI^3} J^2 (J + 1)^2 \hbar^2 = -hDJ^2 (J + 1)^2 \tag{7.34}$$

While our treatment is only approximate, the functional dependence of the centrifugal distortion constant D upon J, k, μ, and I is precisely correct. Hence it is seen that centrifugal distortion effects are greatest for molecules with small moments of inertia and small force constants, and that they increase rapidly for the higher rotational states. Some typical centrifugal distortion constants are listed in Table 7.2.

The final expression for the rotational energy of a linear molecule may now be written by combining Eqs. (7.16), (7.26), and (7.34).

$$E_r = hB_v J (J + 1) - hDJ^2 (J + 1)^2 \tag{7.35}$$

More exact and sophisticated treatments have been performed for diatomic molecules,[11] but they differ from our result only by the inclusion of very small, higher-order effects. For linear polyatomic molecules there are other important energy terms which couple rotation with degenerate bending modes via a Coriolis-type interaction. This effect is of less general interest and will not be considered here. The interested reader is urged to consult the work of Townes and Schawlow.[12]

B.   Selection Rules and Intensities

In chapter 5 the general principles of the absorption and emission of electromagnetic radiation were presented. It was shown that the largest contribution to the transition moment matrix element would usually be that arising from the electric dipole moment, except in cases where this term might go identically to zero, in which cases the higher-order magnetic dipole or electric quadrupole terms would be the most important. For the

---

[11] See, for example, footnote 6, chapters 1 and 2.
[12] See footnote 6.

present case of rotational spectra, nonvanishing electric dipole matrix elements do exist, so we investigate their properties.

For unpolarized radiation the probability of an electric dipole transition from rotational state n to state m is given by

$$P \propto \sum_{F=X,Y,Z} \langle n \mid \tilde{\mu}_F \mid m \rangle^2 \qquad (7.36)$$

where $\tilde{\mu}_F$ is the electric dipole moment operator expressed in the laboratory or space-fixed axis system. For radiation polarized in the X, Y, or, Z directions, the transition probabilities would be proportional to $\mid \langle n \mid \tilde{\mu}_X \mid m \rangle \mid^2$, $\mid \langle n \mid \tilde{\mu}_Y \mid m \rangle \mid^2$ or $\mid \langle n \mid \tilde{\mu}_Z \mid m \rangle \mid^2$, respectively. Microwave radiation is normally polarized, but the electric vector is usually so small as to produce no appreciable orientations of the molecules in space. Only when a strong electric field is superimposed (a large dc field, say) is it possible to distinguish the individual polarizations.

We may express the dipole moment vector for a system of point charges in a molecule by

$$\mathbf{\mu}^{(F)} = \sum_i Q_i e \mathbf{R}_i \qquad (7.37)$$

where $Q_i e$ is the algebraic charge of each particle and $\mathbf{R}_i$ its position vector in the space-fixed axis system. An expression similar to (7.37) can be written in the molecule-fixed axis system,

$$\mathbf{\mu}^{(g)} = \sum_i Q_i e \mathbf{r}_i \qquad (7.38)$$

with $\mathbf{r}_i$ being the position vector in the molecule-fixed axis system. The two expressions for $\mathbf{\mu}$ may, of course, be related by means of the direction cosines between the two axis systems as described in chapter 4. For example, the component of $\mathbf{\mu}^{(F)}$ along the F axis may be written in terms of the components along the g axes

$$\mu_F = \sum_g \Phi_{Fg} \mu_g \qquad (7.39)$$

where $F = X$, Y or, Z; $g = x$, y, and z; and Φ_{Fg} is the cosine of the angle between the space-fixed F axis and the molecule-fixed g axis.

Neither of the above expressions for the dipole moment is particularly useful for evaluation of the matrix elements of (7.36) since $\mathbf{\mu}^{(F)}$ (or $\mathbf{\mu}^{(g)}$) varies in a generally complex fashion as the molecule vibrates. Furthermore, since we are interested here only in nuclear motions in the Born-Oppenheimer approximation, the summations over *particles i* must be considered as summations over *atoms i* with *effective charge* $Q_i e$. Then as the position $\mathbf{r}_i$ of the atoms change during a vibration, the effective charges also change due to the fact that the electrons do not follow precisely the nuclear motions. The net result is that we do not know the analytical form which (7.37) or (7.38) takes although the general behavior for a diatomic molecule must certainly be something like that shown in Fig. 7.9. This figure indicates that the dipole moment must go to zero at $r = 0$ where the two atoms are superimposed, and must go to zero at $r = \infty$ when the molecule becomes two separated, neutral atoms. Of course, the behavior in the intermediate region is uncertain, and may even involve a change in sign, that is, a cross over at $\mu = 0$ may occur at some point. Only for a few diatomic molecules has any extensive investigation of the dipole-moment function been made, and this over a limited range.[13]

Because of the previously mentioned uncertainties, it is most convenient to expand the components of the dipole moment in a power series about the equilibrium positions. For the z component this would give

$$\mu_z = \mu_z^0 + \sum_{k=1}^{3N-6} \frac{\partial \mu_z}{\partial q_k} q_k + \cdots \tag{7.40}$$

with the expansion being carried out in terms of the mass-weighted displacement coordinates, although it is more convenient later to use another set of coordinates. μ_z^0 is the equilibrium dipole moment $(\sum_i Q_i e z_{ei})$ in the z direction, which for the linear molecule would be the well-known permanent electric dipole moment.[14]

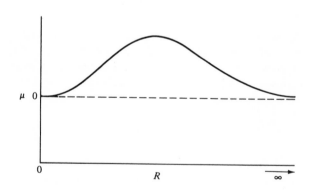

FIG. 7.9. Dipole moment of diatomic molecule as function of internuclear distance.

[13] See for example: J. Goodisman, *J. Chem. Phys.* **38,** 2597 (1963); M. Kaufman, L. Wharton, and W. Klemperer, *ibid.* **43,** 943 (1965); T. C. James, W. G. Norris, and W. Klemperer, *ibid.* **32,** 728 (1950).

[14] This is not identical to the dipole moment measured by dielectric constant techniques, however, as is made clear later in this chapter.

If we introduce expressions of the form of (7.39) into Eq. (7.36) for the transition probability, we obtain

$$P \propto \left| \sum_g \langle n \mid \Phi_{X_g} \tilde{\mu}_g \mid m \rangle \right|^2 + \left| \sum_g \langle n \mid \Phi_{Y_g} \tilde{\mu}_g \mid m \rangle \right|^2 + \left| \sum_g \langle n \mid \Phi_{Z_g} \tilde{\mu}_g \mid m \rangle \right|^2$$

$$(7.41)$$

Equation (7.40) may now be substituted with the further understanding that μ^0_x and μ^0_y must be zero for a linear molecule (choosing the z axis as the molecular axis), and that the matrix elements in (7.41) are independent of the second and higher terms in (7.40) since these terms do not involve any angular variables. Equation (7.41) becomes therefore,

$$P \propto \mu^2 \{ \left| \langle JM \mid \Phi_{Xz} \mid J'M' \rangle \right|^2 + \left| \langle JM \mid \Phi_{Yz} \mid J'M' \rangle \right|^2$$

$$+ \left| \langle JM \mid \Phi_{Zz} \mid J'M' \rangle \right|^2 \quad (7.42)$$

in which the definition $\mu_z{}^0 = \mu$ has been used and the states n and m have been rewritten as JM and $J'M'$, respectively. The determination of transition probabilities, or *selection rules*, is seen to resolve itself into the evaluation of the direction cosine matrix elements. These matrix elements have been described and tabulated in chapter 4. From Table 4.2, recalling that $K = 0$ for the linear molecule, it is seen immediately that the elements are all zero unless

$$\Delta J = J' - J = \pm 1$$

and

$$\Delta M = M' - M = 0, \pm 1$$

$$(7.43)$$

which represent the *selection rules* for the linear molecule. More specifically, the matrix elements are, for $J + 1 \leftarrow J$, (which is the case of interest in absorption spectroscopy)

$$\left| \langle JM \mid \Phi_{Zz} \mid J + 1M \rangle \right|^2 = \frac{(J + 1)^2 - M^2}{(2J + 1)(2J + 3)}$$

$$\left| \langle JM \mid \Phi_{Xz} \mid J + 1M + 1 \rangle \right|^2 = \left| \langle JM \mid \Phi_{Yz} \mid J + 1M + 1 \rangle \right|^2$$

$$= \frac{(J + M + 2)(J + M + 1)}{4(2J + 1)(2J + 3)} \quad (7.44)$$

$$\left| \langle JM \mid \Phi_{Xz} \mid J + 1M - 1 \rangle \right|^2 = \left| \langle JM \mid \Phi_{Yz} \mid J + 1M - 1 \rangle \right|^2$$

$$= \frac{(J - M + 1)(J - M + 2)}{4(2J + 1)(2J + 3)}$$

Finally, by summing P in Eq. (7.42) over all permitted M' values, the total transition intensity for $J + 1 \leftarrow J$ becomes

$$P \propto \mu^2 \frac{(J + 1)}{(2J + 1)} \tag{7.45}$$

It should be emphasized that this expression gives the sum of the probabilities of the three transitions $JM \rightarrow J + 1, M$; $JM \rightarrow J + 1, M + 1$; and $JM \rightarrow J + 1, M - 1$ in the presence of unpolarized radiation, but it applies to the common experimental case in which polarized radiation is used and the M degeneracy is unbroken. Note especially that electric-dipole pure-rotational transitions are not permitted for molecules with vanishing dipole moments. This means that only $C_{\infty v}$ linear molecules have microwave pure-rotational spectra.

Returning briefly to the energy levels for the linear molecule given in (7.35), the transition frequency for $J + 1 \leftarrow J$ may be found to be

$$\nu = \frac{E_r(J + 1) - E_r(J)}{h} = 2B_v(J + 1) - 4D(J + 1)^3 \tag{7.46}$$

Aside from the small centrifugal distortion contribution, it is seen that the transition frequencies are equally spaced in any vibrational state, occurring at $2B_v$, $4B_v$, $6B_v$. ... These transitions are indicated in Fig. 7.8 by the vertical arrows. In Table 7.3 we have tabulated the observed microwave transitions for a typical diatomic molecule.

A complete expression giving the fraction of microwave power absorbed at any frequency includes not only the transition probability [Eq. (7.45)] but also terms involving the number of molecules in the energy state of interest. It will depend further upon the transition linewidth, or more precisely upon the broadening mechanism. The Van Vleck-Weisskopf equation,[15] which has been discussed in Sec. 5.5 is the accepted expression for microwave absorption line shapes at low to medium pressures. This *pressure broadening* theory leads to the fractional absorption coefficient per cm pathlength,

$$\gamma = \frac{8\pi^2 Nf}{3ckT} |\mu_{nm}|^2 \nu^2 \frac{\Delta\nu}{(\nu - \nu_0)^2 + (\Delta\nu)^2} \tag{7.47}$$

where N equals the total number of molecules per cm^3, f equals fraction in the state $\psi_{e,v,r} \equiv \psi_n$ of interest, $|\mu_{nm}|^2$ equals the square of the dipole moment matrix element [which for the linear molecule is given by the right-hand side of Eq. (7.45)], $\Delta\nu$ equals the half-width of the line at half-maximum intensity, ν_0 equals the center of the absorption line [given by (7.46) for linear molecules] and ν is the frequency of the applied electromagnetic radiation. The peak intensity of this *Lorentzian* line (See Fig.

[15] J. H. Van Vleck and V. F. Weisskopf, *Rev. Mod. Phys.* **17**, 227 (1945).

TABLE 7.3

Experimental microwave spectrum of cesium bromide.[a]

Isotope	$J \rightarrow J'$	v	Frequency[b] (MHz)
$Cs^{133}Br^{79}$	$9 \rightarrow 10$	0	$21\ 588.57 \pm 0.10$
		1	$21\ 514.48$
		2	$21\ 440.65$
		3	$21\ 366.36$
		4	$21\ 292.40$
		5	$21\ 218.66$
	$10 \rightarrow 11$	0	$23\ 747.17 \pm 0.10$
		1	$23\ 665.60$
		2	$23\ 583.87$
		3	$23\ 502.95$
		5	$23\ 340.26$
		6	$23\ 259.19$
		7	$23\ 178.25$
		8	$23\ 097.97$
	$11 \rightarrow 12$	1	$25\ 816.53$
		2	$25\ 648.95$
		3	$25\ 550.22$
$Cs^{133}Br^{81}$	$9 \rightarrow 10$	0	$21\ 254.44 \pm 0.10$
	$10 \rightarrow 11$	0	$23\ 379.53 \pm 0.10$
		1	$23\ 299.79$
		2	$23\ 220.22$
		3	$23\ 140.61$
		4	$23\ 061.38$
	$11 \rightarrow 12$	0	$25\ 504.69 \pm 0.10$

[a] A. Honig, M. Mandel, M. L. Stitch, C. H. Townes, *Phys. Rev.* **93,** 953 (1954).
[b] All uncertainties are ± 0.20 unless otherwise noted.

5.3) is obtained by setting $\nu = \nu_0$. As in Eq. (5.75) we obtain

$$\gamma_{max} = \frac{8\pi^2 Nf}{3ckT} | \mu_{nm} |^2 \nu_0^2 \qquad (7.48)$$

According to the theory, $\Delta\nu$ is equal to $1/2\pi\tau$, where τ is the time between collisions. Since τ is proportional to pressure according to kinetic theory, we might expect $N/\Delta\nu$ to be independent of pressure, a conclusion which is experimentally verified at moderate pressures.

The fractional population f, is easily obtained by using the results of equilibrium statistical thermodynamics.[16] First, note that nearly all

[16] See, for example, Malcolm Dole, *Introduction to Statistical Thermodynamics* (Prentice-Hall, New York, 1954).

molecules are in the ground electronic state, so we may divide f into two parts, one for vibrational population and the other for rotational population.

$$f = f_v \cdot f_r \qquad (7.49)$$

The fraction of all molecules in a vibrational state having only one mode v_i excited by v_i quanta is

$$f_v = \frac{g_i \exp(-h\nu_i v_i / kT)}{Z_{\mathrm{vib}}} \qquad (7.50)$$

where g_i is the degeneracy of the v_ith state, ν_i is the vibrational frequency for the mode of interest, and the vibrational partition function Z_{vib} is given by

$$Z_{\mathrm{vib}} = \prod_i^{3N-5} [1 - \exp(-h\nu_i/kT)]^{-g_i} \qquad (7.51)$$

Note that if all ν_i are greater than 600 cm^{-1}, Z_{vib} is within a few percent of unity for small molecules near room temperature.[17] For the ground vibrational state (all $v = 0$), f_v is close to unity for this case.

The rotational distribution is obtained similarly. For a linear molecule in the state J,

$$f_r = \frac{(2J + 1) \exp(-hBJ(J + 1)/kT)}{Z_{\mathrm{rot}}} \qquad (7.52)$$

where

$$Z_{\mathrm{rot}} = \sum_{J=0}^{\infty} (2J + 1) \exp(-hBJ(J + 1)/kT) \qquad (7.53)$$

and the $2J + 1$ represents the M degeneracy. When $hB \ll kT$, the summation may be replaced by an integral over dJ, which yields

$$Z_{\mathrm{rot}} = \frac{kT}{hB} \qquad (7.54)$$

The approximation is quite useful since B is often on the order of 1 cm^{-1} or less, so that $hB \sim (1/200) kT$ at room temperature. For the same reason the numerator in (7.52) becomes approximately $2J + 1$ for low values of J, so (7.52) becomes

$$f_r = \frac{(2J + 1)hB}{kT} \qquad (7.55)$$

Substituting (7.55), (7.45), (7.49), and $\nu_0 = 2B(J + 1)$ into Eq. (7.48),

[17] $k = 0.695$ cm^{-1}/deg, so $kT \sim 200$ cm^{-1} at room temperature.

we obtain

$$\gamma_{max} = \frac{4\pi^2 h N \mu^2 \nu_0^3 f_v}{3ck^2T^2\Delta\nu} \tag{7.56a}$$

In Table 7.4 we have listed several of the observed transitions of OCS and their intensities computed from Eq. (7.56a). For these computations we use: $\mu = 0.71$ Debye, $\Delta\nu = 6$ MHz at a pressure of 1 Torr, $T = 300°$K. For OCS the vibrational frequencies are $\nu_1 = 860$ cm^{-1}, $\nu_2 = 530$ cm^{-1}, and $\nu_3 = 2000$ cm^{-1}. The ν_2 mode is doubly degenerate ($g_2 = 2$) but a Coriolis-type interaction splits the transition into a doublet so this degeneracy may be neglected in the intensity calculation. Finally, a natural isotopic mixture of OCS contains approximately 95% $O^{16}C^{12}S^{32}$, 4% $O^{16}C^{12}S^{34}$, 0.95% $O^{16}C^{13}S^{32}$, and 0.4% $O^{16}C^{13}S^{34}$, so these factors are included in the intensities of Table 7.4. It should be remembered in using Eq. (7.56a) that cgs units are required. If the physical constants are substituted and appropriate conversion factors are used, Eq. (7.56a) becomes

$$\gamma_{max} = 5.5 \times 10^{-17} \frac{\mu^2\nu_0^3}{\Delta\nu} \cdot f_v \tag{7.56b}$$

at $T = 300°$K. Also μ is in debye, ν_0 in MHz, and $\Delta\nu$ has units of MHz/Torr, that is, $\Delta\nu$ is the half-width in MHz at a pressure of 1 Torr.

TABLE 7.4
Some observed transition frequencies and computed intensities of OCS ($T = 300°$K).

Molecule	Vibrational state $v_1v_2v_3$	Transition $J \to J+1$		Frequency (MHz)	Intensity (cm^{-1})
$O^{16}C^{12}S^{32}$	0 0 0	1	2	24 326	5.2×10^{-5}
	0 0 0	2	3	36 489	1.8×10^{-4}
	0 0 0	3	4	48 652	4.2×10^{-4}
	1 0 0	1	2	24 254	8.5×10^{-7}
	2 0 0	1	2	24 180	1.4×10^{-8}
	0 1 0^a	1	2	24 356	4.2×10^{-6}
	0 1 0^a	1	2	24 381	4.2×10^{-6}
$O^{16}C^{12}S^{34}$	0 0 0	1	2	23 731	2.0×10^{-6}
	0 0 0	3	4	47 462	1.6×10^{-5}
	1 0 0	1	2	23 661	3.3×10^{-8}
$O^{16}C^{13}S^{32}$	0 0 0	1	2	24 248	5.1×10^{-7}
	1 0 0	1	2	24 176	8.3×10^{-9}
$O^{16}C^{13}S^{34}$	0 0 0	1	2	23 647	2.0×10^{-8}

^{a} Doublet for $v_2 = 1$ state is caused by vibration-rotation interaction and is known as l-type doubling. See Ref. 6.

Note particularly the strong dependence of $\gamma_{\max}$ upon ν_0 and T. For this reason it is desirable to observe microwave spectra with samples at reduced temperatures. It is also obvious that high-frequency operation is desirable for molecules having small dipole moments. It should be pointed out that typical microwave spectrometers can detect lines having γ on the order 10^{-8} cm^{-1} or greater, so the size of μ often becomes the real limiting factor in the observation of microwave spectra.

As previously mentioned, most microwave spectra exhibit rotational transitions in one or more excited vibrational states. Since μ, ν_0, and $\Delta\nu$ change only slightly in different vibrational states, the ratio of intensities of excited and ground state lines (for given J) is given by

$$\frac{\gamma_{\text{ex}}}{\gamma_0} = \frac{(f_v)_{\text{ex}}}{(f_v)_0} = g_{\text{ex}} \exp(-\Delta E/kT) \qquad (7.57)$$

where ΔE is the energy difference between the two states and g_{ex} is the degeneracy of the excited state. Since for a given mode of vibration the vibrational states are nearly harmonically spaced, the intensities fall off exponentially as $g_{\text{ex}} \exp[-(\Delta E)_0/kT]$, where $(\Delta E)_0$ is the separation between the lowest ($v = 0$) and first ($v = 1$) excited vibrational state. As shown in chapter 8, the energy separation of the lowest states is approximately equal to the fundamental vibrational frequency ω, so the intensities vary as $g_{\text{ex}} \exp(-\omega v/kT)$. This is shown schematically in Fig. 7.10 for a diatomic molecule having ω equal to 200 cm^{-1} and $g_{\text{ex}} = 1$. The equal spacings of the "satellite" lines is a direct consequence of Eqs. (7.46) and (7.25).

For a linear triatomic molecule the situation becomes more complex since now there are three distinct vibrational frequencies (or modes), one of which is doubly degenerate ($g_{\text{ex}} = 2$), so the spectrum of a given rotational transition consists of three sets of vibrational satellite lines. Nevertheless, the ratios of intensities are given by Eq. (7.57) as before.

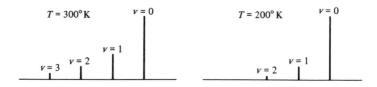

FIG. 7.10.   Satellite pattern for a $J{\to}J{+}1$ transition of a diatomic molecule at two different temperatures.

TABLE 7.5

Rotational transitions of fluoroacetylene in ground vibrational state.[a]

Molecule	Transition	Frequency (MHz)
$F^{19}C^{12}C^{12}H^1$	$J = 1 \leftarrow J = 0$	19 412.37
	$J = 2 \leftarrow J = 1$	38 824.64
	$J = 4 \leftarrow J = 3$	77 648.58
$F^{19}C^{12}C^{12}H^2$	$J = 1 \leftarrow J = 0$	17 472.08
	$J = 2 \leftarrow J = 1$	34 944.07
$F^{19}C^{13}C^{12}H^1$	$J = 1 \leftarrow J = 0$	19 401.12
	$J = 2 \leftarrow J = 1$	38 802.53
$F^{19}C^{12}C^{13}H^1$	$J = 1 \leftarrow J = 0$	18 747.94
	$J = 2 \leftarrow J = 1$	37 495.67

[a]   See Ref. 18 in text.

Before closing our discussion of linear molecules we analyze the spectrum of the fluoroacetylene molecule as a practical example. Later in this chapter we shall return to a consideration of some further interesting details concerning linear molecules, and look further into the problem of determining molecular structures from experimental rotational constants.

EXAMPLE 7-2

In Table 7.5 some of the observed microwave transitions of $FC{\equiv}CH$ have been listed.[18] Because of low intensities, it was possible to observe vibrational satellites for only two of the vibrational modes (five distinct modes exist, two being doubly degenerate) so no satellite frequencies have been listed. Note the clear effect of isotopic substitution for the four different isotopic species.

Considering the data for the normal species ($FC^{12}C^{12}H$), we can calculate B_0 and D in the following manner. From Eq. (7.46)

$$\nu_{0 \to 1} = 2B_0(1) - 4D(1) = 19\ 412.37 \text{ MHz}$$

$$\nu_{1 \to 2} = 2B_0(2) - 4D(8) = 38\ 824.64 \text{ MHz}$$

(7.58)

Solving Eqs. (7.58), we obtain

$$B_0 = 9706.20 \text{ MHz}$$

$$D = 0.0041 \text{ MHz}$$

[18] J. K. Tyler and J. Sheridan, *Trans. Faraday Soc.* **59,** 2661 (1963).

TABLE 7.6

Rotational constants obtained by analysis of data in Table 7.5.

Molecule	B_0 (MHz)	D (MHz)	I (amu A²)
$F^{19}C^{12}C^{12}H^1$	9706.20	0.0041	52.08331
$F^{19}C^{12}C^{12}H^2$	8736.04	0.0038	57.86729
$F^{19}C^{13}C^{12}H^1$	9700.58	−0.012[a]	52.11311
$F^{19}C^{12}C^{13}H^1$	9373.94	0.0087	53.92940

[a] This negative value is surprising and suggests the possibility of a slight experimental error.

Using these values[19] we may predict $\nu_{3\to4}$ and compare it to the measured value. This gives

$$\nu_{3\to4}(\text{calc}) = 77\ 648.55\ \text{MHz}$$

which is quite satisfactory agreement with the measured value. Note that the centrifugal distortion term is not negligible, amounting to $256D$ or 1.05 MHz for the $J = 4 \leftarrow 3$ transition. Analysis of the other data in Table 7.5 has been similarly carried out to yield the results of Table 7.6. Included in Table 7.6 are the moments of inertia (in amu A²) calculated by using the conversion factor 505 531 MHz amu A². For consistency with the work of Ref. 18, we use the old physical constants and O^{16} atomic-mass scale rather than the newer data of Appendixes F and G.

The fluoroacetylene molecule shown in Fig. 7.11 contains three unknown structural parameters. If the origin of the molecular coordinate system is placed at the left-hand carbon atom and the z axis lies along the molecular symmetry axis, the moment of inertia about the x (or y) axis can be calculated from Eq. (3.54).

$$I_{xx} = m_F r_1^2 + m_{C_2} r_2^2 + m_H (r_2 + r_3)^2 \tag{7.59}$$

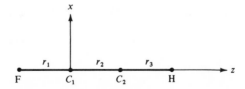

FIG. 7.11. Fluoroacetylene coordinate system used in structural determination of Example 7–2.

[19] Our values differ somewhat from those of Ref. 18 since these workers used a least-squares fit of the three transitions.

The z coordinate of the c.m. is obtained by using Eq. (7.1),

$$\bar{Z} = \frac{1}{M} \left(-m_F r_1 + m_{C_2} r_2 + m_H (r_2 + r_3) \right) \tag{7.60}$$

so the moment of inertia in the principal axis system is obtained directly by means of Eq. (7.2). Because $\bar{Y} = 0$, we obtain

$$I^0_{xx} = I_{xx} - M\bar{Z}^2 \tag{7.61}$$

If Eqs. (7.59) and (7.60) are substituted into Eq. (7.61), I^0_{xx} is seen to be a function of r_1, r_2, r_3, and atomic masses. By using any three of the experimental moments of inertia and the appropriate atomic masses, three equations in the three unknown bond-length parameters are obtained. Solution of these equations would lead to the effective bond lengths in the ground vibrational state.

While this procedure would yield legitimate ground-state parameters, we illustrate another method suggested by Costain.[20] This "substitution coordinate" method is further described in Sec. 7.9, but we simply mention here that the resulting parameters are usually closer to the equilibrium values than are those obtained by direct solution of the ground-state moment of inertia equations as suggested above.

We begin by writing an expression for the moment of inertia of the normal isotopic species ($F^{19}C^{12}C^{12}H^1$) in the principal-axis system.

$$I = I_{xx} = I_{yy} = \sum_{i \neq j}^{4} m_i z_i^2 + m_j z_j^2 \tag{7.62}$$

where the atomic masses and coordinates are labeled one through four, and the jth atom has been separated from the remaining sum. Suppose the jth atom is now isotopically substituted so that its mass becomes $m_j + \Delta m$. If we assume that all the atomic coordinates (or bond distances) are unchanged by this substitution the moment of inertia of the substituted molecule *in the same axis system* used for the normal molecule is

$$I' = \sum_{i \neq j} m_i z_i^2 + (m_j + \Delta m) z_j^2 - (M + \Delta m)$$

$$\times \left\{ \frac{1}{M + \Delta m} \left(\sum_{i \neq j} m_i z_i + (m_j + \Delta m) z_j \right) \right\}^2 \tag{7.63}$$

by direct application of Eq. (7.2), with M being the mass of the normal molecule. By substitution of Eq. (7.62), and remembering that

$$\sum_{i \neq j} m_i z_i + m_j z_j = 0$$

since the coordinate system was situated at the c.m. of the normal species,

[20] C. C. Costain, *J. Chem. Phys.* **29**, 8641 (1958).

we get

$$I' = I + \Delta m z_j^2 - \frac{1}{M + \Delta m}(\Delta m)^2 z_j^2$$

or   (7.64)

$$I' = I - \left(\frac{M \Delta m}{M + \Delta m}\right) z_j^2$$

Equation (7.64) shows that it is possible to calculate the z coordinate of the jth atom with respect to the c.m. of the normal molecule by utilizing the experimental moments of inertia of the normal molecule and the molecule having only the jth atom isotopically substituted. Although we derived this equation with the tetratomic linear molecule in mind, it should be clear that *the result applies generally to any linear molecule*.

Applying Eq. (7.64) to the data in Table 7.6 for the normal species and the $F^{19}C^{12}C^{12}H^2$ species,[21] we obtain

$$57.867\ 29 = 52.083\ 39 + \frac{(44.020\ 206)(1.006\ 593)}{45.026\ 799} z_H^2$$

or   (7.65)

$$|z_H| = 2.4244\ \text{Å}$$

The sign of the atomic coordinate is not given by the calculation but it can normally be chosen by inspection. In the present case the $+$ sign is evident from Fig. 7.11. In an entirely similar manner, the z coordinates of the two C^{13} species can be obtained. All of these results are collected in Table 7.7.

Since the fluorine atom cannot be isotopically substituted this procedure is not appropriate for finding its position. Instead, we utilize the c.m. condition as follows.

$$\sum_i m_i z_i = 0 = 19.004\ 456\ z_F + 12.003\ 804(0.1743 + 1.3716)$$

$$+ 1.008\ 142(2.4244)$$

or   (7.66)

$$z_F = -1.1050\ \text{Å}$$

TABLE 7.7

Experimental atomic coordinates and bond distances in fluoroacetylene.

	F	C_1[a]	C_2	H
z(Å)	-1.1050	0.1743	1.3716	2.4244
	CF	$C\!\equiv\!C$	CH
Bond distance (Å)	1.279	1.197	1.053

[a] Numbering of carbon atoms follows Fig. 7.11.

[21] See Appendix F for atomic-mass values.

The internuclear or bond distances are now obtained by simple subtraction and are given in Table 7.7.

7.4 MICROWAVE SPECTRUM OF THE SYMMETRIC ROTOR MOLECULE

A. Energy Levels

A *symmetric rotor* molecule is one which has two identical principal moments of inertia, the third moment being either larger or smaller. In the former case the molecule is known as an *oblate* symmetric rotor while in the latter case it is termed a *prolate* symmetric rotor. With these definitions it is clear that the linear molecule may be considered to be a special case of the prolate symmetric rotor. Exact symmetric rotors are easily distinguished by their point-group symmetry, since it is easily verified that any molecule having threefold or higher axes of symmetry has at least two equal moments of inertia. Examples are NH_3, CH_3Cl, SF_5Cl, C_6H_6, cyclopropane and chlorobicyclo[1.1.1]pentane. If the molecule has tetrahedral or octahedral symmetry it has three equal moments of inertia and is known as a *spherical* rotor. In addition, molecules having S_2 (rotary-reflection) axes are symmetric rotors, for example, allene ($H_2C\!=\!C\!=\!CH_2$).

The classical rigid-rotor Hamiltonian was given in Sec. 3.3 B by Eq. (3.60), which in operator form is

$$\tilde{\mathfrak{K}}_r = \frac{\tilde{P}_x^{\,2}}{2I_{xx}} + \frac{\tilde{P}_y^{\,2}}{2I_{yy}} + \frac{\tilde{P}_z^{\,2}}{2I_{zz}} \qquad (7.67a)$$

Since the z axis is the uniquely chosen one, a symmetric rotor has $I_{xx} = I_{yy}$ so that Eq. (7.67a) may be rewritten

$$\tilde{\mathfrak{K}}_r = \frac{1}{2I_{xx}}\,(\tilde{P}^2 - \tilde{P}_z^{\,2}) + \frac{1}{2I_{zz}}\,\tilde{P}_z^{\,2} \qquad (7.67b)$$

where $\tilde{P}^2$ has also been introduced by use of Eqs. (4.53) and (4.59). As in Sec. 7.3 A the Hamiltonian matrix elements are easily evaluated in the $|JKM\rangle$ representation. From Table 4.1 we find that both $\tilde{P}_z$ and $\tilde{P}^2$ are diagonal so the Hamiltonian matrix is immediately diagonal. Substituting the matrix elements of $\tilde{P}^2$ and $\tilde{P}_z$ and rearranging terms we obtain

$$E_r = \langle JKM \mid \tilde{\mathfrak{K}}_r \mid JKM \rangle = \frac{\hbar^2}{2I_{xx}}J(J+1) + \frac{\hbar^2 K^2}{2}\left(\frac{1}{I_{zz}} - \frac{1}{I_{xx}}\right) \qquad (7.68)$$

K is the value of the projection of the total angular momentum upon the molecular z axis and has values of $J, J-1\ldots 0 \ldots -J+1, -J$.

Before discussing the energy levels further it is convenient to relabel the principal moments of inertia I_a, I_b, I_c with the convention

$$I_a \leqq I_b \leqq I_c \qquad (7.69)$$

For the symmetric rotor the two equal moments are always identified with I_b. We also define rotational constants

$$A = \frac{h}{8\pi^2 I_a}$$

$$B = \frac{h}{8\pi^2 I_b} \tag{7.70}$$

$$C = \frac{h}{8\pi^2 I_c}$$

For a prolate rotor $(B = C)$ Eq. (7.68) becomes

$$E_r = hBJ(J+1) + h(A-B)K^2 \tag{7.71a}$$

while for an oblate rotor $(A = B)$ it becomes

$$E_r = hBJ(J+1) + h(C-B)K^2 \tag{7.71b}$$

Note that in Eq. (7.71a) the second term is positive while in Eq. (7.71b) the second term is negative.

The structure of the symmetric-rotor energy levels is shown in Fig. 7.12. For illustration purposes the prolate rotor levels are calculated for $A = 1.5B$ and the oblate rotor levels for $C = 0.75B$. Note that all levels with $K > 0$ are doubly degenerate because of the occurrence of the K^2 terms. As for the linear molecule, each state still has a $2J + 1$ degeneracy in M.

The real symmetric rotor shows the effects of vibration-rotation interaction, so the effective rotational constants are defined as in Eq. 7.18. These are related approximately to the equilibrium quantities by

$$A_v = A_e - \sum_i^{3N-6} \alpha_i(v_i + \tfrac{1}{2})$$

$$B_v = B_e - \sum_i^{3N-6} \beta_i(v_i + \tfrac{1}{2}) \tag{7.72}$$

$$C_v = C_e - \sum_i^{3N-6} \gamma_i(v_i + \tfrac{1}{2})$$

but the expressions are of little practical significance in most cases since the large number of vibrational modes makes the determination of the α's, β's and γ's prohibitive.

Centrifugal distortion occurs to about the same extent as for linear molecules. Again the terms which occur are of the fourth power in angular momenta, but are more complicated since they involve both quantum numbers J and K. When centrifugal distortion terms are included Eqs. (7.71) become

$$E_r = hB_vJ(J + 1) + h(A_v - B_v)K^2 - D_JJ^2(J + 1)^2$$
$$- D_{JK}J(J + 1)K^2 - D_KK^4 \quad (7.73a)$$

for the prolate rotor and

$$E_r = hB_vJ(J + 1) + h(C_v - B_v)K^2 - D_JJ^2(J + 1)^2$$
$$- D_{JK}J(J + 1)K^2 - D_KK^4 \quad (7.73b)$$

for the oblate rotor. Table 7.8 gives a list of rotational constants, centrifugal

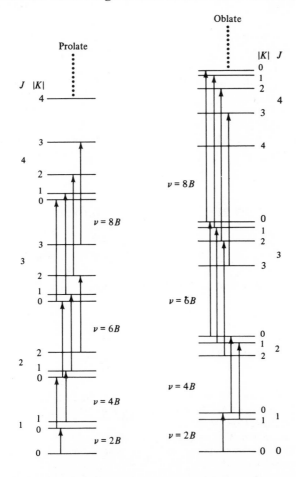

FIG. 7.12.   Energy levels for the rigid prolate and oblate symmetric rotors. For the prolate rotor the energy levels are calculated for the case $A = 1.5B$, while for the oblate rotor $C = 0.75B$.

TABLE 7.8
Rotational parameters for some symmetric rotor molecules.

Molecule	$B_0(MHz)$	$D_J(MHz)$	$D_{JK}(MHz)$	Structure
NH_3 [a]	298 000	19	-28	NH = 1.014 Å, ∠HNH = 106.8°
PH_3 [b]	133 478	3.7	-4.6	PH = 1.421 Å, ∠HPH = 93.5°
PF_3 [c]	7 819.90	...	...	PF = 1.55 Å, ∠FPF = 102°
NF_3 [d]	10 680.96	...	...	NF = 1.371 Å, ∠FNF = 102.1°
CH_3Cl [e]	13 292.95	0.0181	0.189	CH = 1.113 Å, CCl = 1.781 Å, ∠HCH = 110.5°
F_3C—C≡CH [f]	2 877.95	0.00024	0.0063	CF = 1.33 Å, C—C = 1.49 Å, C≡C = 1.21 Å, C—H = 1.06 Å
BH_3CO [g]	8 657.22	...	0.00036	BH = 1.19, BC = 1.54 Å, C≡O = 1.13 Å, ∠HBH = 114°
$XeOF_4$ [h]	2 785.86	0.000576	...	XeO = 1.703 Å, XeF = 1.900 Å, ∠OXeF = 91.8°
$C_8H_{13}Cl$ [i]	1 090.90	...	...	...
C_5H_7Cl [j]	2 161.23	...	...	CC = 1.54 Å, CCl = 1.76 Å

[a]   M. T. Weiss and M. W. P. Strandberg, *Phys. Rev.* **83,** 567 (1951).
[b]   C. C. Loomis and M. W. P. Strandberg, *Phys. Rev.* **81,** 798 (1951).
[c]   P. Kisliuk, *J. Chem. Phys.* **22,** 86 (1954).
[d]   J. Sheridan and W. Gordy, *Phys. Rev.* **79,** 513 (1950).
[e]   W. Gordy, J. W. Simmons, and A. G. Smith, *Phys. Rev.* **74,** 243 (1948).
[f]   Reference 22 in text.
[g]   W. Gordy, H. Ring, and A. B. Burg, *Phys. Rev.* **78,** 512 (1950).
[h]   J. F. Martins and E. B. Wilson, Jr., *J. Molec. Spec.* **26,** 410 (1968).
[i]   A. H. Nethercot and A. Javan, *J. Chem. Phys.* **21,** 363 (1953).
[j]   K. W. Cox and M. D. Harmony, *J. Molec. Spec.* **36,** 34 (1970).

distortion parameters, and structural results for several representative symmetric rotors.

B. Selection Rules and Intensities

The intensities or transition probabilities of symmetric rotor molecules are obtained in the same manner as for linear molecules. Hence the probability is proportional to the sum of the squares of the dipole moment matrix elements as given by Eq. (7.41). Since the symmetric rotor can have a permanent dipole moment only along the unique axis, we require $\mu_x = \mu_y = 0$, so Eq. (7.42) is appropriate for the symmetric rotor as well as for the linear molecule, except that now we must not suppress the K label. By referring to the direction cosine matrix elements in Table 4.2 we find nonzero values only when

$$\Delta J = 0, \pm 1$$

$$\Delta K = 0 \tag{7.74}$$

$$\Delta M = 0, \pm 1$$

Equations (7.74) are therefore the selection rules for the symmetric rotor. Evaluating the right-hand side of Eq. (7.42) by use of Table 4.2 and summing over all permitted M' values, the total dipole matrix element for the transition $JK \to J + 1, K$ is

$$| \mu_{nm} |^2 = \mu^2 \frac{(J + 1)^2 - K^2}{(J + 1)(2J + 1)} \tag{7.75}$$

Since states having $K \neq 0$ are doubly degenerate it is appropriate for these cases to multiply the right-hand side of Eq. (7.75) by two to account for the two coincident transitions.

Returning to Eqs. (7.73), we may now find an expression for the frequency of the transition $JK \to J + 1, K$. For either the prolate or oblate rotor the result is

$$\nu = \frac{E_r(J + 1) - E_r(J)}{h} = 2B_v(J + 1)$$

$$- 2D_{JK}(J + 1)K^2 - 4D_J(J + 1)^3 \tag{7.76}$$

Note the rather unfortunate result that knowledge of the A or C rotational constant cannot be obtained from microwave spectra because of the $\Delta K = 0$ selection rule. We see also that the linear [Eq. (7.46)] and symmetric rotor spectra are of identical form if centrifugal distortion is neglected. Figure 7.12 shows these transitions for the oblate and prolate symmetric

TABLE 7.9

Microwave spectrum of trifluoromethyl acetylene.[a]

Transition		Frequency
$J \rightarrow J'$	$K \rightarrow K$	(MHz)
4–5	0–0	27 779.31
	1–1	
	2–2	27 779.14
	3–3	27 778.76
	4–4	27 778.32
8–9	0–0	51 802.26
	1–1	
	2–2	51 801.90
	3–3	51 801.32
	4–4	51 800.54
	5–5	51 799.56
	6–6	51 798.26
	7–7	51 796.78
	8–8	51 795.10

[a] See Ref. 22.

rotors. Table 7.9 lists a portion of the observed spectrum[22] of trifluoromethyl acetylene in which the effects of centrifugal distortion are clearly visible.

We can obtain an expression for the maximum value of the absorption coefficient of a symmetric-rotor transition if we substitute Eq. (7.75) into Eq. (7.48). If in addition we substitute

$$f_r = (2J + 1) \left(\frac{B^2 C h^3}{\pi (kT)^3} \right)^{1/2} \tag{7.77}$$

for the fraction of molecules in the state JK, and let $h\nu_0 = 2B(J + 1)$, we obtain for an oblate symmetric rotor transition

$$\gamma_{max} = \frac{4\pi N f_v h^2 \mu^2}{3ck^2 T^2 \Delta\nu} \sqrt{\frac{\pi Ch}{kT}} \left(1 - \frac{K^2}{(J + 1)^2} \right) \nu_0^3 \tag{7.78a}$$

For a prolate rotor we use A in place of C. The expression (7.77) for f_r if actually not quite complete as it stands since it does not take into account statistical weight factors due to nuclei with nonzero spin angular momentum. Townes and Schawlow[23] have shown for the common case of three equivalent nuclei of spin I that Eq. (7.78a) should be multiplied by $S(I, K)/(4I^2 + 4I + 1)$, where $S(I, K) = 4I^2 + 4I + 3$ when

[22] W. E. Anderson, R. Trambarulo, J. Sheridan, and W. Gordy, *Phys. Rev.* **82,** 58 (1951).

[23] See Ref. 6, p. 75.

$K = 0$, $S(I, K) = 2(4I^2 + 4I + 3)$ for K a multiple of 3, and $S(I, K) = 2(4I^2 + 4I)$ for K not a multiple of 3. The factors of 2 in the latter two $S(I, K)$ account for the double degeneracy of the $K \neq 0$ states. For the common case in which $I = \frac{1}{2}$ (as in CH_3F), $S(I, K)/(4I^2 + 4I + 1)$ becomes $\frac{3}{2}$, 3, $\frac{3}{2}$, respectively.

Since it very often occurs that the centrifugal distortion constant D_{JK} is quite small, the various K components for a given $J \rightarrow J + 1$ transition are not resolved. In this case the intensity would be gotten by summing (7.78a) over all K from $-J$ to $+J$ using the appropriate statistical weight factors mentioned above. We can perform this summation approximately in closed form if we take an average statistical weighting factor of 2, hence

$$\sum_{K=-J\cdots+J} 2\left(1 - \frac{K^2}{(J+1)^2}\right)$$

$$= 2\left\{(2J+1) - \sum_K \frac{K^2}{(J+1)^2}\right\}$$

$$= 2\left\{(2J+1) - \frac{1}{(J+1)^2}\right.$$

$$\times [J^2 + (J-1)^2 + \ldots 0 \ldots + (-J+1)^2 + (-J)^2]\Big\}$$

$$= 2\left\{(2J+1) - \frac{2}{(J+1)^2} \sum_{n=1\cdots J} n^2\right\}$$

$$= 2\left\{(2J+1) - \frac{J(2J+1)}{3(J+1)}\right\}$$

$$= \frac{2(2J+1)(2J+3)}{3(J+1)} \tag{7.79}$$

Then Eq. (7.78a) becomes for the total intensity

$$\gamma_{max} = \frac{8\pi N f_v h^2 \mu^2}{9ck^2T^2\Delta\nu} \sqrt{\frac{\pi Ch}{kT}} \frac{(2J+1)(2J+3)}{3(J+1)} \nu_0^3 \tag{7.78b}$$

Evaluating the constant factor in Eq. (7.78b) leads to (at $T = 300°K$)

$$\gamma_{max} = 2.73 \times 10^{-21} \sqrt{C}\,\mu^2 \frac{(2J+1)(2J+3)}{3(J+1)} \frac{\nu_0^3}{\Delta\nu} \tag{7.78c}$$

where C (or A) and ν_0 are in MHz, μ is in debye, and $\Delta\nu$ is the half-width in MHz/mm.

For a molecule such as CH_3Cl, $A \sim 1.5 \times 10^5$ MHz, $\mu = 1.9$ debye, $\Delta \nu \sim 20$ MHz/mm so the $J = 1 \rightarrow J = 2$ transition at 53 200 MHz has an absorption coefficient of approximately 4×10^{-5} cm^{-1} which compares favorably in magnitude with the values shown for OCS in Table 7.4. On the other hand, molecules which have heavy atoms off-axis (such as CF_3H) have considerably smaller rotational constants so that γ_{max} may be 10 to 50 times smaller than the above value.

7.5 MICROWAVE SPECTRUM OF THE ASYMMETRIC ROTOR MOLECULE

Most molecules do not fall into either of the previous two classes which have been discussed. Rather, they are of the type for which

$$I_a \neq I_b \neq I_c$$

Those molecules for which $I_a \sim I_b \neq I_c$ are known as *near-oblate* asymmetric rotors while those having $I_a \neq I_b \sim I_c$ are known as *near-prolate* asymmetric rotors. The chief complicating factor that arises for asymmetric molecules in general is that the $\pm K$ degeneracy which exists for the symmetric rotor is broken; that is, $+K$ states have different energies than $-K$ states. This produces approximately a doubling of the total number of energy states and has associated with it a considerable increase in the number of allowed transitions. It might be said that the chief feature of an asymmetric rotor spectrum is that it has no simple regularity but instead seems to have lines positioned in a random fashion.

Before looking into some of the details of the energy levels it is valuable to consider their qualitative features. We have noted already that the prolate and oblate symmetric rotors represent the two natural extreme limits of an asymmetric rotor. Consider a molecule somewhere intermediate to these two cases, that is, $A \neq B \neq C$. We could imagine calculating the energy levels of this molecule by use of perturbation theory. For example, we might write the Hamiltonian as

$$\widetilde{\mathfrak{K}}_r = \widetilde{\mathfrak{K}}(B = C) + \widetilde{\mathfrak{K}}(B - C)$$

where the first term represents the Hamiltonian for a prolate symmetric rotor, while the second term contains the parts which exist because $B \neq C$ and represents the perturbation term. Then the energy is obtained by the usual perturbation theory techniques using prolate rotor basis functions. On the other hand, the same calculation could be performed by rearranging the Hamiltonian so it is of the form

$$\widetilde{\mathfrak{K}}_r = \widetilde{\mathfrak{K}}(A = B) + \widetilde{\mathfrak{K}}(A - B)$$

In this case the first term is the oblate symmetric-rotor Hamiltonian and the second term represents the parts arising because $A \neq B$. The energy levels are now obtained using the oblate rotor basis functions. If properly performed, the two calculations must yield the same result. We see then that the energy state of any asymmetric rotor must *correlate* with the two limiting cases, and that the deviation from the two limiting cases is determined by its B value.

In view of this discussion Fig. 7.12 can be turned into an energy-level correlation diagram as shown in Fig. 7.13 in which the abscissa is some parameter which measures the asymmetry. One such parameter is

$$\kappa = \frac{(B - A) + (B - C)}{A - C} \tag{7.80}$$

which has the value -1 for a prolate rotor and $+1$ for an oblate rotor. All asymmetric rotors must therefore have κ values between -1 and $+1$. Note in Fig. 7.13 that J remains a good quantum number for the asymmetric rotor, but that K is not (except for $J = 0$), since the various energy levels correlate with different prolate and oblate K values. The levels have been labeled as $J_{K_{-1}K_{+1}}$ where K_{-1} is the prolate K value and K_{+1} is the oblate K value. That the correlations shown in Fig. 7.13 are correct may

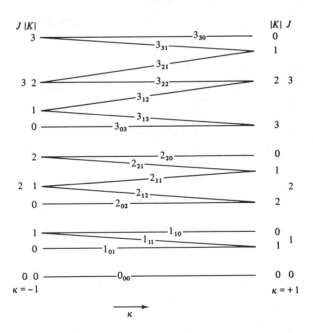

FIG. 7.13.   Correlation diagram for the asymmetric rotor energy levels.

be verified later by use of the "no-crossing rule"[24] for states of the same symmetry. We should stress that no attempt has been made to show correct energies in the correlation diagram.

A. Energy Levels

The Hamiltonian of Eq. (7.67a) can be rewritten as

$$\widetilde{\mathfrak{K}}_r = \frac{2\pi}{\hbar} (A\tilde{P}_x{}^2 + C\tilde{P}_y{}^2 + B\tilde{P}_z{}^2) \tag{7.81}$$

if we make the correlation $x \to a$, $y \to c$, and $z \to b$ followed by substitution of the rotational constants of Eq. (7.70). The axis identification that has been chosen is arbitrary and any of the other five permutations would be acceptable. We always make the experimental choice such that

$$A \geqq B \geqq C$$

In order to set up the Hamiltonian matrix, we need the matrix elements of the $\tilde{P}_g{}^2$. Again, our well-known $|JKM\rangle$ basis set is convenient. Since Table 4.1 shows $\mathbf{P}_z$ to be diagonal, $\mathbf{P}_z{}^2$ is obviously

$$\langle JK | \tilde{P}_z{}^2 | JK \rangle = \langle JK | \tilde{P}_z | JK \rangle^2 = \hbar^2 K^2 \tag{7.82}$$

where we have suppressed the M label. Also, from Table 4.1, we observe that $\mathbf{P}_x$ and $\mathbf{P}_y$ may have $\langle JK | JK \pm 1 \rangle$ elements only. By the rule of matrix multiplication,

$$\langle JK | \tilde{P}_x{}^2 | JK' \rangle = \sum_{K''} \langle JK | \tilde{P}_x | JK'' \rangle \langle JK'' | \tilde{P}_x | JK' \rangle \tag{7.83}$$

gives all the elements of $\mathbf{P}_x{}^2$. Since $\tilde{P}_x$ can only step K up or down by one, it is clear that the only nonzero elements of $\mathbf{P}_x{}^2$ are $\langle JK | JK \rangle$ or $\langle JK | JK \pm 2 \rangle$.

We find for the $\langle JK | JK \rangle$ element,

$$\langle JK | \tilde{P}_x{}^2 | JK \rangle = + \langle JK | \tilde{P}_x | JK + 1 \rangle \langle JK + 1 | \tilde{P}_x | JK \rangle$$
$$+ \langle JK | \tilde{P}_x | JK - 1 \rangle \langle JK - 1 | \tilde{P}_x | JK \rangle \tag{7.84}$$

Substituting the values of Table 4.1 we get

$$\langle JK | \tilde{P}_x{}^2 | JK \rangle = \frac{\hbar^2}{4} [J(J + 1) - K(K + 1)]$$

$$+ \frac{\hbar^2}{4} [J(J + 1) - (K - 1)K]$$

$$= \frac{\hbar^2}{4} [J(J + 1) - K^2] \tag{7.85}$$

---

[24] See, for example, H. Eyring, J. Walter and G. E. Kimball, *Quantum Chemistry* (Wiley, New York, 1944), p. 206.

For the $\langle JK \,|\, JK \pm 2 \rangle$ element we find

$$\langle JK \,|\, \tilde{P}_x{}^2 \,|\, JK \pm 2 \rangle = \langle JK \,|\, \tilde{P}_x \,|\, JK \pm 1 \rangle \langle JK \pm 1 \,|\, \tilde{P}_x \,|\, JK \pm 2 \rangle \quad (7.86)$$

or using Table 4.1 again,

$$\langle JK \,|\, \tilde{P}_x{}^2 \,|\, JK \pm 2 \rangle = \frac{\hbar^2}{4} \{ [J(J+1) - K(K+1)]$$

$$\times [J(J+1) - K(K \pm 1)(K \pm 2)] \}^{1/2} \quad (7.87)$$

In an entirely similar manner the elements of $\mathbf{P}_y{}^2$ are found to be

$$\langle JK \,|\, \tilde{P}_y{}^2 \,|\, JK \rangle = \langle JK \,|\, \tilde{P}_x{}^2 \,|\, JK \rangle$$

and $\quad (7.88)$

$$\langle JK \,|\, \tilde{P}_y{}^2 \,|\, JK \pm 2 \rangle = - \langle JK \,|\, \tilde{P}_x{}^2 \,|\, JK \pm 2 \rangle$$

All the elements needed to compute $\mathfrak{JC}_r$ are now available. Before proceeding it is useful to observe the general form of the Hamiltonian

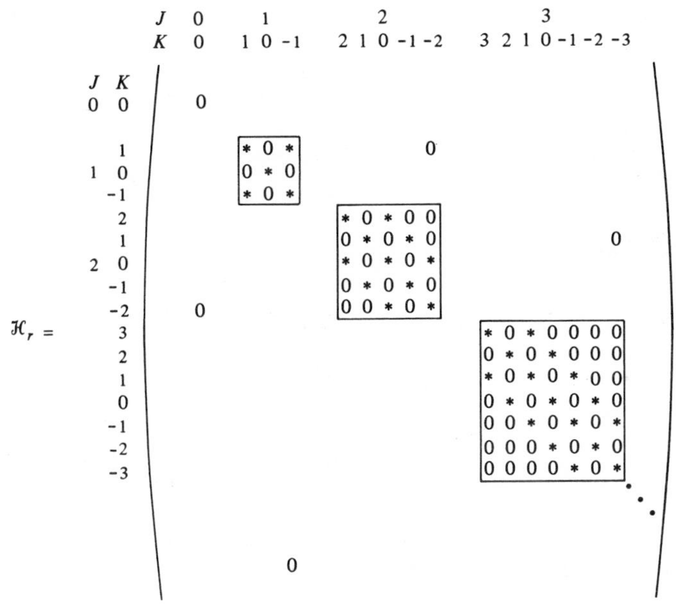

FIG. 7.14. Form of the Hamiltonian matrix for the asymmetric rotor.

matrix. Figure 7.14 shows this matrix with stars (*) occupying the non-zero positions. The ordering of the rows and columns has been chosen in the most convenient way, but of course, the choice is arbitrary and does not affect the final result. Note that since there are no elements off-diagonal in J, the infinite dimensional $J \times J$ matrix reduces to an infinite number of smaller finite blocks each of dimension $2J + 1$.

To obtain the energy levels we proceed as outlined in Sec. 2.2. According to this procedure the asymmetric-rotor energies are obtained from

$$\mathbf{E}_r = \mathbf{A}^{-1} \mathfrak{IC}_r \mathbf{A} \qquad (7.89)$$

where $\mathbf{A}$ is the unitary matrix which brings $\mathfrak{IC}_r$ into diagonal form. Since $\mathfrak{IC}_r$ is blockwise diagonal each block may be treated separately[25] so the energy states of given J are obtained from

$$\mathbf{E}^J{}_r = (\mathbf{A}^J)^{-1} \mathfrak{IC}^J{}_r \mathbf{A}^J \qquad (7.90)$$

We return now to obtain the explicit forms of the Hamiltonian matrix elements. These are easily obtained by appropriate substitutions of Eqs. (7.82), (7.85), (7.87), and (7.88) into the matrix form of Eq. (7.81). For example

$$\langle JK \mid \tilde{\mathfrak{IC}}_r \mid JK \rangle = \frac{2\pi}{\hbar} \{ \langle JK \mid A\tilde{P}_x{}^2 + C\tilde{P}_y{}^2 + B\tilde{P}_z{}^2 \mid JK \rangle$$

$$= \frac{2\pi}{\hbar} \left\{ \frac{\hbar^2}{2} [J(J+1) - K^2][A+C] + \hbar^2 K^2 B \right\} \qquad (7.91)$$

$$= \frac{h}{2} [J(J+1) - K^2][A+C] + hK^2 B$$

Similarly, the off-diagonal elements are

$$\langle JK \mid \tilde{\mathfrak{IC}}_r \mid JK \pm 2 \rangle = \frac{h}{4} \{ [J(J+1) - K(K \pm 1)]$$

$$\times [J(J+1) - (K \pm 1)(K \pm 2)]\}^{1/2} (A - C) \qquad (7.92)$$

In setting up the complete Hamiltonian matrix it is useful to recall that the matrix is Hermitian. Then since all elements are real it is clear that $\mathfrak{IC}_{ij} = \mathfrak{IC}_{ji}$, so the matrix is symmetric about the main diagonal.

It is useful now to illustrate Eq. (7.90) for the low-J blocks. Labeling the Hamiltonian matrix elements as

$$H^J{}_{KK'} \equiv \langle JK \mid \tilde{\mathfrak{IC}}_r \mid JK' \rangle \qquad (7.93)$$

---

[25] See Appendix C.

we obtain the values given in Table 7.10 by use of Eqs. (7.91) and (7.92). From these results we see immediately that

$$E(0_{00}) = 0$$

where the energy level has been labeled in the general format

$$E(J_{K_{-1}K_{+1}})$$

From Table 7.10 we find the $J = 1$ matrix to be

$$\mathcal{3C}^1{}_r = h \begin{pmatrix} \frac{1}{2}(A + C) + B & 0 & \frac{1}{2}(A - C) \\ 0 & A + C & 0 \\ \frac{1}{2}(A - C) & 0 & \frac{1}{2}(A + C) + B \end{pmatrix} \qquad (7.94)$$

It has been shown that a useful first step in the diagonalization process is to apply the Wang transformation[26]

$$\mathbf{X}^{-1} \mathcal{3C}^J{}_r \mathbf{X} \qquad (7.95a)$$

where

$$\mathbf{X} = \mathbf{X}^{-1} = \frac{1}{\sqrt{2}} \begin{pmatrix} \ddots & & & & & \cdot \\ & -1 & & 0 & & 1 \\ & & -1 & & 1 & \\ 0 & & \sqrt{2} & & 0 \\ & & 1 & & 1 & \\ & 1 & & & & 1 \\ \cdot & & & & & \ddots \end{pmatrix} \qquad (7.95b)$$

This transformation does not generally bring about complete diagonalization of $\mathcal{3C}^J{}_r$, but it has been shown that it will always reduce $\mathcal{3C}^J{}_r$ to four smaller blocks of dimension $\sim\frac{1}{2}J$. The diagonalization of these smaller blocks must then be continued by other means. For the $J = 1$ block, the

---

[26] See, for example, M. W. P. Strandberg, *Microwave Spectroscopy* (Methuen and Co., London, 1954), p. 9.

Wang transformation brings about a complete diagonalization, which we show as follows.

$$
\mathbf{X}^{-1}\mathfrak{IC}^1_r\mathbf{X} = \frac{1}{2}
\begin{pmatrix}
-1 & 0 & 1 \\
0 & \sqrt{2} & 0 \\
1 & 0 & 1
\end{pmatrix}
$$

$$
\times
\begin{pmatrix}
\tfrac{1}{2}(A+C)+B & 0 & \tfrac{1}{2}(A-C) \\
0 & A+C & 0 \\
\tfrac{1}{2}(A-C) & 0 & \tfrac{1}{2}(A+C)+B
\end{pmatrix}
\begin{pmatrix}
-1 & 0 & 1 \\
0 & \sqrt{2} & 0 \\
1 & 0 & 1
\end{pmatrix} h
$$

$$
= \frac{1}{2}
\begin{pmatrix}
-(B+C) & 0 & B+C \\
0 & \sqrt{2}(A+C) & 0 \\
A+B & 0 & A+B
\end{pmatrix}
\begin{pmatrix}
-1 & 0 & 1 \\
0 & \sqrt{2} & 0 \\
1 & 0 & 1
\end{pmatrix} h
$$

$$
= \frac{1}{2}
\begin{pmatrix}
2(B+C) & 0 & 0 \\
0 & 2(A+C) & 0 \\
0 & 0 & 2(A+B)
\end{pmatrix} h
$$

$$
=
\begin{pmatrix}
B+C & 0 & 0 \\
0 & A+C & 0 \\
0 & 0 & A+B
\end{pmatrix} h
\tag{7.96}
$$

Therefore the values in the last matrix are the stationary state energies for the $J = 1$ state.

In order to correlate these energies with the state labels of Fig. 7.13, we may consider the energies of the limiting prolate and oblate rotors for $J = 1$ and $K = 0, \pm 1$. For example, from Eqs. (7.71) the energy of the 1_{01} state varies from $h(2B)$ in the prolate case to $h(B+C)$ in the oblate case. From this we deduce that

$$
E(1_{01}) = h(B+C)
\tag{7.97a}
$$

since this agrees immediately with the oblate result and reduces to the prolate result for which $C = B$. In a similar manner we identify the other

TABLE 7.10

Asymmetric rotor matrix elements for $J = 0, 1,$ and 2 states. All values given should be multiplied by h to get energy in ergs.

H^0_{00}	0
$H^1_{11} = H^1_{-1,-1}$	$\frac{1}{2}(A + C) + B$
H^1_{00}	$A + C$
$H^1_{1,-1} = H^1_{-1,1}$	$\frac{1}{2}(A - C)$
$H^2_{22} = H^2_{-2,-2}$	$A + C + 4B$
$H^2_{11} = H^2_{-1,-1}$	$\frac{5}{2}(A + C) + B$
H^2_{00}	$3(A + C)$
$H^2_{02} = H^2_{20}$	$(3/2)(A - C)$
$H^2_{-1,+1} = H^2_{1,-1}$	$(3/2)(A - C)$
$H^2_{-2,0} = H^2_{0,-2}$	$\frac{1}{2}\sqrt{6}(A - C)$

two $J = 1$ states as

$$E(1_{11}) = h(A + C)$$

$$E(1_{10}) = h(A + B)$$ (7.97b)

The $5 \times 5, J = 2$ submatrix can be treated in exactly the same way. Now, however, after application of the Wang transformation we are left with three 1×1 matrices (that is, the eigenvalues directly) plus a 2×2 matrix. This two-dimensional matrix can be diagonalized by use of the two-dimensional rotation matrix Eq. (2.13), or by converting the matrix to a secular determinant and solving as in Eq. (2.10).

This same approach may be followed for the higher J blocks, but it rapidly becomes impractical when three- and higher-dimensional matrices are left after the Wang transformation. Table 7.11 lists all those energy levels which arise from linear or quadratic secular equations. Note that beginning with the $J = 4$ block cubic equations appear so analytical solutions are not readily obtained for these cases.

The general approach requires the use of numerical matrix diagonalization. It is easily shown that the asymmetric-rotor energy levels may be written as[27]

$$E(J_{K_{-1}K_{+1}}) = \frac{1}{2}h(A + C)J(J + 1) + \frac{1}{2}(A - C)E^J_{K_{-1}K_{+1}}(\kappa)$$ (7.98)

where $E^J_{K_{-1}K_{+1}}(\kappa)$ is a dimensionless numerical quantity which depends

[27] G. W. King, R. M. Hainer and P. C. Cross, *J. Chem. Phys.* **11**, 27 (1943).

TABLE 7.11

Asymmetric rotor energy levels which arise from linear or quadratic secular equations.[a]

$J_{K_{-1}K_{+1}}$	*Energy*
0_{00}	0
1_{10}	$A + B$
1_{11}	$A + C$
1_{01}	$B + C$
2_{20}	$2A + 2B + 2C + 2[(B - C)^2 + (A - C)(A - B)]^{1/2}$
2_{21}	$4A + B + C$
2_{11}	$A + 4B + C$
2_{12}	$A + B + 4C$
2_{02}	$2A + 2B + 2C - 2[(B - C)^2 + (A - C)(A - B)]^{1/2}$
3_{30}	$5A + 5B + 2C + 2[4(A - B)^2 + (A - C)(B - C)]^{1/2}$
3_{31}	$5A + 2B + 5C + 2[4(A - C)^2 - (A - B)(B - C)]^{1/2}$
3_{21}	$2A + 5B + 5C + 2[4(B - C)^2 + (A - B)(A - C)]^{1/2}$
3_{22}	$4A + 4B + 4C$
3_{12}	$5A + 5B + 2C - 2[4(A - B)^2 + (A - C)(B - C)]^{1/2}$
3_{13}	$5A + 2B + 5C - 2[4(A - C)^2 - (A - B)(B - C)]^{1/2}$
3_{03}	$2A + 5B + 5C - 2[4(B - C)^2 + (A - B)(A - C)]^{1/2}$
4_{41}	$10A + 5B + 5C + 2[4(B - C)^2 + 9(A - C)(A - B)]^{1/2}$
4_{31}	$5A + 10B + 5C + 2[4(A - C)^2 - 9(A - B)(B - C)]^{1/2}$
4_{32}	$5A + 5B + 10C + 2[4(A - B)^2 + 9(A - C)(B - C)]^{1/2}$
4_{23}	$10A + 5B + 5C - 2[4(B - C)^2 + 9(A - C)(A - B)]^{1/2}$
4_{13}	$5A + 10B + 5C - 2[4(A - C)^2 - 9(A - B)(B - C)]^{1/2}$
4_{14}	$5A + 5B + 10C - 2[4(A - B)^2 + 9(A - C)(B - C)]^{1/2}$
5_{42}	$10A + 10B + 10C + 6[(B - C)^2 + (A - B)(A - C)]^{1/2}$
5_{24}	$10A + 10B + 10C - 6[(B - C)^2 + (A - B)(A - C)]^{1/2}$

[a] See G. W. King, R. M. Hainer, and P. C. Cross, *J. Chem. Phys.* **11**, 27 (1943).

on the state, $J_{K_{-1}K_{+1}}$, and the asymmetry parameter κ defined in Eq. (7.80). This reduced energy $E(\kappa)$ has been obtained by numerical diagonalization and is tabulated in Townes and Schawlow[28] for $0 \leqslant \kappa \leqslant 1$ in increments of 0.01. For negative values of κ the same tables may be used since it has been shown that

$$-E^J_{K_{+1}K_{-1}}(\kappa) = E^J_{K_{-1}K_{+1}}(-\kappa) \tag{7.99}$$

[28] See footnote 6.

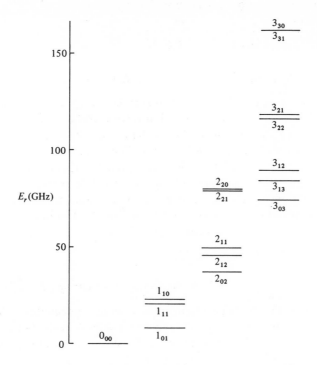

FIG. 7.15. Low-J energy states of cyclopropylamine. Calculated with the data of D. K. Hendricksen and M. D. Harmony, *J. Chem. Phys.*, **51**, 700 (1969).

Finally in Fig. 7.15 we show the energy levels for the low-J states of a typical prolate asymmetric rotor.

As with the linear and symmetric rotor molecules the asymmetric rotor energy states exhibit contributions from nonrigid behavior. These include centrifugal distortion and the variation of rotational constants with vibrational state. In addition, vibration-rotation interactions are often prominent in cases where hindered internal rotation occurs or when certain kinds of low frequency vibrational modes are strongly coupled to the rotational motion. Because of their complexity these cases are not considered here but a brief discussion of internal rotation is given in Sec. 7.8.

B. Selection Rules and Intensities

The dipole moment matrix elements for the asymmetric rotor are obtained by methods similar to, but more complex than, those used for the linear and symmetric-rotor molecules. Beginning with Eq. (7.36) or (7.41), and noting now that $\mu_x{}^0$, $\mu_y{}^0$, and $\mu_z{}^0$ are in general nonzero for an asymmetric molecule, we obtain an expression similar to Eq. (7.42) in the JKM representation.

$$| \mu_{nm}' |^2 = \sum_g \mu_g^2 \{ | \langle JKM | \Phi_{gX} | J'K'M' \rangle |^2$$

$$+ | \langle JKM | \Phi_{gY} | J'K'M' \rangle |^2 + | \langle JKM | \Phi_{gZ} | J'K'M' \rangle |^2 \} \quad (7.100)$$

where $g = x$, y, and z, or in order to correlate with our asymmetric-rotor axis notation, $g = a$, c, and b; n and m refer to JKM and $J'K'M'$, respectively. We have put a prime on the left-hand side of Eq. (7.100) to indicate that it is calculated in the $| JKM \rangle$ basis. Since this basis does not yield the stationary states, Eq. (7.100) does not give the dipole matrix elements between the stationary states of the asymmetric rotor. To obtain the correct matrix elements we must perform the same transformation upon the Φ_{gF} matrices as was used to diagonalize the Hamiltonian matrix in Eq. (7.89). Thus we get instead of Eq. (7.100)

$$| \mu_{nm} |^2 = \sum_g \mu_g^2 \{ | (\mathbf{A}^{-1} \Phi_{gX} \mathbf{A})_{nm} |^2$$

$$+ | (\mathbf{A}^{-1} \Phi_{gY} \mathbf{A})_{nm} |^2 + | (\mathbf{A}^{-1} \Phi_{gZ} \mathbf{A})_{nm} |^2 \} \quad (7.101)$$

in which $\mathbf{A}$ is the transformation matrix which diagonalizes the Hamiltonian and the Φ_{gF} are the direction cosine matrices evaluated in the JKM representation. n and m now refer to the stationary states $| JK_{-1}K_{+1}M \rangle$ and $| J'K'_{-1}K'_{+1}M' \rangle$.

Although evaluation of Eq. (7.101) represents a very formidable numerical problem, there are several easily made observations concerning the selection rules. First, we have observed that the Hamiltonian $\mathcal{3C}$ is diagonal in J, therefore the matrix $\mathbf{A}$ is also. Furthermore, $\mathcal{3C}_r$ contains no M dependence so $\mathbf{A}$ is also diagonal in M. For these reasons, the selection rules on J and M remain the same as those determined solely by the Φ_{gF} elements given in Table 4.2. Thus we have

$$\Delta J = 0, \pm 1, \qquad \Delta M = 0, \pm 1 \quad (7.102)$$

as before.

From (7.101) we observe also that a transition moment can arise along any one of the three principal axes a, b, or c of the molecule, so we may classify asymmetric rotor transitions as a, b, or c type. In a molecule having no symmetry, such as propylene oxide (CH_3—CH_2—CH_2—O), all three types of selection rules apply, whereas for molecules with planes or twofold axes of symmetry only one or two types may occur. For a molecule such as formaldehyde (C_{2v} symmetry) only a-type selection rules apply since $\mu_b = \mu_c = 0$.

Because $\mathcal{3C}_r$ and consequently $\mathbf{A}$ contain elements off-diagonal in K, the Φ_{gF} matrices have their K labels scrambled so that there are no strict selection rules on K. Of course, this is not unexpected since our previous discussion has already shown that K is not a good quantum number. Nevertheless, numerical evaluation of Eq. (7.101) shows that in general the most intense transitions are those for which the K_{-1} or K_{+1} labels change by 0 or ± 1. Nonzero values of $|\mu_{mn}|^2$ are obtained for $|\Delta K_{\pm 1}| = 2, 3, 4$, etc., but the magnitudes fall off rather rapidly with increasing $|\Delta K_{\pm 1}|$.

Some methods of evaluating the dipole moment matrix elements using tabulated quantities are described in the literature. One such discussion is given by Townes and Schawlow.[29] Of course with present day digital computers the numerical solution of Eq. (7.101) is not so difficult. We do not go into this topic further, but give below some symmetry classifications that are extremely useful additions to the previously mentioned selection rules for J and M.

Let us look first at the general problem of asymmetric rotor absorption line intensities. Since $|\mu_{nm}|^2$, ν_0, and $\Delta\nu$ for the asymmetric rotor are comparable to the corresponding values for the linear molecule, Eq. (7.48) shows that any large disparity in intensities must be produced by disparities in fractional population f. For the asymmetric rotor the partition function (to a fair approximation) is[30]

$$Z_{\text{rot}} = \left[\frac{\pi}{ABC} \cdot \left(\frac{kT}{h}\right)^3\right]^{1/2} \tag{7.103}$$

so the fraction of molecules in the state $J_{K_{-1}K_{+1}}$ is

$$f_r = \frac{(2J+1)\,\exp[-E(J_{K_{-1}K_{+1}})/kT]}{[(\pi/ABC)\,(kT/h)^3]^{1/2}} \tag{7.104}$$

For the linear molecule the exponential term was neglected since normally $E \ll kT$. For the asymmetric rotor, however, it is not uncommon to observe transitions in the microwave region which originate from very high-energy states. It is not uncommon, for example, to observe transitions between states having energies of $\sim$500 000 MHz ($\sim$17 cm^{-1}) but separated by only 20 000 MHz. For this reason the exponential in the numerator of Eq. (7.104) plays a very important role in determining the intensities of asymmetric-rotor transitions. Furthermore, the rotational partition function for the linear molecule has a typical value of 500 while for the asymmetric rotor its value is typically 50 000. The net result of these two factors is that f_r for the asymmetric rotor may be easily 10^2–10^4 times smaller than for the linear molecule. Consequently, γ_{max} for the asymmetric rotor may be in the range of 10^{-6} cm^{-1} to 10^{-8} cm^{-1} as opposed to $\sim$10^{-4} cm^{-1} for the

[29] See Ref. 6, pp. 95–100.
[30] See Ref. 6, p. 101.

linear molecule. The low intensities of many asymmetric rotor transitions is one of the foremost difficulties in the study of these molecules.

C. Symmetry Classification of Asymmetric Rotor Transitions

While the previously mentioned selection rules on J and K seem to permit a great and confusing abundance of transitions the situation is really not so bad as it first seems. This is because the totality of allowed transitions can be classified easily as to whether they arise from a-, b- or c-type selection rules. This can be done by noting first that the asymmetric rotor Hamiltonian commutes with all the operations of the four-group (D_2 or V) of Table 7.12. The proof of this can be seen as follows. We first rewrite Eq. (7.81) as

$$\widetilde{\mathfrak{IC}}_r = \frac{2\pi}{\hbar} \left(A P_a{}^2 + B P_b{}^2 + C P_c{}^2 \right) \tag{7.105}$$

Clearly $\widetilde{\mathfrak{IC}}_r$ commutes with the identity E. Because of the symmetry in a, b, c of both the group V and the Hamiltonian we need show only that $C_2{}^a$ commutes with $\widetilde{\mathfrak{IC}}_r$ whence it follows that $C_2{}^b$ and $C_2{}^c$ do also. Performing the operation, we obtain

$$\tilde{C}_2{}^a\widetilde{\mathfrak{IC}}_r = \frac{2\pi}{\hbar} \left[\tilde{C}_2{}^a A \cdot \tilde{C}_2{}^a \tilde{P}_a{}^2 + \tilde{C}_2{}^a B \cdot \tilde{C}_2{}^a \tilde{P}_b{}^2 + \tilde{C}_2{}^a C \cdot \tilde{C}_2{}^a \tilde{P}_c{}^2 \right. \tag{7.106}$$

Now A is a function of $c_i{}^2$ and $b_i{}^2$ so we need investigate the effect of $\tilde{C}_2{}^a$ upon the squares of coordinates. Clearly

$$\tilde{C}_2{}^a b = -b$$
$$\tilde{C}_2{}^a c = -c \tag{7.107a}$$

so that

$$\tilde{C}_2{}^a b^2 = b^2$$
$$\tilde{C}_2{}^a c^2 = c^2 \tag{7.107b}$$

which shows that

$$\tilde{C}_2{}^a A = A \tag{7.108a}$$

TABLE 7.12

Character table for D_2 or V in asymmetric rotor notation.

	E	$C_2{}^c$	$C_2{}^b$	$C_2{}^a$		
A	1	1	1	1	ee	
B_c	1	1	-1	-1	oe	Φ_{cF}
B_b	1	-1	1	-1	oo	Φ_{bF}
B_a	1	-1	-1	1	eo	Φ_{aF}

and by inference

$$\tilde{C}_2{}^a B = B$$

$$\tilde{C}_2{}^a C = C \tag{7.108b}$$

Entirely similar reasoning leads to

$$\tilde{C}_2{}^a \tilde{P}_a{}^2 = \tilde{P}_a{}^2$$

$$\tilde{C}_2{}^a \tilde{P}_b{}^2 = \tilde{P}_b{}^2 \tag{7.109}$$

$$\tilde{C}_2{}^a \tilde{P}_c{}^2 = \tilde{P}_c{}^2$$

Substitution of Eqs. (7.108) and (7.109) into Eq. (7.106) gives

$$\tilde{C}_2{}^a \tilde{\mathcal{H}}_r = \tilde{\mathcal{H}}_r \tag{7.110}$$

which by the arguments of Sec. 6.4 implies that

$$\tilde{C}_2{}^a \tilde{\mathcal{H}}_r - \tilde{\mathcal{H}}_r \tilde{C}_2{}^a = 0 \tag{7.111a}$$

The remaining two operators of V also commute with $\tilde{\mathcal{H}}_r$

$$\tilde{C}_2{}^b \tilde{\mathcal{H}}_r - \tilde{\mathcal{H}}_r \tilde{C}_2{}^b = 0$$

$$\tilde{C}_2{}^a \tilde{\mathcal{H}}_r - \tilde{\mathcal{H}}_r \tilde{C}_2{}^c = 0 \tag{7.111b}$$

by similar reasoning.

The importance of this result as stressed in Sec. 6.4 is that it tells us that the stationary states of the asymmetric rotor can be classified by the irreducible representations or symmetry species of the $V(D_2)$ point group. Thus Table 7.12 shows that all the states may be classified in terms of the four representations, E, B_a, B_b, and B_c. The correlation of the asymmetric rotor eigenstates $|JK_{-1}K_{+1}M\rangle$ with these representations requires rather tedious investigation of the transformation properties of the eigenvectors or wave functions. The treatment has been given by many authors[31] and we merely state the results. It is found, for example, that states for which $|JK_{-1}K_{+1}M\rangle = |J$ even even $M\rangle$ transform like the representation A, or symbolically

$$|J \text{ even even } M\rangle \rightarrow A \tag{7.112a}$$

In the above and following descriptions, even and odd refer to the evenness or oddness of the prolate and oblate K values of the asymmetric-rotor states. Similarly,

$$|J \text{ odd even } M\rangle \rightarrow B_c$$

$$|J \text{ odd odd } M\rangle \rightarrow B_b \tag{7.112b}$$

$$|J \text{ even odd } M\rangle \rightarrow B_c$$

---

[31] See, for example, H. Allen and P. C. Cross, *Molecular Vib-Rotors* (Wiley, New York, 1963), p. 22.

These results have been indicated in the character table 7.12 by the usual abbreviations ee, oe, oo, eo.

For the purposes of numerical work we showed in (7.100) and (7.101) how the dipole-matrix-element evaluation proceeds via the $|JKM\rangle$ representation. The correct matrix elements in the $|JK_{-1}K_{+1}M\rangle$ representation are given by Eq. (7.101), of course, but may be written more simply as

$$| \langle JK_{-1}K_{+1}M \,|\, \tilde{\mu} \,|\, J'K_{-1}'K_{+1}'M' \rangle |^2$$
$$= \sum_{g,F} \mu_g{}^2 \{ |\, \langle JK_{-1}K_{+1}M \,|\, \tilde{\Phi}_{gF} \,|\, J'K_{-1}'K_{+1}'M' \rangle |^2 \} \quad (7.113)$$

From our matrix element theorem of Sec. 6.4 B we know that g-type ($g = a, b, c$) transitions occur only if

$$\Gamma \times \Gamma_{\Phi_{gF}} \times \Gamma' \to \Gamma_A \quad\quad\quad (7.114)$$

where Γ, $\Gamma_{\Phi_{gF}}$, Γ', and Γ_A are the species of $|JK_{-1}K_{+1}M\rangle$, Φ_{gF}, $|J'K_{-1}'K_{+1}'M'\rangle$, and the totally symmetrical representation A, respectively. Since the Φ_{gF}, as previously noted, transform like simple vectors the correlation is $\Phi_{aF} \to B_a$, $\Phi_{bF} \to B_b$, and $\Phi_{cF} \to B_c$ as shown in Table 7.12.

The a-type transitions are found by forming all the triple products $\Gamma \times \Gamma_{\Phi_{aF}} \times \Gamma'$. For example, the character of the representation formed by $\Gamma_{ee} \times \Gamma_{aF} \times \Gamma_{eo}$ for the operator $\tilde{R}$ is obtained from the character table by forming

$$\chi(R) = \chi^{(ee)}(R)\chi^{(aF)}(R)\chi^{(eo)}(R) \quad\quad (7.115)$$

which gives

$$\chi(E) = \chi(C_2{}^a) = \chi(C_2{}^b) = \chi(C_2{}^c) = 1 \quad\quad (7.116)$$

This shows that

$$\Gamma_{ee} \times \Gamma_{aF} \times \Gamma_{eo} \to \Gamma_A \quad\quad\quad (7.117a)$$

Investigating another triple-product representation, $\Gamma_{ee} \times \Gamma_{aF} \times \Gamma_{oo}$, we find the characters to be

$$\chi(E) = 1, \quad \chi(C_2{}^a) = -1, \quad \chi(C_2{}^b) = -1, \quad \chi(C_2{}^c) = 1 \quad (7.118)$$

so that

$$\Gamma_{ee} \times \Gamma_{aF} \times \Gamma_{oo} \to \Gamma_{B_a} \quad\quad\quad (7.117b)$$

In a similar manner we find

$$\Gamma_{ee} \times \Gamma_{aF} \times \Gamma_{oe} \to \Gamma_{B_b}$$
$$\Gamma_{oo} \times \Gamma_{aF} \times \Gamma_{eo} \to \Gamma_{B_b} \quad\quad\quad (7.117c)$$
$$\Gamma_{oo} \times \Gamma_{aF} \times \Gamma_{oe} \to \Gamma_A$$
$$\Gamma_{eo} \times \Gamma_{aF} \times \Gamma_{oe} \to \Gamma_{B_c}$$

which completes all the possibilities for the Φ_{aF} matrix elements. Thus

TABLE 7.13

Symmetry selection rules for asymmetric rotor transitions.

Dipole component	Permitted transitions
a	$ee \leftrightarrow eo$
	$oo \leftrightarrow oe$
b	$ee \leftrightarrow oo$
	$eo \leftrightarrow oe$
c	$ee \leftrightarrow oe$
	$oo \leftrightarrow eo$

we see that a-type transitions can occur only when the $K_{-1}K_{+1}$ labels obey the selection rules $ee \leftrightarrow eo$ or $oo \leftrightarrow oe$.

In a similar fashion we can establish the b- and c-type selection rules which have been listed in Table 7.13. Note that the selection rules are exclusive, that is, those permitted by a-type rules are not permitted by b- or c-type rules, etc. The result of these selection rules is that the number of allowed transitions is considerably reduced for molecules having only one or two nonzero dipole moment components. In Table 7.14 we have summarized and classified the permitted transitions involving $J = 0$, 1, and 2

TABLE 7.14

Low-J asymmetric-rotor transitions.

R (or P) branch[a]		Q branch[a]
	a type	
$0_{00} \leftrightarrow 1_{01}$		$1_{11} \leftrightarrow 1_{10}$
$1_{01} \leftrightarrow 2_{02}$ $1_{01} \leftrightarrow 2_{20}$		$2_{02} \leftrightarrow 2_{21}$
$1_{11} \leftrightarrow 2_{12}$		$2_{12} \leftrightarrow 2_{11}$
$1_{10} \leftrightarrow 2_{11}$		$2_{21} \leftrightarrow 2_{20}$
	b type	
$0_{00} \leftrightarrow 1_{11}$		$1_{01} \leftrightarrow 1_{10}$
$1_{01} \leftrightarrow 2_{12}$		$2_{02} \leftrightarrow 2_{11}$
$1_{11} \leftrightarrow 2_{02}$ $1_{11} \leftrightarrow 2_{20}$		$2_{12} \leftrightarrow 2_{21}$
$1_{10} \leftrightarrow 2_{21}$		$2_{11} \leftrightarrow 2_{20}$
	c type	
$0_{00} \leftrightarrow 1_{10}$		$1_{01} \leftrightarrow 1_{11}$
$1_{01} \leftrightarrow 2_{11}$		$2_{02} \leftrightarrow 2_{12}$
$1_{11} \leftrightarrow 2_{21}$		$2_{12} \leftrightarrow 2_{20}$
$1_{10} \leftrightarrow 2_{02}$ $1_{10} \leftrightarrow 2_{20}$		$2_{11} \leftrightarrow 2_{21}$

[a] Double-headed arrows have been drawn to indicate that selection rules permit transitions in either direction. For R branch $\Delta J = +1$, for P branch $\Delta J = -1$.

TABLE 7.15
Rotational transitions of bicyclobutane (C_4H_6) in MHz.[a]

Transition	Frequency
$0_{00} \rightarrow 1_{10}$	26 625.55
$1_{10} \rightarrow 2_{02}$	26 420.63
$2_{11} \rightarrow 2_{21}$	23 995.38
$3_{12} \rightarrow 3_{22}$	22 664.28
$4_{13} \rightarrow 4_{23}$	20 927.71
$2_{12} \rightarrow 2_{20}$	26 830.19

[a] Marlin D. Harmony and Kent Cox, *J. Am. Chem. Soc.* **88**, 5049 (1966).

states. Note that they consist of Q-branch ($\Delta J = 0$), R-branch ($\Delta J = +1$) and under certain circumstances,[32] P-branch ($\Delta J = -1$) transitions.

For illustrative purposes we show in Table 7.15 the observed microwave transition frequencies of bicyclobutane, which is an example of a molecule having only c-type selection rules. In this case, since μ lies entirely along the C_2 axis (the c-principal axis) μ_a and μ_b are both zero by symmetry. We have also indicated in Table 7.16 some of the transitions which have been observed for the SF_4 molecule. This table also gives the experimental rotational constants that have been obtained by analysis of the spectrum.

D. Assignment of Asymmetric Rotor Spectra

We did not spend any time discussing the manner in which the quantum numbers were assigned to the experimentally observed transitions of linear and symmetric-rotor molecules because of the simple $2B$, $4B$, $6B$.... structure of the spectra as a first approximation. For these molecules an approximate value of the rotational constant based on chemical intuition (bond lengths, etc.) is often sufficient to permit assignment. Of course, the observation of any two adjacent ground-state transitions is always sufficient to establish the assignment. For example, measurement of adjacent lines at 9000 MHz and 12 000 MHz leads to $2B = 3000$ or $B = 1500$, since adjacent lines are spaced at $2B$. Clearly then the 9000 MHz line is the $J = 3 \leftarrow 2$ and the 12 000 MHz line the $J = 4 \leftarrow 3$. Of course, one may have to sort out vibrational satellites and various isotopic species but this presents no great problem.

The theory presented thus far in Sec. 7.5 shows that very few simple relationships exist for the spacings or positions of asymmetric rotor transitions, as shown clearly by Tables 7.15 and 7.16. While the assignment of

---

[32] For various combinations of A, B, and C it is possible that a $J + 1$ state may be lower in energy than a J state, so that a P-branch transition is possible in absorption.

TABLE 7.16

Observed transitions and rotational constants of SF_4.[a]

Transition	Observed frequency (MHz)	Calculated[b] Frequency (MHz)
	$S^{32}F_4$	
$0_{0,0} \rightarrow 1_{1,0}$	$10\ 774.39 \pm 0.02$	$10\ 774.39$
$1_{1,0} \rightarrow 2_{0,2}$	$10\ 962.72 \pm 0.02$	$10\ 962.58$
$1_{0,1} \rightarrow 2_{1,1}$	$18\ 947.68 \pm 0.02$	$18\ 947.75$
$1_{1,0} \rightarrow 2_{2,0}$	$23\ 466.10 \pm 0.02$	$23\ 466.00$
$1_{1,1} \rightarrow 2_{2,1}$	$24\ 149.76 \pm 0.02$	$24\ 149.81$
$2_{0,2} \rightarrow 3_{1,2}$	$27\ 608.05 \pm 0.02$	$27\ 608.21$
$2_{1,2} \rightarrow 3_{2,2}$	$32\ 323.02 \pm 0.02$	$32\ 323.18$
$2_{1,2} \rightarrow 2_{2,0}$	$10\ 586.14 \pm 0.02$	$10\ 586.20$
$3_{1,3} \rightarrow 3_{2,1}$	$12\ 674.35 \pm 0.02$	$12\ 674.32$
$3_{2,1} \rightarrow 3_{3,1}$	$14\ 409.86 \pm 0.02$	$14\ 409.86$
$3_{2,2} \rightarrow 3_{3,0}$	$15\ 301.13 \pm 0.02$	$15\ 301.17$
$4_{2,2} \rightarrow 4_{3,2}$	$13\ 376.18 \pm 0.02$	$13\ 376.23$
$4_{3,1} \rightarrow 4_{4,1}$	$20\ 807.82 \pm 0.02$	$20\ 807.96$
$4_{3,2} \rightarrow 4_{4,0}$	$20\ 979.47 \pm 0.02$	$20\ 979.62$
$5_{2,3} \rightarrow 5_{3,3}$	$11\ 794.65 \pm 0.02$	$11\ 794.68$
$5_{3,2} \rightarrow 5_{4,2}$	$20\ 328.58 \pm 0.02$	$20\ 328.72$
$5_{3,3} \rightarrow 5_{4,1}$	$20\ 992.33 \pm 0.02$	$20\ 992.56$
	$S^{34}F_4$	
$0_{0,0} \rightarrow 1_{1,0}$	$10\ 740.22 \pm 0.05$	$10\ 740.22$
$1_{0,1} \rightarrow 2_{1,1}$	$18\ 895.16 \pm 0.1$	$18\ 895.18$
$2_{0,2} \rightarrow 3_{1,2}$	$27\ 531.90 \pm 0.05$	$27\ 531.84$
$2_{1,2} \rightarrow 2_{2,0}$	$10\ 508.3\ \pm 0.1$	$10\ 508.28$
$4_{2,2} \rightarrow 4_{3,2}$	$13\ 298.2\ \pm 0.1$	$13\ 298.35$
$5_{2,3} \rightarrow 5_{3,3}$	$11\ 735.33 \pm 0.05$	$11\ 735.37$

	$S^{32}F_4$	$S^{34}F_4$
$\frac{1}{2}(A + C)$	4953.83 ± 0.03 MHz	4941.40 ± 0.03 MHz
$\frac{1}{2}(A - C)$	1733.88 ± 0.01 MHz	1721.34 ± 0.02 MHz
κ	-0.500120 ± 0.000003	-0.50189 ± 0.00001

[a] Data from W. M. Tolles and W. D. Gwinn, *J. Chem. Phys.* **36,** 1119 (1962).
[b] Calculated frequencies are obtained using the rotational constants which give best fit to the observed transition frequencies.

asymmetric-rotor spectra is often a matter of intuition, experience and chance, there are several specific methods which are useful.

The first method is essentially trial and error, and is most useful in getting some preliminary notions about the spectrum. It consists of assuming a reasonable molecular structure, calculating A, B, and C, and then predicting the energy levels and transitions as described in Secs. 7.5 A, 7.5 B, and 7.5 C. Under appropriate circumstances it is possible to more or less match the observed spectrum (or some parts of it) by appropriate variations in the assumed structure or rotational constants. Nevertheless the method is seldom sufficient by itself since very small variations in rotational constants may cause large frequency variations. There is often the additional problem of deciding which components of the dipole moment are nonzero. Only very recently, with the advent of large digital computers, has it become feasible to consider the method very seriously.

The most valuable technique for asymmetric rotor line assignment is by observation of the "Stark" effect. A limited quantitative discussion is given of this in Sec. 7.6. Qualitatively the Stark effect is the splitting of a rotational absorption line into various M components when the absorbing molecules are subjected to an electric field of several hundred volts per centimeter. That is, the M degeneracy of the energy levels spoken of earlier is broken down. The theory shows that for the normal asymmetric-rotor Stark effect a $\pm M$ degeneracy still exists so that any state with quantum number J splits into $J + 1$ substates as shown in Fig. 7.16.

In Sec. 7.5 B we inferred the selection rules $\Delta M = 0, \pm 1$. When a strong polarized electric field is present the $\Delta M = 0$ rule applies when the electric vector of the microwave field is parallel to the direction of the strong electric field, while the $\Delta M = \pm 1$ rule applies if the two fields are perpendicular.[33] Since the typical microwave spectrometer is designed in such a way that the strong field is parallel to the microwave field, only $\Delta M = 0$ transitions are observed. Therefore a $J \rightarrow J + 1$ transition leads to $J + 1$ components (as shown in Fig. 7.16), while a $J \rightarrow J$ transition results in J components. This latter result arises because the theory shows that the intensity of $M = 0 \rightarrow M = 0$ transitions is zero when $\Delta J = 0$.

Hence the observation of, say, four clearly resolved Stark components for some unassigned asymmetric-rotor line implies that the transition is either $J = 4 \leftarrow 3$ or $J = 4 \leftarrow 4$. This observation alone is usually not sufficient to decide between the R- or Q-branch options or to assign the $K_{-1}K_{+1}$ labels, but the problem has been simplified to a relatively few possibilities.

A third method which has proven useful in line assignment involves the observation and measurement of the hyperfine splittings of the rotational lines caused by the coupling of nuclear quadrupole moments with

---

[33] See also the brief discussion in Sec. 7.3 B.

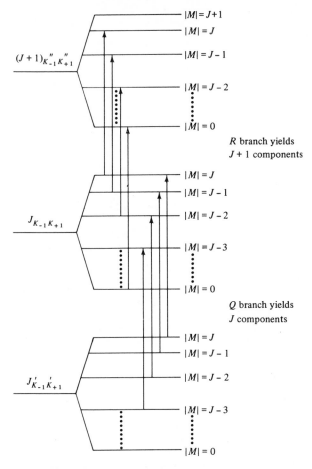

FIG. 7.16.   Qualitative representation of the splitting of asymmetric rotor states by the Stark effect. Arrows indicate $\Delta M = 0$ transitions.

the molecular rotation. We discuss this topic briefly in Sec. 7.7. Here we mention simply that the number and relative spacings of the hyperfine components are distinctly characteristic of the particular transition and therefore may permit immediate assignment of quantum numbers.

EXAMPLE 7-3

The first three observed transitions of bicyclobutane listed in Table 7.15 may be used to determine the rotational constants of the molecule. Using the formulas for the energy levels given in Table 7.11, we find the frequency of the $0_{00} \to 1_{10}$ transition to be

$$\nu = \frac{\Delta E}{h} = A + B = 26\,625.55 \qquad (7.118a)$$

and that for the $2_{11} \to 2_{21}$ to be

$$\nu = \frac{\Delta E}{h} = 3A - 3B = 23\,995.38 \qquad (7.118b)$$

Solving Eqs. (7.118) for A and B we obtain

$$A = 17\,312.00 \text{ MHz}$$
$$B = 9\,313.55 \text{ MHz} \qquad (7.119a)$$

An expression relating A, B, and C is now gotten by use of the $1_{10} \to 2_{02}$ transition which gives

$$\nu = \frac{\Delta E}{h} = A + B + 2C - 2[(B - C)^2 + (A - C)(A - B)]^{1/2} \quad (7.120)$$

$$= 26\,420.63$$

Using the results in Eq. (7.119a), (7.120) may be solved for C with the following result.

$$C = 8393.48 \text{ MHz} \qquad (7.119b)$$

These values of A, B, and C lead to an asymmetry parameter κ of -0.7937.

7.6   STARK EFFECT OF LINEAR AND SYMMETRIC-ROTOR MOLECULES

One of the important electronic properties of a molecule that can be measured by microwave spectroscopy is the electric dipole moment. This property is of great importance in discussing the electronic distribution in molecules and in arriving at valuable concepts of chemical bonding.[34] Dipole-moment measurements have been determined by classical dielectric-constant measurements for many years, but the advent of the microwave method added some distinctly useful features.

In the first place, the classical method serves to determine only a bulk molecular dipole moment which is a complicated average of the dipole moments of the collection of molecules in all the various vibrational states, while the microwave Stark effect method permits measurements of the dipole moment of a molecule in a *particular* vibrational state. In addition, in the classical method only the magnitude of μ can be measured while

---

[34] See for example, L. Pauling, *Nature of the Chemical Bond* (Cornell U.P., Ithaca, N.Y., 1945).

the microwave technique yields also the components $|\mu_a|$, $|\mu_b|$, and $|\mu_c|$, and consequently some information about the orientation of μ in the molecule. Higher accuracy can be obtained by the microwave measurements also, although not as high as that obtained by molecular-beam-resonance experiments.[35]

The microwave dipole-moment determination proceeds via the study of the Stark effect that was discussed qualitatively in Sec. 7.5 D. In the presence of a static electric field $\mathcal{E}$ the energy of an electric dipole is given by $-\mu \cdot \mathcal{E}$ so the total Hamiltonian is

$$\tilde{\mathcal{K}} = \tilde{\mathcal{K}}_r - \tilde{\mu} \cdot \tilde{\mathcal{E}} \tag{7.121a}$$

or

$$\tilde{\mathcal{K}} = \tilde{\mathcal{K}}_r - \tilde{\mathcal{K}}' \tag{7.121b}$$

Because the Stark term is normally small it is possible to treat it as a small perturbation to the rigid-rotor energy. If we choose the direction of $\mathcal{E}$ to be along the Z axis it follows that

$$\tilde{\mathcal{K}}' = -\tilde{\mu}_Z \mathcal{E}_Z \tag{7.122a}$$

or by introduction of the direction cosines

$$\tilde{\mathcal{K}}' = -\mathcal{E}_Z \sum_g \Phi_{gZ} \mu_g \tag{7.122b}$$

where μ_g here is really $\mu_g{}^0$, the leading term in the expansion of μ_g in terms of the vibrational coordinates [see Eq. (7.40)].

For linear and symmetric rotors it is quite easy to evaluate the effect of $\tilde{\mathcal{K}}'$ on the rigid-rotor levels by using first- or second-order perturbation theory with the appropriate $|JKM\rangle$ basis functions. The problem is complicated for the asymmetric rotor since the $|JKM\rangle$ are not the correct zero-order functions.

For the symmetric rotor, the corrections to the rigid-rotor levels $|JKM\rangle$ are found by first-order perturbation theory to be

$$E^{(1)}{}_{JKM} = \langle JKM | \tilde{\mathcal{K}}' | JKM \rangle = -\mu_z \mathcal{E}_Z \langle JKM | \tilde{\Phi}_{zZ} | JKM \rangle \tag{7.123}$$

by use of Eq. (7.122b) and the fact that $\mu_x = \mu_y = 0$. The value of the direction cosine matrix element is found from Table 4.2 to be

$$\langle JKM | \tilde{\Phi}_{zZ} | JKM \rangle = \frac{MK}{J(J+1)} \tag{7.124}$$

The correction to the energy level is then

$$E^{(1)}{}_{JKM} = -\frac{\mathcal{E}_Z \mu_z MK}{J(J+1)} = -\frac{\mu \mathcal{E} MK}{J(J+1)} \tag{7.125}$$

---

[35] See footnote 13, Kaufman et al.

where we have dropped the unnecessary subscripts. Note that the derivation should apply to a linear molecule also, but since $K = 0$ in this case, $E^{(1)} = 0$. It is necessary therefore to investigate the second-order terms for linear molecules.

The second-order correction to the energy levels is given by

$$E^{(2)}{}_{JKM} = \sum_{J'K'M'}{}' \frac{|\langle JKM \mid \tilde{\mathfrak{X}}' \mid J'K'M'\rangle|^2}{E^0{}_{JKM} - E^0{}_{J'K'M'}}$$

$$(7.126)$$

$$= \sum_{J'K'M'}{}' \frac{\mu^2{}_z \mathcal{E}^2{}_Z \mid \langle JKM \mid \tilde{\Phi}_{zZ} \mid J'K'M'\rangle \mid^2}{E^0{}_{JKM} - E^0{}_{J'K'M'}}$$

where the prime on the summation means that $J'K'M' = JKM$ is excluded in the sum. From Table 4.2 we see that Φ_{zZ} is diagonal in M and K so $M' = M$ and $K' = K$ in Eq. (7.126). Also we see that the only nonzero elements occur for $J' = J + 1$ or $J - 1$ so the summation in Eq. (7.126) reduces to only two terms. Evaluating these two matrix elements and again dropping the subscripts on μ and $\mathcal{E}$, Eq. (7.126) becomes

$$E^{(2)} = \mu^2 \mathcal{E}^2 \left\{ \frac{(J + 1)^2 - M^2}{(2J + 1)(2J + 3)} \cdot \frac{1}{E^0{}_{JKM} - E^0{}_{J+1,KM}} \right.$$

$$\left. + \frac{J^2 - M^2}{4J^2 - 1} \cdot \frac{1}{E^0{}_{JKM} - E^0{}_{J-1,KM}} \right\} \quad (7.127)$$

The energy differences in the denominator can be evaluated by use of the expression for the energy levels of a rigid symmetric rotor, Eqs. (7.71). We get

$$E^0{}_{JKM} - E^0{}_{J+1\,KM} = -2B(J + 1)h$$

$$(7.128)$$

$$E^0{}_{JKM} - E^0{}_{J-1\,KM} = 2BJh$$

Substituting Eqs. (7.128) into (7.127) gives the second-order correction,

$$E^{(2)}{}_{JKM} = \frac{\mu^2 \mathcal{E}^2}{2BJ(J + 1)h} \left\{ \frac{J(J + 1) - 3M^2}{(2J - 1)(2J + 3)} \right\} \quad (7.129)$$

In this case a $\pm M$ degeneracy remains but we see that the Stark term does not go to zero for $K = 0$ as it did through first order.

In order to see the magnitudes of the first- and second-order terms, we might assume typical values of $\mu = 1$ debye and $\mathcal{E} = 200$ V/cm. Since Eqs. (7.125) and (7.129) are naturally evaluated in cgs units, μ and $\mathcal{E}$ must be in electrostatic units if E is to be in ergs. The conversion of units is most easily made by using the identity,

$$(0.5035h) \text{ MHz erg sec} = 1 \text{ debye (V/cm)} \qquad (7.130)$$

Then we see that

$$E^{(1)}{}_{JKM} \sim \mu \mathcal{E} \sim (1)(200)(0.5035)(h) \text{ MHz erg sec} \qquad (7.131a)$$

or

$$\frac{E^{(1)}{}_{JKM}}{h} \sim 100 \text{ MHz} \qquad (7.131b)$$

In a similar fashion we find

$$E^{(2)}{}_{JKM} \sim \frac{\mu^2 \mathcal{E}^2}{Bh} \sim \frac{(200)^2 (0.5035)^2 h^2}{Bh} \text{ MHz erg sec} \qquad (7.132a)$$

or using a typical value of $B = 10^4$ MHz we find

$$\frac{E^{(2)}{}_{JKM}}{h} \sim 1 \text{ MHz} \qquad (7.132b)$$

Therefore it is usually found that the second-order terms may be neglected unless high accuracy is desired or unless $K = 0$ in which case $E^{(1)} = 0$.

In Fig. 7.17 we show the dependence of the energy levels upon $\mathcal{E}$ for the $J = 1$ and $J = 2$ states of a linear molecule (or $K = 0$ substates of the symmetric rotor). By use of Eq. (7.129) the slopes of the energy-level plots have been calculated and listed on the diagram. In addition, the two

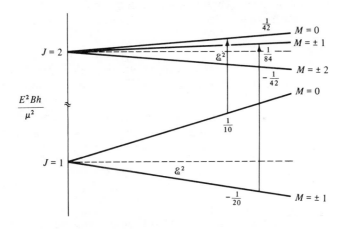

FIG. 7.17.   Stark effect of the linear molecule. Fractional numbers are the slopes of the lines.

allowed transitions for $\Delta M = 0$ have been shown. It is seen that as the field strength is increased the $M = 0 \to M = 0$ line moves lower in frequency while the $M = 1 \to M = 1$ component moves to higher frequencies. If ν_0 is the rigid-rotor frequency of the zero-field line, the two *Stark components* of the $J = 2 \leftarrow 1$ transition have frequencies given by

$$\nu(M = 0) = \nu_0 - \frac{8}{105} \frac{\mu^2 \mathcal{E}^2}{Bh^2} \tag{7.133a}$$

$$\nu(M = 1) = \nu_0 + \frac{13}{210} \frac{\mu^2 \mathcal{E}^2}{Bh^2} \tag{7.133b}$$

A typical dipole-moment measurement resolves itself into the determination of the shift of the absorption lines as a function of electric-field strength. Equations such as those derived above for the $J = 2 \leftarrow 1$ transitions are then used to determine μ^2. The treatment is similar for the first-order Stark effect of the symmetric rotor.

The asymmetric-rotor Stark effect can be similarly treated but, of course, the analysis is more difficult since it involves the direction cosines evaluated in (or transformed to) the representation which diagonalizes $\tilde{\mathcal{K}}_r$. Golden and Wilson[36] have shown that the second-order corrections (the first-order normally vanish) are of the form

$$E^{(2)}{}_{JK_{-1}K_{+1}M} = \mathcal{E}^2 \sum_g (A^{(g)} + B^{(g)}M^2)\mu_g{}^2 \tag{7.134}$$

where the $A^{(g)}$ and $B^{(g)}$ are quantities which can be evaluated theoretically for each state $J_{K_{-1}K_{+1}}$. Note that a minimum of three Stark components must be measured as a function of $\mathcal{E}$ in order to determine μ_a, μ_b, and μ_c.

EXAMPLE 7-4

In a typical experiment on carbonyl sulfide (OCS) the Stark coefficients [the coefficients of $\mathcal{E}^2$ in Eqs. (7.133)] for the $M = 0$ and $M = 1$ components of the $J = 2 \leftarrow 1$ are found to be $-1.619 \times 10^{-6}\,\mathrm{MHz}/(\mathrm{V/cm})^2$ and $1.30 \times 10^{-6}\,\mathrm{MHz}/(V/\mathrm{cm})^2$, respectively. Using the conversion factor given by Eq. (7.130), and the expression for the $M = 0$ component given in Eq. (7.133a) we may write

$$-\frac{8}{105} \frac{\mu^2}{B} (0.5035)^2 = -1.619 \times 10^{-6} \tag{7.135a}$$

with μ in debye and B in MHz. The rotational constant B_0 for OCS is known to have the value 6081.5 MHz, thus Eq. (7.135a) leads to $\mu = 0.714$

---

[36] S. Golden and E. B. Wilson, Jr., *J. Chem. Phys.* **16,** 699 (1948).

debye. The data for the $M = 1$ component gives in a similar fashion,

$$\frac{13}{210} \frac{\mu^2}{B} (0.5035)^2 = 1.308 \times 10^{-6} \qquad (7.135b)$$

or $\mu = 0.710$ debye. The agreement is very good as expected.

7.7 QUADRUPOLE HYPERFINE INTERACTION

A. Energy Levels and Spectra

Hyperfine splittings of rotational lines are very often prominent in microwave spectra, and their analysis can provide information of great importance to chemistry. The most common cause of the hyperfine structure is the interaction of a nuclear electric quadrupole moment with the gradient of the electric field at the nucleus produced by all the charges in the molecule external to the nucleus of interest. Since the topic is a rather complicated one and has been described excellently in several places,[37] we are content here with simply discussing the theory in a qualitative fashion, and illustrating the results for a symmetric-rotor molecule.

In addition to mass, charge, nuclear spin, and magnetic moment (see chapter 9 for additional discussion of nuclear properties), nuclei with spin I greater than $\frac{1}{2}$ always possess an electric quadrupole moment Q. This quadrupole moment is a measure of the deviation of the nuclear charge distribution from spherical symmetry. Depending on whether the charge density is greater or less along the nuclear spin axis the quadrupole moment is positive or negative, respectively. We show the possible nuclear charge distributions in Fig. 7.18 in which the $+$ and $-$ signs refer to regions of

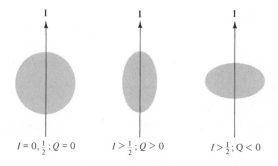

$$I = 0, \tfrac{1}{2}; Q = 0 \qquad I > \tfrac{1}{2}; Q > 0 \qquad I > \tfrac{1}{2}; Q < 0$$

FIG. 7.18.   Nuclear charge distributions.

[37] See Ref. 6, chapter 6.

increased and decreased charge density, respectively. The quadrupole moment Q has units of cm^2 as usually defined and a magnitude of about 1×10^{-25}.

The electric field gradient q along the z-molecular axis may be expressed as the second derivative of the electric potential

$$q = \left\langle \frac{\partial^2 V}{\partial z^2} \right\rangle \tag{7.136}$$

where the average is over the electronic state of the molecule and V is the electric potential at the nucleus. The electric potential V can be simply expressed by

$$V = \sum_i \frac{e}{r_i} + \sum_\alpha \frac{Z_\alpha e}{R_\alpha} \tag{7.137}$$

where the first sum is over all electrons and the second sum is over all nuclei except the one in question. r_i and R_α are the distances of the electrons and nuclei from the nucleus of interest and the Z_α are the charges on the nuclei. It should be clear that if q could be measured experimentally, important information concerning electron distribution might be obtained.

For the symmetric rotor the first-order contribution of the quadrupole interaction energy to the rotational energy is[38]

$$E_Q = E^{(1)}{}_{JKM} = \frac{eqQ[3K^2/J(J+1)-1]}{2I(2I-1)(2J-1)(2J+3)}$$

$$\times [(\tfrac{3}{4})C(C+1) - I(I+1)J(J+1)] \tag{7.138}$$

where

$$C = F(F+1) - I(I+1) - J(J+1) \tag{7.139}$$

F is the quantum number for the total angular momentum, $\mathbf{F} = \mathbf{I} + \mathbf{J}$, and takes the values $I+J, I+J-1 \ldots |I-J|$. Expression (7.138) is valid only for the case where the nucleus of interest lies on the symmetry axis of the molecule, as in CH_3Cl, for example. The dependence of E_Q upon F arises qualitatively because the interaction energy depends upon the orientation of the nuclear quadrupole moment in the molecule and hence upon the orientation of $\mathbf{I}$ with respect to $\mathbf{J}$. The quantum number F is therefore a satisfactory label for the possible hyperfine states. The quantity eqQ in Eq. (7.138) is usually referred to as the *quadrupole coupling constant*, and has values typically in the range 1–100 MHz. Note that Eq. (7.138) applies to a linear molecule by setting $K = 0$.

---

[38] See Ref. 6, p. 154.

In Fig. 7.19 we have shown the quadrupole levels for the lowest two J states of a prolate symmetric rotor having a nucleus of spin $\frac{3}{2}$ on the symmetry axis. CH_3Cl would be an example of this type of molecule. Figure 7.19 also shows the three transitions permitted by the selection rules[39]

$$\Delta F = 0, \pm 1 \tag{7.140}$$

and the previous ones for J and K. The $J = 1 \leftarrow 0$ is therefore seen to be a triplet with frequencies (neglecting centrifugal distortion) given by

$$\nu_{3/2} = 2B + 0.20eqQ$$

$$\nu_{5/2} = 2B - 0.05eqQ \tag{7.141}$$

$$\nu_{1/2} = 2B - 0.25eqQ$$

Multiplets involving higher J states of symmetric rotors are much more complex (more components) as can be verified easily by continuing the energy-level diagram of Fig. 7.19 for higher J states. For asymmetric rotors the theory is considerably more complicated but the quadrupole hyperfine splittings can be analyzed nevertheless.[40] In this case it is gen-

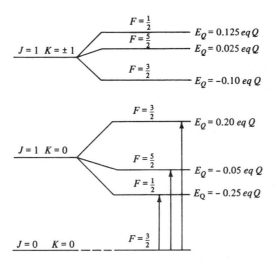

FIG. 7.19. Prolate symmetric rotor states with quadrupole hyperfine interaction.

[39] See footnote 6 or J. K. Bragg, *Phys. Rev.* **74**, 533 (1948).
[40] See footnote 6.

erally found that E_Q depends upon the field gradient along two principal axes, say

$$q_a = \frac{\partial^2 V}{\partial a^2}$$

(7.142)

$$q_b = \frac{\partial^2 V}{\partial b^2}$$

instead of just one as for the symmetric rotor or linear molecule. It is useful to note that Laplace's relation gives the field gradient along the third principal axis.

$$\frac{\partial^2 V}{\partial a^2} + \frac{\partial^2 V}{\partial b^2} + \frac{\partial^2 V}{\partial c^2} = 0$$

(7.143)

For linear and symmetric-rotor molecules the two directions perpendicular to the symmetry axis are equivalent, therefore

$$\frac{\partial^2 V}{\partial x^2} = \frac{\partial^2 V}{\partial y^2} = -\frac{1}{2}\frac{\partial^2 V}{\partial z^2}$$

(7.144)

This condition is spoken of as *cylindrical* symmetry. In Table 7.17 some representative quadrupole coupling constants have been listed for several nuclei in a variety of molecules.

B. Interpretation of Quadrupole Coupling Constants

A variety of schemes for interpreting quadrupole coupling constants have been devised by several workers.[41] These schemes have attempted to use the coupling constants to describe chemical bonds in terms of the well-known concepts of hybridization, ionic character, multiple-bond character, etc., and have been found to be reasonably successful in many cases. Nevertheless, the results have never been entirely satisfactory because of the emphasis put upon such concepts as hybridization and ionic character. In recent years, though, *ab initio* quantum-mechanical calculations of the field gradients have been undertaken, so the experimental results provide a stringent test of electronic wave functions which have (usually) been generated by minimizing the energy in a variational approach.[42]

[41] C. H. Townes and B. P. Dailey, *J. Chem. Phys.* **17**, 782 (1949); W. Gordy, *Disc. Faraday Soc.* **19**, 14 (1955); J. H. Goldstein, *J. Chem. Phys.* **24**, 106 (1956); L. Pierce, R. G. Hayes, J. F. Beecker, *J. Chem. Phys.* **46**, 4352 (1967).

[42] C. T. O'Konski and T. K. Ha, *J. Chem. Phys.* **49**, 5354 (1968).

TABLE 7.17

Representative quadrupole coupling constants.

Molecule	Nucleus	Coupling constant(s) (MHz)
FCl	Cl^{35}	-146.0
ICl	Cl^{35}	-82.5
	I^{127}	-2930
KCl	Cl^{35}	0.04
RbCl	Cl^{35}	0.77
CH_3Cl	Cl^{35}	-74.8
LiI	I^{127}	-198.2
CH_3CN	N^{14}	-4.21
CH_3NC	N^{14}	0.48
N_2O	N^{14}	$(eqQ)_{end} = -0.79$
		$(eqQ)_{central} = -0.24$
$\underline{CH_2-NH-CH_2}$	N^{14}	$(eq_aQ) = 0.69$
		$(eq_bQ) = 2.17$
		$(eq_cQ) = -2.86$
CH_2Cl_2	Cl^{35}	$(eq_aQ) = -41.8$
		$(eq_bQ) = 2.6$
		$(eq_cQ) = 39.2$
CD_2O	D	$(eq_aQ) = -0.013$
		$(eq_bQ) = 0.098$
		$(eq_cQ) = -0.085$

In order to put the discussion on a firmer base, suppose that we have available a set of *one-electron* molecular orbitals ψ_i formed from linear combinations of atomic orbitals φ_j centered at each nucleus,

$$\psi_i = \sum_j c_j^{(i)} \varphi_j \qquad (7.145)$$

We leave unspecified the computational procedure in obtaining the ψ_i except to note that the c_j have been obtained by a variational procedure. Also, the range of j (that is, the size of the basis set) is not specified. By the usual procedure, the expectation value of q_z, say, would be obtained as

$$\left\langle \frac{\partial^2 V}{\partial z^2} \right\rangle = \sum_i \left\langle \psi_i \left| \frac{\partial^2 V}{\partial z^2} \right| \psi_i \right\rangle (2)$$

$$= 2 \sum_i \left\{ \sum_j \sum_k c_j^{(i)*} c_k^{(i)} \left\langle \varphi_j \left| \frac{\partial^2 V}{\partial z^2} \right| \varphi_k \right\rangle \right\} \qquad (7.146)$$

where we have summed over all doubly occupied molecular orbitals.

In any nonempirical calculation the approach would be to simply substitute Eq. (7.137) for V, perform the indicated integrations over atomic orbitals and then sum over all occupied molecular orbitals. The integral evaluations are not easy and invariably have involved numerous approximations. Still, several useful generalizations can be arrived at.

(1) The contributions to $\langle q \rangle$ from the nuclei are often small compared to the contribution from the electrons.

(2) Contributions from terms of the form $\langle \varphi_j \mid \tilde{q} \mid \varphi_k \rangle$, $k \neq j$, are often small compared to those from $\langle \varphi_j \mid \tilde{q} \mid \varphi_j \rangle$.

(3) Spherically distributed electrons give no contribution to $\langle q \rangle$, hence terms such as $\langle 1s_1 \mid \tilde{q} \mid 1s_1 \rangle$ contribute nothing if the nucleus in question is nucleus 1.

(4) Again considering nucleus 1, terms involving d orbitals are usually small compared to those involving p orbitals, that is $\langle 2p_1 \mid \tilde{q} \mid 2p_1 \rangle \gg \langle 3d_1 \mid \tilde{q} \mid 3d_1 \rangle$.

The qualitative result is that if p-orbital contributions are present they tend to dominate the field gradient. In this case Eq. (7.146) becomes

$$\left\langle \frac{\partial^2 V}{\partial z^2} \right\rangle = 2 \sum_i \sum_j^p c_j^{(i)*} c_j^{(i)} q_j \tag{7.147}$$

where $q_j = \langle \varphi_j \mid \tilde{q} \mid \varphi_j \rangle$ and the sum is over p orbitals centered on the nucleus of interest. If we further consider twice the summation over i to be the electron populations N_j of the various orbitals we arrive at

$$q = \left\langle \frac{\partial^2 V}{\partial z^2} \right\rangle = \sum_j^p N_j q_j \tag{7.148a}$$

$$= N_x q_x + N_y q_y + N_z q_z$$

where the latter equality assumes that the p orbitals are of only one principal quantum number and are oriented in the x, y, and z directions.

Expression (7.148a) is one of the common forms used for qualitative or semiempirical considerations. Note that it requires a knowledge of only three matrix elements, namely q_x, q_y, and q_z. These are easily evaluated (neglecting the nuclear terms) for atomic hydrogenlike p orbitals and it is found that

$$q(np_x) = q(np_y) = -\tfrac{1}{2}q(np_z) \tag{7.149}$$

for the case under consideration, that is, when the operator is $\dfrac{\partial^2 V}{\partial z^2}$. Using

(7.149), Eq. (7.148a) becomes

$$q = \{-\tfrac{1}{2}(N_x + N_y) + N_z\}q(np_z) \qquad (7.148b)$$

or

$$q_{mol} = \{-\tfrac{1}{2}(N_x + N_y) + N_z\}q_{at} \qquad (7.148c)$$

where in (7.148c) we clearly distinguish the molecular and atomic field gradients.

To illustrate Eq. (7.148c) let us consider CH_3Cl. In a simple picture of covalent bonding the chlorine $3p_x$, $3p_y$, and $3p_z$ orbitals would be occupied by 2, 2, and 1 electron(s), respectively, where the z axis is the bond axis. Therefore by use of Eq. (7.148c)

$$q_{mol} = -q_{at} \qquad (7.150a)$$

or

$$eQq_{mol} = -eQq_{at} \qquad (7.150b)$$

For atomic chlorine it is known that $eQq_{at} = 109.7$ MHz, so we predict $eQq_{mol} = -109.7$ MHz. The observed value for CH_3Cl is actually -74.70 MHz, which can be taken as an indication that the p electron density in the z direction relative to the x- and y-directions is greater than that predicted by the very simple bonding picture described above.

The picture can be modified in several ways. If, for example, the chlorine bonding orbital were 20% hybridized with Cl $3s$, we would predict $(N_x + N_y)$ and N_z to be 4 and 1.20, respectively, where the latter value arises from 0.8 of a bonding p_z electron and $2(0.2) = 0.4$ of a nonbonding p_z electron. Hence eQq_{mol} is now predicted to be $-0.8\ eQq_{at} = -88$ MHz which is in considerably better agreement with the experimental value but still not outstanding. By adding additional parameters such as ionic character it is possible, of course, to fit theory with experiment exactly.

7.8 HINDERED INTERNAL ROTATION

As previously mentioned there are many occasions when nonrigid-rotor behavior must be considered in the analysis of pure rotational spectra. In earlier sections of this chapter we have described two of these, namely, centrifugal distortion and the variation of moments of inertia with vibrational state. A much more important case occurs typically in molecules having methyl groups which are able to undergo a torsional oscillation, as in CH_3–CH_2F for example. These torsional vibrations occur at frequencies of a few hundred wavenumbers or less. The torsional potential function is very anharmonic and consists of an appropriate series of maxima and minima separated in energy by a few to several hundred or thousand wavenumbers. Because the methyl protons have such small mass they can tunnel through the potential barrier rather easily, giving rise to *hindered internal rotation*. For a symmetrical methyl group undergoing internal rotation the potential function has the general appearance of that shown in Fig. 7.20. It consists of maxima every 120°, that is every time the methyl protons eclipse the atoms of the other half of the molecule. We use the angle α to measure the torsional motion, and indicate the height of the barrier between maxima and minima by V_3. While the exact shape of the potential function is generally not known, it is usually represented to a first approximation by

$$V(\alpha) = \tfrac{1}{2}V_3(1 - \cos 3\alpha) \tag{7.151}$$

which has the correct behavior for threefold situations. Hindered internal rotation occurs for molecules with other than threefold barriers also, such as CH_3-NO_2 (sixfold) and H_2O_2 (chiefly twofold). In these cases the function $V(\alpha)$ is modified appropriately.

The quantum-mechanical treatment for a molecule which is rigid, except for the hindered internal rotation, may be obtained by returning to the general Hamiltonian discussed in chapter 3. The usual approximation is to neglect all the vibrations other than the torsion and, in the case of threefold barriers, to use the vibrational potential function given in Eq. (7.151). When the barrier height is not too high, the angular momentum produced by the torsional motion about the internal rotation axis becomes strongly coupled to the over-all angular momentum, so it becomes necessary to include the appropriate vibration-rotation coupling term from Eq. (3.44). The detailed treatments of this problem have been given in

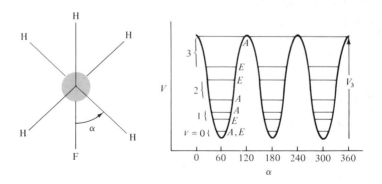

FIG. 7.20.   Potential function for hindered internal rotation of a methyl group.

several places.[43] We merely look at the problem in a qualitative way in order that the spectral effects and results may be discussed.

In the most common cases, V_3 is relatively high (greater than a few hundred cm^{-1}) so the vibration-rotation coupling terms are small. The rotational and torsional problems can then be treated separately and the coupling term treated as a small perturbation.

The torsional wave equation that arises for the threefold case is a well-known one and can be solved with no great difficulty. The result is that each torsional state is split into two substates, one of them doubly degenerate and the other nondegenerate. The torsional wave functions must transform like the irreducible representations of the C_3 point group because of the threefold symmetry of the internal rotor, therefore the states can be labeled as A or E, which are the two possible symmetry species for this group. As shown in Fig. 7.20, the A-E splitting is small in the lower states but becomes larger as the top of the barrier is approached.

When the rotational problem is treated and the perturbation terms have been added, it is found that the rotational energy levels are essentially like those of a rigid rotor except that the level spacings are different in each torsional state v and in each substate E or A. The difference in the rotational energy-level spacings in the E and A substates of a given torsional state v can be shown to be related directly to the barrier height V_3.[44] When the rotational energy levels are plotted for an asymmetric rotor with a threefold barrier the result is as shown schematically in Fig. 7.21.

The selection rules for a hindered asymmetric rotor are those of the rigid molecule plus the additional requirement that

$$A \leftrightarrow A$$
$$\text{(7.152a)}$$
$$E \leftrightarrow E$$

are allowed but

$$A \not\leftrightarrow E \qquad\qquad \text{(7.152b)}$$

Then as shown in Fig. 7.21 each rigid-asymmetric-rotor transition splits into a doublet, one component arising from the A torsional substate and one from the E substate. These doublet splittings are typically a few tenths to several 10's or 100's of MHz in the ground ($v = 0$) torsional state, and analysis of these splittings leads directly to V_3.

For low-barrier molecules ($V < 100$ cm^{-1}) the method outlined above is modified such that the zero-order problem is that of free internal rotation. The barrier-dependent terms are then added in by perturbation theory or by a complete matrix-mechanics approach.

---

[43] A good review article is: C. C. Lin and J. D. Swalen, *Rev. Mod. Phys.* **31**, 841 (1959).

[44] See footnote 43.

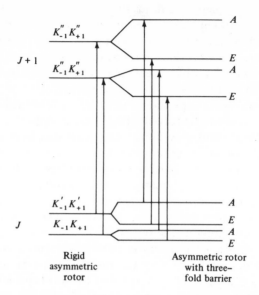

FIG. 7.21.   A portion of the rotational energy-level diagram for an
asymmetric rotor with threefold hindered internal
rotation.

Table 7.18 lists some representative barriers to internal rotation that
have been determined by microwave techniques. While there is considerable experimental evidence now available there is still no solid theoretical
explanation for the origin of these barriers. A variety of theories[45] have
been proposed to account for this phenomenon, including nonbonded interactions, steric effects, σ-bond properties and many others, but the answer
seems most likely to be found only in complete quantum-mechanical calculations of electronic energies.[46] At the present time these efforts are still
beset by serious computational difficulties even for medium-sized molecules.

7.9   STRUCTURAL DETERMINATION

In previous sections of this chapter we have spoken in some detail
about the determination of bond lengths in simple molecules. It was emphasized that the fundamentally most meaningful parameters are those
calculated from the equilibrium moments of inertia. These r_e parameters are,
unfortunately, almost impossible to determine for polyatomic molecules.

[45] See, for example, E. B. Wilson, Jr., *Advan. Chem. Phys.* **2,** 367 (1959).
[46] R. M. Pitzer and W. N. Lipscomb, *J. Chem. Phys.* **39,** 1995 (1963); W. H. Fink
and L. C. Allen, *J. Chem. Phys.* **46,** 2261 (1967).

TABLE 7.18

Barriers to internal rotation from microwave spectroscopy.[a]

Molecule	Barrier (cal/mole)
CH_3OH	$V_3 = 1070$
CH_3NH_2	$V_3 = 1976$
CH_3CHO	$V_3 = 1150$
CH_3CH_2Cl	$V_3 = 3560$
$CH_3CH_2{=\!=}CH_2$	$V_3 = 1978$
CH_3COOH	$V_3 = 483$
$CH_3{-}CH{-}CH_2{-}O$	$V_3 = 2560$
$CH_3{-}CH{-}CH_2{-}NH$	$V_3 = 2608$
CH_3NO_2	$V_6 = 6.03$
CH_3BF_2	$V_6 = 13.77$
H_2O_2	$V_2 = 323$
C_6H_5OH	$V_2 = 3140$

[a] Adapted from C. C. Lin and J. D. Swalen, *Rev. Mod. Phys.* **31**, 841 (1959).

The easiest structural parameters to obtain are the r_0 parameters which satisfy the ground-state moments of inertia. To determine an r_0 structure for a nonplanar molecule we therefore require a knowledge of as many independent moments of inertia as there are parameters. Since the spectrum of a given isotopic species produces only three moments of inertia it is possible to obtain from this data only three relations involving the molecular parameters. The additional necessary equations have traditionally been gotten by determining the moments of inertia of other isotopic species of the same molecule, with the additional assumption being that the relative atomic positions are independent of the isotopic change. This assumption is not too bad, but leads to discrepancies and inconsistencies which limit the accuracies of bond distances obtained in this manner to about 0.01–0.02 Å. This can be seen from the OCS data of Table 7.19, where various pairs of isotopic data have been used to determine the two parameters of the linear molecule. The error in locating hydrogen atoms is often considerably worse than for the heavier atoms because of the very small contribution of these light nuclei to the moments of inertia and because of their large zero-point vibrational amplitude. The chief reason for the discrepancies in r_0 parameters is due to the neglect of vibration-rotation interactions which are small but not negligible in the ground vibrational state.

TABLE 7.19

r_0 and r_s structures for carbonyl sulfide (OCS).

r_0 Structures[a]

Pair of molecules	C—O (Å)	C—S (Å)
$O^{16}C^{12}S^{32}$, $O^{16}C^{12}S^{34}$	1.1647	1.5576
$O^{16}C^{12}S^{32}$, $O^{16}C^{13}S^{32}$	1.1629	1.5591
$O^{16}C^{12}S^{34}$, $O^{16}C^{13}S^{34}$	1.1625	1.5594
$O^{16}C^{12}S^{32}$, $O^{18}C^{12}S^{32}$	1.1552	1.5653
Mean	1.1614	1.5604
Range	0.0095	0.0077

r_s Structures[b]

Reference molecule	C—O (Å)	C—S (Å)
$O^{16}C^{12}S^{32}$	1.16012	1.56020
$O^{18}C^{12}S^{32}$	1.15979	1.56063
$O^{16}C^{13}S^{32}$	1.16017	1.56008
$O^{16}C^{12}S^{34}$	1.16075	1.55963
Mean	1.16021	1.56014
Range	0.00096	0.00100

[a]   C. H. Townes, A. N. Holden and F. R. Merritt, *Phys. Rev.* **74**, 1113 (1948).
[b]   See Ref. 20 in text.

Although the r_0 structural method has its limitations it is widely used. For molecules with 10 or more parameters it usually is impossible to get sufficient isotopic date to calculate all parameters so the common practice is to fix as many parameters as possible from other evidence. For example, C–H bond distances in methyl and methylene groups are always about 1.08 or 1.09 Å, so unless some fine detail is being looked for it is reasonable to fix these parameters initially.

The most satisfactory alternative to the determination of r_e parameters is the r_s substitution method of Costain[47] which we illustrated for the linear molecule in Example 7-2. The method can be applied to any type of molecule although the treatment becomes more complex than for the

[47] See footnote 20.

linear molecule. For an asymmetric rotor with no symmetry, it has been shown that the x coordinate of any atom measured in the principal-axis system of the reference molecule is given by[48]

$$x^2 = (1/2\mu)\{(I_y' - I_y) + (I_z' - I_z) - (I_x' - I_x)\}$$

$$\cdot \left\{1 + \frac{(I_x' - I_x) - (I_y' - I_y) + (I_z' - I_z)}{2(I_x - I_y)}\right\}$$

$$\cdot \left\{1 + \frac{(I_x' - I_x) + (I_y' - I_y) - (I_z' - I_z)}{2(I_x - I_z)}\right\} \quad (7.153)$$

where I_x, I_y, I_z are the moments of inertia of a reference molecule and I_x', I_y', I_z' the moments of inertia of the molecule having the atom of interest isotopically substituted. μ is given by

$$\mu = \frac{M \Delta m}{M + \Delta m} \quad (7.154)$$

where M is the mass of the reference molecule and Δm is the mass of the substituted molecule minus that of the reference molecule. Expressions analogous to Eq. (7.153) can be written for y^2 and z^2 by cyclic permutation of the subscripts.

The principal feature of this method is that only differences of moments of inertia occur, so there is considerable cancellation of nonrigid-rotor effects even though the moments of inertia are evaluated in the ground vibrational state. Costain[49] has shown, in addition, that these r_s parameters are much better approximations to the desirable r_e values. That is,

$$r_e < r_s < r_0 \quad (7.155)$$

is thought to be generally true.

The primary drawbacks to the substitution method are immediately obvious: (1) Only atoms having more than one stable isotope can be located by this procedure, which eliminates its application to such atoms as F or P. (2) For a complete structural determination the number of isotopic substitutions or the number of *sets* of moments of inertia required is equal to the number of atoms in the molecule. For large molecules the chemical synthesis problems usually become the limiting factor since the natural abundances of many isotopes are very low, for example, $C^{13} = 1\%$, $H^2 = .01\%$, $O^{18} = 0.20\%$. If all but one atom is located by the substitution

[48] J. Kraitchman, *Am. J. Phys.* **21**, 17 (1953).
[49] See footnote 20.

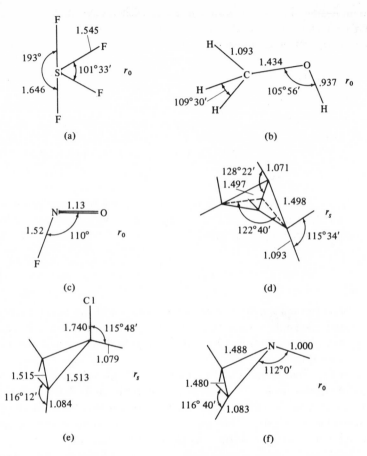

FIG. 7.22.   Structures of some asymmetric rotors. (a) W. M. Tolles
and W. D. Gwinn, *J. Chem. Phys.* **36,** 1119 (1962). (b)
K. V. Ivash and D. M. Dennison, *J. Chem. Phys.* **21,**
1804 (1953). (c) D. W. Magnuson, *J. Chem. Phys.* **19,** 1071
(1951). (d) K. W. Cox, M. D. Harmony, G. Nelson, and
K. B. Wiberg, *J. Chem. Phys.* **50,** 1976 (1969). (e) R. H.
Schwendeman, G. D. Jacobs, and T. M. Krigas, *J. Chem.
Phys.* **40,** 1022 (1964). (f) T. E. Turner, V. C. Fiora, and
W. M. Kendrick, *J. Chem. Phys.* **23,** 1966 (1955).

method, it is common practice to locate the last atom by use of the first
moment equations as was done for F in Example 7-2.

Although the task is often a difficult one the structures of many interesting asymmetric rotors have been determined by microwave spectroscopy in the manner described above. When r_s structures are obtained the

self-consistency for overdetermined systems is always very good, but is usually poorest for hydrogen parameters. Table 7.19 shows the OCS r_s parameters for comparison to the r_0 results previously discussed. It is clear that the consistency is excellent. In addition, the evidence is strong that the r_s parameters differ from the r_e values by only a few thousandths of an angstrom in favorable cases.

Finally in Fig. 7.22 are listed a few of the many asymmetric rotor structures that have been determined in recent years. We have identified the principal method (r_0, r_s or combination) used for each determination and have given the references to the original work. In several of the examples only the most interesting parameters have been listed.

SUPPLEMENTARY REFERENCES

W. Gordy, W. V. Smith, and R. F. Trambarulo, *Microwave Spectroscopy* (Wiley, New York, 1953).

M. W. P. Strandberg, *Microwave Spectroscopy* (Methuen, London, 1954).

T. M. Sugden and C. N. Kenney, *Microwave Spectroscopy of Gases* (Van Nostrand, London, 1965).

C. H. Townes and A. L. Schawlow, *Microwave Spectroscopy* (McGraw-Hill, New York, 1955).

J. E. Wollrab, *Rotational Spectra and Molecular Structure* (Academic, New York, 1967).

CHAPTER 8

INFRARED SPECTROSCOPY

8.1 GENERAL CHARACTERISTICS AND EXPERIMENTAL ASPECTS OF INFRARED SPECTROSCOPY

The most common region of the infrared spectrum (the region covered by common commercial instruments) is 600 to 4000 cm^{-1} or 16 μ to 2.5 μ on the less useful but still common wavelength scale. It is in this frequency region that the bulk of the molecular spectra attributable to molecular vibrations occur, although some vibrational spectroscopy is performed in the *near* infrared (up to 10 000 cm^{-1}) and more recently technological advances have made it possible to observe vibrational spectra in the *far* infrared down to frequencies as low as 20 cm^{-1}.

Infrared spectra may originate from causes other than primarily intramolecular vibrational motions. For example, light molecules in the gas phase may exhibit pure rotational transitions at frequencies up to and beyond the far infrared. More importantly, solids may exhibit absorption bands that are not attributable entirely to intramolecular vibrations. These include lattice vibrational motions and electronic transitions in semiconductors. Although there is considerable interest in these solid-state infrared spectra[1] we consider—in keeping with our general theme of molecular spectra—only infrared molecular vibrational spectra.

One of the major limitations of microwave (rotational) spectroscopy,

[1] See, for example, C. Kittel, *Introduction to Solid State Physics* (Wiley, New York, 1956).

the requirement that the substance be in the gas phase, does not occur in vibrational spectroscopy since the spectra can be observed in all three states of matter: solid, liquid, and gas. The general character of the spectra differ, however, particularly in the latter case in which the vibrational transitions have superimposed on them simultaneous rotational transitions. Consequently, gas-phase spectra consist of resolved or unresolved rotation-vibration *bands* consisting of many relatively sharp lines, which in the former case have the potential of providing some of the same information as obtained from microwave spectra. On the other hand, infrared bands in condensed phases are relatively broader due to the stronger intermolecular interactions, and show little if any of the rotation-vibration fine structure except in the case of solid matrix isolation spectra (see Ref. 1, chapter 7) or very small molecules in liquid solutions. Figure 8.1 shows the behavior of the HCl spectrum in liquid solution and in the gas phase.

Nearly any size or kind of molecule exhibits an infrared spectrum; indeed, as we see later, only homonuclear diatomic molecules lack such a spectrum under normal conditions. Thus an infrared spectrum of polyethylene can be studied, as well as that of the hydrogen chloride molecule. In general, the complexity of the spectrum increases as the number of atoms increase, and we find that it often becomes impossible to do complete quantitative analyses of infrared spectra when as few as five or six atoms are present, although relatively simple spectra may occur for much larger molecules. Nevertheless, qualitative utilization of infrared spectra is possible for nearly all molecules and thus infrared spectroscopy is one of the most widely used tools in the physical sciences.

In comparison to microwave spectroscopy, the resolution of normal infrared spectroscopy is orders of magnitude less. Except in the case of high-resolution studies of gases, some 20 or 30 distinctly visible absorption

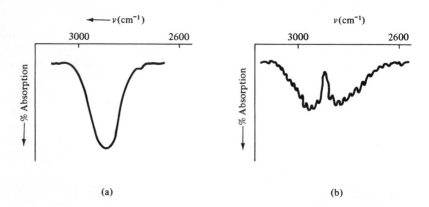

FIG. 8.1.   Sketch of the appearance of the intense infrared band of HCl in (a) solution (inert solvent) and (b) gas phase.

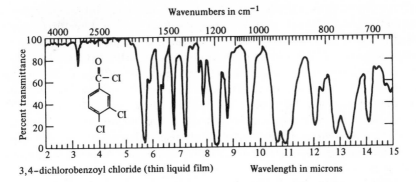

FIG. 8.2.   Infrared spectrum of 3,4-dichlorobenzoyl chloride. From
Robert T. Conley, *Infrared Spectroscopy*, Fig. 544, p. 152.
© Copyright 1966 by Allyn and Bacon, Inc., Boston.
Reprinted by permission of the publisher.

peaks often cover the entire standard infrared range as illustrated in Fig.
8.2 for 3,4 dichlorobenzoyl chloride. Yet the great acceptance and use of
the technique is attributable partly to the fact that only a relatively small
number of absorption features are present in even very complicated molecules.

Figure 8.3 shows a block diagram of a typical double-beam infrared
spectrometer. Continuous infrared radiation from the source is passed
alternately through a sample and reference cell by means of the chopper.
The two light beams then pass into a monochromator and then onto the
detector. The output of the detector is amplified at the chopper frequency
which is typically 100 Hz. The amplifier compares the signal appearing
from the reference and sample cycles and by means of a servo-mechanism

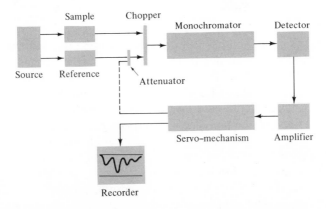

FIG. 8.3.   Simplified block diagram of infrared spectrometer.

adjusts the infrared radiation energy in the reference beam by means of an optical attenuator such that the energy falling on the detector from the two beams is always the same, that is, the spectrometer operates by the null technique. If, for example, the sample is absorbing at the wavelength being passed by the monochromator, the amplifier system senses the imbalance of the two beams and causes the servo-mechanism to attenuate the reference beam. Of course, this process goes on continuously as the monochromator is being scanned so when absorption bands are passed over the optical attenuator always maintains a null condition by decreasing or increasing the reference beam intensity as needed. A recorded spectrum giving per cent transmission (or some similar absorption or transmission measure) is easily obtained by driving a recorder with a signal that is synchronous with that driving the attenuator.

Common sources of infrared radiation are the Nernst glower and the Globar. The former consists of a rare-earth oxide electrode while the latter is a piece of silicon carbide. When heated to temperatures of 1000–2000°K by passing relatively large currents through them, they emit useable amounts of radiant energy over the near infrared region. While these sources produce less radiant energy than would an ideal black body at the same temperature, it is useful to consider briefly the ideal limiting case.

The Planck black-body radiation formula is written in terms of frequency as

$$E_\nu d\nu = \frac{8\pi h\nu^3}{c^2} \frac{1}{\exp(h\nu/kT) - 1} d\nu \tag{8.1a}$$

where $E_\nu d\nu$ is the total radiant power in erg cm^{-2} sec^{-1}, in the frequency range ν to $\nu + d\nu$ at temperature T. This formula can be recast in terms of wavelengths λ, so the power between λ and $\lambda + d\lambda$ is

$$E_\lambda d\lambda = \frac{8\pi hc^2}{\lambda^5} \frac{1}{\exp(hc/\lambda kT) - 1} d\lambda \tag{8.1b}$$

In Fig. 8.4 this function has been plotted for two temperatures. It is seen that the total power output increases at all wavelengths as T increases, but the wavelength of maximum energy moves to shorter wavelengths. Consequently, in the major part of the infrared region the radiant power is relatively low and represents only a small fraction of the total emitted light. Nevertheless, the power available from practical sources is sufficient for spectroscopic purposes, and with sensitive detection apparatus it has been possible to use them at frequencies down into the far-infrared region.

Monochromators for infrared spectroscopy are of either prism or grating types. In general, the optical material problems are among the most difficult to deal with in infrared spectroscopy. Windows and prisms must be transparent to infrared radiation and the latter must have simultaneously good dispersion properties. Let us be content here to simply mention that such materials as NaCl, KBr, LiF, AgCl and others prove to be satisfactory over various portions of the near-infrared spectral region. Again the interested reader is urged to consult the general references for detailed discussions.

Detectors of infrared radiation are of two chief types, thermal and nonthermal. The former are simply "black" absorbers of radiation; the temperature increase of the absorber is then measured commonly by a thermocouple or bolometer (temperature-sensitive resistor). Thermocouple and bolometer detectors are the most common since their broad-banded nature makes them useful over the common near-infrared region.

The nonthermal detectors include chiefly the semiconductors which show photoconductivity or photodiode behavior. Included here would be germanium, lead sulfide, and lead selenide detectors. The general characteristics of these types of detectors are their very high sensitivity, relatively sharp long-wavelength limit, and fast response times. Because of their limited frequency range they are seldom used for general purpose infrared spectroscopy.

Finally, we might mention that gas-phase spectra are studied typically at pressures on the order of 10 Torr with path lengths of 10 cm or longer. Pure liquids or solutions are studied as rather thin films on the order of 0.1 mm thickness. Solid polycrystalline samples present somewhat more

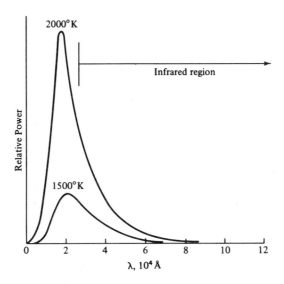

FIG. 8.4. Planck blackbody distribution of radiant energies.

difficulty due to scattering of the incident radiation. To avoid this problem solid samples are usually powdered and dispersed in either a solid or liquid supporting medium. The two most common supporting agents are Nujol (a viscous organic liquid) and KBr. In the first case the sample is "mulled" into the Nujol and the suspension squeezed between NaCl windows. In the second, the sample is ground with KBr and then pressed into a pellet which has little scattering if properly made. Solids may also be studied as amorphous thin films and *reflectance* spectra are obtained less commonly.

8.2   INFRARED SPECTRA OF DIATOMIC MOLECULES

A.   Harmonic Oscillation in Classical Mechanics

In Sec. 3.3 we have shown that the classical Hamiltonian for the harmonically oscillating diatomic molecule is given by

$$\mathcal{K}_v = \frac{p^2}{2\mu} + \tfrac{1}{2}k\xi^2 \tag{3.89}$$

with μ being the reduced mass, p the linear momentum, $\mu\dot{\xi}$ and ξ being the variations in bond length from equilibrium $R - R_e$. By using Hamilton's equation (see Appendix A), we find

$$\frac{\partial \mathcal{K}_v}{\partial \xi} = -\mu\ddot{\xi} \tag{8.2a}$$

which becomes by use of Eq. (3.89)

$$k\xi + \mu\ddot{\xi} = 0 \tag{8.2b}$$

This is readily recognized as Newton's equation $f = \mu a$ since $\mu\ddot{\xi} = \mu a$ and $k\xi = -f$ are the restoring force produced by the Hooke's law potential. Equation (8.2b) describes the time dependence of the bond length of the harmonically oscillating molecule.

It is easily verified that the general solution of the differential equation (8.2b) is simply

$$\xi = A \cos(2\pi\nu_e t + \varphi) \tag{8.3}$$

where

$$\nu_e = \frac{1}{2\pi} \left(\frac{k}{\mu}\right)^{1/2} \tag{8.4}$$

and A and φ are the two constants required for the second-order equation. Hence we see that the variation of the bond distance from equilibrium ξ, is given at any time t by Eq. (8.3); and further we note that ξ varies periodically with a frequency given by Eq. (8.4). This periodic oscillation of ξ

with time is the characteristic behavior of simple harmonic oscillation. We henceforth consider ν_e to be the classical *fundamental* vibrational frequency of the diatomic molecule.

The phase angle φ need be of no concern to us since it merely reflects the choice of the time-scale axis, but it is perhaps useful to indicate how the vibrational amplitude A is determined classically. If we substitute Eq. (8.3) into Eq. (3.89) for the total energy $\mathcal{3C}_v$ we obtain

$$E_v \equiv \mathcal{3C}_v = \tfrac{1}{2}kA^2 \tag{8.5}$$

Thus, knowledge of the total energy of the oscillator at any time permits the amplitude to be found if the force constant k is known. As we see shortly, a typical diatomic molecule may have an energy of 2×10^{-13} erg and a force constant of 4×10^5 dyne cm^{-1}. This leads to a classical vibrational amplitude of 1×10^{-9} cm or 0.1 Å.

B.   Energy Levels of the Diatomic Vibrator

Classical mechanics does not place any restriction upon the permitted energy of a vibrating diatomic molecule. Our interest now is to see what quantum mechanics tells us about the allowed energies of the diatomic vibrator. In fact, the answer has already been given in Eq. (4.30), which applied to a particle of mass m undergoing harmonic oscillation in the x direction. Thus, these results apply here merely by replacing m with μ and using ν_e instead of ν. The result is

$$E_v = \langle v \mid \tilde{\mathcal{3C}}_v \mid v \rangle = H_{vv} = h\nu_e(v + \tfrac{1}{2}) \tag{8.6}$$

with ν_e given by Eq. (8.4) and the vibrational quantum number taking on values of $v = 0, 1, 2, 3, \ldots$. Figure 4.3 has already presented an energy-level diagram in graphical form and this is applicable to the present case if $\xi = R - R_e$ is substituted for x.

Before comparing Eq. (8.6) to experimental results let us review briefly the assumptions which entered the derivation. First, the classical vibrational Hamiltonian Eq. (3.89) was obtained from the general molecular Hamiltonian by separating it from the rotational energy term with neglect of the vibration-rotation interaction term. We have seen in Sec. 7.3 A that the variation of rotational constant with vibrational state is very small, thus the separation of rotational and vibrational energy is very good for the diatomic molecule. The small vibration-rotation interaction that does occur has been adequately described by means of the effective rotational constant B_v.

The second major assumption was that of harmonic oscillation. This entered the derivation through the truncation of the potential energy at the quadratic terms as shown in Eq. (3.73). It is this assumption which is the most serious, since we know that it is physically unreasonable in the

TABLE 8.1

Energy levels of HCl[35]. Harmonic predictions used $\omega_e = 2885.9$; anharmonic predictions used data given in text.

	Measured[a] (cm^{-1})	Calculated (harmonic) (cm^{-1})	Calculated (anharmonic) (cm^{-1})
$E_1 - E_0$	2 885.9	(2 885.9)	2 885.7
$E_2 - E_0$	5 668.0	5 771.8	5 668.0
$E_3 - E_0$	8 347.0	8 657.7	8 347.0
$E_4 - E_0$	10 923.1	11 543.6	10 922.8
$E_5 - E_0$	13 396.5	14 429.5	13 395.3

[a] Data from footnote 2, this chapter, p. 55.

limit of high vibrational energies. This is obvious since the harmonic potential predicts the atoms of the molecule to be always bound together, whereas we know that the diatomic molecule can be dissociated into atoms. Thus we must expect that any major deficiencies of Eq. (8.6) arise from our choice of potential function.

In Table 8.1 we have listed the experimental energies of the first six vibrational states of HCl[35]. It is seen that the energy-level spacings are becoming smaller rather than remaining constant as predicted by Eq. (8.6). Indeed, this phenomenon continues until eventually the vibrational levels coalesce into a continuum in much the same fashion as do the energy levels of the hydrogen atom.

In order to account for this behavior let us add one more term in the potential energy expansion, namely a cubic term of the form

$$V' = k_3 \xi^3 \tag{8.7}$$

If this potential energy term is small we may expect to find its effect on the energy levels by means of perturbation theory. Thus, to second order, the correction terms are

$$E_v^{(1)} + E_v^{(2)} = k_3 \langle v \mid \xi^3 \mid v \rangle + k_3^2 \sum_{v'} \frac{\mid \langle v \mid \xi^3 \mid v' \rangle \mid^2}{E_v^0 - E_{v'}^0} \tag{8.8}$$

As shown by Eq. (4.38), ξ can have only $\xi_{v,v\pm1}$ elements; thus ξ^3 may have only $\xi^3_{v,v\pm1}$ or $\xi^3_{v,v\pm3}$ nonzero elements. For this reason the first-order correction in Eq. (8.8) goes to zero. Using the ordinary rules of matrix multiplication and the values of Eq. (4.38) it is found that the nonzero matrix elements are of the form

$$\xi^3_{v,v\pm3} = \xi_{v,v\pm1}\xi_{v\pm1,v\pm2}\xi_{v\pm2,v\pm1} \tag{8.9a}$$

$$\xi^3_{v,v\pm1} = \xi_{v,v+1}\xi_{v+1,v}\xi_{v,v\pm1} + \xi_{v,v-1}\xi_{v-1,v}\xi_{v,v\pm1} + \xi_{v,v\pm1}\xi_{v\pm1,v\pm2}\xi_{v\pm2,v\pm1} \tag{8.10a}$$

Somewhat tedious algebra leads after substitution of the appropriate values to

$$\xi^3_{v,v+3} = \alpha^{3/2}[(v + 1)(v + 2)(v + 3)]^{1/2} \tag{8.9b}$$

$$\xi^3_{v,v-3} = \alpha^{3/2}[v(v - 1)(v - 2)]^{1/2} \tag{8.9c}$$

$$\xi^3_{v,v+1} = 3\alpha^{3/2}(v + 1)^{3/2} \tag{8.10b}$$

$$\xi^3_{v,v-1} = 3\alpha^{3/2}v^{3/2} \tag{8.10c}$$

with

$$\alpha = \frac{\hbar}{2(k\mu)^{1/2}} \tag{8.11}$$

Further tedious algebra shows that when Eqs. (8.9) and (8.10) are substituted into Eq. (8.8) the correction $E^{(2)}$ becomes

$$E^{(2)} = -\frac{30\hbar^2 k_3^2}{8k^2\mu}\left(v + \frac{1}{2}\right)^2 - \frac{105k_3^2\hbar^2}{8k^2\mu} \tag{8.12}$$

The key point here is the dependence upon v which has been written in the form $(v + \frac{1}{2})^2$. Since we have gone to second order in our perturbation theory, one might ask about the effect of the still higher powers of ξ in the potential energy. If we add a term $k_4\xi^4$ we find that the energy is corrected to first order by an amount

$$E^{(1)} = \frac{3k_4\hbar^2}{2k\mu}\left(v + \frac{1}{2}\right)^2 + \frac{3k_4\hbar^2}{2k\mu} \tag{8.13}$$

Thus again we find the same functional dependence upon the vibrational quantum number.

If the power-series expansion of the potential energy is valid, then we expect $|k_4| \ll |k_3| \ll |k|$. The behavior of the vibrational energy levels as illustrated in Table 8.1 is then reasonable since the $(v + \frac{1}{2})^2$ term of Eq. (8.12) causes the energy level spacing to become smaller as v increases. The contribution produced by the leading term of Eq. (8.13) may be positive or negative in general, but seems to be of such a magnitude that the total coefficient of $(v + \frac{1}{2})^2$ is always negative.

Extension of these arguments leads one to believe that the *anharmonic* oscillator energy levels for the diatomic molecule are given by

$$E_v = h\nu_e(v + \frac{1}{2}) - h\nu_e x_e(v + \frac{1}{2})^2 + h\nu_e y_e(v + \frac{1}{2})^3 \ldots \tag{8.14a}$$

where the coefficients of the powers of $v + \frac{1}{2}$ have been written in the conventional spectroscopic manner and the sign choice is such that $\nu_e x_e > 0$. If all units in Eq. (8.14) are cgs, E_v is in ergs. It is conventional in much infrared literature to express the energy in wavenumbers and to introduce the wave-number frequency $\omega_e = \nu_e/c$. We then write Eq. (8.14a) as

$$E_v = \omega_e(v + \tfrac{1}{2}) - \omega_e x_e(v + \tfrac{1}{2})^2 + \omega_e y_e(v + \tfrac{1}{2})^3 + \cdots \quad (8.14b)$$

It is found in many cases that the first several vibrational energy states are well represented by the first two terms of Eqs. (8.14). For example, in Table 8.1 we have listed the predicted energy spacings using the best values of ω_e and $\omega_e x_e$ for HCl^{35}, namely, $\omega_e = 2988.95$ cm^{-1} and $\omega_e x_e = 51.65$ cm^{-1}. It should be noted that ω_e (or ν_e) is the appropriate quantity to relate to the classical oscillation frequency[2]; thus

$$k_e = 4\pi^2 \nu_e^2 \mu \quad (8.15)$$

gives the classical force constant for infinitesimal displacements about $R = R_e$.

Table 8.2 lists the most important vibrational parameters for a few diatomic molecules that have been spectroscopically investigated. We have identified the isotopes involved, since it is clear that an isotope effect enters through the reduced mass. Since the potential-energy function remains unchanged to a high approximation, the force constant is also expected to be unchanged by isotopic substitution. The principal effect therefore appears in ω_e and consequently in the energy-level spacing. For hydrogen chloride, the reduced mass ratio for the two isotopic species of Table 8.2 is

$$\frac{\mu(DCl)}{\mu(HCl)} \cong \frac{70/37}{35/36} = \frac{72}{37} = 1.95 \quad (8.16)$$

so $\omega_e(DCl)$ should be $(1.95)^{1/2}$ times smaller than $\omega_e(HCl)$ which is the result shown in the table.

TABLE 8.2
Vibrational parameters for diatomic molecules.

Molecule	ω_e (cm^{-1})	$\omega_e x_e$ (cm^{-1})	k_e (10^5 dyne cm^{-1})	D_0^0 (eV)
$C^{12}O^{16}$	2170.21	13.46	19.06	11.09
H_2	4395.2	117.9	5.74	4.476
HD	3817.1	95.0	5.78	4.511
HCl^{35}	2988.95	51.65	5.17	4.430
DCl^{35}	2143.7	26.6	5.16	4.481
HI^{127}	2309.5	39.7	3.15	3.056
N_2	2359.6	14.5	23.02	9.75
O_2	1580.4	12.0	11.79	5.080
KCl	280	0.9	0.86	4.42
NaI	286	0.8	0.94	3.16

[2] G. Herzberg, *Spectra of Diatomic Molecules* (Van Nostrand, Princeton, N.J., 1950), p. 98.

One of the most important applications of the experimental study of the energy levels of the diatomic molecule is the determination of the precise shape of the potential-energy curve. It should be recalled that this potential energy arose via the Born–Oppenheimer approximation for the separation of electronic and nuclear motions, and that it really represents the sum of kinetic and potential energies of the electrons. Thus if a sufficient number of the potential-energy derivatives of Eq. (3.72) could be measured the entire potential curve could be drawn out. This is not usually possible since in most cases the vibrational levels are not accurately known when v is large. One does know the general shape of the potential curve from theoretical considerations (see chapter 11). It has the form shown in Fig. 8.5. The shape of the curve is usually known experimentally near the bottom from measurements of ω_e and $\omega_e x_e$; the energy difference D_e between the minimum of the curve and the limiting potential energy when $R \rightarrow \infty$ is also known.

Morse[3] has shown that this general behavior may be well represented in many instances by a potential function of the form

$$V(\xi) = D_e[1 - \exp(-\beta\xi)]^2 \tag{8.17}$$

D_e is the dissociation energy measured from the minimum of the potential function [the (imaginary) $v = -\frac{1}{2}$ state], and is related to the ground-state ($v = 0$) dissociation energy $D_0{}^0$ by

$$D_e = D_0{}^0 + \tfrac{1}{2}h\nu_e - \tfrac{1}{4}h\nu_e x_e + \cdots \tag{8.18}$$

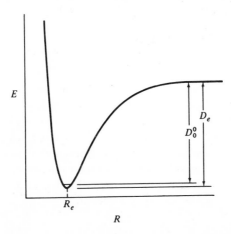

FIG. 8.5.   Typical potential function for a diatomic molecule.

[3] P. M. Morse, *Phys. Rev.* **34**, 57 (1929).

as can be seen easily by use of Eq. (8.14a) and Fig. 8.5. The parameter β is mainly responsible for determining the steepness of the potential function. The "Morse" curve Eq. (8.17) is physically reasonable except when $\xi = R - R_e = -R_e$, that is for $R = 0$. Here we should expect $V(\xi)$ to become infinite due to the $1/R$ Coulombic repulsion of the two nuclei. The Morse curve predicts a large but finite value, but since the curve is of no interest for $R \sim 0$ this is no serious defect.

Morse has shown that Schrödinger's wave equation can be solved using the potential of Eq. (8.17), the resulting energies being

$$E_v = h \left(\frac{D_e}{2\pi^2\mu} \right)^{1/2} \beta \left(v + \frac{1}{2} \right) - \frac{h^2\beta^2}{8\pi^2\mu} \left(v + \frac{1}{2} \right)^2 \tag{8.19}$$

where D_e and E_v are in ergs, β in cm^{-1}, and μ in g molecule^{-1}. This result is similar to the form previously inferred in Eq. (8.14a), and comparison shows that

$$\beta = \left(\frac{2\pi^2\mu}{D_e} \right)^{1/2} \nu_e \ cm^{-1} \tag{8.20a}$$

If D_e is in cm^{-1}, μ in amu, and $\nu_e = c\omega_e$ is introduced,

$$\beta = 0.1218\omega_e \left(\frac{\mu}{D_e} \right)^{1/2} \tag{8.20b}$$

where β is in reciprocal Angstrom units. The force constant can be expressed in terms of β and D_e by means of Eqs. (8.15) and (8.20a), or equivalently, by differentiation of $V(\xi)$ in Eq. (8.17). This gives

$$k_e = \left(\frac{\partial^2 V}{\partial \xi^2} \right)_{\xi=0} = 2\beta^2 D_e \ \text{(cgs units)} \tag{8.20c}$$

For our H^1Cl^{35} example, D_e is $3.723 \times 10^4 \ cm^{-1}$ so β has the value 1.87 Å^{-1} from Eq. (8.20b). The force constant given by Eq. (8.20c) is of course in agreement with the value of Table 8.2 calculated from Eq. (8.15).

The only major point remaining with respect to the vibrational energy levels is to emphasize that in the gas phase each vibrational state v has associated with it a manifold of rotational energy states as described in chapter 7. In fact, using the result given by Eq. (7.35), the energy of any rotation-vibration state $\langle vJ |$ may be written

$$E(v, J) = h\nu_e(v + \tfrac{1}{2}) - h\nu_e x_e(v + \tfrac{1}{2})^2 + hB_v J(J + 1) - hDJ^2(J + 1)^2$$

$$\tag{8.21}$$

through terms in $(v + \tfrac{1}{2})^2$.

C. Selection Rules and Intensities

In Sec. 7.3 B the rotational selection rules for the linear molecule were obtained by expanding the dipole moment in powers of the mass-weighted displacement coordinates as in Eq. (7.40). It was shown that the first term (μ_z^0 for the linear molecule) was the one which determined the pure rotational-transition selection rules. In the present case it is the second term in Eq. (7.40) that determines the selection rules between different vibrational states. Thus the probability of a transition between vibrational states is determined by

$$P \propto | \langle v | \left(\frac{\partial \mu_z}{d\xi}\right)_0 \xi | v' \rangle |^2 \tag{8.22}$$

which arises from Eq. (7.40) by expanding in terms of ξ instead of q and by noting that the sum reduces to one term for the diatomic molecule. If we wish to account for the simultaneous changes in rotational quantum numbers during the vibrational transition we need to introduce the direction cosine relations and sum over all three space-fixed axes which gives

$$P \propto | \langle vJM | \Phi_{Xz} \left(\frac{\partial \mu_z}{\partial \xi}\right)_0 \xi | v'J'M' \rangle |^2$$

$$+ | \langle vJM | \Phi_{Yz} \left(\frac{\partial \mu_z}{\partial \xi}\right)_0 \xi | v'J'M' \rangle |^2 \tag{8.23a}$$

$$+ | \langle vJM | \Phi_{Zz} \left(\frac{\partial \mu_z}{\partial \xi}\right)_0 \xi | v'J'M' \rangle |^2$$

Our convention is, as usual, that the molecular axis is the z axis for the diatomic molecule.

Separating the vibrational and rotational parts (no coupling of vibration and rotation) we obtain

$$P \propto P_{JM} \left| \left\langle v \left| \left(\frac{\partial \mu_z}{\partial \xi}\right)_0 \xi \right| v' \right\rangle \right|^2 \tag{8.23b}$$

in which P_{JM} is simply that part of Eq. (7.42) involving the direction cosines.

The derivative evaluated at $\xi = 0$ is a constant for a particular molecule, so we need to evaluate simply the matrix element $\langle v | \xi | v' \rangle$ in order to determine the vibrational selection rules. Using the results of Sec. 4.2 again we find the only nonzero matrix elements of the *harmonic oscillator* to be those of Eq. (4.38). Hence we have the selection rule

$$v' - v = \Delta v = \pm 1 \qquad (8.24)$$

where the plus sign is appropriate for absorption. In the harmonic oscillator approximation we have the result that transitions occur only between adjacent vibrational levels. The frequency of a vibrational transition $v \to v + 1$ is found to be

$$\nu = \frac{E_{v+1} - E_v}{h} = \nu_e - 2\nu_e x_e(v + 1) + \cdots \qquad (8.25)$$

by using Eq. (8.14a). Thus the transition frequency, aside from the anharmonicity terms, is simply the classical oscillator frequency ν_e. The small added terms, $-2\nu_e x_e(v + 1)$, etc., cause the $v = 0 \to 1$, $v = 1 \to 2$, $v = 2 \to 3$, etc., transitions to not coincide exactly in frequency, although they are rather close.

The actual intensity of any vibrational transition can be obtained by appropriate substitution into Eq. (5.69). For the transition $v \to v + 1$, using the matrix elements of Eq. (4.38), we find the intensity to be given by

$$\gamma_{v \to v+1} = \frac{8\pi^3(N_v - N_{v+1})}{3hc} \; \nu_{v \to v+1}(v + 1) \; \frac{\hbar}{2(k\mu)^{1/2}} \left(\frac{\partial \mu_z}{\partial \xi}\right)_0^2 \qquad (8.26)$$

Since vibrational energy levels are separated typically by greater than 500 cm^{-1}, the usual Boltzmann distribution shows that $N_v - N_{v+1} \sim N_v$ at temperatures below about 300°K. Thus

$$\frac{\gamma_{v+1 \to v+2}}{\gamma_{v \to v+1}} = \frac{N_{v+1}}{N_v} \frac{(v + 2)}{(v + 1)} \qquad (8.27a)$$

in which we have used the additional approximation $\nu_{v \to v+1} \sim \nu_{v+1 \to v+2}$. Then introducing the Boltzmann equation for N_{v+1}/N_v,

$$\frac{\gamma_{v+1 \to v+2}}{\gamma_{v \to v+1}} = \exp(-h\nu_e/kT) \left(\frac{v + 2}{v + 1}\right) \qquad (8.27b)$$

where the energy difference between $v = 0$ and $v = 1$ has been approximated by $h\nu_e$.

For a vibration having $\omega_e \sim 1000$ cm^{-1},

$$\exp(-1000/kT) = e^{-5} \sim 0.007$$

when $kT \cong 200$ cm^{-1}. The ratio of intensities of the

$$v = 0 \to 1 \quad \text{and} \quad v = 1 \to 2$$

is given from Eq. (8.27b) as

$$\frac{\gamma_{1\to2}}{\gamma_{0\to1}} = 0.007\,(2/1) = 0.014.$$

In a similar manner we find

$$\frac{\gamma_{2\to3}}{\gamma_{0\to1}} = (0.007)^2(3/1) = 1.5 \times 10^{-4}.$$

Thus we see that the intensities of the transitions permitted by the harmonic oscillator selection rules fall off very rapidly as v increases because of the exponential Boltzmann factor. Since the exponential term becomes larger as T gets larger, the higher v transitions become more intense at high temperatures. The transitions $v \to v + 1$ with $v > 1$ are therefore commonly called "hot bands," since they are most easily observed with heated samples.

Before proceeding with a more detailed discussion of infrared spectra, let us return briefly to Eq. (8.23b) for the transition probability. In addition to the harmonic oscillator selection rule $\Delta v = \pm 1$, we see that $P = 0$ if $(\partial\mu_z/\partial\xi)_0 = 0$. Figure 7.9 showed the general behavior of $\mu(R)$ for a molecule having a nonzero dipole moment. Clearly, except for the *accidental* case when $\xi = 0$ occurs at the maximum of the curve, $(\partial\mu_z/\partial\xi)_0$ is not zero unless μ_z is zero for any value of R. We expect, therefore, that *heteronuclear* diatomic molecules exhibit infrared spectra, which consist of at least the fundamental absorption bands, $v = 0 \to v = 1$. On the other hand, if $\mu(R) = 0$ always, such as for a *homonuclear* diatomic molecule, the first derivative is zero also. The result is that *homonuclear diatomic molecules have no ordinary infrared spectra.* Table 8.3 lists the observed fundamental transition frequencies in the infrared for several diatomic molecules.

TABLE 8.3

Vibrational fundamentals of diatomic molecules observed in the infrared spectrum.

Molecule	Frequency (cm^{-1})
HF	3958
HCl^{35}	2886
HBr^{79}	2559
HI^{127}	2230
CO	2143
NO	1876
NaCl	378
$Br^{79}F$	665
$I^{127}Cl^{35}$	380

D. Effects of Anharmonicity on Selection Rules

So far we have considered the vibrational selection rules which arise from the harmonic oscillator approximation and the approximation that only the dipole-moment derivatives through the first power [see Eq. (7.40)] need be considered. Thus it might be expected that the selection rules could be modified by either or both of two effects: *vibrational anharmonicity* or *electrical anharmonicity*. It can be seen readily that even if the molecule vibrates harmonically, a term

$$\left(\frac{\partial^2 \mu_z}{\partial \xi^2}\right)_0 \xi^2$$

in Eq. (7.40) leads to selection rules of $\Delta v = \pm 2$. This follows because the matrix elements of ξ are $v, v \pm 1$ in form, so ξ^2 may have $v, v \pm 2$ elements.

Even though the electrical anharmonicity is unquestionably involved in the breakdown of the selection rules, it is most likely the vibrational anharmonicity which accounts for the major intensity of infrared *overtone* bands. The first, second, third, etc., overtones obey the rules

$$\Delta v = \pm 2, \pm 3, \pm 4 \ldots$$

with decreasing intensities as Δv increases. Often the first overtone ($v = 0 \rightarrow v = 2$) has an intensity of a few percent of the fundamental. It occurs at a frequency of approximately twice the fundamental, but is actually somewhat lower because of the $-\omega_e x_e$ term in the energy expression.

The reason for the occurrence of overtones (aside from electrical anharmonicity) and the quantitative treatment of their intensities, can be seen readily from perturbation theory. If a term

$$\tilde{\mathcal{K}}' = \frac{1}{6}\left(\frac{\partial^2 v}{\partial \xi^2}\right)_0 \xi^3 = k_3 \xi^3 \tag{8.28}$$

is added as a perturbation to the harmonic oscillator Hamiltonian, the wave functions, correct to first order, are of the form[4]

$$\langle v \mid = \langle v \mid^0 + a \langle v + 1 \mid^0 \mid \langle v \mid \tilde{\mathcal{K}}' \mid v + 1 \rangle \mid^2$$
$$+ b \langle v - 1 \mid^0 \mid \langle v \mid \tilde{\mathcal{K}}' \mid v - 1 \rangle \mid^2 + c \langle v + 3 \mid^0 \mid \langle v \mid \tilde{\mathcal{K}}' \mid v + 3 \rangle \mid^2$$
$$+ d \langle v - 3 \mid^0 \mid \langle v \mid \tilde{\mathcal{K}}' \mid v - 3 \rangle \mid^2 \tag{8.29}$$

---

[4] From ordinary first-order perturbation theory the wave functions, correct to first order, are

$$\langle n \mid = \langle n^0 \mid + \sum_m{}' \frac{H_{mn}^{(1)}}{E_n^0 - E_m^0} \langle m^0 \mid$$

where the $\langle m^0 \mid$ and E_m^0 are the unperturbed eigenfunctions and eigenvalues, respectively.

The result is that small amounts of the $v - 1, v + 1, v - 3,$ and $v + 3$ wave functions are mixed in with the unperturbed state v. If the matrix elements of the dipole-moment operator of Eq. (8.22) are evaluated using the functions of the form of Eq. (8.29), it can be shown that the transition probability is nonzero for the first five and the seventh overtone. Inclusion of still higher terms in Eq. (8.28) obviously causes all overtones to be allowed. It should be emphasized, however, that the intensities of the overtones fall off quite rapidly and therefore those higher than the first are seldom observed.

EXAMPLE 8-1

As a more specific example we can predict the intensity of the first overtone relative to that of the fundamental of HCl. We represent the potential energy as a power series in ξ in the usual way:

$$V(\xi) = \frac{1}{2}\left(\frac{\partial^2 V}{\partial \xi^2}\right)_0 \xi^2 + \frac{1}{6}\left(\frac{\partial^3 V}{\partial \xi^3}\right)_0 \xi^3 + \cdots \tag{8.30}$$

The force constant for HCl is $k = (\partial^2 V/\partial \xi^2)_0 = 5.2 \times 10^5$ dyne cm^{-1} from Table 8.2. The cubic coefficient can be obtained by differentiating Eq. (8.17) which gives

$$\left(\frac{\partial^3 V}{\partial \xi^3}\right)_0 = -3\beta k_e \tag{8.31}$$

Using the value of β that was previously found for HCl we get $(\partial^3 V/\partial \xi^3)_0 = -2.92 \times 10^{14}$ dyne cm^{-2}.

With the potential function now completely described through the cubic term we can proceed to evaluation of the transition probabilities using Eq. (8.22). For the fundamental

$$P_{0\to1} \propto (\partial \mu_z/\partial \xi)_0^2 \,|\, \langle 0 \,|\, \xi \,|\, 1 \rangle \,|^2 = (\partial \mu_z/\partial \xi)_0^2 \alpha \tag{8.32}$$

where $\alpha = \hbar/2(k_e\mu)^{1/2}$ as before.

In order to find $P_{0\to2}$, the overtone transition probability, we use equations of the form of (8.29) to find the wave functions in the presence of the ξ^3 term in Eq. (8.30). Straightforward evaluation of matrix elements leads to

$$\langle 0 \,| = \langle 0^\circ \,| + A \langle 1^\circ \,| + B \langle 3^\circ \,| \tag{8.33a}$$

$$\langle 2 \,| = \langle 2^\circ \,| + C \langle 1^\circ \,| + D \langle 3^\circ \,| + E \langle 5^\circ \,| \tag{8.33b}$$

where

$$A = \frac{-3\alpha^{3/2}k_3}{h\nu_e} \qquad (8.34a)$$

$$B = \frac{-\sqrt{6}\,\alpha^{3/2}k_3}{3h\nu_e} \qquad (8.34b)$$

$$C = \frac{6\alpha^{3/2}k_3}{h\nu_e} \qquad (8.34c)$$

$$D = \frac{-9\sqrt{3}\,\alpha^{3/2}k_3}{h\nu_e} \qquad (8.34d)$$

$$E = \frac{-2\sqrt{15}\,\alpha^{3/2}k_3}{3h\nu_e} \qquad (8.34e)$$

and

$$k_3 = (1/6)(\partial^3 V/\partial\xi^3)_0 \qquad (8.34f)$$

Using Eq. (8.22) with the first-order functions $\langle 0 \mid$ and $\langle 2 \mid$ of Eqs. (8.33), the transition probability is

$$P_{0\to2} \propto (\partial\mu_z/\partial\xi)_0^2 \mid \langle 0 \mid \xi \mid 2 \rangle \mid^2 = (\partial\mu_z/\partial\xi)_0^2(C\xi_{01} + A\xi_{12} + B\xi_{23})^2 \qquad (8.35a)$$

But the well-known matrix elements are

$$\xi_{01} = \alpha^{1/2}$$

$$\xi_{12} = \sqrt{2}\,\alpha^{1/2} \qquad (8.36)$$

$$\xi_{23} = \sqrt{3}\,\alpha^{1/2}$$

so Eq. (8.35a) becomes

$$P_{0\to2} \propto (\partial\mu_z/\partial\xi)_0^2 \frac{8\alpha^4 k_3^2}{h^2\nu_e^2} \qquad (8.35b)$$

Except for one further factor the ratio of Eq. (8.35b) to Eq. (8.32) is the desired intensity ratio. But as shown by the full equation for the transition probability [see Eq. (5.69)] an extra factor of 2 should be included since the overtone occurs at a frequency of essentially twice the fundamental. Thus we find finally

$$\frac{P_{0\to2}}{P_{0\to1}} = \frac{16\alpha^3 k_3^2}{h^2\nu_e^2} \qquad (8.37a)$$

Using Eqs. (8.31), (8.34f), $\nu_e^2 = (1/4\pi^2)(k_e/\mu)$ and the definition of α we find

$$\frac{P_{0\to2}}{P_{0\to1}} = \frac{\beta^2\hbar}{2(k_e\mu)^{1/2}} = \beta^2\alpha \tag{8.37b}$$

Substituting the known values of β, k_e, $\hbar$, and μ ($\mu = 0.980\ g/\text{mole}$) the intensity ratio is

$$\frac{P_{0\to2}}{P_{0\to1}} = 0.021 \tag{8.37c}$$

This result is in excellent agreement with the observed intensity ratio of approximately 0.02 reported by Herzberg.[5]

E. Rotation-Vibration Spectra

The gas-phase infrared spectrum of a diatomic molecule consists, in general, of a *band* of absorption lines. This occurs because any given vibrational transition, say the $v = 0 \to v = 1$, may occur with a variety of different initial and final rotational energies. Thus, the question is, between which initial and final rotational states can the transition $\Delta v = +1$ occur? The answer is given by P_{JM} of Eq. (8.23b), which was shown in Sec. 7.3 B to be nonzero only when $\Delta J = +1$ and $\Delta M = 0, \pm1$. Thus the complete set of selection rules for heteronuclear diatomic molecule rotation-vibration infrared spectra[6] is

$$\Delta v = \pm1$$

$$\Delta J = \pm1 \tag{8.38}$$

$$\Delta M = 0, \pm1$$

The last selection rule is of little consequence in usual infrared spectroscopy.

In contrast to the situation for pure rotational spectra, the $\Delta J = +1$ (R branch) and the $\Delta J = -1$ (P branch) transitions lead to different absorption frequencies for the general $v'' \to v'$ transition shown in Fig. 8.6. In the figure the transitions have been labeled by $P(J)$ and $R(J)$ where J refers to the rotational quantum number of the lower (J'') state. Note that there is no $P(0)$ transition. The frequencies of the P- and R-branch tran-

---

[5] See footnote 2, p. 54.

[6] As for most rules, there is an exception here. For molecules having nonsigma electronic ground states, $\Delta J = 0$ is also permitted. This occurs because the electronic orbital angular momentum causes the molecule to behave like a symmetric rotor molecule for which $\Delta J = 0$ is permitted. See footnote 2, p. 119.

sitions for a fundamental vibrational mode are given by

$$\nu_P = P(J) = \nu_0 + B_v'J(J - 1) - B_v''J(J + 1) \tag{8.39a}$$

$$\nu_R = R(J) = \nu_0 + B_v'(J + 1)(J + 2) - B_v''J(J + 1) \tag{8.39b}$$

where $J = J''$ and

$$\nu_0 = \nu_e - 2\nu_e x_e(v + 1) \tag{8.40}$$

is the frequency of the *pure* vibrational transition $v \rightarrow v + 1$. Equations (8.39) are obtained directly from Eq. (8.21) by neglecting the centrifugal distortion term. If the rotational constants in the upper and lower vibrational states were identical, Eqs. (8.39) would predict a series of lines spaced at $2B$, with a gap of $4B$ at the band center due to the absence of the $P(0)$ transition. The fundamental rotation-vibration band of HCl is shown in Fig. 8.7. Reference to Fig. 8.6 and Eqs. (8.39) shows that the R branch lies to higher frequency and the P branch to lower frequency, as shown for the HCl spectrum.

The intensity distribution of the components of the rotation-vibra-

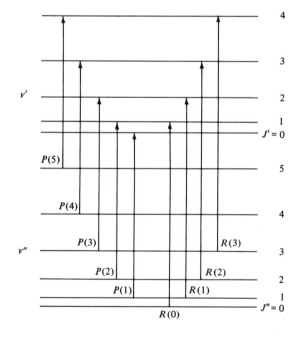

FIG. 8.6.   Rotation-vibration transitions of a diatomic molecule.

tion band of Fig. 8.7 is typical. The intensity is low in both the P and R branches for small J values, reaches a maximum for some intermediate values of J and then decreases rather rapidly as J becomes large. The relative intensities are obtained simply by considering dipole matrix elements for rotational transitions along with the Boltzmann distribution among rotational states. Thus for the R-branch lines, the intensity is given by

$$I(R) \propto \mu^2 \frac{(J+1)}{(2J+1)} f_r \qquad (8.41a)$$

where f_r is given by (7.52) and the preceding factor is the transition moment for $J \to J + 1$ transitions given by Eq. (7.45). In Sec. 7.3 B we did not derive the transition moment for the $J \to J - 1$ case, but it can be found easily by the same techniques. The result for the intensities of the P-branch transitions is

$$I(P) \propto \mu^2 \frac{J}{2J+1} f_r \qquad (8.41b)$$

Note that analysis of rotation-vibration bands leads to rotational constants and hence internuclear distances R_v. Much of our knowledge of bond distances of heteronuclear diatomics comes from infrared determinations of this type. When, however, the rotational constant becomes less than ca. 1 cm^{-1} (such as for NaCl), the experimental difficulties of resolving the various components makes the method less suitable than the microwave techniques of chapter 7.

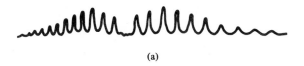

(a)

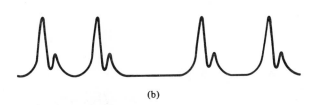

(b)

FIG. 8.7.   Rotation-vibration band of HCl(g) under (a) low resolution conditions and (b) high resolution conditions showing Cl^{35} and Cl^{37} isotopes.

EXAMPLE 8-2

Analysis of rotation-vibration bands of diatomic molecules is easily carried out. The most efficient means is by use of "combination difference" relations of the form

$$R(J) - P(J) = 2B_v'(2J + 1) \qquad (8.42a)$$

and

$$R(J) - P(J + 2) = 2B_v''(2J + 3) \qquad (8.42b)$$

which are easily derived from Eqs. (8.39). Reference to Fig. 8.6 shows clearly why the first relation involves only the B value of the lower state while the second relation involves only the upper-state B value.

To illustrate the theory of this section we have listed a few of the measured components of the HCl^{35} $v = 0 \rightarrow v = 1$ band in Table 8.4. In this case $B_v'' = B_0$ and $B_v' = B_1$. Using the values of $P(J)$ and $R(J)$ of Table 8.4, and Eqs. (8.42), B_0 and B_1 have been calculated and tabulated in Table 8.4. The average values are seen to be

$$B_1 = 10.12 \text{ cm}^{-1}$$

$$B_0 = 10.44 \text{ cm}^{-1}$$

TABLE 8.4

A portion of the infrared rotation-vibration band for the **v** $= 0 \rightarrow$ **v** $= 1$ transition of HCl^{35}.
[Data from C. F. Meyer and A. A. Levin, *Phys. Rev. 84*, 44 (1929).]

$P(J)$	J	$R(J)$ (cm^{-1})
	0	2906.25
2865.09	1	2925.78
2843.56	2	2944.89
2821.49	3	2963.24
2798.78	4	

$R(J) - P(J)$	J	B_1 (cm^{-1})
60.69	1	10.11
101.33	2	10.13
141.75	3	10.12

$R(J) -$ $P(J + 2)$	J	B_0 (cm^{-1})
62.69	0	10.45
104.29	1	10.43
146.11	2	10.44

The band origin ν_0 can now be found from the $P(J)$ and $R(J)$ using Eqs. (8.39). For example,

$$R(0) = 2906.25 = \nu_0 + 10.12(2) - 10.44(0)$$

or

$$\nu_0 = 2886.01 \text{ cm}^{-1}.$$

Similarly,

$$P(1) = 2865.09 = \nu_0 + 10.12(0) - 10.44(2)$$

or

$$\nu_0 = 2885.97 \text{ cm}^{-1}.$$

The average value of ν_0 which is known as the *band origin*, is then, 2885.99 cm^{-1}. The fundamental vibrational frequency ν_e cannot be found from the data on the $v = 0 \rightarrow v = 1$ transition alone, since, as shown by Eq. (8.40), ν_0 depends upon both ν_e and $\nu_e x_e$.

The rotation-vibration constant α_e and the equilibrium rotational constant B_e can be found by means of Eq. (7.26). We find

$$B_0 = B_e - \tfrac{1}{2}\alpha_e$$

and

$$B_1 = B_e - \tfrac{3}{2}\alpha_e$$

Thus our previous results for B_0 and B_1 lead to

$$B_e = 10.59 \text{ cm}^{-1}$$

$$\alpha_e = +0.32 \text{ cm}^{-1}$$

Finally R_e is easily calculated from B_e using the relation [see Eq. (7.24)]

$$R_e = \left(\frac{h}{8\pi^2 c \mu B_e}\right)^{1/2} \tag{8.43}$$

with B_e in cm^{-1} and μ in g/molecule. Inserting the appropriate values for HCl^{35}, the equilibrium internuclear distance is calculated to be $R_e = 1.275$ Å.

8.3 INFRARED SPECTRA OF POLYATOMIC MOLECULES

A. Classical Mechanics, Normal Modes, and Normal Coordinates

In this section we treat the polyatomic molecule as a harmonic oscillator. Thus the general Hamiltonian is that of Eq. (3.74)

$$\mathcal{H}_v = \frac{1}{2} \sum_i^{3N} \dot{q}_i^2 + \frac{1}{2} \sum_{i,j}^{3N} f_{ij} q_i q_j \tag{3.74}$$

where the q_i are the mass-weighted coordinates. This equation does include $3N - 6$ (or $3N - 5$) redundant variables which can, in principle, be removed by the redundancy conditions described previously. However, we carry along the redundant variables for the time being, since there is no generally simple way of removing them as in the case of the diatomic molecule.

If we now proceed with the classical problem in the same fashion as for the diatomic molecule, we use first Hamilton's equations of motion which give (since $p_j = \dot{q}_j$)

$$\frac{\partial \mathcal{K}_v}{\partial q_j} = -\ddot{q}_j \tag{8.44}$$

Using Eq. (3.74) we find the time dependence of the $3N$ coordinates to be given by the set of coupled differential equations

$$\ddot{q}_j + \sum_i^{3N} f_{ji} q_i = 0 \qquad j = 1 \ldots 3N \tag{8.45}$$

Since each of these equations is of the form of that encountered previously we expect a solution to be

$$q_i = A_i \cos (\lambda^{1/2} t + \varphi) \qquad i = 1 \ldots 3N \tag{8.46}$$

where for simplicity we have defined

$$\lambda^{1/2} = 2\pi\nu \tag{8.47}$$

Substitution of Eqs. (8.46) into Eqs. (8.45) leads to the equations

$$\cos(\lambda^{1/2} t + \varphi) \left[\sum_i f_{ji} A_i - \lambda A_j \right] = 0 \qquad j = 1 \ldots 3N \tag{8.48}$$

These equations would be immediately satisfied if the first factor were zero, but this case corresponds to all $q_i = 0$ which is physically uninteresting. Thus we set the second factor equal to zero, which gives

$$A_1(f_{11} - \lambda) + A_2 f_{12} + A_3 f_{13} + \cdots A_n f_{1n} = 0$$

$$A_1 f_{21} + A_2(f_{22} - \lambda) + A_3 f_{23} + \cdots A_n f_{2n} = 0 \tag{8.49}$$

$$A_1 f_{n1} + A_2 f_{n2} + \qquad \cdots A_n(f_{nn} - \lambda) = 0$$

where $n = 3N$ for simplicity. More compactly, Eqs. (8.49) may be written

$$\sum_i^{3N} (f_{ji} - \lambda \delta_{ji}) A_i = 0 \qquad j = 1 \ldots 3N \tag{8.50}$$

Except for the trivial case when all $A_i = 0$, the homogeneous Eqs.

(8.49) or (8.50) have a solution (for the A_i) only when

$$
\begin{vmatrix}
f_{11} - \lambda & f_{12} & f_{13}\cdots \\
& & \\
f_{21} & f_{22} - \lambda & f_{23}\cdots \\
\cdot & \cdot & \\
\cdot & \cdot & \\
\cdot & \cdot & \\
f_{3N,1} & \cdots & f_{3N,3N} - \lambda
\end{vmatrix} = 0 \qquad (8.51)
$$

When expanded, the determinant of Eq. (8.51) yields a polynomial of order $3N$ in λ. If each of the $3N$ (not necessarily different) values of λ is inserted into Eqs. (8.49) a set of A_i's is obtained for each eigenvalue λ. Thus the $3N$ q_i can be calculated for each λ value by use of Eq. (8.46).

Note that for a given eigenvalue λ there corresponds a vibrational frequency ν and a set of coordinates q_i which vary sinusoidally in time with constant phase φ. Thus all the atoms oscillate harmonically and reach their maximum amplitudes simultaneously. The $3N$ frequencies ν are known as the *normal* or *fundamental* frequencies, and the vibrational motions specified by the set of $3N$ q's accompanying each ν are known as the *normal modes* of vibration.

We see that our results for the polyatomic molecule have been similar to those for the diatomic molecule, except now the relation between force constants and vibrational frequencies is much more complicated. The only remaining question concerns the six (or five) redundant coordinates that have been carried along. These correspond to the nonvibrational motions and consequently must correspond to normal modes having no restoring forces, that is, they are modes of zero frequency. Wilson *et al.*,[7] have shown in general that six (or five) of the λ values of Eq. (8.51) must go to zero. The remaining $3N - 6$ (or 5) eigenvalues correspond to the genuine vibrational modes of the molecule. We see shortly the more practical ways for determining the normal vibrational modes and frequencies of any molecule. For now, we simply illustrate in Fig. 8.8 the normal modes and frequencies for several simple molecules.

EXAMPLE 8-3

While the general proof is lengthy and specific examples are usually tedious, we can, without too much difficulty, show that five of the roots of Eq. (8.51) are indeed zero for a diatomic molecule. The point is that any correct (harmonic) vibrational potential function automatically leads to five (or six) zero roots if the mathematics is carried out properly.

[7] E. B. Wilson, Jr., J. C. Decius, and P. C. Cross, *Molecular Vibrations* (McGraw-Hill, New York, 1955), p. 22.

Consider a diatomic molecule having q_1, q_2, q_3 situated on m_1 and q_4, q_5, q_6 situated on m_2; further, we assume that the two sets of mass-weighted displacement coordinates are similarly oriented with q_1 and q_4 being along the molecular axis. We have seen previously that $\frac{1}{2}k(R - R_e)^2$ is a proper potential function for the diatomic molecule. In order to stay with mass-weighted coordinates we imagine that $R - R_e$ is measured in mass-weighted length units rather than the conventional length units. Hence the force constant k will differ numerically and dimensionally from that of Eq. (3.89) but is related by a simple scale factor. Then without explicitly intro-

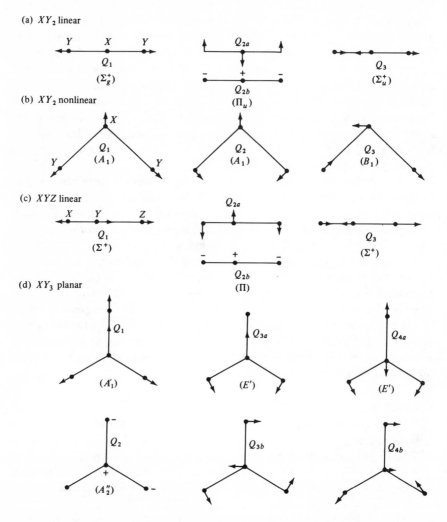

FIG. 8.8. Normal modes for some common molecules (schematic).

ducing any redundancy relations, we know that the potential energy must be of the form

$$V = \tfrac{1}{2}k(R - R_e)^2$$

$$= \tfrac{1}{2}k\{[(R_e + q_4 - q_1)^2 + (q_5 - q_2)^2 + (q_6 - q_3)^2]^{1/2} - R_e\}^2 \qquad (8.52)$$

where the term in square brackets is simply the internuclear separation R at any time during the vibration. The right-hand side of Eq. (8.52) follows by simple geometrical considerations. Note that the potential function (8.52) does not in any way prohibit the full $3N$ degrees of freedom, but of course, the potential affects only the genuine vibrational motion.

In a trivial fashion it is possible to evaluate all the force constants of Eq. (8.51). For example,

$$\frac{1}{\tfrac{1}{2}k} \cdot \frac{\partial^2 V}{\partial q^2{}_1} = 2 - 2R_eR^{-1} + 2R_eR^{-3}(R_e + q_4 - q_1)^2 \qquad (8.53)$$

where R is the first term in brackets in Eq. (8.52). Evaluating the derivative at the equilibrium position $q_i = 0$ we find

$$f_{11} = (\partial^2 V/\partial q_1{}^2)_0{}^2 = k \qquad (8.54)$$

In a similar fashion we find:

$$f_{44} = k, \qquad f_{14} = f_{41} = -k$$

$$f_{ii} = 0, \qquad f_{ij} = 0, \qquad i \text{ and } j \neq 1 \text{ or } 4$$

The secular equation now takes the form

$$\begin{vmatrix} k - \lambda & 0 & 0 & -k & 0 & 0 \\ 0 & \lambda & 0 & 0 & 0 & 0 \\ 0 & 0 & \lambda & 0 & 0 & 0 \\ -k & 0 & 0 & \lambda & 0 & 0 \\ 0 & 0 & 0 & 0 & \lambda & 0 \\ 0 & 0 & 0 & 0 & 0 & \lambda \end{vmatrix} = 0 \qquad (8.55)$$

This equation is easily expanded to give (first interchange rows 2 and 4 and then columns 2 and 4)

$$\lambda^4[(k - \lambda)^2 - k^2] = 0 \qquad (8.56)$$

which obviously has the roots

$$\lambda = 0 \qquad (5 \text{ times})$$

$$\lambda = 2k$$

$$(8.57)$$

Thus, as anticipated, five of the roots of the secular equation are non-genuine modes and have zero frequency. Note that this solution to the diatomic vibrator problem proceeds in a quite different manner from that outlined in Example 3-2 and culminating in Eq. (8.4). In this earlier solution the redundancy relations were used explicitly to remove the five redundant variables. In the present case the redundant variables were not explicitly removed, but a vibrational potential having the correct functional dependence upon the single nonredundant variable was chosen.

For those who wish the correct answer in Eq. (8.57), the scaling factor is easily shown to be 2μ. That is,

$$k[\text{Eq. (8.4)}] = 2\,\mu k[\text{Eq. (8.57)}] \qquad (8.58)$$

Before proceeding to the quantum-mechanical problem it is useful to introduce the concept of a *normal coordinate*. First we recognize that the *normal modes* and *normal frequencies* are real physical results and do not depend upon the coordinate system with which we describe the system. That is, if we set up the vibrational problem [Eq. (3.74)] with a different set of $3N$ coordinates (non–mass-weighted, different orientations, linear combinations, etc.) we always arrive at a determinant of the general form of Eq. (8.51), whose solutions (λ) are identical to those obtained with the mass-weighted coordinates.

Of particular interest to us are coordinates that are related to the mass-weighted coordinates by linear (or real orthogonal) transformations. The *normal coordinates*, $Q_1, Q_2 \ldots Q_k \ldots Q_{3N}$, are such a set, that is,

$$\mathbf{Q} = \mathbf{lq} \qquad (8.59a)$$

or

$$Q_k = \sum_i l_{ki} q_i \qquad (8.59b)$$

The special additional properties of the normal coordinates are that the kinetic and potential energies are of the form

$$2T = \sum_k^{3N} \dot{Q}_k{}^2 \qquad (8.60a)$$

$$2V = \sum_k^{3N} \lambda_k Q_k{}^2 \qquad (8.60b)$$

Thus the *normal coordinates* are those coordinates which eliminate the cross terms in the potential energy, and leave the kinetic energy as a sum of squared terms. The additional feature shown by Eq. (8.60b) is that each of the $3N$ normal coordinates corresponds to *one* normal frequency ($\nu_k = \lambda_k^{1/2}/2\pi$) and mode only.

We do not deal at this time with the general methods of determining the normal coordinates (in terms of the q's, say). We simply point out that the pictorial representations of the normal modes (Fig. 8.8 for example) serve as graphical descriptions of the normal coordinates.

EXAMPLE 8-4

Using the diatomic molecule as a simple example again, it can be shown that the normal coordinates are given by

$$
Q = \begin{pmatrix} T_x \\ T_y \\ T_z \\ 0 \\ R_x \\ R_y \end{pmatrix} = \begin{pmatrix} 0 & 1/\sqrt{2} & 0 & 0 & 1/\sqrt{2} & 0 \\ 0 & 0 & 1/\sqrt{2} & 0 & 0 & 1/\sqrt{2} \\ 1/\sqrt{2} & 0 & 0 & 1/\sqrt{2} & 0 & 0 \\ -1/\sqrt{2} & 0 & 0 & 1/\sqrt{2} & 0 & 0 \\ 0 & 0 & 1/\sqrt{2} & 0 & 0 & -1/\sqrt{2} \\ 0 & 1/\sqrt{2} & 0 & 0 & -1/\sqrt{2} & 0 \end{pmatrix} \begin{pmatrix} q_1 \\ q_2 \\ q_3 \\ q_4 \\ q_5 \\ q_6 \end{pmatrix}
$$

$$(8.61)$$

where the $q_1 \ldots q_6$ were defined in Example 8-3 and the x, y, and z directions are made clear below. If the normal coordinates are sketched, as below, the T and R coordinates are seen to represent translation and rotation, respectively, while the coordinate O corresponds to the vibrational motion. We have assumed in the drawings that $m_1 = m_2$.

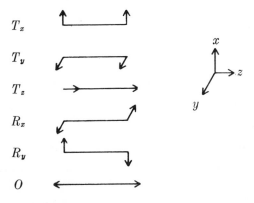

It is clear from the sketches that the normal modes T, R, and O are as described above; the translational, rotational and vibrational motions are evident. To verify the correctness of the mathematical form of the Q's, we transform the Hamiltonian of Eq. (3.74) to the normal coordinates. If the transformation matrix of Eq. (8.61) is l, the reverse transformation is

$$\mathbf{q} = \mathbf{l'Q} \tag{8.62}$$

since the inverse of a real orthogonal matrix is its transpose. Thus we can express all of the mass-weighted coordinates in terms of the normal coordinates. Using the results of Eqs. (8.61) or (8.62) it can be shown that

$$\sum_i q_i^2 = \sum_i Q_i^2 \tag{8.63a}$$

so

$$\sum_i \dot{q}_i^2 = \sum_i \dot{Q}_i^2 \tag{8.63b}$$

The result in Eq. (8.63a) is an example of a general property of orthogonal transformations; namely, the sum of squared coordinates is constant. We see then that the kinetic energy is of the form of Eq. (8.60a).

In order to investigate the transformation of the potential energy we recall first that most of the f_{ij} in Eq. (3.74) are zero (see Example 8-3), the result being

$$2V = f_{11}q_1^2 + f_{44}q_4^2 + 2f_{14}q_1q_4 = f_{11}(q_1^2 + q_4^2) + 2f_{14}q_1q_4 \tag{8.64}$$

since $f_{11} = f_{44}$ when $m_1 = m_2$. From Eq. (8.62) we find

$$q_1 = (1/\sqrt{2})(T_z - O)$$

$$q_4 = (1/\sqrt{2})(T_z + O) \tag{8.65}$$

which gives

$$2V = 2f_{11}O^2 \tag{8.66}$$

But, as shown in Example 8-3, $2f_{11} = 2k = \lambda$, so

$$2V = \lambda O^2 \tag{8.67}$$

which is precisely the result required for the potential energy given by Eq. (8.60b). We have shown therefore that the normal coordinates for the diatomic molecule are those of Eq. (8.61).

B. Vibrational Energy Levels of Polyatomic Molecules

The energy levels for the harmonically oscillating polyatomic molecule are easily obtained using the kinetic and potential energies expressed in normal coordinates. Furthermore, an exact quantum-mechanical solution cannot be obtained with any other set of coordinates due to the cross terms in the potential energy. If a wave-mechanical approach is used, it can be shown that the variables of Schrödinger's equation can be separated if $\tilde{\mathcal{3C}}$ is formulated from Eqs. (8.60). Thus one finds that if the wave function is written

$$\psi_v = \psi(Q_1) \cdot \psi(Q_2) \cdots \psi(Q_i) \cdots \psi(Q_{3N-6}) \qquad (8.68)$$

the variables separate, giving equations of the form

$$\tilde{\mathcal{3C}}_i \psi(Q_i) = E(Q_i)\psi(Q_i) \qquad (8.69)$$

with the total energy being

$$E_v = E(Q_1) + E(Q_2) + \cdots E(Q_i) + \cdots E(Q_{3N-6}) \qquad (8.70)$$

As we shall see immediately, the solutions to Eq. (8.69) are well known.

In the matrix formulation we simply evaluate the matrix elements in the representation $\langle v_1 v_2 \ldots v_i \ldots v_{3N-6} |$, where the v_i are quantum numbers associated with each normal coordinate. The quantum-mechanical Hamiltonian is easily gotten from Eqs. (8.60) since the conjugate momentum is $P_k = \dot{Q}_k$. Thus

$$\tilde{\mathcal{3C}} = \sum_k \tfrac{1}{2}(\tilde{P}_k^2 + \lambda_k Q_k^2) \qquad (8.71)$$

It is verified easily by a little reflection that the Hamiltonian matrix is diagonal in the above representation, so the energies are

$$E_v = \langle v_1 \cdots v_i \cdots | \tfrac{1}{2} \sum_k (\tilde{P}_k^2 + \lambda_k Q_k^2) | v_1 \cdots v_i \cdots \rangle$$

$$= \sum_i \langle v_i | \tfrac{1}{2}(\tilde{P}_i^2 + \lambda_i Q_i^2) | v_i \rangle \langle v_1 \cdots v_{3N-6} | v_1 \cdots v_{3N-6} \rangle$$

$$= \sum_i E(Q_i) \qquad (8.72a)$$

The energies $E(Q_i)$ are obtained from the well-known harmonic oscillator matrix elements of Sec. 4.1, since each $E(Q_i)$ merely represents harmonic oscillation along the normal coordinate Q_i. The results are most quickly obtained by noting that the Hamiltonian operator of Eq. (8.71) looks like that of a harmonic oscillator of mass unity and force constant λ. Thus from Eq. (4.30)

$$E(Q_i) = \hbar\lambda^{1/2}(v_i + \tfrac{1}{2}) = h\nu_i(v_i + \tfrac{1}{2}) \qquad (8.73)$$

where we have used Eq. (8.47). From Eq. (8.72a) we find the vibrational energy states of the polyatomic molecule to be

$$E_v = \sum_i^{3N-6} h\nu_i(v_i + \tfrac{1}{2}) \qquad (8.72b)$$

where the ν_i are the fundamental (classical) vibrational frequencies and

$v_i = 0, 1, 2\ldots$ are the vibrational quantum numbers. The vibrational frequencies have been related to force constants in Sec. 8.3 A, and are discussed further in a later section. Note that in this section we have treated explicitly only the $3N - 6$ (or 5) genuine vibrational modes.

In Fig. 8.9 we have drawn the energy-level diagram for the ammonia molecule using the four experimentally known vibrational frequencies. The states are labeled by the quartet of numbers, $(v_1v_2v_3v_4)$, which correspond respectively to v_1, v_2, v_3, and v_4.[8] Note that there are $3N - 6 = 6$ vibrational modes for ammonia, but v_3 and v_4 are each doubly degenerate. Since a mode with degeneracy d produces d identical terms in Eq. (8.72b), it is common to sum over only the distinct v_α values and to rewrite (8.72b) as

$$E_v = \sum_{\alpha}^{\text{distinct } v_\alpha} h\nu_\alpha(v_\alpha + \tfrac{1}{2}d_\alpha) \qquad (8.74)$$

In Fig. 8.9 it can be seen that there are three basic types of energy levels; the

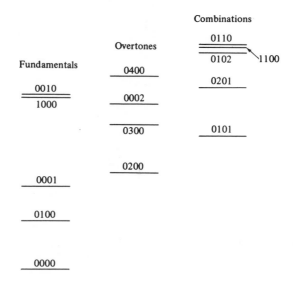

FIG. 8.9. Energy-level diagram for ammonia (neglecting inversion). Calculated using $v_1 = 3337$ cm^{-1}, $v_2 = 950$ cm^{-1}, $v_3 = 3414$ cm^{-1}, and $v_4 = 1628$ cm^{-1}.

[8] Usual spectroscopic convention is such that the *most* symmetrical vibrational modes have the lowest numbers, the symmetrical stretching modes being lower than the bending. Less symmetrical modes have the higher numbers. The specific conventions need not concern us too much.

fundamentals of the form $(\cdots 1 \cdots)$, the *overtones* of the form $(\cdots v_i > 1 \cdots)$, and the *combinations* of the form $(v_1 v_2 v_3 v_4)$ with at least two of the $v_i \neq 0$.

Just as with diatomic molecules, anharmonic terms in the potential energy cause the vibrational energy levels of the modes of polyatomic molecules to deviate from the simple equally spaced harmonic behavior. This anharmonic behavior is very difficult to treat theoretically, even for triatomic molecules, but the general behavior is similar to that for diatomic molecules, namely the energy expression contains terms proportional to $(v_i + \frac{1}{2})^2$, etc. In addition, due to the presence of several normal modes, cross terms in $(v_i + \frac{1}{2})(v_j + \frac{1}{2})$ appear also. It is often not possible to determine experimentally these anharmonic corrections but when it is, the fundamental frequencies are usually corrected for these contributions.

There are other perturbations which produce very marked effects on the energy levels. The most important of these is that known as "Fermi resonance", which occurs when two nearly degenerate energy levels are connected by a nonvanishing perturbation matrix element. For example, if ψ_n^0 and ψ_m^0 are the wave functions of the nearly degenerate levels, E_n^0 and E_m^0, the usual first-order degenerate perturbation theory (neglecting all other levels) leads to

$$\begin{vmatrix} E_n^0 - E & H_{nm}' \\ H_{mn}' & E_m^0 - E \end{vmatrix} = 0 \qquad (8.75)$$

where

$$H_{mn}' = \langle m \mid \widetilde{\mathfrak{3C}}' \mid n \rangle \qquad (8.76)$$

The solution of the quadratic equation leads to the well-known "repulsion" of the unperturbed energy levels if H_{mn}' is not zero.

We need not worry here about the mathematical form of $\widetilde{\mathfrak{3C}}'$, but we note simply that it consists of the cubic, quartic, etc., terms in the potential energy. Since the Hamiltonian belongs to the totally symmetrical representation of the group of Schrödinger's equation, our group-theory results [Eq. (6.65)] state that the integral of Eq. (8.76) is nonzero only when $\Gamma^{(m)} \to \Gamma^{(n)}$, that is, the symmetries of the two unperturbed states must be the same. The most commonly occurring "Fermi resonances" are between fundamental and overtone states of different vibrational modes. For example, the (100) state of CO_2 is observed at 1388 cm^{-1}, the (010) at 667 cm^{-1}, and the (020) at 1286 cm^{-1}. In the simple harmonic approximation, (020) is expected to occur at about $2 \times 667 = 1334$ cm^{-1}, which shows the perturbation very clearly. Dennison[9] has shown that the separation of the two unperturbed levels, $\delta = E^0_{100} - E^0_{020}$, should be 17 cm^{-1}, and that the perturbation term coupling the two states (which have the same sym-

[9] D. M. Dennison, *Rev. Mod. Phys.* **12**, 175 (1940).

metry) is 50 cm^{-1}. Expanding Eq. (8.75) we find

$$E = \frac{E_n{}^0 + E_m{}^0}{2} \pm \tfrac{1}{2}(\delta^2 + 4 \mid H_{nm}{}' \mid^2)^{1/2} \tag{8.77}$$

so the separation between the perturbed levels (100) and (020) is predicted to be

$$\Delta E = (\delta^2 + 4 \mid H_{nm}{}' \mid^2)^{1/2} \tag{8.78}$$

Using Dennison's values for CO_2 we find from Eq. (8.78), $\Delta E = 101$ cm^{-1}. This agrees well with the experimental value of $1388 - 1286 = 102$ cm^{-1}.

Because of the large number of modes in polyatomic molecules it should not be surprising to find accidental degeneracies leading to Fermi resonances of this type. They are relatively easy to detect experimentally in many cases since, as we have seen, the energy-level shifts are quite large. Note that the phenomena may occur with combination levels also. The only requirements are (a) near degeneracies and (b) states of the same symmetries. We see shortly that the symmetry properties of the vibrational levels are easily determined, so the possibility of Fermi resonance can be easily predicted. It should be mentioned also that the perturbation $H_{mn}{}'$ destroys the "pure" character of the unperturbed states, that is, the states $\psi_m{}^0$ and $\psi_n{}^0$ become mixed. Thus the selection rules and intensities may be markedly affected, depending on the strength of the perturbation. The common result is that overtone (or combination) bands involving states in Fermi resonance are often much stronger than they would otherwise be, since they have "borrowed" some of the character of the more strongly allowed transition.

Finally, it should be clear that in the gas phase the energy levels of each vibrational mode is accompanied by the usual manifold of rotational energy states. Thus the energy of any state $\langle J\tau v \mid$, where we use τ to indicate the rotational quantum number K for the linear or symmetric rotor, or the labels K_{-1} and K_{+1} for the asymmetric rotor (see chapter 7), is given by

$$E(J\tau v) = \sum_i^{3N-6} h\nu_i(v_i + \tfrac{1}{2}) + E(J\tau) \tag{8.79}$$

Depending upon whether the molecule is linear, or a symmetric or asymmetric rotor, the appropriate expressions for $E(J\tau)$ are (7.35), (7.73), or (7.98), respectively.

C. Selection Rules and Spectra

The expression for the probability of a vibrational transition for a polyatomic molecule follows by a very simple extension of the results for the diatomic molecule. Instead of Eq. (8.22) we have

$$P \propto \sum_{g=x,y,z} | \langle v_1 v_2 \cdots v_{3N-6} | \sum_i^{3N-6} (\partial \mu_g / \partial Q_i)_0 \, Q_i \, | \, v_1' v_2' \cdots v_{3'N-6} \rangle |^2 \quad (8.80a)$$

where it is noted that we sum over all genuine normal modes and all three molecular axes. The matrix element of the sum can be rewritten as a sum of matrix elements,

$$P \propto \sum_g | \sum_i \langle v_1 | v_1' \rangle \langle v_2 | v_2' \rangle \cdots \langle v_i | (\partial \mu_g / \partial Q_i)_0 Q_i | v_i' \rangle \cdots \langle v_{3N-6} | v_{3'N-6} \rangle |^2$$

$$(8.80b)$$

where the scalar products have been factored out since they are unaffected by the operator Q_i. Because of the orthogonality of the harmonic oscillator eigenstates, a particular term j in the sum over i goes to zero unless $v_i = v_i'$ for all states but the jth. From the well-known harmonic oscillator matrix elements, we know that $\langle v_j | Q_j | v_j' \rangle$ vanishes unless $v_j' = v_j \pm 1$. Thus the jth term in the sum is nonvanishing only if $v_i = v_i'$ for $i \neq j$, and $v_j' = v_j \pm 1$. If the jth term vanishes we then repeat the identical analysis for the remaining $(3N - 7)$ terms in the sum. Note that, at most, only one term can be nonvanishing. If, for example, the jth is nonvanishing, this means $v_j' = v_j \pm 1$. But then in all other terms in the summation the factor $\langle v_j | v_j \pm 1 \rangle = 0$ appears.

Our resulting selection rule is that $\Delta v_i = 0$ for all modes but one, which must have $\Delta v = \pm 1$. For example, for H_2O the following transitions are included among the allowed ones;

$$(000) \rightarrow (010)$$
$$(000) \rightarrow (100)$$
$$(000) \rightarrow (001)$$
$$(120) \rightarrow (220)$$
$$(120) \rightarrow (130)$$
$$(120) \rightarrow (121)$$

while the following are among those not permitted;

$$(000) \rightarrow (020)$$
$$(000) \rightarrow (110)$$
$$(000) \rightarrow (021)$$
$$(120) \rightarrow (230)$$
$$(120) \rightarrow (030)$$
$$(120) \rightarrow (111)$$

In the harmonic approximation we see from Eq. (8.72b) that the only possible transition frequencies are just the $3N - 6$ (or 5) classical vibrational frequencies of the molecule.

Equation (8.80b) contained three other extremely important factors, the changes in the dipole moment components with normal coordinate. Hence we see that the transition probability goes to zero unless

$$(\partial\mu/\partial Q_i)_0 \neq 0$$

regardless of whether or not the selection rules on v_i are satisfied. When the form of the normal modes is known, it is relatively easy to see whether or not $(\partial\mu/\partial Q_i)_0$ is zero. For example, a molecule such as CO_2 in Fig. 8.8 has two modes for which the derivative is nonzero, namely ν_2 and ν_3. For these two modes it is clear that the dipole moment of the molecule in its equilibrium configuration is different from that of the molecule when it has undergone its maximum oscillation as shown by the arrows. On the other hand the ν_1 mode of CO_2 has a zero value of $(\partial\mu/\partial Q_1)$ since the symmetrical nature of the molecule is not destroyed. Thus for CO_2, vibrational transitions are possible for the ν_2 and ν_3 modes only. These modes are described as being *infrared active*, while ν_1 is *infrared inactive*. Figure 8.8 labels the infrared activity of the modes of the various molecules illustrated. It is seen that all three distinct modes of the water molecule are infrared active, so the illustrative transitions above would all be permitted.

Since it is clear that the vanishing or nonvanishing behavior of the dipole derivatives is strongly dependent upon the molecular symmetry it should not be surprising that the selection rules can be predicted by group theory. Indeed, this is an extremely important application of group theory and it makes it unnecessary to puzzle over the normal-mode diagrams in

TABLE 8.5

Some observed infrared bands of N_2O.[a]

$\nu\,(cm^{-1})$	Upper state numbers[b] $(v_1v_2v_3)$	Band intensity[c]
588.8	(010)	s
1167.0	(020)	m
1285.0	(100)	vs
1867.5	(110)	w
2223.5	(001)	vs
2461.5	(120)	m
2563.5	(200)	m
2798.3	(011)	w
3365.6	(021)	w

[a] Adapted from footnote 10, Herzberg, Table 58.
[b] Lower state is (000) in all cases.
[c] w = weak, m = medium, s = strong, v = very.

TABLE 8.6

Some observed infrared bands of CS_2.[a]

ν (cm^{-1})	Upper state[b] $(v_1 v_2 v_3)$	Band intensity[c]
397.7	(010)	s
1523	(001)	s
2184	(101)	w
2329	(021)	vw
878	(001)	vw

[a] Adapted from footnote 10, Herzberg, Table 57.
[b] Lower state is (000) in all cases except the last in which case it is (100).
[c] w = weak, s = strong, v = very.

order to decide upon the selection rules. In addition, we find that anharmonic contributions cause a breakdown of the rule $\Delta v = \pm 1$ just as with diatomics, so *overtone* and *combination* bands are found in the infrared spectra of polyatomic molecules. But the group-theory rules [or the observations about the behavior of $(\partial \mu / \partial Q_i)$] are still valid and are found to place stringent requirements upon the possible transitions.

To place the spectra of polyatomic molecules on a somewhat more practical footing we show in Tables 8.5 and 8.6 the observed spectral bands of N_2O and CS_2 and their assignments. It should be pointed out that for larger polyatomic molecules it becomes increasingly difficult to perform complete spectroscopic assignments due to the large number of bands which occur in the relatively short infrared region. Even so, certain simplifications to be discussed later make it possible to obtain qualitative, if not quantitative, information.

D. Rotation-Vibration Spectra

In the gas phase the vibrational transitions are accompanied by changes in rotational quantum numbers. For rather small (light) polyatomics it is often possible to resolve, at least partially, the rotation-vibration bands and hence to obtain some knowledge of the rotational constants. The selection rules for the rotation-vibration bands are obtained by appropriate combination of the vibrational and rotational selection rules. There are, however, a number of restrictions on the ways in which the rotational and vibrational selection rules are combined.[10] This topic is rather lengthy and since the analysis of rotation-vibration bands is of limited and specialized utility, we do go into the treatment in this work. Example 8-5 below does present one illustration of a simple rotation-vibration analysis.

[10] G. Herzberg, *Infrared and Raman Spectra* (Van Nostrand, Princeton, N.J., 1945), chap. IV.

EXAMPLE 8-5

The infrared spectrum of the NF_2 free radical has been studied by Harmony *et al.*[11] The fundamentals of this bent triatomic were identified as $\nu_1 = 1074$ cm^{-1}, $\nu_2 = 573$ cm^{-1}, and $\nu_3 = 931$ cm^{-1}. The ν_1 symmetrical stretching band has been resolved in the gas phase as shown in Fig. 8.10. Since NF_2 is an asymmetric rotor, its rotation-vibration selection rules should be predicted on this basis. However, it turns out that the molecule can be treated (to the accuracy of the infrared data) as a nearly prolate symmetric rotor. In this case, the transition moment is perpendicular to the a principal axis, which is the approximate axis of the symmetric rotor. The rotational selection rules[12] for this perpendicular ($\perp$) band are $\Delta J = 0$, ± 1 and $\Delta K = \pm 1$. The latter rule is clearly different from that applicable to the pure rotational spectrum for which only $\Delta K = 0$ is permitted. The reason is that in the present case the transition moment is perpendicular, rather than parallel to the top axis.

The subbands in Fig. 8.10 have been assigned to the $\Delta J = 0$, $\Delta K = +1$ and $\Delta K = -1$ transitions. The assignments and measured frequencies are listed in Table 8.7, where ^PQ_K and ^RQ_K refer to the $\Delta K = -1$ and $\Delta K = +1$ transitions, respectively, and the subscript K labels the state from which the transition originated. The $\Delta J = \pm 1$ transitions are not observed since they form a weak, unresolved background. For the prolate rotor, Eq. (7.73a) gives the energy levels, except it is perhaps best to write $\frac{1}{2}(B_v + C_v)$ in place of just B_v since we are effectively averaging the B and C asymmetric rotor rotational constants due to our symmetric-rotor approximation.

It is easy to see then that the following combination differences are

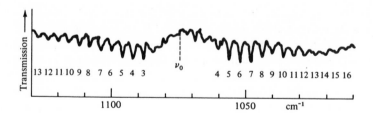

FIG. 8.10.   The ν_1 band of the NF_2 radical at 1074 cm^{-1}. (See footnote 11.) From M. D. Harmony, R. J. Myers, L. J. Schoen, D. R. Lide, and D. E. Mann, *J. Chem. Phys.*, **35**, 1129 (1961). (With permission.)

[11] M. D. Harmony, R. J. Myers, L. J. Schoen, D. R. Lide, and D. E. Mann, *J. Chem. Phys.* **35**, 1129 (1961).
[12] See footnote 10.

valid:

$$^RQ_{K-1} - {}^PQ_{K+1} = 4\left(A_0 - \frac{B_0 + C_0}{2}\right)K \qquad (8.81a)$$

$$^RQ_K - {}^PQ_K = 4\left(A_1 - \frac{B_1 + C_1}{2}\right)K \qquad (8.81b)$$

Using the data of Table 8.7 we find

$$\left(A_0 - \frac{B_0 + C_0}{2}\right) = \left(A_1 - \frac{B_1 + C_1}{2}\right) = 1.981 \pm 0.005 \text{ cm}^{-1}$$

Thus we obtain information about only the *combination* rotational constant $[A - \frac{1}{2}(B + C)]$. It is obviously not possible to fit the two structural parameters to this piece of information. However, Harmony *et al.*[13] have shown that $R(NF) = 1.37$ Å and $\angle FNF = 104.3°$ are a reasonable pair of parameters which satisfy the experimental results.

Before leaving this example we might point out that the reason for the missing bands (for $K \leq 3$) is undoubtedly due to a breakdown in the symmetric-rotor approximation. For $K \leq 3$ the $\pm K$ states are split rather widely so the $\Delta J = 0$ transitions no longer fall at the same frequencies for the ^PQ_K branches, but form a smeared out, diffuse background.

TABLE 8.7

Frequencies[a] of the Q branch subbands of ν_1 of NF_2. (See Fig. 8.10)

| | P_{Q_K} (cm^{-1}) | | R_{Q_K} (cm^{-1}) | |
K	Obs.	Calc.[b]	Obs.	Calc.[b]
3		1064.4	1088.2	1088.2
4	1060.6	1060.4	1092.0	1092.1
5	1056.7	1056.5	1096.1	1096.1
6	1052.5	1052.5	1100.1	1100.0
7	1048.4	1048.6	1104..1	1104.0
8	1044.4	1044.6	1107.9	1108.0
9	1040.4	1040.6	1111.9	1111.9
10			1116.0	1115.9
11			1119.9	1119.8

[a] See footnote 11, this chapter.
[b] Calculated using $(A - (B + C)/2) = 1.981$ cm^{-1} and $\nu_0 = 1074.3$ cm^{-1}.

8.4 GROUP-THEORETICAL ASPECTS OF MOLECULAR VIBRATIONS

A. Symmetry Classification of the Normal Modes and Normal Coordinates

The starting point for the determination of the symmetry of the normal modes lies in the fact that the coordinates used to describe the vibrational motions form a basis for a representation of the group of

[13] See footnote 11.

Schrödinger's equation. The appropriate group is clearly the group of covering operations of the *equilibrium* molecule, since the mathematical form of the kinetic and potential energies must be invariant to the interchange (via covering operations) of identical equilibrium nuclei.

The representation generated by the coordinates is in general reducible, as we shall see shortly. Furthermore, the representations generated by any complete sets of $3N$ coordinates are equivalent. Thus the representation generated by the Cartesian displacement coordinates (Δx_i, etc.) is the same as that generated by the mass-weighted coordinates (q_i). This is true even though the two sets of coordinates are not related by a real orthogonal transformation, since the symmetry operations interchange or mix only those coordinates situated on atoms of the same mass.

Since unitary (or real orthogonal) transformations produce equivalent matrices, the representation generated by the $3N$ normal coordinates Q_i is, of course, also equivalent to that generated by the Δx_i's, etc., or the q_i's. The important result to be illustrated is that the representation matrices generated by the normal coordinates are in their most reduced form. That is, *the normal coordinates transform like the irreducible representations of the group*, although this is not meant to imply that *each* irreducible representation is associated with a normal coordinate.

The normal vibrational modes (normal coordinates) given in Fig. 8.8 can be used to illustrate these statements. For H_2O the appropriate group is C_{2v}. By visual inspection we find $\tilde{E}(Q_1) = Q_1$, $\tilde{C}_2(Q_1) = Q_1$, $\tilde{\sigma}_v(Q_1) = Q_1$, $\tilde{\sigma}_v'(Q_1) = Q_1$, so Q_1 transforms like A_1 of C_{2v}. Similarly, Q_2 transforms like A_1, while for Q_3 we find

$$\tilde{E}(Q_3) = Q_3, \qquad \tilde{C}_2(Q_3) = -Q_3, \qquad \tilde{\sigma}_v(Q_3) = Q_3, \qquad \tilde{\sigma}_v'(Q_3) = -Q_3$$

so Q_3 transforms like B_1. Of course, the transformation properties of the rotations and translations are given in the character table, the result being $\Gamma^{(R+T)} = A_1 + A_2 + 2B_1 + 2B_2$. The $3N$-dimensional representation including all the degrees of freedom is therefore $\Gamma^{(R+T+V)} = 3A_1 + A_2 + 3B_1 + 2B_2$. This latter result is that to which *any* complete set of $3N$ coordinates must lead.

In a similar fashion Q_1 of $CO_3^=$ in Fig. 8.8 transforms like A_1' of D_{3h} while Q_2 transforms like A_2'' since it changes sign under C_2, σ_h, and S_3. On the other hand, Q_{3a} and Q_{3b} form a two-dimensional basis for E'. For example,

$$\tilde{C}_3 \begin{pmatrix} Q_{3a} \\ Q_{3b} \end{pmatrix} = \begin{pmatrix} -\frac{1}{2} & -\sqrt{3}/2 \\ +\sqrt{3}/2 & -\frac{1}{2} \end{pmatrix} \begin{pmatrix} Q_{3a} \\ Q_{3b} \end{pmatrix} \tag{8.82a}$$

and

$$\tilde{\sigma}_h \begin{pmatrix} Q_{3a} \\ Q_{3b} \end{pmatrix} = \begin{pmatrix} -1 & 0 \\ 0 & 1 \end{pmatrix} \begin{pmatrix} Q_{3a} \\ Q_{3b} \end{pmatrix} \tag{8.82b}$$

Equation (8.82a) can be verified graphically using the diagrams of Fig. 8.8. Similarly Q_{4a} and Q_{4b} form a basis for a representation of E' also.

It is evident then, that the normal coordinates do form a basis for a completely reduced representation of the group for the examples cited. To prove that the representation is in completely reduced (block-diagonal) form, we need merely show that the symmetry operations do not mix coordinates corresponding to different vibrational frequencies. Consider two nondegenerate modes Q_m and Q_n, and assume $\lambda_m \neq \lambda_n$. If the operation $\tilde{R}$ mixes the coordinates, we must get

$$\tilde{R}Q_m = aQ_m + bQ_n \tag{8.83a}$$

$$\tilde{R}Q_n = a'Q_m + b'Q_n \tag{8.83b}$$

Let us assume for simplicity that all other modes are nondegenerate also,[14] which means

$$\tilde{R} \sum_{i \neq m,n} Q_i^2 = \sum_i (\pm Q_i)^2 = \sum_i Q_i^2 \tag{8.84a}$$

and

$$\tilde{R} \sum_{i \neq m,n} \dot{Q}_i^2 = \sum_i \dot{Q}_i^2 \tag{8.84b}$$

Thus we see that

$$\tilde{R}(2T) = \left[\sum_{i \neq m,n} \dot{Q}_i^2 \right] + \dot{Q}_m^2(a^2 + a'^2) + \dot{Q}_n^2(b^2 + b'^2) + 2\dot{Q}_m\dot{Q}_n(ab + a'b')$$

$$\tag{8.85}$$

and

$$\tilde{R}(2V) = \left[\sum_{i \neq m,n} \lambda_i Q_i^2 \right] + Q_m^2(\lambda_m a^2 + \lambda_n a'^2)$$

$$+ Q_n^2(\lambda_m b^2 + \lambda_n b'^2) + 2Q_mQ_n(\lambda_m ab + \lambda_n a'b') \tag{8.86}$$

by operating upon $2T$ and $2V$ expressed in terms of normal coordinates and using Eqs. (8.83) and (8.84). But the kinetic and potential energies must be invariant to $\tilde{R}$, so comparison to Eqs. (8.60) shows that

$$a^2 + a'^2 = 1 \tag{8.87a}$$

$$b^2 + b'^2 = 1 \tag{8.87b}$$

$$ab + a'b' = 0 \tag{8.87c}$$

---

[14] A little thought shows that this assumption is not necessary, but we leave the reasoning to the reader.

and

$$\lambda_m = \lambda_m a^2 + \lambda_n a'^2 \qquad (8.87\text{d})$$

$$\lambda_n = \lambda_m b^2 + \lambda_n b'^2 \qquad (8.87\text{e})$$

$$\lambda_m ab + \lambda_n a'b' = 0 \qquad (8.87\text{f})$$

Plugging (8.87a) into (8.87d) and (8.87b) into (8.87e) gives

$$a'^2 = \frac{\lambda_m}{\lambda_n} a'^2 \qquad (8.88\text{a})$$

and

$$b^2 = \frac{\lambda_n}{\lambda_m} b^2 \qquad (8.88\text{b})$$

Because $\lambda_n \neq \lambda_m$ by assumption, we find that the equations are satisfied only if $a'^2 = b^2 = 0$, which according to Eqs. (8.83) means that no mixing can occur.

Our real interest is in deducing the symmetry of the normal modes before their precise form is known, as in the preceding examples. Since the preliminary discussion led to the conclusion that the $3N$-dimensional representation of the normal modes could be generated with any complete basis set, we simply need to find the representation matrices (or their characters) and then reduce the representation under the symmetry group. The irreducible representations of which the reducible representation is composed give the symmetry classifications of the normal modes or coordinates.

It is most convenient to begin with the $3N$ displacement coordinates, $\Delta x_1, \Delta y_1, \Delta z_1, \Delta x_2 \ldots \Delta z_N$, which we relabel as $\xi_1, \xi_2, \xi_3 \ldots \xi_{3N}$ for ease of notation. This basis set is shown for H_2O in Fig. 8.11, where ξ_1, ξ_4, and ξ_7 are out of the plane. Performing the operations of the C_{2v} group upon the set

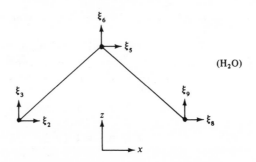

FIG. 8.11.   Displacement coordinates for H_2O (or any bent symmetrical XY_2 molecule). By convention, y axis is out of plane and z axis lies along C_2 axis.

of ξ's we find

$$\tilde{E}\boldsymbol{\xi} = \begin{pmatrix} 1 & & & & & & & & \\ & 1 & & & & & & & \\ & & 1 & & & & & & \\ & & & 1 & & 0 & & & \\ & & & & 1 & & & & \\ & & & & & 1 & & & \\ & & 0 & & & & 1 & & \\ & & & & & & & 1 & \\ & & & & & & & & 1 \end{pmatrix} \boldsymbol{\xi} \qquad (8.89a)$$

$$\tilde{C}_2\boldsymbol{\xi} = \begin{pmatrix} 0 & 0 & 0 & 0 & 0 & 0 & -1 & 0 & 0 \\ 0 & 0 & 0 & 0 & 0 & 0 & 0 & -1 & 0 \\ 0 & 0 & 0 & 0 & 0 & 0 & 0 & 0 & 1 \\ 0 & 0 & 0 & -1 & 0 & 0 & 0 & 0 & 0 \\ 0 & 0 & 0 & 0 & -1 & 0 & 0 & 0 & 0 \\ 0 & 0 & 0 & 0 & 0 & 1 & 0 & 0 & 0 \\ -1 & 0 & 0 & 0 & 0 & 0 & 0 & 0 & 0 \\ 0 & -1 & 0 & 0 & 0 & 0 & 0 & 0 & 0 \\ 0 & 0 & 1 & 0 & 0 & 0 & 0 & 0 & 0 \end{pmatrix} \boldsymbol{\xi} \qquad (8.89b)$$

$$\tilde{\sigma}_v'\boldsymbol{\xi} = \begin{pmatrix} 0 & 0 & 0 & 0 & 0 & 0 & 1 & 0 & 0 \\ 0 & 0 & 0 & 0 & 0 & 0 & 0 & -1 & 0 \\ 0 & 0 & 0 & 0 & 0 & 0 & 0 & 0 & 1 \\ 0 & 0 & 0 & 1 & 0 & 0 & 0 & 0 & 0 \\ 0 & 0 & 0 & 0 & -1 & 0 & 0 & 0 & 0 \\ 0 & 0 & 0 & 0 & 0 & 1 & 0 & 0 & 0 \\ 1 & 0 & 0 & 0 & 0 & 0 & 0 & 0 & 0 \\ 0 & -1 & 0 & 0 & 0 & 0 & 0 & 0 & 0 \\ 0 & 0 & 1 & 0 & 0 & 0 & 0 & 0 & 0 \end{pmatrix} \boldsymbol{\xi} \qquad (8.89c)$$

$$\tilde{\sigma}_v\boldsymbol{\xi} = \begin{pmatrix} -1 & 0 & 0 & 0 & 0 & 0 & 0 & 0 & 0 \\ 0 & 1 & 0 & 0 & 0 & 0 & 0 & 0 & 0 \\ 0 & 0 & 1 & 0 & 0 & 0 & 0 & 0 & 0 \\ 0 & 0 & 0 & -1 & 0 & 0 & 0 & 0 & 0 \\ 0 & 0 & 0 & 0 & 1 & 0 & 0 & 0 & 0 \\ 0 & 0 & 0 & 0 & 0 & 1 & 0 & 0 & 0 \\ 0 & 0 & 0 & 0 & 0 & 0 & -1 & 0 & 0 \\ 0 & 0 & 0 & 0 & 0 & 0 & 0 & 1 & 0 \\ 0 & 0 & 0 & 0 & 0 & 0 & 0 & 0 & 1 \end{pmatrix} \boldsymbol{\xi} \qquad (8.89d)$$

The characters of the representations are therefore

$$\chi(E) = 9, \qquad \chi(C_2) = -1, \qquad \chi(\sigma_v') = 1, \qquad \chi(\sigma_v) = 3 \qquad (8.90)$$

Using the reduction formula given by Eq. (6.41c) we find the 9-dimensional representation, $\Gamma^{(RTV)}$, to be composed of

$$\Gamma^{(RTV)} = 3A_1 + A_2 + 3B_1 + 2B_2 \qquad (8.91)$$

Equation (8.91) gives the irreducible representations or the symmetry *species* of the $3N$ normal modes, which include rotation and translation in addition to vibration. The symmetry species of the former two types of motions are given in the character table and can be subtracted out leaving only the vibrational species. Thus,

$$\Gamma^{(RT)} = A_1 + A_2 + 2B_1 + 2B_2 \qquad (8.92)$$

so

$$\Gamma^{(V)} = \Gamma^{(RTV)} - \Gamma^{(RT)} = 2A_1 + B_1 \qquad (8.93)$$

This result is the same as previously obtained and shows clearly that the *symmetry* of the normal modes can be found without actually solving the classical vibrational problem.

The process of setting up the complete transformation matrices is obviously laborious, even for $N = 3$. A bit of astute observation shows that it is an unnecessary step. First, note that a particular coordinate provides a contribution to the character (that is, provides a transformation coefficient on the diagonal) only when it is not shifted by the symmetry operation from the atom to which it is attached. That is, we need focus attention only on those atoms which would not be shifted by the covering operations of the group.

Secondly, the contributions to the character from each unshifted atom is always the same for a given type of operation. Clearly, for $\tilde{E}$, the contribution per unshifted atom is $+3$; while for $\tilde{\sigma}$ it is always $+1$ per unshifted atom since a reflection always changes the sign of one coordinate and leaves the other two alone. Similarly, $\tilde{C}_2$ must always change the sign of two coordinates (since they rotate by 180°) while leaving the third unchanged, which provides a -1 contribution to the character per unshifted atom.

Similar considerations can be carried out for any $\tilde{C}_n$ rotation, and for $\tilde{\imath}$ and $\tilde{S}_n$. The general results for $\tilde{C}_n$ and $\tilde{S}_n$ are obtained by recalling the well-known 2×2 rotational transformation. Table 8.8 summarizes the character contributions per unshifted atom. In the table $C_n{}^k$ means $(C_n)^k$, for example, $C_3{}^2 = (C_3)^2$. It is seen that the general $C_n{}^k$ result of the table reduces to our special result of -1 when $n = 2$ and $k = 1$.

With the results of Table 8.8 it is an easy task to find the symmetry species of any molecule. For $CO_3{}^=$ (see Fig. 8.8) we find the characters of the D_{3h} operations to be

$$\chi(E) = 12, \qquad \chi(C_3) = 0, \qquad \chi(C_2) = -2,$$

$$\chi(\sigma_h) = 4, \qquad \chi(S_3) = -2, \qquad \chi(\sigma_v) = 2$$

which gives by reduction

$$\Gamma^{(RTV)} = A_1' + A_2' + 2A_2'' + 3E' + E'' \qquad (8.94)$$

From the D_{3h} character table

$$\Gamma^{(RT)} = A_2' + A_2'' + E' + E''$$

so

$$\Gamma^{(V)} = A_1' + A_2'' + 2E' \qquad (8.95)$$

which agrees again with our previous result.

We see shortly the several important advantages that accrue merely by knowing the symmetry species of the normal modes. We can point out immediately that in many cases this knowledge alone provides us with a qualitative picture of what the normal vibrations look like. For H_2O as an example, we know that two modes are totally symmetric (A_1). Thus, the only two vibrational pictures one can draw having this symmetry are the symmetrical stretch (Q_1) and the bend (Q_2), as shown in Fig. 8.8. Likewise, a B_1 mode must change sign under C_2 and σ_v and thus one is led to the antisymmetrical stretch (Q_3). Of course, the symmetry does not tell the lengths or precise orientations of the displacement vectors.

The symmetry species determination can be carried out in other coordinate systems also, to good advantage. One such system which is extremely useful later is the *internal coordinate* system. For example, H_2O can have $3N - 6 = 3$ internal (vibrational) coordinates only. The manner in which internal coordinates are chosen is flexible, but the most common choice is that of so-called *valence* coordinates, namely, bond stretches and angle deformations. The important criterion is that the internal coordinates must span the complete ($3N - 6$) space of interest. In general, for ease of manipulation, complete sets of *equivalent* internal coordinates should be used for any group-theory analysis. By equivalent we mean coordinates which transform into each other under the group operations. For H_2O the most suitable internal (valence) coordinates would be: (1) stretch of the left-hand bond, (2) stretch of the right-hand bond, and (3) increase of the

TABLE 8.8

Contribution to the character of the 3N-dimensional representation per unshifted atom.

Operation	Character contribution per unshifted atom
E	3
σ	1
i	-3
$C_n{}^k$	$1 + 2\cos(2\pi k/n)$
$S_n{}^k$	$-1 + 2\cos(2\pi k/n)$

HOH bond angle. In this case the two stretches form an equivalent set, while the bending coordinate is a set in itself. One could, of course, specify the vibrational motions with other internal coordinates, for example, (1) an $H \cdots H$ (non-bonded) stretch, (2) an $OH \cdots H$ angle deformation, and (3) OH stretch. This latter set of coordinates is, however, unsatisfactory from a group-theory point of view since it does not consist of equivalent sets.

To be more definite we may consider $CO_3^=$ again. We require a minimum of 6 internal coordinates. One set of equivalent coordinates consists of the 3 CO stretches shown in Fig. 8.12, while another is composed of the 3 OCO bond-angle deformations. Note that since $\alpha_1 + \alpha_2 + \alpha_3 = 0$ is required because of the planarity of the ion, one of these angle coordinates is *redundant*. Our group theory proceeds easily if we carry along all the angular coordinates and remove the redundancy at the end. Finally, we require one coordinate to specify out-of-plane motion. We call this I, for inversion. Figure 8.12 shows all of our choices. The coordinates that are drawn in the figure need not be given a quantitative significance. We simply require that those within each set be equivalent. As is shown later, it is possible to relate quantitatively the internal coordinates to the Cartesian displacement coordinates.

Since the R_i's transform only into each other they must form a basis for a three-dimensional representation. We find the characters of the D_{3h} group operators to be $\chi(E) = 3, \chi(C_3) = 0, \chi(C_2) = 1, \chi(\sigma_h) = 3, \chi(S_3) = 0$,

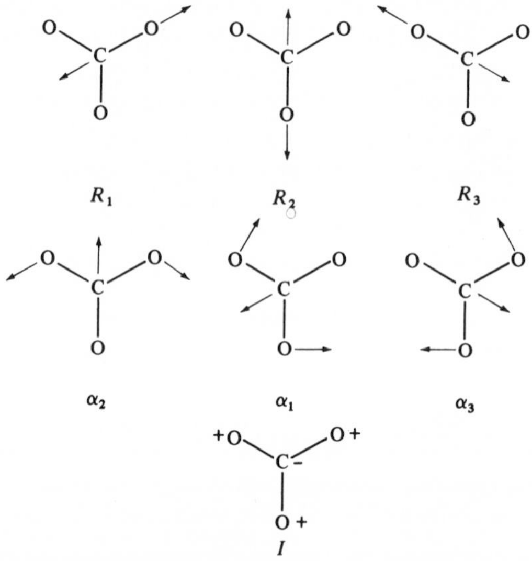

FIG. 8.12.   Internal coordinates for $CO_3^=$.

and $\chi(\sigma_v) = 1$. These results arise easily; for example, for one of the C_2 axes,

$$\tilde{C}_2 \begin{pmatrix} R_1 \\ R_2 \\ R_3 \end{pmatrix} = \begin{pmatrix} 0 & 0 & 1 \\ 0 & 1 & 0 \\ 1 & 0 & 0 \end{pmatrix} \begin{pmatrix} R_1 \\ R_2 \\ R_3 \end{pmatrix} \qquad (8.96)$$

thus $\chi(C_2) = 1$. Reducing the representation formed by the R_i we find

$$\Gamma^{(R)} = A_1' + E' \qquad (8.97)$$

where the rather unfortunate notation $\Gamma^{(R)}$ means the representation formed by the R_i coordinates, and has nothing to do with rotation.

In an entirely similar manner we find the characters for the α set of coordinates to be $\chi(E) = 3, \chi(C_3) = 0, \chi(C_2) = 1, \chi(\sigma_h) = 3, \chi(S_3) = 0$, and $\chi(\sigma_v) = 0$, which gives

$$\Gamma^{(\alpha)} = A_1' + E' \qquad (8.98)$$

The coordinate I is seen easily to transform like A_2'', thus

$$\Gamma^{(I)} = A_2'' \qquad (8.99)$$

The sum of Eqs. (8.97)–(8.99) gives the total representation formed by the *seven* internal coordinates. However, we desire the six-dimensional representation of the six (independent) internal coordinates. This is gotten by subtracting the representation which gives the symmetry of the redundancy relation. That is, we need the transformation properties of $\alpha_1 + \alpha_2 + \alpha_3$. This turns out to be of A_1' symmetry. Thus the correct symmetry species of the independent α's is

$$\Gamma^{(\alpha)} = E' \qquad (8.100)$$

instead of Eq. (8.98). Note now that the sum of Eqs. (8.97), (8.99), and (8.100) gives

$$\Gamma^{(V)} = \Gamma^{(R)} + \Gamma^{(\alpha)} + \Gamma^{(I)} = A_1' + A_2'' + 2E' \qquad (8.101)$$

which is identical to our previous result [Eq. (8.95)] using Cartesian displacement coordinates.

This internal-coordinate treatment has, however, provided us with more detailed knowledge of the vibrational motions than was available from the Cartesian displacement coordinate analysis. Specifically, we have found that the stretching modes are of A_1' and E' symmetry, the in-plane (doubly-degenerate) bending mode is of E' symmetry, and the out-of-plane motion is A_2''. It is useful therefore to perform both types of analysis. The Cartesian displacement treatment gives the total number and types of symmetry species, while the internal-coordinate method provides the breakdown in terms of equivalent sets and also provides a check on the former method.

EXAMPLE 8-6

Once the symmetry of the internal coordinates is known it is an easy matter to use the methods of Sec. 6.4 C to obtain the symmetry coordinates in terms of the internal coordinates. Considering $CO_3^=$ again, let us look first for the A_1' stretching coordinate. Using the operations of D_{3h} and Fig. 8.12, we can apply Eq. (6.68) taking $\phi_j = R_1$. This gives

$$S_1^{A_1'} = N\{R_1 + (R_2 + R_3) + (R_1 + R_2 + R_3) + R_1$$
$$+ (R_2 + R_3) + (R_1 + R_2 + R_3)\}$$
$$= (1/\sqrt{3})(R_1 + R_2 + R_3) \quad (8.102a)$$

In a similar manner we find *one* member of the E' degenerate stretching mode to be

$$S_2^{E'} = N\{2R_1 - (R_2 + R_3) + 0(R_1 + R_2 + R_3) + 2R_1$$
$$- (R_2 + R_3) + 0(R_1 + R_2 + R_3)\}$$
$$= (1/\sqrt{6})(2R_1 - R_2 - R_3) \quad (8.102b)$$

The second member of the degenerate pair is chosen to be orthogonal, so we may write

$$S_3^{E'} = (1/\sqrt{2})(R_2 - R_3) \quad (8.102c)$$

One member of the degenerate pair of E' bending coordinates is now found by using $\phi_j = \alpha_2$. This gives

$$S_4^{E'} = N\{2\alpha_2 - (\alpha_1 + \alpha_3) + 0(\alpha_1 + \alpha_2 + \alpha_3) + 2\alpha_2$$
$$- (\alpha_1 + \alpha_3) + 0(\alpha_1 + \alpha_2 + \alpha_3)\}$$
$$= N'(2\alpha_2 - \alpha_1 - \alpha_3) \quad (8.102d')$$

But this contains a redundant variable, which we eliminate by writing $\alpha_2 = -\alpha_1 - \alpha_3$. Thus we get

$$S_4^{E'} = (1/\sqrt{2})(\alpha_1 + \alpha_3) \quad (8.102d)$$

The second bending coordinate is most easily chosen as

$$S_5^{E'} = (1/\sqrt{2})(\alpha_1 - \alpha_3) \quad (8.102e)$$

Finally we note that I is already a symmetry coordinate:

$$S_6^{A_2''} = I \quad (8.102f)$$

Thus we have explicit expressions for the symmetry coordinates in terms of internal coordinates. The six symmetry coordinates of Eqs. (8.102) have, in addition, the same symmetry as the normal coordinates, although they are not mass weighted. Thus we expect the normal coordinates (or modes) to have the same general appearance as the symmetry coordinates. While the nondegenerate symmetry coordinates look qualitatively like the corresponding normal modes, the four normal modes of E' symmetry are linear combinations of the E' symmetry coordinates so the resemblance may be somewhat clouded. We see later that symmetry coordinates have an extremely important role to play in practical calculations relating force constants and fundamental vibrational frequencies.

B. Symmetry Classification of the Vibrational Energy States

In this section we develop the means of determining the symmetry species for the vibrational states of polyatomic molecules. Our explicit treatment is for the harmonic-oscillator approximation, but in fact, the results are applicable to a real anharmonic molecule. This is because an anharmonic oscillator wave function can always be expanded in terms of a complete set of harmonic-oscillator functions. Furthermore, since the eigenstates must have the symmetry of the molecular point group, the harmonic-oscillator functions making up a given anharmonic eigenfunction must all be of the same symmetry. Therefore, classification of a harmonic state may also be considered to be a classification of the anharmonic state to which it is most similar.

The result of this brief argument is that our symmetry classifications are really more inclusive than just the harmonic oscillator. When we use these symmetry arguments to predict selection rules, we have results which are correct even for an anharmonic oscillator. For example, the group-theory selection rules permit certain transitions not allowed by the harmonic-oscillator considerations of Sec. 8.3 C. On the other hand we find that transitions forbidden by symmetry are usually correct as long as the molecular symmetry is maintained.

To begin the discussion it is perhaps best to revert to the wave-mechanics formalism, in which case the wave function for the state $| v_1 v_2 \ldots v_{3N-6} \rangle$ is given by Eq. (8.68) as

$$\psi_v = \psi_{v_1}(Q_1) \cdot \psi_{v_2}(Q_2) \cdots \psi_{v_{3N-6}}(Q_{3N-6}) \tag{8.103}$$

where the $\psi_{v_i}(Q_i)$ are the well-known harmonic-oscillator functions, namely

$$\psi_{v_i}(Q_i) = \mathfrak{N}_i \exp(-\tfrac{1}{2}\gamma_i Q_i{}^2) H_{v_i}(\gamma_i{}^{1/2}Q_i) \tag{8.104}$$

In Eq. (8.104), $\mathfrak{N}_i$ is a normalizing constant, $\gamma_i = 4\pi^2 \nu_i/h$, and $H_{v_i}(\gamma_i{}^{1/2}Q_i)$ is a Hermite polynomial.[15]

---

[15] See for example, L. Pauling and E. B. Wilson, *Introduction to Quantum Mechanics* (McGraw-Hill, New York, 1937).

Ground-state symmetry $(\mid 000\cdots0\rangle)$:
In this case $H_{v_i} = 1$ for all i so

$$\psi_{\text{Ground}} = \Re \exp\left(-\tfrac{1}{2}\sum_i \gamma_i Q_i^2\right) \tag{8.105}$$

The transformation properties of ψ_{Ground} are then gotten from

$$\tilde{R}\psi_G = \Re \exp\left(-\tfrac{1}{2}\sum_i \gamma_i \tilde{R}Q_i^2\right) \tag{8.106}$$

We have already seen that the Q_i form bases for the irreducible representations, thus for the nondegenerate modes $\tilde{R}Q_i^2 = (\pm Q_i)^2 = Q_i^2$. For degenerate modes, $\tilde{R}Q_i^2 = (a_{ii}Q_i + a_{ij}Q_j + \cdots)^2$ that is, the degenerate coordinates are mixed. But since the a_{ij} are unitary or real orthogonal, the sum $\sum_i \tilde{R}Q_i^2$ over the degenerate set is invariant. For example, suppose Q_2 and Q_3 are a degenerate pair. Then

$$\begin{aligned}
\sum_{i=2,3} \tilde{R}Q_i^2 &= (a_{ii}Q_i + a_{ij}Q_j)^2 + (a_{jj}Q_j + a_{ji}Q_i)^2 \\
&= Q_i^2(a_{ii}^2 + a_{ji}^2) + Q_j^2(a_{ij}^2 + a_{jj}^2) \\
&\quad + 2Q_iQ_j(a_{ii}a_{ij} + a_{jj}a_{ji}) \\
&= Q_i^2 + Q_j^2
\end{aligned} \tag{8.107}$$

by use of the unitary properties of a. We conclude that ψ_G is always *totally symmetric*, that is,

$$\tilde{R}\psi_G = \psi_G \tag{8.108}$$

Fundamental symmetry $(\mid 00\cdots v_j\cdots00\rangle)$:
Now we have all $H_{v_i} = 1$, except for the mode Q_j which has $v_j = 1$. Then the Hermite polynomial is $H_{v_j} \propto Q_j,$[16] and the fundamental-state wave function becomes

$$\psi_F = \Re Q_j \exp\left(-\tfrac{1}{2}\sum_i \gamma_i Q_i^2\right) \tag{8.109}$$

Because the exponential part is totally symmetric we find that ψ_F transforms like the normal coordinate which is excited with one quanta, that is

$$\tilde{R}\psi_F = \Re \exp\left(-\tfrac{1}{2}\sum_i \gamma_i Q_i^2\right)\tilde{R}Q_j \tag{8.110}$$

The problem is therefore solved, since we have already found in Sec. 8.4 A how the Q_j transform according to the irreducible representations of the group.

[16] See footnote 15.

Overtone level symmetry $(\,|\,00\cdots v_j>1\cdots 0\rangle)$:

For nondegenerate modes the analysis is easy. We need simply investigate the symmetry of H_{v_j} where Q_j is the multiply excited mode. That is,

$$\tilde{R}\psi_0 = \mathfrak{N} \exp\left(-\tfrac{1}{2}\sum_i \gamma_i Q_i^2\right)\tilde{R}H_{v_j} \qquad (8.111)$$

where ψ_0 is the nondegenerate overtone state. We have seen that $\psi_G \sim Q^0$ and $\psi_F \sim Q^1$. The properties of the Hermite polynomials are such that *all* v = even states are even functions of Q while *all* v = odd states are odd functions of Q.[17] Hence for even overtones (v_j = even)

$$\tilde{R}\psi_0 = \mathfrak{N} \exp\left(-\tfrac{1}{2}\sum_i \gamma_i Q_i^2\right)(+H_{v_j}) \qquad (8.112)$$

$$= \psi_0$$

so *even overtones are of the totally symmetric species.*

On the other hand the odd overtones lead to

$$\tilde{R}\psi_0 = \mathfrak{N} \exp\left(-\tfrac{1}{2}\sum_i \gamma_i Q_i^2\right)(\chi_j H_{v_j})$$

$$= \chi_j\psi_0 \qquad (8.113)$$

where χ_j is simply ± 1 as given by

$$\tilde{R}Q_j = \chi_j Q_j$$

That is, the *odd overtones* (v_j = odd) *have the symmetry of the corresponding fundamental or normal coordinate.*

When the overtones belong to degenerate fundamentals the analysis is considerably more complex because of the larger number of basis functions. For example, when the v = 3 state of a doubly degenerate vibration is being considered, the four degenerate functions $|\,v_a v_b\rangle$ are $|\,03\rangle$, $|\,12\rangle$, $|\,21\rangle$, and $|\,30\rangle$ where v_a and v_b refer to the two degenerate normal modes.

It can be shown that the characters $\chi_v(R)$ for the vth quantum state for an overtone of a *doubly degenerate* vibration[18] are given by

$$\chi_v(R) = \sum_{|l|} \chi_{|l|}(R) \qquad (8.114)$$

where $|\,l\,| = 1, 3, 5\ldots v$ if v is odd or $|\,l\,| = 0, 2, 4\ldots v$ if v is even. $\chi_{|l|}(R)$ is obtained by the following rules:

(1) If $|\,l\,| = 0$, $\chi_{|0|}(R) = 1$

(2) If $|\,l\,| \neq 0$, and if $\chi^{(j)}(R) = 0$ and $\chi^{(j)}(R^2) > 0$, then $\chi_{|l|}(R) = 0$. In this rule $\chi^{(j)}(R)$ is the character of the Rth operation of the

---

[17] See footnote 15.

[18] J. C. Decius, *J. Chem. Phys.* **17**, 504 (1949).

irreducible representation to which the doubly degenerate mode belongs. Similarly $\chi^{(j)}(R^2)$ is the character for the R^2 operation.

(3) If rules (1) and (2) do not apply, $\chi_{|u|}(R) = 2\cos(l\alpha_R)$ where α_R is obtained from $\chi_R^{(j)} = 2\cos(\alpha_R)$.

After the characters are found the representation of the overtone level is reduced in the usual manner.

Similar but more complicated, results can be obtained for finding the symmetry species for overtones of triply degenerate normal modes.[19]

Symmetry of Combination or Any General Vibrational State.

Here we consider any state $|v_1 v_2 \cdots v_{3N-6}\rangle$ where at least two v_i are greater than zero. Since each nondegenerate normal coordinate, or each set of degenerate normal coordinates,[20] forms a basis for an irreducible representation of the group, the representation of any general vibrational state is obtained by forming the direct product of the representations of the individual states which are excited. These results are easily obtained since the symmetries of the individual states have been obtained in the preceding sections. An example should aid now in the understanding of these principles.

EXAMPLE 8-7

We consider again the $CO_3^=$ ion of D_{3h} symmetry, which has been shown previously to have normal modes of symmetry $\Gamma_1 = A_1'$, $\Gamma_2 = A_2''$, $\Gamma_3 = E'$, and $\Gamma_4 = E'$. The rules for the ground state and fundamentals lead to the following symmetries:

$$\Gamma(0000) = A_1', \qquad \Gamma(0100) = A_2'', \qquad \Gamma(0001) = E'.$$

$$\Gamma(1000) = A_1', \qquad \Gamma(0010) = E',$$

That is, the fundamentals have the symmetries of the normal coordinates themselves. Consider now the $v_1 = 2$ overtone state of the symmetrical stretching mode. Our previous theory states that even overtones of nondegenerate states are totally symmetric, thus

$$\Gamma(2000) = A_1'.$$

For the (3000) state the symmetry is that of the normal coordinate, so

$$\Gamma(3000) = A_1' \text{ also.}$$

---

[19] See footnote 18.

[20] The statement is correct also, if we substitute for "normal coordinate(s)" the words "normal mode wave function(s)."

However, the identical principles applied to the (0200) and (0300) states give

$$\Gamma(0200) = A_1'$$

$$\Gamma(0300) = A_2''.$$

In order to find the symmetry of the doubly degenerate state (0020) we apply Eq. (8.114) which for our double excitation becomes

$$\chi_2(R) = \chi_{|0|}(R) + \chi_{|2|}(R) \tag{8.115}$$

By rule (1), $\chi_{|0|}(R) = 1$ for all R. To obtain the second term we write the characters for the E' representation with the aid of the D_{3h} character table. This gives

R	E	C_3	C_2	σ_h	S_3	σ_v
$\chi(R)$	2	-1	0	2	-1	0
$\chi(R^2)$	2	-1	2	2	-1	2

where the second line is obtained by observing that $\chi(C_3^2) = \chi(C_3)$, $\chi(C_2^2) = \chi(E)$, $\chi(\sigma_h^2) = \chi(E)$, $\chi(S_3^2) = \chi(C_3)$, $\chi(\sigma_v^2) = \chi(E)$. Thus by rule (2), $\chi_{|2|}(C_2) = \chi_{|2|}(\sigma_v) = 0$. We apply rule (3) to the remaining $\chi_{|2|}(R)$ which gives $\chi_{|2|}(E) = \chi_{|2|}(\sigma_h) = 2$ and $\chi_{|2|}(C_3) = \chi_{|2|}(S_3) = 0$. Putting all of our intermediate results into Eq. (8.115) we find the characters to be

R	E	C_3	C_2	σ_h	S_3	σ_v
$\chi_2(R)$	3	0	1	3	0	1

Reducing this in the usual manner we find

$$\Gamma(0020) = A_1' + E'$$

The degenerate overtone (0020) is seen to be composed of two species of total dimension three.

It is easy now to determine the symmetries of a number of combination states. For example, the symmetry of the (1100) state is, by our previous results,

$$\Gamma(1100) = \Gamma(1000) \times \Gamma(0100)$$

$$= A_1' \times A_2'' = A_2''$$

that of the (0110) is

$$\Gamma(0110) = \Gamma(0100) \times \Gamma(0010)$$

$$= A_2'' \times E' = E''$$

while that of the (0011) is

$$\Gamma(0011) = \Gamma(0010) \times \Gamma(0001)$$

$$= E' \times E'$$

$$= A_1' + A_2' + E'$$

This latter result arises, for example, by decomposition of the representation $E' \times E'$ whose characters are 4, 1, 0, 4, 1, 0 running left to right in the D_{3h} character table. Multiply excited combination states are analyzed similarly, thus for the (1320) state we have

$$\Gamma(1320) = \Gamma(1000) \times \Gamma(0300) \times \Gamma(0020)$$

$$= A_1' \times A_2'' \times (A_1' + E')$$

$$= A_2'' \times (A_1' + E')$$

$$= A_2'' + E''$$

C. Symmetry Selection Rules for Infrared Vibrational Spectra

Sections 8.2 C and 8.3 C have presented the general quantitative means for determining vibrational selection rules and spectral line intensities. As shown in Sec. 6.4 B, group theory provides a powerful means of specifying selection rules with a minimum of effort. In no area of spectroscopy is this more evident than in infrared (vibrational) spectroscopy.

Neglecting the rotational parts of the matrix elements, it has been shown in Secs. 7.3 B, 8.1 C and 8.3 C that the intensity of a vibrational transition is given by

$$P \propto \sum_g | \langle v | \tilde{\mu}_g | v' \rangle |^2 \tag{8.116}$$

where v and v' refer to all the vibrational quantum numbers and $\tilde{\mu}_g$ is the dipole moment along the g axis of the molecule. Although for quantitative evaluation of transition probabilities it is useful to expand $\tilde{\mu}_g$ in a power series in the normal coordinates, our present application requires only the original definition of the electric dipole moment given by Eq. (7.38). Thus we have

$$\mu_x = \sum_i Z_i e x_i \tag{8.117a}$$

$$\mu_y = \sum_i Z_i e y_i \tag{8.117b}$$

$$\mu_z = \sum_i Z_i e z_i \tag{8.117c}$$

where Z_i is the charge on the ith particle.

Group theory tells us that the transition probability of Eq. (8.116) can be nonzero only if

$$\Gamma^{(v)} \times \Gamma^{(\mu_g)} \times \Gamma^{(v')} = \Gamma^{(A)} \tag{8.118}$$

for at least one of the μ_g $(g = x, y, z)$. Since we have already determined the symmetry properties (irreducible representations) of the vibrational states $|v\rangle$ in Sec. 8.4 B, we need merely ask now about the symmetry properties of the μ_g. By Eqs. (8.117) it is clear that μ_g is simply a linear combination of Cartesian coordinates. Thus the transformation properties of μ_x, for example, are just the same as the transformation properties of any x_i, that is, of the general coordinate x. But we have discussed this problem in our general discussion of group theory and the results were trivially obtained. For practical purposes, of course, the representations to which x, y, and z belong are given for each group in the appropriate character table.

We consider first the *fundamental* infrared transitions. Then $|v\rangle$ is the ground state and $|v'\rangle$ the fundamental state so

$$\Gamma^{(v)} = \Gamma^{(A)}$$

and

$$\Gamma^{(v')} = \Gamma^{(Q_i)}$$

where Q_i refers to the mode having $v_i = 1$ and $\Gamma^{(A)}$ is our usual notation for the totally symmetric representation. Equation (8.118) becomes, for fundamental bands,

$$\Gamma^{(A)} \times \Gamma^{(g)} \times \Gamma^{(Q_i)} = \Gamma^{(A)} \tag{8.119a}$$

or by our earlier group-theory results,

$$\Gamma^{(g)} \times \Gamma^{(Q_i)} = \Gamma^{(A)} \tag{8.119b}$$

or more simply,

$$\Gamma^{(g)} = \Gamma^{(Q_i)} \tag{8.119c}$$

Equation (8.119c) states that a fundamental transition, $(00\ldots0) \rightarrow (00, v_i = 1, \ldots0)$, may occur, that is, be infrared active, only if at least one of the x, y, or z belongs to the same irreducible representation as the normal mode being excited.

Thus for the H_2O molecule we have found the normal modes and hence the fundamental states to have the symmetries A_1, A_1, and B_1 under the C_{2v} group. Looking at the character table we find that z transforms like A_1 and x like B_1, so we conclude that all three fundamentals of H_2O are infrared active. This is the same result that we deduced earlier by looking at the form of the $(\partial\mu/\partial Q_i)$ terms.

As a less trivial example consider the $CO_3^=$ ion. The symmetries of the fundamental states were

$$\Gamma^{(Q_1)} = A_1', \qquad \Gamma^{(Q_2)} = A_2'', \qquad \Gamma^{(Q_3)} = E', \qquad \text{and} \qquad \Gamma^{(Q_4)} = E'$$

Looking at the D_{3h} character table we find that x and y transform like E' while z transforms like A_2''. Therefore, all $CO_3^=$ fundamental modes are IR active except ν_1, the totally symmetric mode. Again the result agrees with what one would infer about $(\partial\mu/\partial Q_i)$ for the four modes of Fig. 8.8. Note, however, that our present results are obtained by a process which bypasses entirely the actual construction of a normal-mode diagram. The group-theory method is trivial to apply no matter how complicated the molecule, while normal-mode diagrams for complicated molecules are very difficult to analyze in terms of $(\partial\mu/\partial Q_i)$ without the use of symmetry.

The selection rules for overtones and combinations are obtained by the same process. For all transitions originating from the ground state, the requirement of Eq. (8.118) becomes

$$\Gamma^{(g)} \times \Gamma^{(v')} = \Gamma^{(A)} \tag{8.120a}$$

or

$$\Gamma^{(g)} = \Gamma^{(v')} \tag{8.120b}$$

Note that Eq. (8.119c) is a special case of this. Equation (8.120b) states that the only allowed transitions (originating from the ground state) are those which terminate in states belonging to the same irreducible representations as either x, y, or z.

For the $CO_3^=$ example we have previously found from the D_{3h} table that E' and A_2'' are the species of x, y, and z. Therefore transitions are permitted only to states of either of these two symmetries. For all the example states treated in Example 8-7 we find the following results:

$$
\begin{array}{lll}
(0000) &\rightarrow (1000) & \text{Inactive} \\
&\rightarrow (0100) & \text{Active} \\
&\rightarrow (0010) & \text{Active} \\
&\rightarrow (0001) & \text{Active} \\
&\rightarrow (2000) & \text{Inactive} \\
&\rightarrow (3000) & \text{Inactive} \\
&\rightarrow (0200) & \text{Inactive} \\
&\rightarrow (0300) & \text{Active} \\
&\rightarrow (0020) & \text{Active} \\
&\rightarrow (1100) & \text{Active} \\
&\rightarrow (0110) & \text{Inactive} \\
&\rightarrow (0011) & \text{Active} \\
&\rightarrow (1320) & \text{Active} \tag{8.121}
\end{array}
$$

It should be emphasized that these are strong symmetry arguments, and that the predictions of inactive transitions are rigorously followed except under unusual conditions, such as high pressure, in which case forbidden transitions might be induced. On the other hand, an active transition may or may not be experimentally observable. We know that active fundamentals should be observable (since $\Delta v = +1$). But overtones and combinations violate the rule that Δv must be zero for all but one mode for which $\Delta v = \pm 1$; thus these transitions occur only because of electrical and mechanical anharmonicity and are expected to be considerably less intense than the fundamentals.

It is perhaps of value to introduce a common nomenclature for the vibrational transitions. For any transition originating from the ground state and ending in the state $| v_1 v_2 \cdots v_{3N-6} \rangle$, the symbol used is $v_1 \nu_1 + v_2 \nu_2 + \cdots v_{3N-6} \nu_{3N-6}$. This nomenclature for the transitions of Eq. (8.121) gives (listed in order)

$$\nu_1$$

$$\nu_2$$

$$\nu_3$$

$$\nu_4$$

$$2\nu_1$$

$$3\nu_1$$

$$2\nu_2$$

$$3\nu_2$$

$$2\nu_3$$

$$\nu_1 + \nu_2$$

$$\nu_3 + \nu_4$$

$$\nu_1 + 3\nu_2 + 2\nu_3 \qquad (8.122)$$

The symbolism is natural, since it represents the frequency of the transition in the harmonic-oscillator approximation.

The group-theoretical method may be used to predict selection rules for transitions not originating in the ground state. Consider, for example $(0010) \rightarrow (0100)$ of the $CO_3^=$ ion. Forming the triple products, we find for the μ_x or μ_y matrix element

$$E' \times E' \times A_2'' = A_1'' + A_2'' + E''$$

and for the μ_z matrix element

$$E' \times A_2'' \times A_2'' = E'$$

Thus this transition, which would be termed $\nu_2 - \nu_3$, is inactive, since the triple product does not contain A_1'. On the other hand, for the $(0010) \rightarrow (1000)$ transition, the μ_x or μ_y element leads to

$$E' \times E' \times A_1' = A_1' + A_2' + E'$$

The triple product contains A_1' in this case and hence $\nu_1 - \nu_3$ is an active transition.

In Tables 8.9 and 8.10 the stronger infrared bands have been listed for NH_3 and C_2H_2. The measured frequencies and assignments are given, including the point-group symmetry classification for the upper states of the transitions.

8.5 NORMAL COORDINATE ANALYSIS

A. FG Matrix Theory; A Simple Example

Up to now we have avoided the actual practical calculations and procedures involved in setting up and solving the classical secular equation which relates the fundamental vibrational frequencies and force constants. This process and that of determining the exact form of the normal vibrational modes is usually referred to as *normal coordinate analysis*.

The normal coordinate-analysis problem has been avoided since the classical problem as formulated in Sec. 8.3 A is not a very convenient method to use. For one thing, the secular equation still contains the $3N - 6$ (or 5) zero roots corresponding to translation and vibration. Secondly, the mass-weighted (or displacement) coordinates are not very convenient for physical applications. This is because one would prefer to formulate the potential energy in terms of bonding or nonbonding parameters, that is, internuclear distances and bond angles. Nevertheless, the problem can be treated by the original method as has been outlined for the diatomic molecule in Example 8-3.

A final reason for avoiding normal coordinate analysis until now is that group theory plays an important role in simplifying the problem. We see shortly that a knowledge of the normal coordinate symmetries leads to a reduction in the size of the secular equations which must be solved.

Wilson[21] devised a method based on internal coordinates that is convenient for any molecule, and that is universally used for normal coordinate analyses. It begins by defining a matrix **B** which relates a set of $3N - 6$

---

[21] E. B. Wilson, Jr., *J. Chem. Phys.* **7**, 1047 (1939); See also footnote 7 for a complete discussion.

TABLE 8.9

Some observed infrared bands of NH₃ in the gas phase.[a]

$\nu \, (cm^{-1})$	Assignment	Intensity
$\left.\begin{array}{l} 931.6 \\ 968.1 \end{array}\right\}$[b]	$\nu_2 (A_1)$	s
1627.5	$\nu_4 (E)$	vs
$\left.\begin{array}{l} 2440 \\ 2473 \end{array}\right\}$	$\nu_3 - \nu_2$ or $\nu_2 + \nu_4$	w
$\left.\begin{array}{l} 3335.9 \\ 3337.5 \end{array}\right\}$	$\nu_1 (A_1)$	s
3414	$\nu_3 (E)$	—
$\left.\begin{array}{l} 4176 \\ 4216 \end{array}\right\}$	$\nu_2 + 2\nu_4 (A_1 + E)$	m
$\left.\begin{array}{l} 4269 \\ 4302 \end{array}\right\}$	$\nu_1 + \nu_2 (A_1)$	s
$\left.\begin{array}{l} 6595 \\ 6624 \end{array}\right\}$	$2\nu_1 (A_1)$	m

[a] Adapted from footnote 10, Herzberg, Tables 72 and 73.
[b] Parentheses indicate bands split by the inversion motion of the molecule.

TABLE 8.10

Some observed infrared bands of gaseous acetylene (C₂H₂).[a]

$\nu \, (cm^{-1})$	Assignment	Intensity[b]
729.1	$\nu_5 (\Pi_u)$	vs
1328.1	$\nu_4 + \nu_5 (\Sigma_u{}^+)$	s
2701.5	$\nu_2 + \nu_5 (\Pi_u)$	m
3287	$\nu_3 (\Sigma_u{}^+)$	vs
3897	$\nu_3 + \nu_4 (\Pi_u)$	m
4091	$\nu_1 + \nu_5 (\Pi_u)$	m
8512	$\nu_1 + \nu_2 + \nu_3 \; (\Sigma_u{}^+)$	—
9640	$3\nu_3 (\Sigma_u{}^+)$	vs

[a] Adapted from footnote 10, Herzberg, Table 68.
[b] m = medium, s = strong, v = very.

(or 5) internal coordinates S_i to the $3N$ Cartesian displacement coordinates ξ_k as shown in Eqs. (8.123).

$$S_i = \sum_k^{3N} B_{ik}\xi_k \tag{8.123a}$$

or

$$\mathbf{S} = \mathbf{B}\boldsymbol{\xi} \tag{8.123b}$$

Note that the $\mathbf{B}$ matrix is rectangular, having $3N - 6$ (or 5) rows and $3N$ columns as shown by Eq. (8.124)

$$\mathbf{B} = \qquad\qquad\qquad\qquad\qquad 3N - 6 \text{ (or 5)} \tag{8.124}$$

$$3N$$

The internal coordinates may be any set which are physically convenient; for example, for H_2O they might be the extensions of the two OH bonds and the angle deformation between the two bonds. The matrix $\mathbf{B}$ can always be worked out by geometrical considerations as we presently illustrate.

Then a very important square matrix of dimension $3N - 6$ (or 5) is defined by

$$G_{ij} = \sum_k^{3N} \frac{1}{m_k} B_{ik}B_{jk} \tag{8.125}$$

In Appendix H it is shown that the $\mathbf{G}$ matrix is closely related to the kinetic energy of the vibrating molecule. Note that $\mathbf{G}$ is symmetric (that is, $G_{ij} = G_{ji}$) since the equation is unchanged by interchanging B_{ik} and B_{jk}. Clearly, the $\mathbf{G}$ matrix elements are obtained directly once the $\mathbf{B}$ matrix elements are known.

We now write the potential energy in a form which makes physical sense, that is, we write it in terms of internal coordinates as,

$$2V = \sum_{i,j}^{3N-6(\text{or }5)} F_{ij}S_iS_j \tag{8.126}$$

The form is a general quadratic one as used previously, for example in Eq. (3.74). The elements F_{ij} form what we call the $\mathbf{F}$ matrix. If we define one further matrix $\mathbf{L}$, we are in a position to write down the secular equation for the classical problem. $\mathbf{L}$ is defined as the matrix which relates the internal coordinates to the normal coordinates.

$$\mathbf{S} = \mathbf{LQ} \tag{8.127a}$$

Of course, we are usually interested in finding the Q's from the S's, which is accomplished by the inverse transformation

$$\mathbf{Q} = \mathbf{L^{-1}S} \tag{8.127b}$$

As emphasized in Appendix H, the matrix $\mathbf{L}$ is not unitary or real orthogonal in general.

The relation between $\mathbf{G}$, $\mathbf{F}$, and $\mathbf{L}$ is that the classical vibrational eigenvalue matrix is given by

$$\mathbf{L^{-1}GFL} = \mathbf{\Lambda} \tag{8.128}$$

where $\mathbf{\Lambda}$ is diagonal and contains the $3N - 6$ (or 5) λ_i, where $\lambda_i = 4\pi^2\nu_i^2$. The relation given by Eq. (8.128) is derived in Appendix H. We simply note here that if the $(\mathbf{GF})$ matrix is formed and then diagonalized in an appropriate fashion, the vibrational eigenvalues are obtained. Further, the transformation matrix $\mathbf{L}$ which brings about the similarity transformation is the same $\mathbf{L}$ of Eq. (8.127a). The computational approach can be further clarified by writing the ikth element of Eq. (8.128):

$$\sum_j \left\{ (\mathbf{GF})_{ij} - \delta_{ij}\lambda_k \right\} L_{jk} = 0 \qquad k = 1, 2, 3 \ldots 3N - 6 \tag{8.129}$$

Thus the λ_k are obtained from the secular determinant

$$| \mathbf{GF} - \mathbf{E}\lambda_k | = 0 \tag{8.130}$$

Equations (8.129) and (8.130) are entirely analogous to our earlier Eqs. (8.50) and (8.51). Equation (8.128) leads to other useful forms by means of appropriate manipulations, for example,

$$\sum_j \left\{ (\mathbf{FG})_{ij} - \delta_{ij}\lambda_k \right\} L^{-1}_{kj} = 0 \qquad k = 1 \ldots 3N - 6 \tag{8.131a}$$

$$| \mathbf{FG} - \mathbf{E}\lambda_k | = 0 \tag{8.131b}$$

These last two forms are useful since the solution of the $3N - 6$ equations of the form Eq. (8.131a) leads directly to the $\mathbf{L^{-1}}$ matrix elements and consequently the normal coordinates by use of Eq. (8.127b).

It is best now to proceed with a simple example, the linear XY_2 molecule of Fig. 8.13. The required four $(3N - 6)$ internal coordinates are taken to be the two XY bond extensions and the two $\angle YXY$ bends, one in the

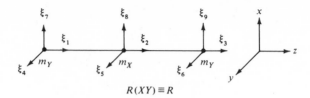

$$R(XY) \equiv R$$

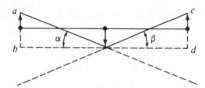

FIG. 8.13.   Displacement coordinates and other parameters for the
linear symmetrical XY_2 molecule. In (a) the displacement
coordinates are defined, while in (b) parameters are
defined for the exaggerated bent molecule.

yz and another in the xz plane. We write immediately by reference to
Fig. 8.13,

$$S_1 = \xi_1 - \xi_2 \tag{8.132a}$$

$$S_2 = \xi_2 - \xi_3 \tag{8.132b}$$

$$S_3 = \theta_{xz} \tag{8.132c}$$

$$S_4 = \theta_{yz} \tag{8.132d}$$

The functional forms of S_1 and S_2 are obvious by reference to Fig. 8.13(a).
In order to determine S_3 and S_4 in terms of the displacement coordinates
we refer to Fig. 8.13(b), which shows the molecule bent in the xz plane.
θ_{xz} is the change in the angle from the 180°, linear configuration. Thus θ_{xz} is
given by

$$S_3 = \theta_{xz} = \alpha + \beta \tag{8.133a}$$

Now for very small amplitude vibrations, which is the case of physical
interest,

$$\alpha \sim \sin \alpha \sim \frac{ab}{R}$$

or   (8.134a)

$$\alpha = \frac{\xi_7 - \xi_8}{R}$$

where R is the equilibrium XY bond length. Similarly, we find for β,

$$\beta = \frac{\xi_9 - \xi_8}{R} \tag{8.134b}$$

Then defining $\rho = 1/R$, Eq. (8.133a) may be written

$$S_3 = \rho(\xi_7 - 2\xi_8 + \xi_9) \tag{8.133b}$$

In a similar fashion,

$$S_4 = \rho(\xi_4 - 2\xi_5 + \xi_6) \tag{8.135}$$

The **B** matrix is now written down easily by inspection of Eqs. (8.132a), (8.132b), (8.133b), and (8.135). To see this most easily, we write out Eq. (8.123b) completely for our example.

$$\begin{pmatrix} S_1 \\ S_2 \\ S_3 \\ S_4 \end{pmatrix} = \begin{pmatrix} 1 & -1 & 0 & 0 & 0 & 0 & 0 & 0 & 0 \\ 0 & 1 & -1 & 0 & 0 & 0 & 0 & 0 & 0 \\ 0 & 0 & 0 & 0 & 0 & 0 & \rho & -2\rho & \rho \\ 0 & 0 & 0 & \rho & -2\rho & \rho & 0 & 0 & 0 \end{pmatrix} \begin{pmatrix} \xi_1 \\ \xi_2 \\ \xi_3 \\ \xi_4 \\ \xi_5 \\ \xi_6 \\ \xi_7 \\ \xi_8 \\ \xi_9 \end{pmatrix} \tag{8.136}$$

The **G** matrix is now obtained by using Eq. (8.125). Note that the multiplication of the **B** matrix elements does not follow the ordinary matrix multiplication, Row times Column, but instead is Row times Row. Thus G_{11} is given by

$$G_{11} = \frac{1}{m_Y} (1)^2 + \frac{1}{m_X} (-1)^2 + (0 \text{ seven times})$$

$$= \mu_Y + \mu_X \tag{8.137a}$$

where $\mu_Y = 1/m_Y$ and $\mu_X = 1/m_X$. Note that one must associate the proper mass with each internal coordinate. As a second example,

$$G_{12} = \mu_Y(1)(0) + \mu_X(-1)(1) + \mu_Y(0)(-1) + (0 \text{ six times}) = -\mu_X \tag{8.137b}$$

In the same manner the remaining **G** matrix elements are found to be:

$$G_{22} = \mu_X + \mu_Y$$

$$G_{33} = 2\rho^2(\mu_Y + 2\,\mu_X)$$

$$G_{44} = 2\rho^2(\mu_Y + 2\,\mu_X) \tag{8.137c}$$

$$G_{13} = G_{31} = G_{34} = G_{43} = G_{23} = G_{32} = G_{14} = G_{41} = G_{42}$$

$$= G_{24} = 0$$

We now are in a position to choose a reasonable potential function. The simplest choice satisfying our physical intuition is

$$2V = k_R(S_1^2 + S_2^2) + k_\theta(S_3^2 + S_4^2) \tag{8.138}$$

The first term involves stretching of the two identical bonds while the second involves bending the molecule in each of the two planes. This potential energy function is known as a *valence-bond* function since it expresses the potential energy in terms of force constants for chemical bonds. Comparing Eqs. (8.138) and (8.126), the **F** matrix elements are

$$F_{11} = F_{22} = k_R$$

$$F_{33} = F_{44} = k_\theta \tag{8.139}$$

$$F_{ij} = 0 \qquad i \neq j$$

Using the elements of the **G** and **F** matrices we can now form **GF** (=**FG** in this case), which leads to the secular equation

$$\begin{vmatrix} k_R(\mu_X+\mu_Y)-\lambda & -k_R\mu_X & 0 & 0 \\ -k_R\mu_X & k_R(\mu_X+\mu_Y)-\lambda & 0 & 0 \\ 0 & 0 & 2k_\theta\rho^2(\mu_Y+2\mu_X)-\lambda & 0 \\ 0 & 0 & 0 & 2k_\theta\rho^2(\mu_Y+2\mu_X)-\lambda \end{vmatrix}$$

$$= 0 \quad (8.140)$$

Two identical roots are immediately obtained, and in order to be in agreement with accepted conventions we label them λ_2.

$$\lambda_2 = 2k_\theta\rho^2(\mu_Y + 2\,\mu_X) \tag{8.141}$$

This result is expected, as we verify shortly, since one vibrational mode is doubly degenerate. Solution of the quadratic equation leads to

$$\lambda_1 = k_R\mu_Y \tag{8.142}$$

$$\lambda_3 = k_R(\mu_Y + 2\,\mu_X) \tag{8.143}$$

We now have explicit formulas relating the three fundamental vibrational frequencies of bent XY_2 to two valence-bond force constants. If the function of Eq. (8.138) were a true representation of the potential field, it should be possible to fit the experimental λ_i (or ν_i) for any XY_2 molecule to two parameters, k_R and k_θ. In fact we usually find that this potential function is too simple, and since in the present case three λ_i values are experimentally available, it is sensible to introduce a third force constant.

Intuitively it is evident that when one bond stretches, the potential energy of the other bond is affected. Thus it is appropriate to write a potential function reflecting this interaction. Instead of Eq. (8.138) we write

$$2V = k_R(S_1^2 + S_2^2) + 2k_{12}S_1S_2 + k_\theta(S_3^2 + S_4^2) \qquad (8.144)$$

where k_{12} is the bond-stretching *interaction* force constant. This introduces two new elements in the **F** matrix, $F_{12} = F_{21} = k_{12}$, so the 2×2 part of the secular determinant becomes

$$\begin{vmatrix} k_R(\mu_X + \mu_Y) - k_{12}\mu_X - \lambda & -k_R\mu_X + k_{12}(\mu_X + \mu_Y) \\ -k_R\mu_X + k_{12}(\mu_X + \mu_Y) & k_R(\mu_X + \mu_Y) - k_{12}\mu_X - \lambda \end{vmatrix} = 0 \qquad (8.145)$$

Solving this equation we obtain

$$\lambda_1 = \mu_Y(k_R + k_{12}) \qquad (8.146)$$

$$\lambda_3 = (k_R - k_{12})(2\mu_X + \mu_Y) \qquad (8.147)$$

Note that the result for the bending vibration [Eq. (8.141)] remains unchanged.

It should be emphasized that the valence-bond function is not the only type of potential energy function which might be used to describe the vibrational motions. The other very common type is the *central force* field. Here one assumes the primary forces are those between atoms, whether bonded or *nonbonded*. For linear XY_2 this field is rather unrealistic; it is probably better illustrated for nonlinear XY_2. In this case, we would choose the internal coordinates as the two *bond* extensions S_1 and S_2 and the nonbond distance extension between the two end atoms. Then the central-field potential would be

$$2V = k_r(S_1^2 + S_2^2) + k_{YY}S_3^2 \qquad (8.148)$$

An interaction term $2k_{12}S_1S_2$ might also be added. In fact, this type function is rather poor for XY_2 molecules, leading even in many cases to imaginary values of the force constants when the experimental frequencies are used for their evaluation. On the other hand a modified central-force potential function seems to be very useful for molecules containing carbon in tetrahedral bonding situations.[22]

---

[22] H. C. Urey and C. A. Bradley, *Phys. Rev.* **38**, 1969 (1931).

EXAMPLE 8-8

The fundamental vibrational frequencies of $C^{12}O_2$ are 1337, 667, and 2349 cm^{-1} for ν_1, ν_2, and ν_3, respectively. We first evaluate the valence-force constants using Eqs. (8.141)–(8.143). The reduced masses are

$$\mu_Y = 6.02 \times 10^{23}/16.0 = 3.76 \times 10^{22} \, g$$

$$\mu_X = 6.02 \times 10^{23}/12.0 = 5.02 \times 10^{22} \, g$$

and the λ values are

$$\lambda_1 = 4(9.87)(1337)^2(3.00 \times 10^{10})^2 = 6.35 \times 10^{28} \sec^{-2}$$

$$\lambda_2 = 4(9.87)(667)^2(3.00 \times 10^{10})^2 = 1.58 \times 10^{28}$$

$$\lambda_3 = 4(9.87)(2349)^2(3.00 \times 10^{10})^2 = 19.60 \times 10^{28}$$

Then we find from Eq. (8.141)

$$k_\theta \rho^2 = 0.572 \times 10^5 \text{ dyne/cm}$$

$$= 0.572 \text{ mdyne/Å}$$

It is customary to not quote values of k_θ since as defined in Eq. (8.138) it does not have the conventional force/length units [its units are erg/(radian)^2]. From Eqs. (8.142) and (8.143) we calculate

$$k_R = 16.9 \text{ mdyne/Å}$$

and

$$k_R = 14.2 \text{ mdyne/Å}$$

respectively. Thus we see quite clearly that the two parameter potential function gives only a fair fit to the three fundamental frequencies.

On the other hand, we can obviously fit the three frequencies exactly if we use the three-parameter potential function including k_{12}. Our previous result for $k_\theta \rho^2$ is of course unchanged, but we now get from Eqs. (8.146) and (8.147)

$$k_R = 15.6 \text{ mdyne/Å}$$

$$k_{12} = 1.35 \text{ mdyne/Å}$$

The small size of the interaction constant relative to the stretching force constant is physically reasonable, since one would hope to be explaining the dominant part of the energy with the bond stretching and bending force constants. The validity of the three-parameter potential is perhaps best assessed by predicting ν_3 for $C^{13}O_2$, which has the value 2284 cm^{-1}. Since the potential energy is expected to be invariant to isotopic substitution to a good approximation, we simply use k_R and k_{12} as given above, but of course

$$\mu_X = \frac{6.02 \times 10^{23}}{13.0} = 4.63 \times 10^{22} \, g$$

Using these values we find $\nu_3 = 2280$ cm^{-1} which is rather good agreement.

Although we are in a position to evaluate the normal coordinates in terms of the S_i and hence in terms of the ξ_i, we put this off until we describe the application of group theory to normal coordinate analysis.

B. Group Theory and Normal Coordinate Analysis

It might have been guessed by now that the form of the secular determinant depends upon the choice of internal coordinates. Indeed, it occurs that internal coordinates having the symmetry of the normal modes lead to a secular determinant (or **FG** matrix) with maximum factoring. This is because *there can be no* **F** *or* **G** *matrix elements connecting symmetry coordinates belonging to different irreducible representations.* Furthermore, there are no matrix elements connecting symmetry coordinates belonging to different rows or columns within a degenerate representation, or belonging to different rows or columns of different degenerate representations of the same symmetry species if the symmetry coordinates are properly chosen.

The general proof of the above statements has been given by Wilson, *et al.*[23] We merely prove the result for the case of nondegenerate representations. Suppose S_i and S_j are symmetry coordinates belonging to the ith and jth one-dimensional irreducible representations of the group of interest. Then since the representations are different there must exist an operation $\tilde{R}$ of the group, such that

$$\tilde{R}S_i = +S_i \tag{8.149a}$$

while

$$\tilde{R}S_j = -S_j \tag{8.149b}$$

Thus it follows that

$$\tilde{R}(S_iS_j) = (RS_i)(RS_j) = -S_iS_j \tag{8.150}$$

Now consider the potential energy as written, for example, in Eq. (8.126) (using S's instead of S's). In general it may contain terms of the form $F_{ij}S_iS_j$. But we have shown already that V (and also T) must be invariant to any group operation, that is

$$\tilde{R}V = +V \tag{8.151}$$

The only way that Eq. (8.151) can be compatible with Eq. (8.150) is if $F_{ij} = 0$. This proves, then, that there can be no nonzero **F** matrix elements connecting symmetry coordinates belonging to different irreducible repre-

[23] See footnote 7, Sec. 6.3 and Appendix XII.

sentations. Similar considerations apply to the **G** matrix, since, as shown in Appendix H, the kinetic energy contains products of the form $\dot{s}_i \dot{s}_j$ when expressed in the **G** matrix formalism.

We can state one more useful result concerning the maximum number of independent quadratic force constants that exist for any molecule. Suppose there are n_i symmetry coordinates of the nondegenerate species $\Gamma^{(i)}$. The block in the **F** matrix involving these coordinates then has a number of independent force constants equal to the sum of those on the diagonal and all those on *one* side of the diagonal. This is

$$n_i + (n_i - 1) + (n_i - 2) + \cdots 1 = \tfrac{1}{2} n_i(n_i + 1) \qquad (8.152)$$

using the well-known algebraic result for the sum of the first n numbers. Thus each set of coordinates of a given species $\Gamma^{(i)}$ permits $\tfrac{1}{2} n_i(n_i + 1)$ force constants, so the total number of independent force constants is

$$\frac{1}{2} \sum_{n_i} n_i(n_i + 1) \qquad (8.153)$$

where the sum is over all sets of n_i symmetry coordinates. If a set of coordinates is degenerate (and thus belongs to the same species), the **F** matrix contains as many *identical* factors as the degeneracy of the set. Equation (8.153) still applies if we count only one member from each degenerate set.

For example, if there were three coordinates having the symmetry of a one-dimensional representation, there would be $\tfrac{1}{2}(3)(4) = 6$ force constants arising from this block. If there were two pairs of doubly degenerate coordinates belonging to the same two-dimensional representation, there would be $\tfrac{1}{2}(2)(3) = 3$ force constants.

We return now to our discussion of the XY_2 linear symmetrical molecule. By the methods of Sec. 8.4 A it is easily found that the species of the vibrational modes are

$$\Gamma^{(V)} = \Sigma_g^+ + \Sigma_u^+ + \Pi_u \qquad (8.154)$$

where the appropriate group is $D_{\infty h}$. We know then, that the secular determinant has the block diagonal form shown in Fig. 8.14 if symmetry coordinates are used. In this case the determinant is completely reduced, since according to our previous discussion, there can be no elements connecting different species or different rows within a degenerate species. Of course, the two Π_u elements are identical, so we really need not clutter things up by including more than one member of the degenerate set. Note that Eq. (8.153) predicts a maximum of three force constants, so we know that our previous potential function [Eq. (8.144)] was complete in this respect.

Let us verify our group theory predictions by applying the **FG** matrix theory to XY_2 using symmetry coordinates. We note first that our original coordinates S_3 and S_4 *were* symmetry coordinates, that is, the operations of $D_{\infty h}$ leave them invariant. This is why they appeared in block diagonal form in our original treatment. For consistency with our earlier choices we write

$$S_2 = S_3 = \rho(\xi_7 - 2\xi_8 + \xi_9) \tag{8.155}$$

The symmetry coordinates of Σ_g^+ and Σ_u^+ type can be found by applying Eq. (6.68) as was done in Example 8-6. More simply, it is easily verified that $S_1 + S_2$ has Σ_g^+ symmetry, while $S_1 - S_2$ has Σ_u^+ symmetry. In order that the transformation between $\mathbf{S}$ and $\mathcal{S}$ be real orthogonal we write

$$\mathcal{S}_1 = \frac{1}{\sqrt{2}} (S_1 + S_2) = \frac{1}{\sqrt{2}} (\xi_1 - \xi_3) \tag{8.156}$$

and

$$\mathcal{S}_3 = \frac{1}{\sqrt{2}} (S_1 - S_2) = \frac{1}{\sqrt{2}} (\xi_1 - 2\xi_2 + \xi_3) \tag{8.157}$$

The symmetry coordinates are concisely related to the original internal coordinates by

$$\begin{pmatrix} \mathcal{S}_1 \\ \mathcal{S}_2 \\ \mathcal{S}_3 \end{pmatrix} = \begin{pmatrix} 1/\sqrt{2} & 1/\sqrt{2} & 0 \\ 0 & 0 & 1 \\ 1/\sqrt{2} & -1/\sqrt{2} & 0 \end{pmatrix} \begin{pmatrix} S_1 \\ S_2 \\ S_3 \end{pmatrix} \tag{8.158}$$

The reader should recognize simply by the definition of the ξ_i in Fig. 8.13 that $\mathcal{S}_1$, $\mathcal{S}_2$, and $\mathcal{S}_3$ have precisely the symmetry of the normal modes in Fig. 8.8, as expected.

Using the symmetry coordinates as expressed in Eqs. (8.155)–(8.157), we may write down the $\mathbf{B}$ matrix and hence the $\mathbf{G}$ matrix. The $\mathbf{G}$ matrix

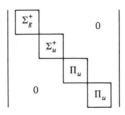

FIG. 8.14. Form of secular determinant for linear XY_2 molecule using symmetry coordinates.

elements are easily found to be

$$G_{11} = \mu_Y$$

$$G_{22} = 2\rho^2(\mu_Y + 2\mu_X) \tag{8.159a}$$

$$G_{33} = \mu_Y + 2\mu_X$$

$$G_{ij} = 0, \ i \neq j$$

Our result for **G** verifies our group theory expectations since there are no off-diagonal elements.

If we write the three-constant potential energy function of Eq. (8.144) in terms of symmetry coordinates it becomes

$$2V = (k_R + k_{12})S_1^2 + (k_R - k_{12})S_3^2 + k_\theta(S_2^2 + S_{2'}^2) \tag{8.160}$$

where the reverse transformation of Eq. (8.158) has been used. Also we have written $S_{2'} = S_4$. Then the **F** matrix elements are

$$F_{11} = k_R + k_{12}$$

$$F_{33} = k_R - k_{12} \tag{8.159b}$$

$$F_{22} = k_\theta$$

$$F_{ij} = 0, \ i \neq j$$

Thus **F** has the expected diagonal form also.

Using (8.159a) and (8.159b) the secular equation is

$$\begin{vmatrix} \mu_Y(k_R + k_{12}) - \lambda & 0 & 0 \\ 0 & 2\rho^2(\mu_Y + 2\mu_X)k_\theta - \lambda & 0 \\ 0 & 0 & (\mu_Y + 2\mu_X)(k_R - k_{12}) - \lambda \end{vmatrix} = 0 \tag{8.161}$$

The determinant is completely factored and the roots are those previously obtained.

While this is a rather simple example, the same principles may be applied to more complicated molecules with a corresponding simplification of the secular determinant, although it is seldom entirely factored. The other very useful result of the symmetry coordinate approach is that it usually permits at least a partial correlation of eigenvalues with symmetry coordinates (or normal modes). Thus in this example, we know immediately that λ_1, λ_2, and λ_3 correspond to the normal coordinates Q_1, Q_2 (two of these), and Q_3, as depicted in Fig. 8.8.

EXAMPLE 8-9

Once the eigenvalues of the secular determinant have been determined, it is always possible to find $\mathbf{L}^{-1}$, which provides an explicit relationship between the normal coordinates and the internal coordinates as shown by Eq. (8.127b). The $\mathbf{L}^{-1}$ elements are found by solving the linear homogeneous equations of the form of Eq. (8.131a). This is particularly easy to do for linear symmetrical XY_2 using the symmetry coordinate system just described. For a three-dimensional secular determinant (as in the present case since we have recognized that the two degenerate coordinates have identical forms in the xz and yz planes), Eqs. (8.131a) are in general

$$\{(\mathbf{FG})_{11} - \lambda_i\}L^{-1}{}_{i1} + (\mathbf{FG})_{12}L^{-1}{}_{i2} + (\mathbf{FG})_{13}L^{-1}{}_{i3} = 0$$

$$(\mathbf{FG})_{21}L^{-1}{}_{i1} + \{(\mathbf{FG})_{22} - \lambda_i\}L^{-1}{}_{i2} + (\mathbf{FG})_{23}L^{-1}{}_{i3} = 0 \quad (8.162)$$

$$(\mathbf{FG})_{31}L^{-1}{}_{i1} + (\mathbf{FG})_{32}L^{-1}{}_{i2} + \{(\mathbf{FG})_{33} - \lambda_i\}L^{-1}{}_{i3} = 0$$

We simply substitute the $\mathbf{FG}$ matrix elements of Eq. (8.161) and find the set of L^{-1} elements which satisfy Eq. (8.162) for each of the three λ_i values. We find trivially the following results. For $\lambda_1 = \mu_Y(k_R + k_{12})$:

$$L^{-1}{}_{12} = L^{-1}{}_{13} = 0 \neq L^{-1}{}_{11} \quad (8.163a)$$

For $\lambda_2 = 2\rho^2(\mu_Y + 2\mu_X)k_\theta$:

$$L^{-1}{}_{21} = L^{-1}{}_{23} = 0 \neq L^{-1}{}_{22} \quad (8.163b)$$

For $\lambda_3 = (\mu_Y + 2\mu_X)(k_R - k_{12})$:

$$L^{-1}{}_{31} = L^{-1}{}_{32} = 0 \neq L^{-1}{}_{33} \quad (8.163c)$$

Note that the homogeneous nature of the equations prohibits determination of $L^{-1}{}_{11}$, $L^{-1}{}_{22}$, and $L^{-1}{}_{33}$. It should also be emphasized that only when the secular determinant is entirely diagonal is the $\mathbf{L}^{-1}$ matrix diagonal.

The actual values of the nonzero elements can be obtained by equating the potential energy expressed in internal (symmetry) coordinates to that expressed in normal coordinates. This becomes

$$\lambda_1Q_1{}^2 + \lambda_2Q_2{}^2 + \lambda_3Q_3{}^2 = (k_R + k_{12})S_1{}^2 + (k_R - k_{12})S_3{}^2 + k_\theta S_2{}^2 \quad (8.164a)$$

where again we have omitted $S_{2'}$. Using Eq. (8.127b) and the elements given in Eqs. (8.163) we rewrite Eq. (8.164a) as

$$\mu_Y(k_R + k_{12})(L^{-1}{}_{11})^2S_1{}^2 + 2\rho^2(\mu_Y + 2\mu_X)k_\theta(L^{-1}{}_{22})^2S_2{}^2$$
$$+ (\mu_Y + 2\mu_X)(k_R - k_{12})(L^{-1}{}_{33})^2S_3{}^2$$
$$= (k_R + k_{12})S_1{}^2 + (k_R - k_{12})S_3{}^2 + k_\theta S_2{}^2 \quad (8.164b)$$

where we have also substituted λ_1, λ_2, and λ_3 as previously obtained. But this last equation can only be true if the coefficients of the S_i^2 are equal on both sides of the equation. Thus we find

$$(L^{-1}{}_{11})^2 = \mu_Y^{-1}$$

$$(L^{-1}{}_{22})^2 = [2\rho^2(\mu_Y + 2\,\mu_X)]^{-1} \qquad (8.165a)$$

$$(L^{-1}{}_{33})^2 = (\mu_Y + 2\,\mu_X)^{-1}$$

or

$$L^{-1}{}_{11} = \mu_Y^{-1/2}$$

$$L^{-1}{}_{22} = [2\rho^2(\mu_Y + 2\,\mu_X)]^{-1/2} \qquad (8.165b)$$

$$L^{-1}{}_{33} = (\mu_Y + 2\,\mu_X)^{-1/2}$$

Using Eqs. (8.155)–(8.157) we finally arrive at the useful forms for the normal coordinates,

$$Q_1 = (2\,\mu_Y)^{-1/2}(\xi_1 - \xi_3) \qquad (8.166a)$$

$$Q_2 = [2(\mu_Y + 2\,\mu_X)]^{-1/2}(\xi_7 - 2\xi_8 + \xi_9) \qquad (8.166b)$$

$$Q_3 = [2(\mu_Y + 2\,\mu_X)]^{-1/2}(\xi_1 - 2\xi_2 + \xi_3) \qquad (8.166c)$$

It still remains to specify the relative values of the ξ_i. This is easily done by using the requirement that the c.m. of the molecule does not move during the vibrations. Thus we have for Q_1, $\xi_1 = -\xi_3$; for Q_2, $\xi_7 = \xi_9 = -(m_X/2m_Y)\xi_8$; for Q_3, $\xi_1 = \xi_3 = -(m_X/2m_Y)\xi_2$. The absolute (maximum) classical displacements are gotten by equating $\lambda_i Q_i^2$ to $h\nu_i(v_i + \frac{1}{2})$ for any excitation v_i of the ith normal mode.

Returning to the general symmetry considerations, we recall that a plane symmetrical XY_3 species (such as $CO_3^=$) has six normal modes with species $A_1' + A_2'' + 2E'$. If *appropriate* symmetry coordinates are used to form the **FG** matrix, we know it must have the form shown in Fig. 8.15. Notice that the only off-diagonal elements are those connecting the first and second E' coordinates of the R set with the first and second coordinates of the α set, respectively. The 4×4 E' block can be rearranged into two identical 2×2 submatrices, and by retaining one we obtain Fig. 8.15b.

Some care must be exercised in choosing the degenerate symmetry coordinates in any actual calculation if one desires the maximum factoring shown in Fig. 8.15. The rule for obtaining the maximum factoring has been given by Wilson, et al.[24] as follows. When more than one set of degenerate coordinates of the same symmetry exists, *the first member of each degenerate set of coordinates must be chosen so that its transformation prop-*

---

[24] See footnote 7, Sec. 6.4.

erties are identical to those of the first member of each other set. If this rule is followed, the maximum factoring is obtained automatically.

In Example 8-6 we determined a set of E' coordinates for the stretches (R) and a set for the bends (α). It is easily shown that $S_4{}^{E'}$ *does not* have the same transformation properties as $S_2{}^{E'}$. Thus considering the D_{3h} group, $S_2{}^{E'}$ transforms into itself under the σ_h and under *one* of the C_2's and *one* of the σ_v's; but $S_4{}^{E'}$, under the three identical operations, transforms into itself only under σ_h. Thus the maximum factoring would not be obtained if the symmetry coordinates of Example 8-6 were used. The situation is easily remedied by generating $S_4{}^{E'}$ with the function α_1 rather than α_2. One then gets

$$S_4{}^{E'} = N(2\alpha_1 - \alpha_2 - \alpha_3) = \frac{1}{\sqrt{2}} (\alpha_2 + \alpha_3) \qquad (8.102d'')$$

This coordinate has the same transformation properties as $S_2{}^{E'}$ and thus would be the appropriate choice along with $S_2{}^{E'}$ for evaluating the $2 \times 2\ E'$ block of Fig. 8.15(b).

Two other examples should finally make the symmetry arguments clear. The reader may readily verify that the vibrational species of benzene (D_{6h}) are

$$\Gamma^{(V)} = 2A_{1g} + A_{2g} + A_{2u} + 2B_{1u} + 2B_{2g} + 2B_{2u}$$
$$+ 4E_{2g} + 2E_{2u} + E_{1g} + 3E_{1u}$$

Appropriately chosen symmetry coordinates must then lead to the **FG** matrix block diagonal form of Fig. 8.16(a). Note we have retained only

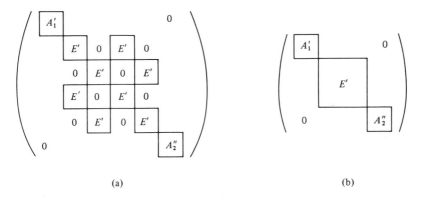

(a)   (b)

FIG. 8.15. Structure of the **FG** matrix of plane symmetrical XY_3 using symmetry coordinates. In (a) the complete matrix is shown while in (b) only one of the 2×2 identical E' blocks is represented.

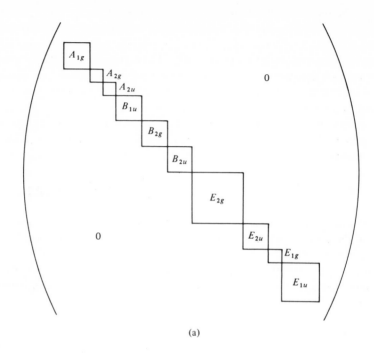

(a)

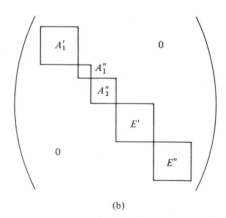

(b)

FIG. 8.16. Structure of the **FG** matrices for (a) benzene (D_{6h}) and (b) ethane (D_{3d}) if symmetry coordinates are used.

one of the two identical E_{2g} blocks, one of the two E_{2u} blocks, one member of E_{1g}, and one of the two E_{1u} blocks. Thus a 30×30 matrix ($3N - 6$ with $N = 12$) factors into three 1×1 matrices, five 2×2 matrices, one 3×3, and one 4×4 matrix. Note in general, *that n symmetry species of degeneracy d lead to d identical matrices of dimension n.*

For the stable, staggered form of ethane (D_{3d}) the vibrational modes are

$$\Gamma^{(V)} = 3A_1' + A_1'' + 2A_2'' + 3E' + 3E''$$

as can be verified by the usual techniques. Retaining only the unique matrix blocks, the **FG** matrix has the form shown in Fig. 8.16(b). Again we emphasize that this maximum factorization occurs only if properly constructed symmetry coordinates are used.

While we have not attempted to give a complete treatment of normal coordinate analysis in this and the preceding section, it is hoped that the discussion provides at least a starting point for understanding the practical methods. The reader who becomes deeply involved in such endeavors should consult the works of Herzberg[25] and Wilson, Decius and Cross.[26] In particular, the latter authors point out numerous practical shortcuts in the utilization of **FG** matrix theory.

EXAMPLE 8-10

As another simple example, consider the nonlinear symmetrical XY_2 molecule of Fig. 8.17(a). Fig. 8.11 could have been used but the displacement coordinate orientations are not as convenient. If we choose S_1 and S_2 as the stretching coordinates for the left-hand and right-hand bonds, respectively, and S_3 as the angle deformation coordinate, we find by use of Fig. 8.17(a) that

$$S_1 = \xi_4 + s\xi_6 + c\xi_9$$

$$S_2 = \xi_5 - s\xi_6 + c\xi_9 \qquad (8.167)$$

$$S_3 = \rho(\xi_7 + \xi_8 - 2s\xi_9)$$

where

$$s = \sin \tfrac{1}{2}\alpha$$

$$c = \cos \tfrac{1}{2}\alpha$$

$$\rho = 1/R$$

Note that S_3 can be found by a method similar to that used in Fig. 8.13(b) to obtain the bending coordinate of the linear molecule.

[25] See footnote 10.
[26] See footnote 7.

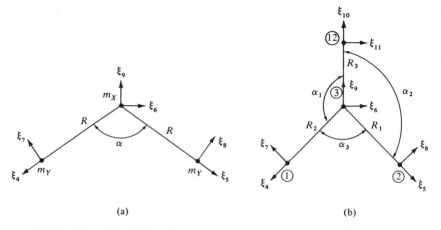

FIG. 8.17.   (a) XY_2 molecule; R and α are equilibrium bond length
and angle. (b) XY_3 molecule; R_i and α_i represent the
stretching and bending coordinates; equilibrium bond
lengths and angles are used as R and α in text. $\rho = 1/R$
and $\mu_i = 1/m_i$.

Now we have previously found the vibrational species of nonlinear
XY_2 to be $2A_1 + B_1$, so we know that symmetry coordinates of this form
lead to an **FG** matrix composed of a 2×2 A_1 block and a one-dimensional
B_1 "block." The appropriate symmetry coordinates are easily found to be

$$S_1 = \frac{1}{\sqrt{2}}\,(S_1 + S_2) = \frac{1}{\sqrt{2}}\,(\xi_4 + \xi_5 + 2c\xi_9)$$

$$S_3 = \frac{1}{\sqrt{2}}\,(S_1 - S_2) = \frac{1}{\sqrt{2}}\,(\xi_4 - \xi_5 + 2s\xi_6) \tag{8.168}$$

$$S_2 = S_3 = \rho(\xi_7 + \xi_8 - 2s\xi_9)$$

where S_1 and S_2 are of A_1 symmetry and S_3 is of B_1 symmetry.

If we write the **G** matrix in terms of the nonsymmetry internal coordinates of Eqs. (8.167) we get

$$\mathbf{G} = \begin{pmatrix} \mu_Y + \mu_X(s^2 + c^2) & \mu_X(c^2 - s^2) & -2\,\mu_X sc\rho \\ & \mu_Y + \mu_X & -2\,\mu_X sc\rho \\ & & 2\,\mu_Y \rho^2 + 4\,\mu_X s^2\rho^2 \end{pmatrix} \tag{8.169}$$

while the use of the symmetry coordinates of Eqs. (8.168) leads to

$$
\mathsf{G} = \begin{pmatrix} \mu_Y + 2c^2\mu_X & -2\sqrt{2}\ sc\rho\mu_X & 0 \\ -2\sqrt{2}\ sc\rho\mu_X & 2\rho^2(\mu_Y + 2s^2\mu_X) & 0 \\ 0 & 0 & \mu_Y + 2c^2\mu_X \end{pmatrix} \tag{8.170}
$$

which shows the expected factoring.

Since G and $\mathbf{G}$ differ only in the linearly related coordinate sets used in their evaluation, they must be related by a similarity transformation

$$
\mathsf{G} = \mathbf{C}'\mathbf{G}\mathbf{C} \tag{8.171}
$$

In fact it can be shown that $\mathbf{C}$ is the real orthogonal matrix which transforms symmetry coordinates to internal, that is,

$$
\mathbf{S} = \mathbf{C}\mathbf{s} \tag{8.172}
$$

The reader may easily verify that Eqs. (8.169) and (8.170) satisfy Eq. (8.171). It is often useful to find the G matrix from $\mathbf{G}$ by use of this equation.

Choosing the simplest valence potential function, we write

$$
2V = k_R(S_1^2 + S_2^2) + k_\theta S_3^2 = k_R(s_1^2 + s_3^2) + k_\theta s_2^2 \tag{8.173}
$$

so the $\mathfrak{F}$ matrix is

$$
\mathfrak{F} = \begin{pmatrix} k_R & 0 & 0 \\ 0 & k_\theta & 0 \\ 0 & 0 & k_R \end{pmatrix} \tag{8.174}
$$

Combining Eqs. (8.170) and (8.174) and solving the secular determinant the three roots λ_1, λ_2, and λ_3 are directly obtained. We leave this final step to the reader.

EXAMPLE 8-11

We consider a somewhat more complicated example now, namely the plane symmetrical XY_3 species of Fig. 8.17(b). For internal coordinates we choose the previously discussed sets of equivalent coordinates, (R_1, R_2, R_3), (α_2, α_3), and I. The appropriate symmetry coordinates would be those described in Example 8-6 and Eqs. (8.102). We could now proceed, as in the previous example, by writing the $\mathbf{S}$ and $\mathbf{s}$ in terms of $\boldsymbol{\xi}$ of Fig. 8.17(b), from which the $\mathbf{G}$ and G matrices could be obtained.

Some effort can be saved by noting that in $\mathbf{G}$, the elements within the R block are of the identical form to those obtained in the previous example.

Thus, the R_1R_1 element is just $\mu_Y + \mu_X$, which is the **G** matrix element for any bond stretch. This is obviously true for R_1R_1 since the coordinate system of Fig. 8.17(b) is identical to that of Fig. 8.17(a). Furthermore, the R_2R_2 and R_3R_3 elements have the same value, as must be true by symmetry. In a similar manner, we know that the R_1R_2, R_1R_3, R_2R_3 elements all have the value $\mu_X(c^2 - s^2)$ as in the previous example. The R part of **G** is then

$$
\mathbf{G}(R) = \begin{array}{c} \\ R_1 \\ R_2 \\ R_3 \end{array}
\overset{\displaystyle \begin{array}{ccc} R_1 & R_2 & R_3 \end{array}}{
\begin{pmatrix}
\mu_X + \mu_Y & -\tfrac{1}{2}\mu_X & -\tfrac{1}{2}\mu_X \\
-\tfrac{1}{2}\mu_X & \mu_X + \mu_Y & -\tfrac{1}{2}\mu_X \\
-\tfrac{1}{2}\mu_X & -\tfrac{1}{2}\mu_X & \mu_X + \mu_Y
\end{pmatrix}}
\qquad (8.175)
$$

where we have used $s = \tfrac{1}{2}\sqrt{3}$ and $c = \tfrac{1}{2}$ since $\alpha = 120°$. The two diagonal $\alpha\alpha$ elements are also like those of Example 8-10, and have the value $\rho^2(2\,\mu_Y + 3\,\mu_X)$.

By the same reasoning, the **G** elements $R_1\alpha_3$, $R_2\alpha_3$, $R_3\alpha_2$, $R_1\alpha_2$ all have the value $-2\,\mu_X s c \rho = -(\tfrac{1}{2}\sqrt{3})\rho\mu_X$ as in Eq. (8.169). It must be noted now that there are two additional types of elements involving the R and α that did not arise in the XY_2 case. These elements are of the form $\alpha_2\alpha_3$ and $R_2\alpha_2$. To obtain these we write down a portion of the transformation relating **R** and $\boldsymbol{\alpha}$ to $\boldsymbol{\xi}$. By means of Fig. 8.17(b) we find

$$
\begin{pmatrix} R_2 \\ \\ \alpha_2 \\ \\ \alpha_3 \end{pmatrix} =
\begin{pmatrix}
0 & 0 & 0 & 1 & 0 & \tfrac{1}{2}\sqrt{3} & 0 & 0 & \tfrac{1}{2} & 0 & 0 & 0 \\
0 & 0 & 0 & 0 & 0 & \tfrac{3}{2}\rho & 0 & -\rho & (\tfrac{1}{2}\sqrt{3})\rho & 0 & -\rho & 0 \\
0 & 0 & 0 & 0 & 0 & 0 & \rho & \rho & -\sqrt{3}\rho & 0 & 0 & 0
\end{pmatrix}
\begin{pmatrix} \xi_1 \\ \cdot \\ \cdot \\ \cdot \\ \xi_{12} \end{pmatrix}
$$

$$(8.176)$$

Note that R_2 and α_3 have the same form as in the previous example (R_2 corresponds to S_1 and α_3 to S_3), while α_2 is derived in the usual manner. We now find from the **B** matrix elements of Eq. (8.176) that

$$
G_{\alpha_2\alpha_3} = -\rho^2(\mu_Y + \tfrac{3}{2}\,\mu_X)
$$

$$
G_{R_2\alpha_2} = \sqrt{3}\,\rho\mu_X
$$

$$(8.177)$$

This provides all the elements of **G** except those arising from I, which can be defined as[27]

$$
I = (\xi_1 + \xi_2 - 3\xi_3 + \xi_{12})\cdot\frac{\rho}{\sqrt{3}}
\qquad (8.178)
$$

[27] See footnote 7, p. 59.

The G_{II} element is therefore $3\,\mu_Y + \mu_X$, and $G_{I\alpha} = G_{IR} = 0$ since the displacement coordinates of the R and α are orthogonal to those of I. Finally then, the **G** matrix is

$$
\mathbf{G} =
\begin{array}{c}
\\
R_1 \\
R_2 \\
R_3 \\
\alpha_2 \\
\alpha_3 \\
I
\end{array}
\begin{array}{c}
\begin{array}{cccccc}
R_1 & R_2 & R_3 & \alpha_2 & \alpha_3 & I
\end{array}\\
\left[
\begin{array}{cccccc}
a & b & b & d & d & 0 \\
b & a & b & c & d & 0 \\
b & b & a & d & c & 0 \\
d & c & d & e & f & 0 \\
d & d & c & f & e & 0 \\
0 & 0 & 0 & 0 & 0 & g
\end{array}
\right]
\end{array}
\tag{8.179}
$$

where

$$a = \mu_X + \mu_Y$$

$$b = -\tfrac{1}{2}\mu_X$$

$$c = \sqrt{3}\,\rho\mu_X$$

$$d = -(\tfrac{1}{2}\sqrt{3})\rho\mu_X \tag{8.180}$$

$$e = (2\,\mu_Y + 3\,\mu_X)\rho^2$$

$$f = -(\mu_Y + \tfrac{3}{2}\mu_X)\rho^2$$

$$g = \mu_Y + 3\,\mu_X$$

The matrix **G** of Eq. (8.179) may now be transformed to symmetry coordinates using Eq. (8.171) where

$$
\mathbf{C'} =
\begin{bmatrix}
1/\sqrt{3} & 1/\sqrt{3} & 1/\sqrt{3} & 0 & 0 & 0 \\
2/\sqrt{6} & -1/\sqrt{6} & -1/\sqrt{6} & 0 & 0 & 0 \\
0 & 1/\sqrt{2} & -1/\sqrt{2} & 0 & 0 & 0 \\
0 & 0 & 0 & 1/\sqrt{2} & 1/\sqrt{2} & 0 \\
0 & 0 & 0 & 1/\sqrt{2} & -1/\sqrt{2} & 0 \\
0 & 0 & 0 & 0 & 0 & 1
\end{bmatrix}
\tag{8.181}
$$

By tedious but straightforward matrix multiplication,

$$\mathcal{G} = \begin{pmatrix} a + 2b & 0 & 0 & 0 & 0 & 0 \\ 0 & a - b & 0 & (1/\sqrt{3})(d - c) & 0 & 0 \\ 0 & 0 & a - b & 0 & c - d & 0 \\ 0 & (1/\sqrt{3})(d - c) & 0 & e + f & 0 & 0 \\ 0 & 0 & c - d & 0 & e - f & 0 \\ 0 & 0 & 0 & 0 & 0 & g \end{pmatrix}$$

(8.182)

which upon rearranging has the form predicted by Fig. 8.15.

The simplest valence potential function is

$$2V = k_R(R_1^2 + R_2^2 + R_3^2) + k_\theta(\alpha_1^2 + \alpha_2^2 + \alpha_3^2) + k_\Delta I^2 \quad (8.183a)$$

which becomes by removal of α_1,

$$2V = k_R(R_1^2 + R_2^2 + R_3^2) + 2k_\theta(\alpha_2^2 + \alpha_2\alpha_3 + \alpha_3^2) + k_\Delta I^2 \quad (8.183b)$$

With the aid of Eq. (8.181), the potential energy in symmetry coordinates is

$$2V = k_R(S_1^2 + S_2^2 + S_3^2) + k_\theta(3S_4^2 + S_5^2) + k_\Delta S_6^2 \quad (8.183c)$$

and the $\mathfrak{F}$ matrix is

$$\mathfrak{F} = \begin{pmatrix} k_R & & & & & \\ & k_R & & & 0 & \\ & & k_R & & & \\ & & & 3k_\theta & & \\ & & & & k_\theta & \\ & 0 & & & & k_\Delta \end{pmatrix}$$

(8.184)

Using Eqs. (8.182) and (8.184) the $\mathfrak{F}\,\mathcal{G}$ matrix is easily formed. Carrying only one of the two E' blocks (which have identical roots), we find

$$\begin{vmatrix} (a + 2b)k_R - \lambda & 0 & 0 & 0 \\ 0 & (a - b)k_R - \lambda & (c - d)k_R & 0 \\ 0 & (c - d)k_\theta & (e - f)k_\theta - \lambda & 0 \\ 0 & 0 & 0 & gk_\Delta - \lambda \end{vmatrix} = 0$$

(8.185)

From the upper left corner we have

$$\lambda_1 = \mu_Y k_R \qquad (A_1') \qquad\qquad (8.186a)$$

and from the lower right

$$\lambda_2 = (\mu_Y + 3\mu_X)k_\Delta \qquad (A_2'') \qquad\qquad (8.186b)$$

This latter root has been labeled $\lambda_2(\nu_2)$ in keeping with convention. The solution to the 2×2 E' block may be written

$$\lambda_3\lambda_4 = 3k_\theta k_R \rho^2 (\mu_Y^2 + 3\mu_X\mu_Y) \qquad\qquad (8.186c)$$

$$\lambda_3 + \lambda_4 = (3\rho^2 k_\theta + k_R)(\mu_Y + \tfrac{3}{2}\mu_X) \qquad\qquad (8.186d)$$

Equations (8.186) specify the four fundamental vibrational frequencies in terms of three potential constants. For a better representation of the potential field, a fourth parameter could be introduced, for example, a term $k_{R\theta}(R_1\alpha_3 + R_1\alpha_2 + R_2\alpha_1 + R_2\alpha_3 + R_3\alpha_2 + R_3\alpha_1)$ could be added to Eq. (8.183).

8.6 SOME QUALITATIVE ASPECTS OF VIBRATIONAL ANALYSES

A. Force Constants and Group Frequencies

If the concept of a valence force potential function is to be generally useful, we would expect that the stretching force constant for a given bond would be essentially constant from molecule to molecule. Thus the force constant for CH stretching in CH_4 should be very similar to that in C_2H_6 or C_2H_4. This is indeed the experimental finding, since $k_R = 4.7–5.1$ mdyne Å^{-1} for the CH stretch in these molecules. The precise values depend upon the specific form of V, that is, the interaction constants included.

This result is in agreement with our chemical intuition since, *as a first approximation*, we expect the chemical bonding between C and H to be the same in all hydrocarbons. On the other hand, we know that the bond energies of CH in CH_4, C_2H_4 and HC≡CH are somewhat different, as are the CH bond lengths and other properties such as acidity, etc. Therefore we are not surprised that the force constants permit this same finer distinction. For example, the force constants for CH stretching are ca. 4.8, 5.1, and 5.8 in CH_4, C_2H_4, and C_2H_2, respectively, an increase which parallels the increase in bond energies and the decrease in bond lengths. Thus the force constants do depend upon the entire molecular environment and may be used to distinguish well-known chemical concepts such as hybridization, resonance, ionic character, inductive effects, etc.

In Table 8.11 are listed the approximate values of the force constants for various bonds and bond angles, for which the above arguments also apply. One generalization that is immediately evident from the table is

TABLE 8.11

Typical bond stretching and bending force constants. All units are mdyne $\mathring{A}^{-1}$. Bending force constants quoted are actually $\rho_1\rho_2 k_\theta$, where $\rho_i = 1/r_i$, and $r_i =$ bond length.

Bond Stretching

≡C—H	5.9	—C—C—	4.5
=C—H	5.1	C=C	9.6
—C—H	4.8	—C≡C—	15.6
N—H	6.4	C=O	12.1
—O—H	7.7	—C≡N	17.7
H—F	9.7	—Si—Si—	1.7
H—Cl	5.2	—C—N	5.2
H—Br	4.1	—O—O—	3.5–5.0
H—I	3.2	—C—O—	5.0–5.8
—C—F	5.6		
—C—Cl	3.4		
—C—Br	2.8		
—C—I	2.3		

TABLE 8.11 (*continued*)

Bond Bending

$\equiv$C$-$H	0.21	C$=$C (with H)	0.55	I$-$C$-$ (with H)	0.45
$=$C (with H, H)	0.30	F$-$C$-$ (with H)	0.76	F$-$C$-$ (with F)	0.71
$-$C (with H, H)	0.46	Cl$-$C$-$ (with H)	0.58	Cl$-$C$-$ (with Cl)	0.33
		Br$-$C$-$ (with H)	0.52	O$=$C (with H)	1.5

that the bending constants (in mdyne Å^{-1} units) are about an order of magnitude smaller than those for stretches. It is clear also that the force constants do indeed impart chemical knowledge; for example, the carbon-carbon bond strengths increase in the order C$-$C < C$=$C < C$\equiv$C, and the C-X bond strengths increase in the order C$-$I < C$-$Br < C$-$Cl < C$-$F, and so forth.

It is found that force constants correlate qualitatively with several other physical properties. One of the most widely known correlations expresses bond length in terms of stretching force constant, and is known as Badger's Rule. It may be written

$$k(r - d_{ij})^3 = 1.86 \qquad (8.187)$$

if k is in mdyne Å^{-1} and r in Å. d_{ij} is a constant which depends upon the rows i and j of the periodic table to which the two atoms belong. Some of these values are: $d_{11} = 0.68$, $d_{22} = 1.25$. The equation seems to hold to about 5% in most cases.

Possibly the most remarkable qualitative feature of vibrational spectra is the occurrence of characteristic *group* frequencies. By group frequencies we mean fundamental vibrations that are attributable primarily to a pair of bonded atoms, or to three atoms, and which are essentially constant from molecule to molecule. For example, all molecules with a C$-$C bond have

TABLE 8.12
Typical group frequencies in cm^{-1}.

Stretching

—CH$_3$	2960 2870	$\diagdown$ —C—F $\diagup$	1100
$\diagdown$ CH$_2$ $\diagup$	2926 2853	$\diagdown$ —C—Cl $\diagup$	650
		$\diagdown$ —C—Br $\diagup$	560
≡CH	3300	$\diagdown$ —C—I $\diagup$	500
=CH$_2$	3020	$\diagdown$ N—H $\diagup$	3350
—C—OH	1100	—S—H	2570
C—C	900	—P—H $\diagup$	2400
C—N	1000		

$\diagdown$C=C$\diagup$ 1650

—C≡C— 2050

—C≡N 2240

$\underset{H}{\overset{R}{\diagdown}}$C=O 1725

R$_2$C=O 1715

NH$_2$—C$\overset{O}{\diagup}$ 1700

—O—H 3630 (no H-bonding)

TABLE 8.12 (*continued*)

Bending

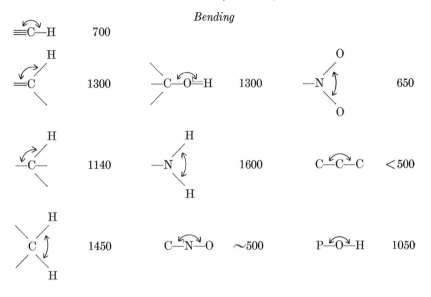

a fundamental frequency near 900 cm⁻¹, those with a C=C bond near 1650 cm⁻¹ and those with a C≡C bond near 2050 cm⁻¹. Similarly, molecules with CH₃ groups normally have stretching vibrations at about 2960 and 2870 cm⁻¹, and bending vibrations near 1460 and 1380 cm⁻¹. A rather brief listing of these and other group frequencies is given in Table 8.12. In most cases we list a single number for a given group, but for some, such as C=O, we list several to illustrate the small changes produced by the particular molecular environment.

The reason for the constancy of group frequencies is certainly associated with the constancy of the bond force constants, but is somewhat more involved than merely that. For CH, NH, or OH stretching motions the reason for group frequencies is straightforward. The hydrogen atom is so light compared to the atom to which it is attached that the normal modes which involve appreciable hydrogen motion have very little motion of the heavier atoms. That is, the hydrogen acts like it is vibrating against an infinitely heavy mass. Thus in these cases, if the force constant remains invariant from molecule to molecule, the CH, OH, and NH stretching motions do also. It is useful here to note that when two or more CH, etc., groups are present in a molecule, two or more stretching frequencies are possible, but they are very similar. We have seen this for the linear XY_2 molecule, whose vibrational frequencies were related to force constants in Eqs. (8.141), (8.146), and (8.147). Looking at Eq. (8.147), we see that if $\mu_X \ll \mu_Y$ (or $m_X \gg m_Y$), the asymmetrical stretching frequency ν_3 is nearly identical to the symmetrical stretching frequency ν_1, differing only by virtue of the small interaction constant k_{12}. The results for more complicated molecules are similar, so we are not really surprised to find that heptane C_7H_{16} shows two overlapped CH bands (at 2944 and 2865 cm^{-1}) in the infrared spectrum under moderate resolution.

The reason why group frequencies occur for heavier bonded atoms is somewhat more complicated. In simple terms, they occur when the group in question has a force constant which is considerably different from that of its neighboring bonds, even though the atoms in question are of approximately the same mass; or they occur if the force constants are similar and the masses quite different, as illustrated for CH, etc. Thus the carbonyl stretch is distinct in formaldehyde (H_2CO) and also in acetaldehyde (CH_3–CHO). In the former case the carbonyl group masses are much different from those of the adjacent atoms, but in the latter the adjacent atom is heavy. Nevertheless, the carbonyl frequency is distinct because the C=O force constant is much larger than that of the C–C bond.

We can get some quantitative feel again by considering an example. The frequencies for the three stretching frequencies of the linear Y–X–X–Y molecule are given by[28]

$$\lambda_3 = (\mu_X + \mu_Y)k_{XY} \tag{8.188a}$$

$$\lambda_2\lambda_1 = 2\,\mu_X\mu_Y k_{XY}k_{XX} \tag{8.188b}$$

$$\lambda_2 + \lambda_1 = (\mu_X + \mu_Y)k_{XY} + 2\,\mu_X k_{XX} \tag{8.188c}$$

First, suppose that $\mu_Y \gg \mu_X$ (as in C_2H_2). Then $\lambda_3 \cong \mu_Y k_{XY}$, $\lambda_2 \sim \mu_Y k_{XY}$, and $\lambda_1 \sim 2\,\mu_X k_{XX}$, where λ_3 and λ_2 are the X–Y stretching eigenvalues (one symmetrical and one antisymmetrical) and λ_1 is for X–X stretching. The latter result is just what one would get by considering X–X to be a diatomic molecule with force constant k_{XX}. Thus we see that a heavy-group frequency does arise if the group is adjacent to only light atoms. Suppose now that $\mu_X \sim \mu_Y$, but $k_{XX} \gg k_{XY}$. Then $\lambda_3 = (\mu_X + \mu_Y)k_{XY}$, $\lambda_2 \sim (\mu_X + \mu_Y)k_{XY}$ and $\lambda_1 \sim 2\,\mu_X k_{XX}$. Again we find that the X–X heavy-atom group has a characteristic frequency that is more or less independent of its environment.

On the other hand it is difficult to distinguish a C–C or C–N group frequency in CH_3–CH_2–NH_2 or a C–C or C–O group frequency in CH_3–CH_2–OH, because both the masses *and* the force constants are very similar. A useful qualitative way of considering the group-frequency concept is to consider the energy of the various bonds. In this view we expect that C–C and C–N bonds have similar total energies. When they are adjacent in a molecule they are strongly coupled and a sort of resonance occurs. This

---

[28] See footnote 10, p. 181.

mixes the individual pure modes and produces new frequencies which are somewhat different. C–C and C–H bonds have different vibrational energies and are therefore only slightly coupled or disturbed if they are adjacent.

The occurrence of group frequencies has been very usefully exploited in infrared spectroscopy for the identification of organic molecules. It is not our purpose to go into this any further, but several excellent books describing this kind of analysis have been included in the list of reference books at the end of the chapter.

SUPPLEMENTARY REFERENCES

H. C. Allen and P. C. Cross, *Molecular Vib-Rotors* (Wiley, New York, 1963).

G. M. Barrow, *Introduction to Molecular Spectroscopy* (McGraw-Hill, New York, 1962).

R. P. Bauman, *Absorption Spectroscopy* (Wiley, New York, 1962).

L. J. Bellamy, *The Infrared Spectra of Complex Organic Molecules* (Wiley, New York, 1958), 2nd ed.

R. T. Conley, *Infrared Spectroscopy* (Allyn and Bacon, Boston, 1966).

G. Herzberg, *Infrared and Raman Spectra of Polyatomic Molecules* (D. Van Nostrand, Princeton, N.J., 1945).

G. Herzberg, *Spectra of Diatomic Molecules* (D. Van Nostrand, Princeton, N.J., 1950).

G. W. King, *Spectroscopy and Molecular Structure* (Holt, Rinehart and Winston, New York, 1964).

R. M. Silverstein and G. C. Bassler, *Spectrometric Identification of Organic Compounds* (Wiley, New York, 1967), 2nd ed.

CHAPTER 9

NUCLEAR MAGNETIC RESONANCE SPECTROSCOPY

9.1 GENERAL CHARACTERISTICS AND EXPERIMENTAL ASPECTS

Nuclear magnetic resonance spectroscopy (NMR) is usually performed in the radiofrequency region between 20 and 200 MHz. It is clear from this fact that the energy changes involved are extremely small; for example, a frequency of 100 MHz corresponds to an energy of about 0.01 cal/mole. The NMR experiment distinguishes itself further by its resonance nature, as described in Sec. 1.3. The resonant absorption of the RF radiation can be observed by varying the radiofrequency at fixed magnetic field, or by varying the magnetic field strength at fixed radiofrequency.

NMR spectra may be observed for appropriate inorganic or organic molecules existing in any of the three states of matter, solid, liquid or gas. Studies in the liquid phase are by far the most common, however, although solid-state studies have their own particular utility. Studies of solids have been generally referred to as "wide-line" or "broad-line" NMR because of the much greater line widths that are observed in these cases (several gauss as opposed to about 0.1 milligauss for liquids). We see later that the types of information available from solid and liquid spectra are somewhat different. Recently, it has been found that molecules dissolved in nematic (high viscosity) liquids produce sharp-line NMR spectra which are, nevertheless, of a nature similar to that of solids. The advantage of the latter method is that the narrow lines permit resolution and analysis of the spectra in a way usually not possible for solids.[1]

The resolving power of an NMR instrument is very high when liquids are being studied. If resolving power is defined as frequency divided by line width, we obtain a value of greater than 10^7, which is larger than that obtained by microwave spectroscopy by about one order of magnitude. From a more practical point of view, however, this high resolving power is misleading, since the majority of NMR (proton) spectra occur in a frequency *range* of about 1 KHz. Thus with typical line widths of 0.1 Hz, it is possible to distinguish no more than about 10^4 resonance lines in the useful region.

The major limitation of NMR spectroscopy is that it can be performed only for molecules which contain nuclei with nonzero spin. This is not such a serious problem, since all hydrocarbons contain protons (spin $\frac{1}{2}$), so NMR is possible with all of these. On the other hand one is seriously limited in seeing nuclear magnetic resonances for several desirable nuclei, such as carbon and oxygen. To see resonances from these nuclei requires the use of C^{13} or O^{17} isotopes, since the abundant isotopes have spin zero. Another limitation occurs with paramagnetic molecules, for which it is often not possible to observe nuclear magnetic resonance even though appropriate nuclei are present.

One other important distinguishing feature of NMR is that it is primarily a nuclear phenomenon, but as we see it provides information concerning the molecule as a whole. This may be contrasted to infrared or microwave spectroscopy, each of which is primarily an over-all molecular phenomenon (molecular vibration and rotation), although each provides information about localized regions (such as chemical bonds) of the molecule. Thus in a sense we may consider that NMR probes molecular properties from the inside while IR and microwave spectroscopy probe them from the outside.

A number of experimental arrangements have been utilized for NMR spectroscopy. In Fig. 9.1 we show a simplified block diagram of the double-coil nuclear magnetic resonance spectrometer which is common for high-resolution studies of liquids. The instrument consists of a high-quality electromagnet with a field-homogeneity on the order of one part in 10^7. The magnet pole faces are typically 12 in. diam. with a gap of $1\frac{1}{2}$ inches. As we see shortly, a magnetic field strength of 14 kilogauss is needed for observing proton resonances at 60 MHz. Radiofrequency radiation (at 60 MHz, say) is put into the sample via a transmitter coil which is part of a tuned circuit. Absorbed radiation is detected by a receiver coil which is appropriately coupled to the transmitting circuit. Both coils have their axes at right angles to each other and to the static magnetic field direction.

---

[1] See, for example, L. C. Snyder and S. Meiboom, *J. Chem. Phys.* **47**, 1480 (1967).

The coils and sample holder are constructed into a compact unit called the "induction head" or "probe."

Practically all NMR studies are performed at a fixed frequency, so the resonant condition is achieved by sweeping the magnetic field. This turns out to be the easiest alternative, since the amplification system is a fixed-frequency one which permits high-sensitivity, narrow-band operation. The necessary small magnetic sweep field is usually obtained by use of small Helmholtz coils built into the NMR probe. The small magnetic field produced by these coils adds to the large field of the electromagnet, so that only the small field is swept in a typical NMR experiment. For proton NMR experiments a sweep range of a few hundred milligauss is sufficient.

Figure 9.1 shows how the signal detected by the receiver coil might be treated. After preamplification (which may occur in the probe itself) the signal can be heterodyned to some lower intermediate frequency (IF) by means of local oscillator and mixer stages. In this manner most of the amplification is carried on in narrow-band IF stages. If the probe frequency

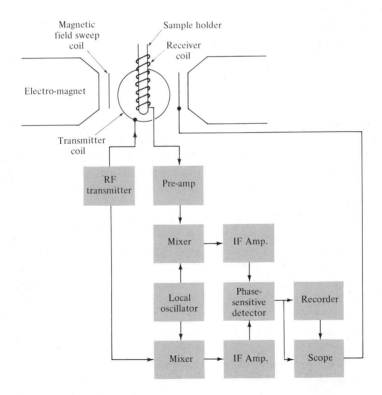

FIG. 9.1.   Simplified block diagram of a crossed-coil high-resolution NMR spectrometer.

is changed (for example, to observe other nuclei) only the local oscillator frequency needs to be changed in order to utilize the same IF stages. Further increase in signal-to-noise ratio is obtained by phase-sensitive (lock-in) detection using as reference the IF signal obtained by heterodyning the original RF transmitting signal. The output of the phase-sensitive detector is then suitable for oscilloscope or recorder presentation.

With properly designed magnets, probes and subsidiary electronics, a resolution of 1 part in 10^8 is achievable. It is somewhat difficult to specify the sensitivity of a conventional NMR spectrometer, but *very* roughly, something like 1 mole % of protons can be detected in a volume of a few tenths of a milliliter at 60 MHz. For other nuclei the sensitivity is usually much less. Recent applications of digital time-averaging techniques have increased the sensitivity by about a factor of at least 10 in practical applications. For a more complete description of NMR instrumentation the reader is urged to consult the excellent work of Pople, Schneider, and Bernstein.[2]

Before leaving this brief discussion we should mention one other important instrumental feature which improves instrument resolution and sensitivity. This is the technique of sample spinning. In this method the sample (typically in a 5-mm-diam glass tube) is rotated at a speed of $\sim$25 rps, the result being that the resonant absorption lines narrow by about a factor of 10 and increase appreciably in amplitude. This remarkable result occurs because the macroscopic rotation of the molecules in the sample tends to average out the field inhomogeneity over the sample.

It is not unusual to have a variation of field strength of 1 mG over the sample volume with the best available magnet. This would produce minimum absorption linewidths of about 1 mG since the nuclei of interest would show continuous absorption over the range of inhomogeneity. Thus at 10 000 G, a resolution of 1 in 10^7 would be the maximum possible without sample spinning.

A simple qualitative explanation of the averaging phenomenon can be made as follows with the aid of Fig. 9.2. If a particular molecule A moves across the field variation range ΔH very slowly, it has its resonance condition at a magnetic field setting somewhat different than that of molecule B, for example, or molecule C. If we assume that the magnet setting for resonance is H_0, then molecule C absorbs when the field strength is H_0. But molecule A does not absorb until the magnet strength is increased by δH_1, and B does not absorb until the field strength is decreased by δH_2. Thus we have a linewidth of approximately $\delta H_1 + \delta H_2 = \Delta H$. On the other hand, if the molecules move across the gradient ΔH rapidly, their potential energy is more nearly that of the average field (H_0 in Fig. 9.2), so the

[2] J. A. Pople, W. G. Schneider and H. J. Bernstein, *High-resolution Nuclear Magnetic Resonance* (McGraw-Hill, New York, 1959).

resonant field is more sharply peaked about H_0 for all molecules in the sample.

The time scales that are involved may be estimated by a simple application of the Heisenberg uncertainty relation in the form

$$\delta E \delta t \sim h \tag{9.1a}$$

If the time τ in traversing ΔH is equated with δt, the uncertainty in the potential energy of the molecule is

$$\delta E = h/\tau \tag{9.1b}$$

As we will see very soon, the potential energy of a nucleus is $g_N \beta_N H$ so that δE is $g_N \beta_N \delta H$. Thus

$$\tau = h/g_N \beta_N \delta H \tag{9.1c}$$

relates the traversal time to the uncertainty in the field felt by the molecule.

Then if the sample is rotated more or less rapidly (τ small), the uncertainty in field strength felt by a given molecule is relatively large, say $\delta H > \Delta H$. In this case the field variation ΔH is of no significance so all molecules behave the same on the average. But if the sample is stationary (τ large), the field strength uncertainty for a given molecule is very small, say $\delta H < \Delta H$, in which case each molecule behaves as an individual in the particular field that it feels. For protons, when $\delta H \sim \Delta H = 10^{-3}$ G, we

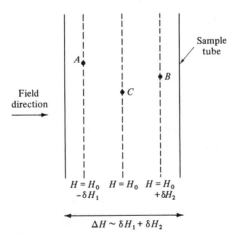

FIG. 9.2. Magnetic field variation across an NMR sample tube. Drawing assumes that field has been adjusted to H_0 (resonant condition for molecules of interest) at center of tube. A, B, and C are representative molecules and $\Delta H \sim 10^{-3}$ G.

find from Eq. (9.1c) that

$$\tau \sim 0.2 \text{ sec}$$

Thus if the sample is rotated at speeds on the order of 1000 rpm ($\tau \sim 0.06$ sec) the field variation will be averaged out since $\delta H > \Delta H$. With this technique NMR resolution has been pushed to 1 part in 10^8 even though the homogeneity of existing magnets is not that good.

9.2 BEHAVIOR OF NUCLEI IN MAGNETIC FIELDS

A. Nuclear Moments

Although mass and charge were among the earliest discernible properties of the nucleus, it was not until much later that the nuclear magnetic dipole moment and the nuclear electric quadrupole moment were found to exist. Closely associated with these two properties is that of intrinsic nuclear spin angular momentum. While magnetic moments and electric quadrupole moments are classically recognized properties, nuclear spin angular momentum is a strictly quantum-mechanical phenomenon. As mentioned in Sec. 2.4, the spin angular momenta ($\tilde{I}_X$, $\tilde{I}_Y$, $\tilde{I}_Z$) satisfy the ordinary commutation relations for space-fixed axes. The eigenvalues and matrix elements are those derived in Sec. 4.3, and listed in Table 4.1. Thus the eigenvalues are given by

$$\tilde{I}^2\psi(n) = \hbar^2 I(I+1)\psi(n) \tag{9.2}$$

and

$$\tilde{I}_Z\psi(n) = \hbar M_I\psi(n) \tag{9.3}$$

where I is the nuclear spin quantum number, $M_I = I, I-1, \ldots, -I+1, -I$, and $\psi(n)$ is the nuclear eigenfunction.

The permissible values of I are zero, integers, or half-integral positive numbers. The value of I for any particular nucleus is essentially an experimental finding. Nuclear spin values for several common nuclei are listed along with other nuclear properties in Table 9.1. From this table we can see three generalizations which arise:

(1) Nuclei with even mass number and even atomic number (even-even nuclei) have $I = 0$.

(2) Nuclei with even mass number and odd atomic number (even-odd nuclei) have integral values of I.

(3) Nuclei with odd mass number (odd-even or odd-odd nuclei) have half-integral values of I.

Since the neutron and the proton each have $I = \frac{1}{2}$ it is easy to see by simple vector addition arguments that even mass number nuclei must have integral or zero values of I, and that odd mass number nuclei must have half-integral values. It is a difficult problem in nuclear-physics theory,

TABLE 9.1

Properties of some common nuclei.[a]

Isotope	NMR frequency in MHz for a 10 kG field	Natural abundance %	Nuclear spin in units of $\hbar$	Nuclear g factor	γ $10^4\ G^{-1}\ sec^{-1}$	Magnetic moment (units of Bohr magnetons)	Quadrupole moment (units of $e \times 10^{-24}\ cm^2$)
H^1	42.576	99.98	$\frac{1}{2}$	5.5854	2.6752	2.7927	0
H^2	6.536	0.0156	1	0.8574	0.4107	0.8574	2.77×10^{-3}
B^{10}	4.575	18.83	3	0.6002	0.2875	1.8005	7.4×10^{-2}
B^{11}	13.660	81.17	$\frac{3}{2}$	1.7920	0.8583	2.6880	3.55×10^{-2}
C^{12}	...	98.9	0	0	0	0	0
C^{13}	10.705	1.11	$\frac{1}{2}$	1.4044	0.6726	0.7022	0
N^{14}	3.076	99.64	1	0.4036	0.1933	0.4036	7.1×10^{-2}
O^{16}	...	99.76	0	0	0	0	0
O^{17}	5.772	0.0037	$\frac{5}{2}$	-0.7572	-0.3627	-1.8930	-4.0×10^{-3}
F^{19}	40.055	100	$\frac{1}{2}$	5.2546	2.5168	2.6273	0
P^{31}	17.236	100	$\frac{1}{2}$	2.2610	1.0829	1.1305	0
Cl^{35}	4.172	75.4	$\frac{3}{2}$	0.5473	0.2621	0.8209	-7.9×10^{-2}
Cl^{37}	3.472	24.6	$\frac{3}{2}$	0.4555	0.2182	0.6833	-6.21×10^{-2}
Br^{79}	10.667	50.6	$\frac{3}{2}$	1.3994	0.6703	2.0991	0.34
Br^{81}	11.499	49.4	$\frac{3}{2}$	1.5084	0.7225	2.2626	0.28

[a] Adapted from Varian Associates Nuclear Magnetic Resonance Table, 4th ed., 1964.

however, to explain why some even mass number nuclei have spin zero while others have spins of one, two, or three; or why some odd mass number nuclei have $I = \frac{1}{2}$, while others have $I = \frac{3}{2}$ or $\frac{5}{2}$, etc.

Any nucleus having $I > 0$ has a magnetic moment whose direction is parallel to that of $\mathbf{I}$ and whose magnitude is proportional to $|\mathbf{I}|$. Thus we may write

$$\boldsymbol{\mu} = \gamma \mathbf{I} \tag{9.4a}$$

where γ is known as the gyromagnetic ratio. γ is an experimentally determined quantity which may be positive or negative. Another common form for Eq. (9.4a) is

$$\boldsymbol{\mu} = \frac{g_N \beta_N}{\hbar} \cdot \mathbf{I} \tag{9.4b}$$

where

$$\beta_n = \frac{e\hbar}{2M_p c} \tag{9.5}$$

is the nuclear Bohr magneton and g_n is known as the nuclear g factor. This latter form arises from theoretical considerations, and β_n is defined in analogy to early considerations (see chapter 10) involving the electron. In practical units, β_n has the value 5.050×10^{-24} erg G^{-1}. g_n is a small dimensionless algebraic number. Some γ and g_n values are listed in Table 9.1.

The quantum-mechanical eigenvalues of the magnetic moment are obtained by substituting Eq. (9.4a) into Eqs. 9.2 and 9.3. We find then,

$$\tilde{\mu}^2 \psi(n) = \hbar^2 \gamma I (I + 1) \psi(n) \tag{9.6}$$

and

$$\tilde{\mu}_Z \psi(n) = \hbar \gamma M_I \psi(n) \tag{9.7}$$

The magnitude of the nuclear magnetic moment is therefore $\hbar[\gamma I(I+1)]^{1/2}$. This result not withstanding, it is common practice to define the maximum value of the Z component of $\tilde{\mu}$ as "the magnetic moment." Thus by reference to Eq. (9.7) (setting $M_I = I$), we define "the magnetic moment" as

$$\mu = \hbar \gamma I \tag{9.8a}$$

or using Eq. (9.4b),

$$\mu = g_n \beta_n I \tag{9.8b}$$

Values of the magnetic moment μ have also been tabulated in Table 9.1 for the common nuclei.

The electric quadrupole moment of a nucleus is a measure of the nonsphericity of the nuclear charge distribution. As customarily defined, the nuclear quadrupole moment is

$$Q = \frac{1}{e} \int \rho(3z^2 - r^2)\,dx\,dy\,dz \tag{9.9}$$

where $\rho = \rho(x, y, z)$ is the charge density at any position (x, y, z) measured relative to the center of mass of the nucleus, and the z axis is in the direction of I_z. The electric quadrupole moment is clearly distinguished from the electric dipole moment by the squared terms.

From Eq. (9.9) it is easily seen that $Q = 0$ if the charge distribution is spherical, that is, $\rho = Cr^2$ where C is some constant. In this case the integral of Eq. (9.9) becomes

$$Q = \frac{C}{e} \int r^2(3z^2 - r^2)\,dx\,dy\,dz \tag{9.10a}$$

or in polar coordinates

$$Q = 4\pi \frac{C}{e} \int r^6(3\cos^2\vartheta - 1)\sin\vartheta\,d\vartheta\,d\varphi\,dr \tag{9.10b}$$

But the integral over $d\vartheta$ vanishes so we obtain $Q = 0$ as stated.

For any nuclear charge distribution other than spherical, Q as defined in Eq. (9.9) is nonzero. Two nonspherical cases are clearly distinguished and have been previously described in Sec. 7.7 A. In the first the charge density is greatest in the z direction, in which case Eq. (9.9) shows that $Q > 0$. If the charge density is greatest in the x and y directions $Q < 0$. These cases have been shown previously in Fig. 7.18 where the quadrupolar ellipsoids are labeled *prolate* and *oblate* in analogy to the definitions for the symmetric rotor molecule.

As is the case with magnetic moments and nuclear spin values, quadrupole moments must be determined experimentally. Theoretical considerations[3] do, however, dictate that Q is nonzero only when $I > \frac{1}{2}$. Table 9.1 gives values for the quadrupole moments in those cases for which data are available.

B. Nuclear Zeeman Effect

In a magnetic field H_0 the potential energy of a magnetic dipole is given by

$$V = -\boldsymbol{\mu}\cdot\mathbf{H}_0 \tag{9.11}$$

where H_0 is the magnetic field strength. If the magnetic dipole is a bare

---

[3] C. H. Townes and A. L. Schawlow, *Microwave Spectroscopy* (McGraw-Hill, New York, 1955), p. 142.

nucleus, the quantum-mechanical Hamiltonian for this nucleus would be simply

$$\tilde{\mathfrak{IC}} = -\tilde{\mu} \cdot \tilde{H}_0 \tag{9.12a}$$

Choosing the static magnetic field direction along the Z axis, Eq. (9.12a) becomes

$$\tilde{\mathfrak{IC}} = -\frac{1}{\hbar} g_n \beta_n \tilde{I}_Z H_0$$

$$= -\gamma \tilde{I}_Z H_0 \tag{9.12b}$$

by use of Eqs. (9.4).

The eigenvalues of $\tilde{\mathfrak{IC}}$ are trivially obtained from those previously written for $\tilde{I}_Z$. This gives

$$E_n = \langle I_Z \mid \tilde{\mathfrak{IC}} \mid I_Z \rangle = -g_n \beta_n M_I H_0$$

$$= -\gamma \hbar M_I H_0 \tag{9.13}$$

Thus the $2I + 1$ degeneracy of the nuclear spin state is lifted in the presence of a magnetic field. This is known as the nuclear Zeeman effect and is shown in Fig. 9.3 for $I = \frac{1}{2}$ and $I = 1$ cases.

We discuss the selection rules in greater detail in Sec. 9.4 A, but since the results are of such simplicity we state them here; namely,

$$\Delta M_I = \pm 1 \tag{9.14}$$

Nuclear magnetic resonance absorption thus occurs at a frequency

$$\nu = g_n \beta_n H_0 / h$$

$$= \gamma H_0 / 2\pi \tag{9.15}$$

These transitions are illustrated in Fig. 9.3. Note that the $2I$ transitions for

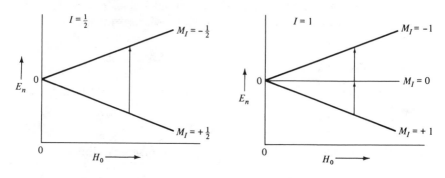

FIG. 9.3.   Nuclear Zeeman effect for $I = \frac{1}{2}$ and $I = 1$ cases.

a nucleus of spin I are all coincident. We are now in a position to evaluate the frequency of an NMR experiment. From Table 9.1 and Eq. (9.15) we find that when $H_0 = 10\ 000$ G, $\nu = 42.58$ MHz for the proton; $\nu = 40.05$ MHz for F^{19}; and $\nu = 10.70$ MHz for C^{13}.

What we are really interested in is a nucleus embedded in a molecule, not just a bare nucleus as previously considered. As described in Sec. 3.4 the Zeeman energy term is a small perturbation, in general, in the total Hamiltonian, and is therefore expected to produce a fine structure on the rotational energy levels. Normally the Zeeman effect that is observed arises from the ground electronic and vibrational states so it is appropriate to consider that the Zeeman energy is averaged over these states as shown in Eq. (3.96). In the gas phase it is possible that the nuclear spin and rotational angular momentum are coupled via spin-rotation or nuclear quadrupole coupling (see Sec. 7.7), in which case M_I is no longer a good quantum number and the nuclear Zeeman energy may not separate as illustrated in Eq. (3.95). In the liquid or solid states, however, the rotational angular momentum is effectively *quenched* since the rotational motions are severely limited. For condensed phases then, we expect that the nuclear Zeeman term is separable from the larger Hamiltonian terms and can therefore be treated separately.

The Zeeman Hamiltonian and energy for a nucleus embedded in a molecule should therefore be of the previously described form with one major exception. The magnetic field strength that is needed in the expression for the resonant frequency is the *field felt by the nucleus*. This is lower in general than the external applied field since the nucleus is screened by the surrounding electrons. The dominant cause of this screening effect is the induced circulation of the atomic electrons, which produces a small magnetic field in the vicinity of the nucleus as shown in Fig. 9.4. This induced magnetic field is in the opposite direction to the applied field and thus causes the *effective* field at the nucleus to be *less* than the applied field. This screening mechanism is known as *diamagnetic* screening since the phenomenon is similar to that producing the diamagnetic susceptibility of atoms. In addition to the dominant diamagnetic screening term, smaller contributions may exist which *add* to the applied field, and consequently are known as *paramagnetic* "screening" terms. We describe all these effects in more detail later.

For now we note simply that $H_{eff} < H_0$ because of the dominant diamagnetic screening. If σ is the dimensionless screening constant, we can write

$$H_{eff} = H_0(1 - \sigma) \tag{9.16}$$

where σ is expected to be a positive number. For light nuclei, σ turns out to be on the order of 10^{-5} to 10^{-3}, and it increases as the atomic number increases. The resonance condition for a nucleus in a molecule is now given by

$$\nu = \frac{g_n\beta_nH_0(1-\sigma)}{h}$$

$$= \frac{\gamma H_0(1-\sigma)}{2\pi} \tag{9.17}$$

rather than by Eq. (9.15).

Now the most remarkable feature of NMR is its ability to distinguish not only between different kinds of nuclei, but also between the same nucleus in different electronic environments due to the changes in σ. For example, in the frequency region of proton absorption, the ethyl chloride molecule (CH_3CH_2Cl) shows two distinct groups of absorption lines as shown in Fig. 9.5. Since this spectrum was obtained with a fixed frequency of 60 MHz, the resonances occur at a magnetic field of 14 100 G. Neglecting for the present the reason for the small triplet and quartet splittings, we can see that the centers of the two resonances are separated by 30.1 mG and that the integrated intensities are in the ratio of 3:2.

This latter result suggests that the more intense multiplet arises from the three methyl protons and the less intense one from the methylene protons. This is, in fact, the correct interpretation. The methyl protons are said to form a chemically *equivalent* set of nuclei, which means that they each have the same σ value. Likewise the methylene protons form an equivalent set having a σ value somewhat smaller than that for the methyl group. In a qualitative sense we may conclude that each member of an equivalent set absorbs at the same frequency, which explains the 3:2 ratio

FIG. 9.4. Origin of chemical shielding of nucleus by electrons. Applied field is H, induced field produced by circulating electrons is H'.

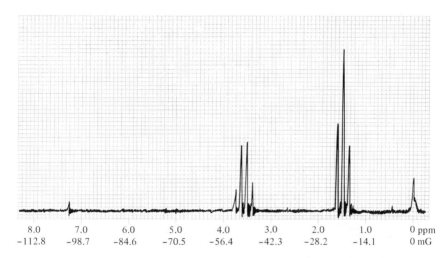

8.0	7.0	6.0	5.0	4.0	3.0	2.0	1.0	0 ppm
−112.8	−98.7	−84.6	−70.5	−56.4	−42.3	−28.2	−14.1	0 mG

FIG. 9.5. Spectrum of ethyl chloride (in CDCl₃) at 60 MHz. Scale readings of zero correspond to 14 080 G. Meaning of scales other than mG are described later. From "High Resolution NMR Spectra Catalogue," Vol. 1 (Palo Alto, Calif.: Varian Associates, 1962). With permission.

in this case. Since Cl^{35} or Cl^{37} have nuclear spins of $\frac{3}{2}$ we might expect that their absorption might be seen also (Cl^{35} should absorb at about 4.17 MHz in a 10 kG field). However, due to a relaxation effect produced by the chlorine quadrupole moment this resonance cannot be observed.

Neglecting the Cl^{35} (or Cl^{37}) nucleus, we could write the Zeeman Hamiltonian for ethyl chloride by a slight extension of the one-nucleus Hamiltonian. This gives

$$\tilde{\mathcal{3C}} = -\{\gamma_H \tilde{I}_Z(H_1)H + \gamma_H \tilde{I}_Z(H_2)H + \gamma_H \tilde{I}_Z(H_3)H$$
$$+ \gamma_H \tilde{I}_Z(H_4)H' + \gamma_H \tilde{I}_Z(H_5)H'\} \quad (9.18)$$

in which H and H' are the effective fields felt by the methyl and methylene nuclei, respectively; $\tilde{I}_Z(X)$ refers to nucleus X and H_1, H_2, H_3 are the methyl protons. There are no terms in $\tilde{\mathcal{3C}}$ coupling the $\tilde{I}_Z$, so each term has the same eigenvalues as previously obtained. This gives for the energy

$$E = -\hbar\gamma_H H \sum_{i=1}^{3} M_I(i) - \hbar\gamma_H H' \sum_{i=4,5} M_I(i) \quad (9.19a)$$

Writing $\sum_i M_I(i) = M_F$ we get

$$E = -\hbar\gamma_H H M_F(CH_3) - \hbar\gamma_H H' M_F(CH_2) \quad (9.19b)$$

Since protons have nuclear spins of $\frac{1}{2}$, all nuclei may have M_I values of $\pm\frac{1}{2}$, which leads to $M_F(CH_3) = \frac{3}{2}, \frac{1}{2}, -\frac{1}{2}, -\frac{3}{2}$ and $M_F(CH_2) = 1, 0, -1$. If the complete energy-level diagram were drawn, Eq. (9.19b) would lead to $(4)(3) = 12$ distinct states. The selection rules permit only $\Delta M_F(CH_3) = \pm 1$, or $\Delta M_F(CH_2) = \pm 1$. This leads to (for the absorption spectra)

$$\nu(CH_3) = \frac{1}{2\pi}(\gamma_H H)$$

$$\nu(CH_2) = \frac{1}{2\pi}(\gamma_H H') \tag{9.20}$$

as the only permitted absorption frequencies. Moreover, since there are three identical transitions arising from $\Delta M_F(CH_3) = 1$ for every two arising from $\Delta M_F(CH_2) = 1$,[4] the methyl resonance should be $\frac{3}{2}$ as intense as the methylene resonance.

At fixed frequency we note that the difference in resonant applied magnetic field is [from Eqs. (9.20)]

$$H_0(CH_3) - H_0(CH_2) = \frac{2\pi\nu}{\gamma_H}\left(\frac{1}{1 - \sigma_{CH_3}} - \frac{1}{1 - \sigma_{CH_2}}\right) \tag{9.21a}$$

Since $\sigma \ll 1$ we may write $(1 - \sigma)^{-1} = 1 + \sigma$, so Eq. (9.21a) becomes

$$H_0(CH_3) - H_0(CH_2) = \frac{2\pi\nu}{\gamma_H}(\sigma_{CH_3} - \sigma_{CH_2}) \tag{9.21b}$$

On the other hand, at fixed magnetic field H_0, the difference in resonant frequencies is

$$\nu(CH_3) - \nu(CH_2) = -\frac{\gamma_H H_0}{2\pi}(\sigma_{CH_3} - \sigma_{CH_2}) \tag{9.22}$$

In the present case, the experiment was carried out at a fixed frequency of 60 MHz. Thus, using Eq. (9.21b) with $\Delta H = 0.00301$ G, we find $\sigma_{CH_3} - \sigma_{CH_2} = +2.14 \times 10^{-6}$. We see then that the changes in σ produced by the chemical environment are sensitively detected and measured.

Perhaps a brief word concerning *chemical equivalence* of protons is in order here. If the CH_3–CH_2–Cl molecule were perfectly rigid the methyl

---

[4] To obtain this result one must account for all degeneracies. For example, $M_F(CH_3) = \frac{1}{2}$ can be obtained in three ways, viz., $\frac{1}{2}, \frac{1}{2}, -\frac{1}{2}; \frac{1}{2}, -\frac{1}{2}, \frac{1}{2}; -\frac{1}{2}, \frac{1}{2}, \frac{1}{2}$. Also, the rules $\Delta M_F(CH_3) = 1$ or $\Delta M_F(CH_2) = 1$ must be supplemented by the restriction that the spin of only *one* nucleus may change at a time. Thus $\frac{1}{2}\frac{1}{2}-\frac{1}{2} \rightarrow -\frac{1}{2}\frac{1}{2}-\frac{1}{2}$ is permitted but $\frac{1}{2}\frac{1}{2}-\frac{1}{2} \rightarrow -\frac{1}{2}-\frac{1}{2}\frac{1}{2}$ is not.

protons would not all be equivalent. In a rigid staggered conformation, for example, two of the methyl protons would be staggered with respect to the chlorine, while the third would be staggered with respect to the methylene protons. In this rigid conformation there would therefore be two types of methyl protons due to the different electronic environments. However, rotation about the carbon-carbon single bond is relatively easy (rapid) in the gas or liquid phase so the screening of the methyl protons is averaged. In situations where bulky substituents are present, or in cases where double-bond character is present, internal rotation about single bonds may not produce an average chemical screening in which case the various methyl protons may have different values of σ. Some quantitative discussions of these time-dependent phenomena are given in Sec. 9.4.

The general results illustrated for ethyl chloride are found to be qualitatively valid for other molecules. We can therefore summarize the chief features of an NMR spectrum that have been elucidated up to this point. They are:

(1) Every nucleus having a magnetic moment ($I \geqq \frac{1}{2}$) may show absorption of electromagnetic radiation at a frequency given by Eq. (9.17).

(2) The same nucleus in different chemical (electronic) environments has slightly different resonant frequencies due to differences in the screening of the nucleus.

(3) If several identical nuclei are chemically equivalent (same σ) their resonant frequencies are identical, and the intensity of the absorption line is directly proportional to the number of equivalent nuclei.

9.3 GENERAL FEATURES OF LIQUID AND GAS PHASE NMR SPECTRA

A. Chemical Shift

At the end of Sec. 9.2 B we summarized the principal features of NMR spectra that originated from the nuclear Zeeman effect. It is evident (particularly for proton magnetic resonances, PMR) that the relative positions of the resonances of nuclei in various molecular environments are a sensitive probe of the electron distribution in the molecule. In order to alleviate the problem of always specifying both the magnetic field and the frequency at which resonance occurs, it has become common to express the *chemical shift* of a nucleus relative to some standard compound containing the same nucleus. If $H_0(s)$ is the applied field at which a certain nucleus in the sample compound is in resonance at a fixed frequency, and $H_0(r)$ is the resonant field of the same nucleus in a reference molecule, the chemical shift in parts per million (ppm) is defined by

$$\delta = \frac{H_0(r) - H_0(s)}{H_0(r)} \times 10^6 \qquad (9.23)$$

Using an equation similar to Eq. (9.21b) we can recast this expression as

$$\delta = \frac{[\sigma(r) - \sigma(s)]2\pi\nu}{H_0(r)\gamma} \times 10^6 \qquad (9.24a)$$

But using Eq. (9.17) with $(1 - \sigma) \sim 1$, we get

$$\delta = [\sigma(r) - \sigma(s)] \times 10^6 \qquad (9.24b)$$

Thus we see that δ is positive when the sample nucleus is *less* screened than that of the reference.

The utility of defining the chemical shift as in Eq. (9.23) is twofold. First, it is a practical relative measurement of absorption line positions. Some kind of relative measurement is necessary since, as mentioned earlier, an accurate measurement of magnetic field is not possible; thus only differences in field can be measured with any great accuracy. Secondly, the definition is such that it is independent of experimental frequency. Suppose that in a proton resonance experiment carried out at 60 MHz, $H_0(r) - H_0(s) = 0.14$ G. At 60 MHz, a proton resonance occurs at 14.1 kG (i.e., $H(r) = 14.100$) so $\delta = 10$ ppm. If the same resonance were studied at 100 MHz, δ would be still 10 ppm; but of course, $H_0(r) - H_0(s)$ would now be $(23\,500)(10)(10^{-6})$ or 0.23 G since the resonant field is $(100/60)(14\,100) = 23\,500$ G. More simply since $H \propto \nu$, it is clear that the chemical shift *in gauss* is directly proportional to ν. Thus if ν is doubled, $H_0(r) - H_0(s)$ is doubled, etc., but the δ value remains unchanged.

Another chemical shift scale that has become popular for protons is the τ scale, where

$$\tau = 10.00 - \delta \qquad (9.25)$$

This scale gives the reference material $(\delta = 0)$ a chemical shift value of $\tau = 10$, while protons with larger values of δ, that is those which occur at lower fields, have smaller τ values. Thus τ and H are parallel scales, whereas δ and H are antiparallel.

Finally, it is common in commercial instruments to record NMR spectra with a scale that is linear in frequency (cps or Hz) *even though the experiment is carried out at fixed frequency.* The connection between the real magnetic field scale and the fictitious frequency scale (valid when $\sigma \ll 1$) is seen by combining Eqs. (9.21b) and (9.22) which shows that

$\Delta \nu \propto -\Delta H_0$ where the proportionality constant is simply $2\pi/\gamma$. For the proton

$$2\pi/\gamma = 4258 \text{ Hz/G}$$

$$= 4.258 \text{ Hz/mG} \tag{9.26}$$

which is simply the factor that we have been making frequent use of right along to convert from frequency to magnetic field or vice versa.

To make some of these points concrete we list on the schematic "spectrum" of Fig. 9.6 the chemical-shift values for various protons using tetramethyl silane $[Si(CH_3)_4]$, TMS, as the reference. This compound is convenient since its 12 equivalent protons give it a single strong, sharp resonance. Moveover its resonance occurs at a higher field than that for most protons so most δ values are positive. Additional experimental advantages are that it is a liquid which can be mixed in low concentration with samples so that it serves as an internal standard. This latter feature makes TMS more suitable than CH_4, for example, since CH_4 is a gas at ordinary temperatures.

The chemical shift values in Fig. 9.6(a) are mostly typical values for various functional groups, but it must be emphasized that all the values may vary somewhat depending on the solvent and the concentration, and upon the particular molecular substituents. In particular, for hydrogen-bonding protons the chemical shift may vary by several chemical shift units (δ or τ). Nevertheless, a great body of experimental information leads to the conclusion that the chemical shifts of protons are determined by their functional group environment.[5] Thus a chloromethyl group ($-CH_2Cl$) always shows proton-resonance absorption near $\delta = 3.4$, an aldehyde proton near $\delta = 9.7$, and so forth. These *characteristic chemical shift* values are the NMR counterparts of the group vibrational frequencies in infrared spectroscopy.

Nuclei other than protons show similar, but usually larger, chemical-shift behavior. C^{13} nuclei, for example, exhibit the chemical shifts shown in Fig. 9.7, where the reference compound is benzene. Note that the chemical-shift range is about 10 times larger than for protons. In a similar manner the chemical shifts of fluorine nuclei (F^{19}) have been extensively studied. The range of chemical shifts is even larger in this case, as exemplified for a simple series by CF_3H, CF_2H_2, CFH_3, CF_3Cl, CF_2Cl_2, and $CFCl_3$, which have chemical shifts δ (relative to CF_4) of -18, -81, -210, 37, 60, and 77 ppm, respectively. Even more spectacular, fluorine gas (F_2) has a chemical shift of $+491$ relative to the same standard.

---

[5] An excellent description of the extent and application of these principles is given by R. M. Silverstein and G. C. Bassler, *Spectrometric Identifications of Organic Compounds* (Wiley, New York, 1967).

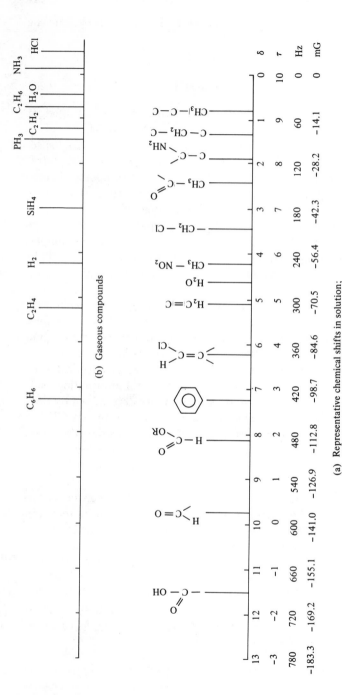

(a) Representative chemical shifts in solution; only absorbing protons shown.

(b) Gaseous compounds

FIG. 9.6. Some observed proton chemical-shift values for simple molecules and various functional groups. Chemical shifts in Hz and mG apply to experiments carried out at 60 MHz. TMS used as reference in (a), gaseous methane as reference in (b). (a) is adapted from W. G. Schneider, H. J. Bernstein, and J. A. Pople, *J. Chem. Phys.* **28**, 601 (1958).

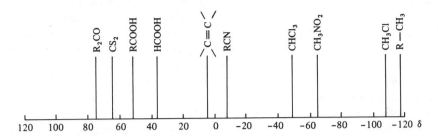

FIG. 9.7.   C¹³ chemical shifts. C₆H₆ used as reference compound. R stands for any alkyl group, for which the chemical shifts given are representative. Data adapted from P. C. Lauterbur, *J. Chem. Phys.* **26,** 217 (1957); and C. H. Holm, *J. Chem. Phys.* **26,** 707 (1957).

It is not our intent in this work to go into chemical shifts in much detail. We do, however, discuss some of the pertinent theoretical aspects later in this chapter. It should be obvious without further discussion that the *chemical shift* can play an important role in the qualitative identification of molecules.

B.   Hyperfine Splitting of Nuclear Resonance Absorption Lines

As has been previously mentioned, and illustrated in Fig. 9.5, NMR absorption lines in solution and in the gas phase often exhibit a hyperfine structure. The hyperfine splitting of some particular resonance is produced by an indirect magnetic interaction of the absorbing nucleus with adjacent nuclei having magnetic moments. For protons these interactions produce splittings in the range of about 0–25 Hz; fluorine-fluorine interactions may produce splittings in the range 0–200 Hz, and fluorine-hydrogen interactions yield splittings of 0–80 Hz.

The indirect interaction mentioned above is usually referred to as a *spin–spin* interaction. While a direct magnetic dipole–dipole interaction may occur (see Sec. 9.7 C), it always averages to zero for rapidly tumbling molecules in liquid or gas phases. On the other hand the spin–spin mechanism is independent of molecular orientation and thus remains. In a simple way, the spin–spin mechanism can be seen as follows. Consider two covalently bonded nuclei of spin ½, A and B, and the two electrons a and b that are primarily responsible for the covalent bond. Fig. 9.8 outlines the problem diagrammatically. Then nuclear dipole A tends to align electron dipole a (or conversely) antiparallel, since this is the lowest state of magnetic energy for two dipoles. The two bonding electrons must have opposite spins (Pauli principle) so their dipoles are oppositely directed. Finally, then, electron dipole b tends to align nuclear dipole B antiparallel also.

Thus we can see that nuclei A and B are coupled energetically by an indirect mechanism which is simply: nucleus $A \to$ electron $a \to$ electron $b \to$ nucleus B. It is clear also that the potential energy of nucleus B in a particular spin state, say $M_I = -\frac{1}{2}$ (or spin down, $\downarrow$), is different for the two possible orientations ($\uparrow$ or $\downarrow$) of the nuclear spin of A.

The same mechanism may be extended over several nuclei by involving more electrons along the route. It is found, as might be expected, that the interaction energy usually falls off rapidly as the chain length increases. For protons the interaction is often negligible if they are separated by more than two atoms.

Similar arguments apply for nuclei with spins greater than $\frac{1}{2}$. A nitrogen nucleus, for example, may present three spin states ($M_I = 1, 0, -1$) for coupling to another nucleus.

The spin–spin mechanism for two nuclei A and B can be represented for spectroscopic purposes by the Hamiltonian term,

$$\widetilde{\mathcal{H}}' = \widetilde{\mathcal{H}}_{ss} = J_{AB}\widetilde{I}_A \cdot \widetilde{I}_B \qquad (9.27)$$

where J_{AB} is the *coupling constant* whose magnitude is a measure of the strength of the interaction. J_{AB} is usually defined such that its units are in Hz. We treat this problem exactly in a later section, but we wish to look first at some of the simple results that occur. First, we note that the spin–spin interaction is independent of magnetic field (or electromagnetic frequency), that is, *the coupling constant is independent of magnetic field*. This should be contrasted to the chemical shift in Hz or gauss which is directly proportional to magnetic field.

We now state several rules which are proved later.

(1) A chemically equivalent set of nuclei produces no spin–spin hyperfine splittings of the resonance line occurring for that set. This means that H_2 and $C^{12}H_4$ each shows only a single sharp

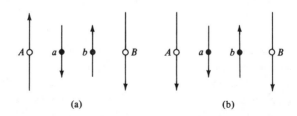

(a)   (b)

FIG. 9.8.   Indirect spin–spin coupling of nuclear spins A and B via electrons a and b. In (a) we show the lowest energy coupling mode, while (b) represents a higher energy interaction due to the parallel spins of A and a.

NMR resonance line, for example. This result does not mean that the protons of an equivalent set do not interact via the spin–spin mechanism. We see later that the absence of splittings is a consequence of the selection rules.

A relatively small number of molecules exhibit what are known as "first-order spectra"; an even larger number of molecules may be approximated by first-order spectra so it proves to be a very useful concept. We see later that the results are derivable from first-order perturbation theory which is the origin of the nomenclature. For our purposes we can state that first-order spectra arise when (a) the relative chemical shifts of all nonequivalent magnetic nuclei are large compared to the coupling constants between the various nuclei, and (b) all the coupling constants of nuclei from one equivalent set with those of a second equivalent set are identical. (a) is always satisfied when the nuclei have different gyromagnetic ratios, for example, in $C^{12}H_3F$ or in HD. However, requirement (a) may be satisfied for the same nucleus in different chemical environments if $\Delta\delta/J$ is large enough. For example, the two equivalent sets of protons of CH_3CH_2Cl are found to satisfy this requirement and to show first-order spectra (at 60 MHz).

In order to have a simple general method for specifying nuclear spin systems we use the letters $A, B, C \ldots$ to signify nonequivalent nuclei of the same species (isotope) whose relative chemical shifts are of the same magnitude. The letters $X, Y, Z \ldots$ signify other such nuclei whose chemical shifts are relatively large compared to $A, B, \ldots$ As in chemical formulas a subscript indicates the number of equivalent nuclei of a given type. Thus HD is an example of an AX system, $C^{12}H_3F$ an example of AX_3 (or A_3X), $C^{13}H_3F$ an example of AMX_3, and the protons of CH_3CH_2Cl an example of A_2X_3. Note in the third example that we use M, N, O, $\ldots$ as a third group of nuclei whose chemical shifts are different from the other two groups. Except for the ethyl chloride case these descriptions are obviously correct since the resonant frequencies of the different nuclear species are much different. In the ethyl chloride case, it is an experimental fact that the observed spectrum is of the A_2X_3 type, rather than the A_2B_3 type. First-order spectra are always of the form $A_aM_mX_x\ldots$.

Requirement (b) is not always satisfied either. Nuclei which do satisfy (b) are said to be *magnetically* equivalent. Requirement (b) is satisfied for HD, of course, since there is only one J value. It is satisfied for $C^{12}H_3F$ also, since all three J_{HF} values are identical by symmetry. Similarly for CH_3CH_2Cl the six J_{HH}, [three methyl protons (H) with two methylene protons (H')] are identical by virtue of the same internal rotation averaging which causes the methyl protons to be chemically equivalent. On the other hand, the two sets of chemically equivalent nuclei of 1,2-difluoroethylene $C^{12}_2H_2F_2$ do not satisfy the requirement since, as shown in Fig. 9.9, there are two distinct, nonidentical coupling constants between the H_2 and the F_2 set. Thus although $C_2H_2F_2$ is properly described as an A_2X_2 system, it does not show first-order behavior.

With this rather lengthy discourse on the definition of "first-order" spectra taken care of, we can add the following rules to the one previously given.

> (2) A chemically equivalent set of nuclei A_n splits the resonance of set X_m into $2F(A) + 1$ components where $F(A)$ is the *total* nuclear spin quantum number of set A_n, that is, $F(A) = nI(A)$ if $I(A)$ is the nuclear spin of nucleus A.
>
> Corollary: For protons (or any nuclei having spin $\frac{1}{2}$) an equivalent statement of (2) is that the set A_n splits the resonance of X_m into $n + 1$ components. The proof of this is left to the reader.
>
> (3) The $2F(A) + 1$ components described in (2) are equally spaced with the spacing equal to J_{AX}. The $2F(X) + 1$ components into which the A_n resonance is split are equally spaced at J_{AX} also.

Rule (2) is understandable on the basis of the simple description of the spin–spin mechanism given earlier. If $F(A)$ is the total nuclear spin of set A_n there must be $2F(A) + 1$ different components which may interact (couple) with another set X_m, thereby causing the X_m resonance to split into $2F(A) + 1$ components. Finally we can describe the relative intensities of the multiplets.

> (4) The intensities of the components of the hyperfine multiplets are proportional to the number of product spin functions which correspond to each of the $2F + 1$ values of M_F, where $M_F = F$, $F - 1, \ldots, -F + 1, -F$.

For example, two magnetically equivalent protons may have $M_F = 1, 0, -1$. If α and β are the functions corresponding to $M_I = +\frac{1}{2}$ and $-\frac{1}{2}$,

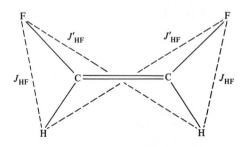

FIG. 9.9. Coupling constants in difluoroethylene.

respectively, the product functions for each M_F value are:

$$M_F = +1 \qquad \alpha(1)\alpha(2)$$
$$M_F = 0 \qquad \alpha(1)\beta(2)$$
$$\beta(1)\alpha(2)$$
$$M_F = -1 \qquad \beta(1)\beta(2)$$

Thus two equivalent protons that are coupled to another nuclear set split this set into three ($n + 1 = 3$) components with relative intensities 1:2:1. Similarly three protons (or any three $I = \frac{1}{2}$ nuclei) yield

$$M_F = \tfrac{3}{2} \qquad \alpha(1)\alpha(2)\alpha(3)$$
$$M_F = \tfrac{1}{2} \qquad \alpha(1)\alpha(2)\beta(3)$$
$$\alpha(1)\beta(2)\alpha(3)$$
$$\beta(1)\alpha(2)\alpha(3)$$
$$M_F = -\tfrac{1}{2} \qquad \beta(1)\beta(2)\alpha(3)$$
$$\beta(1)\alpha(2)\beta(3)$$
$$\alpha(1)\beta(2)\beta(3)$$
$$M_F = -\tfrac{3}{2} \qquad \beta(1)\beta(2)\beta(3)$$

which produces a 1:3:3:1 quartet.

A single nitrogen nucleus ($I = 1$) has $M_F = M_I = 1, 0, -1$. Each of these states has a single wave function, namely, ϕ_1, ϕ_0, or ϕ_{-1}. Thus a nitrogen nucleus has the capability of producing a 1:1:1 triplet for any nuclei to which it is coupled. Two magnetically equivalent nitrogens will yield

$$M_F = 2 \qquad \phi_1(1)\phi_1(2)$$
$$M_F = 1 \qquad \phi_1(1)\phi_0(2)$$
$$\phi_0(1)\phi_1(2)$$
$$M_F = 0 \qquad \phi_1(1)\phi_{-1}(2)$$
$$\phi_0(1)\phi_0(2)$$
$$\phi_{-1}(1)\phi_1(2)$$
$$M_F = -1 \qquad \phi_{-1}(1)\phi_0(2)$$
$$\phi_0(1)\phi_{-1}(0)$$
$$M_F = -2 \qquad \phi_{-1}(1)\phi_{-1}(2)$$

which would produce a 1:2:3:2:1 quintet.

For protons or any nuclei of spin $\frac{1}{2}$, the following corollary to rule (4) applies:

> Corollary: The relative intensities in a multiplet produced by n equivalent protons are given by the number of combinations of n objects taken m at a time, where $m = 0, 1, 2, \ldots, n$. These numbers of combinations are written mathematically as $\binom{n}{m} = n!/m!(n-m)!$.

For one proton this leads to a $1:1$ ratio, for two protons to a $1:2:1$ ratio, for four protons to $1:4:6:4:1$, etc. Note that this corollary obviates the need for actually writing down all the possible spin-product functions.

We can see now why the ethyl chloride spectrum of Fig. 9.5 showed a

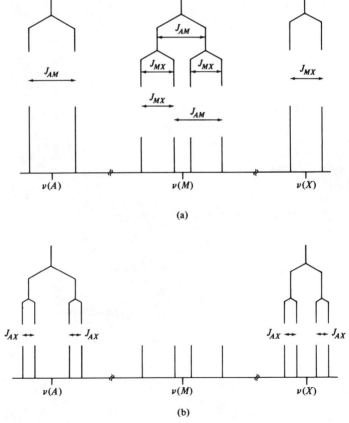

FIG. 9.10. Hypothetical AMX spectra. In (a), $J_{AX} = 0$, in (b), $J_{AX} \neq 0$.

1:2:1 triplet for the methyl resonance and a 1:3:3:1 quartet for the methylene resonance. The triplet arises from the spin–spin coupling of the *two* methylene protons with the methyl protons, while the quartet arises from the coupling of the *three* methyl protons with the methylene protons. We observe also that the spacings of the components of the two multiplets are identical and have the value 1.67×10^{-3} G or 7.1 Hz. Thus the coupling constant is $J_{CH_3-CH_2} = 7.1$ cps.

It may be wondered why no coupling and consequent splitting from the chlorine ($I = \frac{3}{2}$) is observed. This is due to the large quadrupole moment of the Cl^{35} or Cl^{37} nuclei. We discuss this *relaxation* effect later, but we simply note here that under ordinary circumstances no hyperfine splitting is produced by nuclei with spins greater than 1. In most cases $I = 1$ nuclei produce no observable splittings either, although in a few instances deuterium and nitrogen may produce observable hyperfine splittings.

For molecules with more than two sets of equivalent nuclei, perfect first-order spectra tend to be relatively rare. The expected first-order result is that every set of possible couplings produces splittings according to the preceding rules. Thus, an AMX spectrum might appear as in Fig. 9.10 for a system of spin-$\frac{1}{2}$ nuclei. In Fig. 9.10(a) we assume $J_{AX} = 0$. Note that the M resonance is a quartet due to a doublet splitting produced by A and a doublet splitting produced by X. A diagrammatic method of deriving the spectrum is shown in the figure. In Fig. 9.10(b) the AMX spectrum is illustrated with the same J_{AM} and J_{MX} values as in Fig. 9.10(a) but with J_{AX} not equal to zero.

More complicated systems of the type $A_a M_m X_x$ can be considered in an entirely similar manner. In this regard it is useful to emphasize that proton couplings fall off rather rapidly as the number of bonds separating the protons increases. Thus when protons are separated by two carbon atoms as in

the proton coupling constant may be 5–10 Hz, while for separations of three carbon atoms as in

the coupling constant is usually less than 1 Hz. For this reason these latter couplings are often negligible. On the other hand in unsaturated systems such as

$$H-\overset{\diagdown}{\underset{\diagup}{C}}-C\equiv C-\overset{\diagup}{\underset{\diagdown}{C}}-H$$

or

$$H-\overset{\diagdown}{\underset{\diagup}{C}}-\overset{|}{C}=\overset{|}{C}-\overset{\diagup}{\underset{\diagdown}{C}}-H$$

the long-range proton coupling constants are in the range of 1–3 Hz.

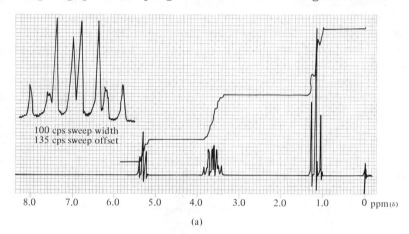

100 cps sweep width
135 cps sweep offset

8.0   7.0   6.0   5.0   4.0   3.0   2.0   1.0   0 ppm(δ)

(a)

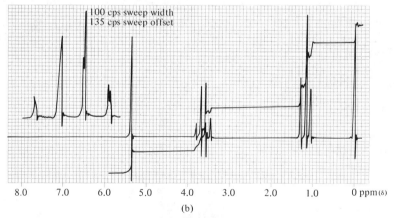

100 cps sweep width
135 cps sweep offset

8.0   7.0   6.0   5.0   4.0   3.0   2.0   1.0   0 ppm(δ)

(b)

FIG. 9.11. Spectra of ethanol. (a) Pure ethanol, 500 Hz sweep width. Inset is the octet expanded by factor of 5. (b) Acidified ethanol, 500 Hz sweep width. From John R. Dyer, *Applications of Absorption Spectroscopy of Organic Compounds*, © 1965. Reprinted by permission of Prentice-Hall, Inc., Englewood Cliffs, N. J.

As an example of an AM_2X_3 first-order system[6] we might consider CH_3CH_2OH. If a pure first-order spectrum occurred we would expect the CH_3 resonance to be a triplet with spacing J_{MX} due to coupling with the CH_2 group (and with neglect of the long-range OH coupling). The OH resonance should be a triplet with spacing J_{AM}, again neglecting the long-range coupling. The CH_2 resonance should be an octet (4×2) due to OH and CH_3 proton coupling. This octet should consist of four lines of relative intensity three, and four of relative intensity one. The ordering of these components depends, of course, upon the magnitudes of J_{AM} and J_{MX}.

Pure dry ethanol exhibits a spectrum in quite good agreement with these predictions as shown in Fig. 9.11(a). The reader can verify that the observed multiplets are satisfied quite well with $J_{AB} = 5.0$ Hz and $J_{BC} = 7.2$. The intensity deviations are indications of the incipient breakdown of the first-order spectral rules.

In Fig. 9.11(b) the spectrum of slightly acidified ethanol is shown. A similar spectrum results simply from "wet" ethanol. We note that the OH resonance is not split and that the CH_2 resonance shows splittings from only the CH_3 protons. This result is common for cases where the protons are exchangeable (that is, exchanging with solvent protons), as in the case of the OH protons. We discuss these *exchange* phenomena in detail in Sec. 9.4 E, but we simply note here that when exchange is rapid the spin-spin splitting of the resonance of the exchangeable proton(s) disappears. Moreover, the spin-spin splitting of other resonances which should be produced by the exchangeable proton(s) is found to disappear. In short, the spectrum behaves as though the exchangeable protons (or other exchangeable nuclei) were not coupled to other nuclei in the molecule.

In Table 9.2 we have listed a sampling of some of the observed spin-spin coupling constants for proton–proton, proton–fluorine, fluorine–fluorine, and proton–carbon 13 interactions. In closing this section we consider several examples which behave in a first-order fashion, but it should be emphasized that most spectra deviate from this simple theory. Nevertheless, the theory is often a valid first approximation, and used in conjunction with the group chemical shifts of Figs. 9.6 and 9.7 it can be very helpful for qualitative functional group analysis of large organic molecules.

EXAMPLE 9-1

The spin–spin interaction produces no hyperfine splitting of the proton resonance of H_2, but the situation is different for HD. Here first-order rules must apply since $\nu_0(H)$ is much different from $\nu_0(D)$. We predict the proton

---

[6] Our convention in complex systems is to label the nuclei left to right according to their left to right appearance on an ordinary spectrum. This means that the first set of nuclei has the *lowest* τ value (or highest δ) value. Thus we label ethanol as AM_2X_3, where $A = $ OH proton, $M = CH_2$ proton, and $X = CH_3$ proton.

TABLE 9.2
Representative coupling constants involving proton, fluorine, and C^{13} nuclei.

Structure	$J(Hz)$	Structure	$J(Hz)$
H_2	286	H—C—F	40–80
H—C—H (geminal)	12–15	H—F	615
CH_3—CH_2—X	6–8	CH—C—F	5–30
HC—OH	5–10	H—C=C—F (cis)	1–8
CH—C(=O)—H	1–3	H—C=C—F (trans)	12–40
H—C=C—H (cis)	7–12	F—C—F	150–225
H—C=C—H (trans)	13–18	F—C=—F	28–87
H—C=C—H (geminal)	1–3	H—C—C^{13}—H	123
H (benzene ring) H	6–10 (ortho) 1–3 (meta) 0–1 (para)	H—C=C^{13}—H	157

TABLE 9.2 (*continued*)

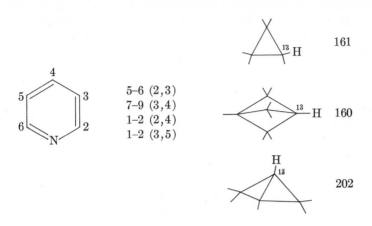

resonance to be a 1:1:1 triplet due to three possible values of $M_I(\text{D})$, namely 1, 0, -1. The deuteron resonance should be a 1:1 doublet and at a magnetic field of 10 000 G should occur at a frequency of

$$\nu = \frac{\gamma_{\text{D}}(10^4)}{2\pi} = 6.54 \text{ MHz}$$

Measurements on the proton resonance show the expected behavior and the spacing between adjacent components is 44 Hz, that is, $J_{\text{HD}} = 44$ Hz. We can calculate J_{HH} for the hydrogen molecule by referring to the theoretical expression for the coupling constant given in Eq. (9.121). As a first, good approximation the electronic structures of HD and H_2 are identical so the only difference in J_{HH} and J_{HD} arises because of the difference in γ_{H} and γ_{D}. Thus

$$\frac{J_{\text{H}_2}}{J_{\text{HD}}} = \frac{\gamma_{\text{H}}}{\gamma_{\text{D}}} \tag{9.28}$$

so

$$J_{\text{H}_2} = (6.51)(44) = 286 \text{ Hz} \tag{9.29}$$

This is the origin of the value given in Table 9.2.

EXAMPLE 9-2

The NMR of the F^{19} and P^{31} nuclei in liquid PF_3 have been studied by Gutowsky, McCall, and Slichter.[7] As expected the P^{31} resonance appears

[7] H. S. Gutowsky, D. W. McCall, and C. P. Slichter, *J. Chem. Phys.* **21,** 279 (1953).

as a 1:3:3:1 quartet, with spacing 1410 Hz. The F^{19} resonance is a 1:1 doublet spaced at 1400 Hz. Thus we conclude that $J_{PF} = 1405 \pm 5$ Hz. Note clearly that this very simple spectrum proves the equivalence of the three fluorines, but does not distinguish between C_{3v} or D_3 symmetries which are both consistent with the observations.

EXAMPLE 9-3

Figure 9.12 shows a sketch of the N^{14} and proton resonances of pure NH_3. Note that the expected 1:1:1 triplet proton resonance is distorted

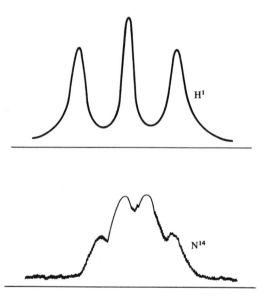

FIG. 9.12. N^{14} and proton resonances of pure NH_3 in liquid phase. From R. A. Ogg and J. D. Ray, *J. Chem. Phys.* **26**, 1515 (1957). With permission.

due to the N^{14} quadrupole moment. The same effect has badly broadened the 1:3:3:1 N^{14} resonance. Nevertheless the spacings of both multiplets are consistent with $J_{NH} = 46$ Hz.

EXAMPLE 9-4

Figure 9.13 shows the F^{19} NMR spectrum of F_2—C=CFCl at 30 MHz. Although considerable intensity distortion occurs, the spectrum is rather well-described as an AMX first-order type. Numbering the fluorine nuclei as in the figure we find $J_{12} = 78$ Hz, $J_{13} = 58$ Hz, and $J_{23} = 115$ Hz.

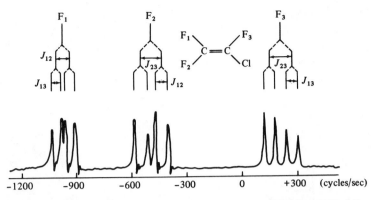

FIG. 9.13. F^{19} spectrum of $CF_2 = CFCl$ at 30 MHz. Chemical shifts are relative to C_4F_8. From J. N. Shoolery, *Varian Technical Information Quarterly* **1**, 3 (1955). With permission.

EXAMPLE 9-5

Finally we illustrate the qualitative structural capabilities of PMR by means of the spectrum (Fig. 9.14) of ethyl N-methyl carbamate,

$$CH_3—N \overset{\displaystyle H}{\underset{\displaystyle \overset{C—O—CH_2}{\underset{\displaystyle O}{\parallel}}}{}} CH_3$$

Four distinct proton resonances are expected and we find four resonances centered at δ values of 5.16, 4.14, 2.78 and 1.23 ppm. The broad low-field absorption at 5.16 ppm is characteristic of quadrupole-broadened NH proton resonances and is so assigned. Note that this broadened peak shows no splitting due to nitrogen or methyl proton coupling. The high-field resonance at 1.23 ppm is appropriate (Fig. 9.6) for an aliphatic methyl group split by two adjacent methylene protons. The methylene resonance must, in turn, be split by the methyl protons into a quartet, which permits us to assign the $\delta = 4.14$ ppm resonance to the CH_2 group. Further confirmation of this assignment arises by noting that the chemical shift of this —OCH_2— group is similar to that in ethanol (Fig. 9.5). The remaining doublet at 2.78 ppm is assignable to the N-methyl protons, the doublet splitting being produced by coupling with the NH group proton. Note that this coupling appears on the N-methyl resonance while it did not for the NH resonance due to the quadrupolar broadening. Thus the spectrum can be satisfactorily described as $AM_2P_3X_3$ where A = NH proton, M_2 = CH_2,

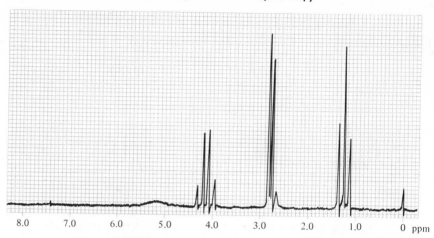

8.0   7.0   6.0   5.0   4.0   3.0   2.0   1.0   0 ppm

FIG. 9.14.   60 MHz spectrum of ethyl N-methyl carbamate (in
CDCl$_3$). 500 Hz sweep width. From "High Resolution
NMR Spectra Catalogue," Vol. 1 (Palo Alto, Calif.:
Varian Associates, 1962). With permission.

$P_3 =$ N—CH$_3$, and $X_3 =$ aliphatic CH$_3$. The small resonances at $\delta = 7.41$
and 2.67 ppm are due to CHCl$_3$ and an unknown impurity. The former is
understandable since the spectrum was run using CDCl$_3$ as the solvent.

9.4   TIME-DEPENDENT PHENOMENA

As with all forms of spectroscopy, the single most important time-
dependent phenomenon is the absorption of electromagnetic radiation,
which leads in NMR spectroscopy to transitions among the magnetic
energy states. However, in addition to this aspect, NMR spectroscopy
shows clearly several other spectral features which may be broadly classed
as time-dependent, or kinetic phenomena. These include relaxation[8] and
exchange phenomena, and in addition, double resonance (spin decoupling)
techniques are included in this category. Because these various processes
play an important role in the experiments, and yield information of chemical interest, we discuss the principles involved in at least a semiquantitative sense.

A.   Selection Rules

As described earlier, the electromagnetic radiation of frequency ν
is applied perpendicularly to the strong static magnetic field H_0. The inter-

[8] Relaxation phenomena may be classed broadly as nonradiative energy-transfer
processes. They must occur in IR and microwave spectral studies also, but their role
is of less general importance to the spectral interpretation.

action Hamiltonian is now of the *magnetic dipole* type, and is entirely analogous to that described in Sec. 5.3 for electric dipole transitions. Thus instead of Eq. (5.30) we have for the magnetic field part of the electromagnetic radiation

$$H = H_X = 2H_1 \cos 2\pi\nu t \tag{9.30}$$

Note that the magnetic field is polarized in the X (or Y) direction since the Z direction is determined by the strong field H_0. The time-dependent perturbation Hamiltonian is then of the form of Eq. (5.31b). For the present case, if we have only one magnetic nucleus per molecule,

$$\tilde{\mathcal{3C}}'(t) = -2H_1\tilde{\mu}_X \cos 2\pi\nu t \tag{9.31}$$

where we have only the X component of the magnetic moment. The transition probability $m \to n$ at the peak of the absorption line ($\nu = \nu_0$) is then similar to Eq. (5.37), namely

$$P_{mn} = \frac{H_1^2}{\hbar^2} \mid \langle m \mid \tilde{\mu}_X \mid n \rangle \mid^2 \tag{9.32}$$

(The factor of three has been omitted since $H_1^2 = H_X^2$). Thus our selection rule problem involves only one simple matrix element. Using Eq. (9.4a), we finally get

$$P_{mn} = \frac{H_1^2\gamma^2}{\hbar^2} \mid \langle m \mid \tilde{I}_X \mid n \rangle \mid^2 \tag{9.33a}$$

From our well-known earlier results (see Table 4.1) we know that the matrix element in Eq. (9.33a) is zero unless $n = m \pm 1$. Our selection rule for a single nucleus is therefore

$$\Delta M_I = \pm 1 \tag{9.34a}$$

This is the result which we have already utilized several times.

Suppose now we have a set of several equivalent nuclei. Then instead of Eq. (9.33a) we have

$$P_{mn} = \frac{H_1^2}{\hbar^2} \gamma^2 \sum_i \mid \tilde{I}_X(i) \mid_{mn}^2 \tag{9.33b}$$

In Sec. 9.5 we shall find the precise form of the wave functions so that the matrix elements in Eq. (9.33b) can be evaluated exactly. As a first approximation we note that if the nuclei are only weakly coupled by spin-spin interaction, each nucleus behaves independently so Eq. (9.34a) applies for each nucleus separately. But since $\tilde{I}_X(i)$ is a one-particle operator, only *one* nucleus at a time may undergo the transition given by Eq. (9.34a).

A more rigorous selection rule can be written down in terms of *total* nuclear spin angular momentum. The total angular momenta are defined for the equivalent nuclei by

$$\tilde{F} = \sum_i \tilde{I}(i)$$

$$\tilde{F}_Z = \sum_i \tilde{I}_Z(i), \text{ etc.} \qquad (9.35)$$

It is shown later that M_F, the eigenvalue of $\tilde{F}_Z$, is rigorously a good quantum number, so

$$\Delta M_F = \pm 1 \qquad (9.34b)$$

is a necessary (but not sufficient) condition for a transition.

If several sets of equivalent nuclei exist, the transition probability is

$$P_{mn} = \frac{H_1^2}{\hbar^2} \{ \gamma_A^2 \sum_i | \tilde{I}_X(A)_i |_{mn}^2 + \gamma_B^2 \sum_i | \tilde{I}_X(B)_i |_{mn}^2$$
$$+ \cdots + \gamma_X^2 \sum_i | \tilde{I}_X(X)_i |_{mn}^2 + \cdots \} \qquad (9.33c)$$

For this situation the selection rule of Eq. (9.34b) applies rigorously if $\tilde{F}$, $\tilde{F}_Z$, etc. are defined as the totals over *all* sets. A less rigorous, but very often applicable rule is that

$$\Delta M_F(X) = \pm 1 \qquad (9.34c)$$

must apply for each set X separately, where now $\tilde{F}$, $\tilde{F}_Z$, etc., are defined for each set. To the approximation that Eq. (9.34c) is true, it is also true that only one set at a time may undergo the change given by that rule.

Later we find that group-theoretical selection rules are an important aid to spectral analyses of molecules having symmetrical nuclear arrangements. For the NMR case, transitions may occur if

$$\Gamma^{(m)} \times \Gamma^{(\mu_X)} \times \Gamma^{(n)} \to \Gamma^{(A)} \qquad (9.36)$$

where $\Gamma^{(\mu_X)}$ is the representation of the X component of the nuclear magnetic moment operator. Whereas the components of the *electric* dipole moment operator transform like X, Y, and Z, the *nuclear* magnetic moment operators have much simpler transformation properties. These magnetic moment operators are strictly nuclear properties, and since the group operations (the group of Schrödinger's equation) simply interchange equivalent nuclei the total magnetic moment operator is invariant to all the point-group operations. Suppose, for example that there were three equivalent nuclei. Then under a C_3 operation

$$\tilde{C}_3 \tilde{\mu}_X = \tilde{C}_3 (\tilde{\mu}_{X1} + \tilde{\mu}_{X2} + \tilde{\mu}_{X3})$$
$$= (\tilde{\mu}_{X2} + \tilde{\mu}_{X3} + \tilde{\mu}_{X1})$$
$$= \tilde{\mu}_X \qquad (9.37)$$

which shows that $\tilde{\mu}_X$ is invariant to the C_3 operation. Thus we conclude that $\Gamma^{(\mu x)} = \Gamma^{(A)}$, the totally symmetrical representation. The requirement of Eq. (9.36) therefore simplifies to

$$\Gamma^{(m)} \times \Gamma^{(n)} \rightarrow \Gamma^{(A)} \qquad (9.38a)$$

or

$$\Gamma^{(m)} = \Gamma^{(n)} \qquad (9.38b)$$

Equation (9.38b) states that *transitions may occur only between states of the same symmetry*.

Thus the selection-rule theory is quite simple for ordinary NMR spectroscopy. We should perhaps mention that the rule which requires that ΔM_F change for only one set of nuclei at a time, is not strictly required if the coupling between sets of nuclei is large compared to the separation of the sets. For our purposes the rule is satisfactory, however. A breakdown of the rule normally results in only very weak additional transitions which are not detected.

B. *Thermal Equilibration and Relaxation Times in NMR Spectroscopy*

For simplicity, in this section we limit our analysis to a two-state system such as exists for a single proton or any other spin-$\frac{1}{2}$ nucleus. In the absence of the strong external field H_0, the two states $M_I = +\frac{1}{2}$ and $M_I = -\frac{1}{2}$ are degenerate and are equally populated at thermal equilibrium. After the field H_0 is applied, the population ratio is given by Boltzmann's equation

$$\frac{n_-}{n_+} = \exp(-g_n\beta_n H_0/kT) \qquad (9.39a)$$

At 300°K the value of the exponent for the proton is

$$\frac{5.585(5.050 \times 10^{-24})(10^4)}{(1.380 \times 10^{-16})(300)} = 6.82 \times 10^{-6}$$

in a field of 10^4 G. Thus the populations differ very little and we may expand Eq. (9.39a) to give

$$\frac{n_-}{n_+} = 1 - \frac{g_n\beta_n H_0}{kT} \qquad (9.39b)$$

The difference in population of the lower and upper states can be expressed as

$$n = n_+ - n_- = \frac{g_n\beta_n H_0}{kT}n_+ \qquad (9.40)$$

Before considering the effect of the absorption of electromagnetic radiation upon the populations it is valuable to ask what happens if the field H_0 is now turned off. The answer is, of course, that the system of α and β nuclei gain or lose energy as necessary to return to the field-free 50:50 distribution. The fact that this occurs demands that there be radiationless energy transfer processes involving transfer of the Zeeman energy into translation, vibration, and rotation modes of the molecular system, and vice versa. These processes are always occurring, so α and β spins are always interconverting when the field H_0 is on. We call these processes *spin-lattice* relaxation.

If W_u and W_d (see Fig. 9.15) are the radiationless transition probabilities up and down, respectively, at thermal equilibrium we require

$$n_+ W_u = n_- W_d \tag{9.41}$$

in the absence of electromagnetic radiation. Since n_+ and n_- are nearly identical, we note that $W_u \sim W_d$. We define therefore

$$W = W_u \cong W_d \tag{9.42}$$

We desire now an expression which gives the rate of change of the population difference under nonequilibrium conditions. This may be written as the rate of decrease of n

$$\frac{dn}{dt} = 2(n_- W_d - n_+ W_u) \tag{9.43}$$

in which the factor of 2 accounts for the fact that any transition changes n

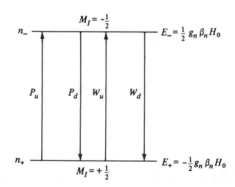

FIG. 9.15.   Energy transfer in a two-level system. P_u and P_d are stimulated absorption and emission probabilities and W_u and W_d are nonradiative transition probabilities.

by 2. By combining Eqs. (9.39b) and (9.41) we obtain

$$W_d = W_u \left(1 + \frac{g_n \beta_n H_0}{kT} \right) \qquad (9.44)$$

which when substituted into Eq. (9.43) yields

$$\frac{dn}{dt} = 2W_u \left(n_- - n_+ + \frac{n_- g_n \beta_n H_0}{kT} \right)$$

$$= 2W \left(-n + \frac{n_- g_n \beta_n H_0}{kT} \right) \qquad (9.45a)$$

In the last equality we have used Eqs. (9.40) and (9.42) also.

If the system were in thermal equilibrium, the population difference would be given by Eq. (9.40). As previously described, $n_- \sim n_+$ so we may write

$$n_{eq} = \frac{n_- g_n \beta_n H_0}{kT} \qquad (9.46)$$

Then Eq. (9.45a) becomes

$$\frac{dn}{dt} = 2W (n_{eq} - n) \qquad (9.45b)$$

Integration of this expression gives

$$n - n_{eq} = (n - n_{eq})_0 \exp(-2Wt) \qquad (9.47)$$

where $(n - n_{eq})_0$ is the value of $n - n_{eq}$ at $t = 0$. Equation (9.47) specifies the rate at which the system would return to equilibrium if it (the system) were initially disturbed from equilibrium. The characteristic time required for the value of $n - n_{eq}$ to become $1/e$ of the equilibrium value is $1/2W$. We call this decay time the spin-lattice relaxation time T_1.

$$T_1 = 1/2W \qquad (9.48)$$

For the liquids, the proton relaxation time, T_1 is in the range 10^{-2} to 10^2 sec. Typically then, it takes about one second for the nuclear spin system to respond to any perturbation which tends to shift the equilibrium population.

Turning to the case where transitions are occurring due to the presence of electromagnetic radiation, the rate of decrease of n_+ is $P_u n_+ - P_d n_-$ and the rate of decrease of n_- is $P_d n_- - P_u n_+$. Since the transition probabilities up and down are identical (see chapter 5), the net rate of decrease of n

due to stimulated transitions is

$$\frac{dn}{dt} = -2Pn \tag{9.49}$$

where $P = P_u = P_d$. It is clear from this expression that the effect of the induced transitions is to cause the normal population difference to go to zero, at which point no more radiation could be absorbed. But of course the relaxation processes tend to work in the opposite fashion so if they are faster than the induced transitions radiation is continuously absorbed. The over-all result is obtained by combining Eqs. (9.45b) and (9.49) to give

$$\frac{dn}{dt} = -\frac{(n - n_{eq})}{T_1} - 2Pn$$

$$= \frac{1 + 2PT_1}{T_1}\left(-n + \frac{n_{eq}}{1 + 2PT_1}\right) \tag{9.50}$$

We can now see clearly the circumstances under which the electromagnetic energy absorption process becomes inefficient. For steady-state conditions $(dn/dt = 0)$, the population difference is

$$n = \frac{n_{eq}}{1 + 2PT_1} \tag{9.51}$$

Hence when $2PT_1 \gg 1$, $n \ll n_{eq}$, which means that the upper and lower states approach equal populations. This condition is known as *saturation*. Since spin-lattice relaxation times are usually fixed properties of the system under study, we can be certain that saturation is avoided by keeping the transition probability sufficiently low. From Eq. (9.33a)

$$P = H_1^2\gamma^2(\tfrac{1}{4})$$

using the well-known matrix element

$$| \langle m | \tilde{I}_X | n \rangle |^2 = \tfrac{1}{4}\hbar^2.$$

From Table 9.1, $\gamma_H = 2.7 \times 10^4$ G^{-1} sec^{-1}, so

$$2PT_1 \sim 2H_1^2(2.7 \times 10^4)^2(\tfrac{1}{4})(1) \sim 3.6 \times 10^8 H_1^2$$

for $T_1 = 1$ sec.[9] Thus saturation becomes important when $H_1 \sim 10^{-4}$ G.

---

[9] As described in chapter 5, P has units of sec^{-2} when defined as in Eq. (5.37). When multiplied by the line shape function $S(\nu)$ as in Eq. (5.57a) we obtain transition probability per second. Hence our calculation is strictly correct for $S(\nu = \nu_0) = 1$ sec, which by Eq. (5.59) corresponds to a linewidth of about 1 Hz.

In some experiments (double-resonance, for example) this limit is intentionally exceeded, but in the ordinary liquid phase NMR spectroscopy this value serves as a practical upper limit.

C. Linewidths and Lineshapes

Since the natural radiative lifetime of the excited Zeeman magnetic energy states is very long[10] ($\tau \gg 1$ sec), this contribution ($\sim 1/\tau$) to the NMR linewidth is unimportant. One might expect, however, that the spin-lattice relaxation time T_1 would be one of the major contributors to the linewidth since $T_1 \sim 1$ sec. Indeed this is the case, at least qualitatively. The theory of NMR linewidths has been beautifully formulated by Bloch.[11] Bloch's theory shows that two distinct relaxation times are involved, one of which is T_1; the second is given the symbol T_2 and is known as the *transverse* or *spin-spin* relaxation time. In Bloch's theory the former relaxation time describes the relaxation of the Z component of the magnetic moment of the macroscopic sample while T_2 describes the same effect for the X and Y directions. The details need not concern us here, but it is useful to point out that for systems in which the molecules tumble (or rotate) at a rate on the order of 10^{10} per sec or faster, T_1 and T_2 become identical. This is the usual situation for (nonviscous) liquids and gases, but not for solids. The rapid tumbling averages out the magnetic interactions so that the relaxation mechanism becomes identical in the X, Y, and Z directions.

In the absence of power saturation effects, the Bloch theory predicts a Lorentzian lineshape function as given by Eq. (5.59), and illustrated in Fig. 5.3. Explicitly, it is found that the line half-width is

$$\delta \nu = 1/2\pi T_2 \tag{9.52}$$

so Eq. (5.59) becomes

$$S(\nu) = 4T_2/(4\pi^2 T_2^2 (\nu - \nu_0)^2 + 1) \tag{9.53}$$

Thus in the absence of other larger broadening effects, the half-width of an NMR absorption line is $1/2\pi T_2 \cong 1/2\pi T_1$.

D. Nuclear Spin Relaxation

The microscopic theories of nuclear spin relaxation are too complex for us to treat in this book. However, the qualitative physical principles involved in the time-dependent perturbation-theory calculations are easily understood. In the absence of other relaxation phenomena, T_1 and T_2 arise because of the well-known magnetic dipole-dipole interaction, which can

---

[10] See, for example, N. Bloembergen, E. M. Purcell, and R. V. Pound, *Phys. Rev.* **73,** 679 (1948).

[11] F. Bloch, *Phys. Rev.* **70,** 460 (1946).

be written

$$\widetilde{\mathfrak{IC}}_D = \frac{\mu_1 \cdot \mu_2}{r^3} - \frac{3(\mu_1 \cdot \mathbf{r})(\mu_2 \cdot \mathbf{r})}{r^5} \tag{9.54}$$

in which $\mathbf{r}$ is the relative position vector of the two dipoles. The important point with regard to Eq. (9.54) is that $\widetilde{\mathfrak{IC}}_D$ is a function of time because of the tumbling, diffusional-type motion of the molecules in the liquid (or gas) phase.

The calculational problem, then, is simply to calculate the transition probability(ies) between the magnetic sublevels in the presence of the perturbation $\widetilde{\mathfrak{IC}}_D(t)$. This calculation proceeds basically in the manner outlined in chapter 5. The major new feature which must be added is the proper stochastic description of the randomly fluctuating nature of $\widetilde{\mathfrak{IC}}_D(t)$. Once these (radiationless) transition probabilities are calculated it is a relatively simple matter to evaluate T_1 (and T_2) by an extension of the methods used in Sec. 9.4 B. Such a calculation may lead in a simple case to an expression which relates T_1 (and T_2) to such measureable quantities as viscosity or diffusion coefficient.[12]

Fluctuating energy terms other than the dipole–dipole type [Eq. (9.54)] can be treated in an identical fashion. Any term which has nonzero matrix elements connecting the magnetic sub-states has the possibility of inducing radiationless transitions and therefore may contribute to T_1 and T_2. One of the important mechanisms is the interaction of the nuclear electric quadrupole moment with the electric field gradient. This is often the dominant linebroadening mechanism for nuclei with spin $I \geqq 1$, such as N^{14} or Cl^{35}. The transition probabilities for this mechanism are in general larger than for the dipole–dipole mechanism, so T_1 and T_2 are smaller. This is the reason why the linewidths of the NMR resonances of N^{14} (see Fig. 9.12) and other quadrupolar nuclei are very large for most molecules.

E. Exchange Processes

In addition to the relaxation processes previously described, there are many interesting exchange processes which have a marked effect on the shapes, intensities, and positions of NMR absorption lines. We have mentioned previously two of the more common results of such processes, namely, (1) the chemical equivalence of the protons of methyl groups in asymmetric molecules,[13] and (2) the apparent lack of spin-spin coupling of the OH proton in acidified alcohols. The qualitative explanation of the former effect is that the methyl group rotates so rapidly that during the measure-

---

[12] A particularly lucid and simple calculation of T_1 and T_2 for water is given by A. Carrington and A. D. McLachlan, *Introduction to Magnetic Resonance* (Harper and Row, New York, 1967), chapter 11.

[13] We describe this loosely as an "exchange" process although there is really no proton exchange occurring. It might be called more properly an "interchange" process.

ment period the protons pass through all environments many times. Thus the protons behave as though they were chemically equivalent; for the same reason they show magnetic equivalence, also. In the alcohol case, the spin-spin splitting disappears because the OH protons are being transferred rapidly from molecule to molecule. In acidic solution the reaction might be written

$$R\!-\!OH + \left[\begin{array}{c} H \quad\quad H \\ \diagdown\quad\diagup \\ O \\ | \\ H \end{array}\right]^{+} \rightleftarrows R\!-\!OH_2^+ + H_2O$$

Thus during the time of a measurement many protons, each with random M_I values ($\tfrac{1}{2}$ or $-\tfrac{1}{2}$ in this case), jump on and off the molecule. The protons in the group R feel no distinct spin orientations of the OH proton and therefore are not split into a doublet as expected. By similar reasoning the OH proton does not stay on a given molecule long enough to feel the fixed spin orientations of the R-group protons and thus the OH multiplet disappears also.

Both of these cases represent examples of *fast exchange*. Experimentally, fast exchange occurs when the exchange rate ν_E is much greater than the separation between the absorption lines which *would exist* if exchange were not occurring. In the case of ethanol [Fig. 9.11(b)], the OH triplet splittings of the neat sample were 5 Hz. We may conclude then that the exchange rate is greater than 5 per sec.

McConnell[14] has developed a quantitative theory for the case where chemical exchange of a nucleus occurs between two sites having different resonant frequencies. The two sites may be in the same molecule or in two different molecules. The results are rather complicated in the general case, but with the following assumptions a useful equation results.

(1) The frequency of exchange of a nucleus from site A to site B, ν_{AB}, is identical to the reverse exchange, ν_{BA}. Thus $\nu_{AB} = \nu_{BA} = \nu_E$.
(2) The transverse relaxation times are very long so that $1/T_{2_A} = 1/T_{2_B} = 0$.

The Bloch equations[15] modified for exchange lead to the following form for the line-shape function

$$S(\nu) = N\left(\frac{\nu_E(\nu_A - \nu_B)^2}{4\nu_E^2(\bar{\nu} - \nu)^2 + 4\pi^2(\nu_A - \nu)^2(\nu_B - \nu)^2}\right) \quad (9.55)$$

N is a normalization constant, ν_A and ν_B are the resonant frequencies of the

[14] H. M. McConnell, *J. Chem. Phys.* **28**, 430 (1958).
[15] See footnote 11.

nuclei on the two sites under conditions of no or slow exchange, $\bar{\nu}$ is the mean of ν_A and ν_B and ν_E is the jump (exchange) frequency. ν_E may also be considered to be the transition probability per second, and ν_E^{-1} is the average lifetime of the nucleus on either site.

The line-shape function can be plotted as a function of the dimensionless quantity

$$\frac{\pi \mid \nu_A - \nu_B \mid}{\nu_E} = \frac{\pi \Delta \nu}{\nu_E}$$

as shown in Fig. 9.16. When the exchange rate is very slow ($\nu_E < 1$), Eq. (9.55) shows that we obtain two Lorentzian lines separated by $\mid \nu_A - \nu_B \mid$. For fast exchange ($\nu_E > 1$) a single Lorentzian occurs at the mean frequency $\bar{\nu}$. At intermediate exchange rates overlapping Lorentzian lines are observed. For analysis purposes it is useful to consider the condition when the two curves merge to produce a flat-topped line shape. This occurs when

$$\frac{\pi \Delta \nu}{\nu_E} = \sqrt{2}$$

so the exchange rate is given by

$$\nu_E = \frac{\pi \Delta \nu}{\sqrt{2}} \tag{9.56}$$

At faster rates a single line is observed while slower rates produce a doublet.

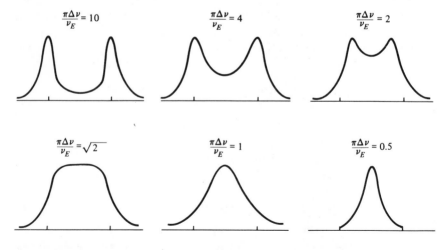

FIG. 9.16.   Lineshape function for different values of the parameter $\pi \mid \nu_A - \nu_B \mid / \nu_E$. Each drawing has been normalized to the same peak height.

The same type of theory may be extended to cases of chemical exchange among more than two sites, but of course the equations become more complex. We are content merely to illustrate the simplest cases.

Some of the more interesting exchange phenomena are those involving hindered internal rotation. Since there is a potential barrier hindering the internal rotation, the exchange rate must follow a simple Arrhenius-type expression

$$\nu_E = \nu_E{}^0 \exp(-E_a/kT) \tag{9.57}$$

By varying the temperature it is often possible to observe the complete range of exchange rates as illustrated in Fig. 9.16. An interesting example of this is N,N dimethylformamide

$$
\begin{array}{ccc}
\text{H} & & \text{CH}_3 \\
\diagdown & & \diagup \\
& \text{C}\!-\!\text{N} & \\
\diagup\!\diagup & & \diagdown \\
\text{O} & & \text{CH}_3
\end{array}
$$

At room temperature the two methyl groups are not equivalent and give rise to sharp resonances at $\delta = 2.95$ and $\delta = 2.80$ ppm. As the temperature is increased the exchange rate increases by virtue of Eq. (9.57), and as a consequence the resonances broaden and begin to coalesce. Finally at 11°C a flat-topped curve characterized by the condition given by Eq. (9.56) is observed. At still higher temperatures a single sharp resonance occurs at the mean frequency ($\delta = 2.875$ ppm). Evaluation of ν_E at several temperatures leads via Eq. (9.57) to a barrier to internal rotation of 15 kcal/mole.

In general, the NMR technique is useful for measuring barriers in the range of about 5 to 25 kcal/mole. This is a rather fortunate development, since the microwave method (Sec. 7.8) is appropriate for the region below 5 kcal/mole. An entirely similar method has been used to study the barrier to inversion of cyclic nitrogen containing molecules, such as 1,2,2-trimethylaziridine,

$$
\begin{array}{ccc}
& \text{H} & \\
& \diagdown & \\
\text{H}\!-\!\!-\!\!-\!\text{C} & & \\
& \diagdown & \\
& & \text{N}\!-\!\text{CH}_3 \\
& \diagup & \\
\text{CH}_3\!-\!\!-\!\!-\!\text{C} & & \\
& \diagup & \\
& \text{CH}_3 &
\end{array}
$$

for which an activation energy of 24 kcal has been obtained.[16]

[16] M. Jautelat and J. D. Roberts, *J. Am. Chem. Soc.* **91,** 642 (1968).

EXAMPLE 9-6

In Fig. 9.17 we show the methyl resonances of N,N-dimethylform-amide as a function of temperature. To determine the barrier to internal rotation we need to determine ν_E at each temperature. This is best done by fitting the experimental curves to Eq. (9.55) by means of digital computer calculations. We can perform the calculation somewhat more crudely in the following way. First we note that below 84°C the separation of the two peaks remains constant at 0.16 ppm or

$$\Delta\nu = \nu_A - \nu_B = 0.16 \times 10^{-6}\ (60 \times 10^6)$$

$$= 9.6\ \text{Hz}$$

We note also that the coalescence temperature is 111°C, thus by use of Eq. (9.56) we find

$$\nu_E(111°\text{C}) = \frac{3.14(9.6)}{1.41} = 21.4\ \text{sec}^{-1}$$

It is more difficult to calculate ν_E for the other temperatures. The coalesced singlet lines can be analyzed rather well, however, in terms of the linewidths. From Eq. (9.55), one-half the peak intensity at $\nu = \bar{\nu}$ is given by

$$\tfrac{1}{2}S(\nu = \bar{\nu}) = N\left\{\frac{2\nu_E}{(\Delta\nu)^2\pi^2}\right\} \tag{9.58}$$

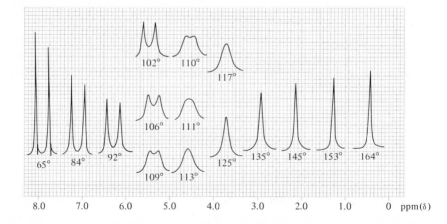

FIG. 9.17. Methyl resonance of N-N dimethyl formamide at various temperatures. From John R. Dyer, *Applications of Absorption Spectroscopy of Organic Compounds*, © 1965. Reprinted by permission of Prentice-Hall, Inc., Englewood Cliffs, N. J.

The half-width of the line for the fast exchange case is then gotten by equating Eq. (9.58) to the right-hand side of Eq. (9.55).

$$N \left\{ \frac{2\nu_E}{(\Delta\nu)^2 \pi^2} \right\} \cong \left\{ \frac{\nu_E (\Delta\nu)^2}{4\nu_E{}^2 (\bar{\nu} - \nu)^2} \right\} N \qquad (9.59)$$

Note we have neglected the second term in the denominator of Eq. (9.55) for the *fast* exchange case. Solving Eq. (9.59) we get

$$\bar{\nu} - \nu = \frac{(\Delta\nu)^2 \pi}{2\sqrt{2}\ \nu_E} = \delta\nu_{1/2} \qquad (9.60)$$

As a first approximation we might apply Eq. (9.60) to the 125°C linewidth. At lower temperatures it is unlikely that the fast exchange approximation holds. From the spectrum, $2\delta\nu_{1/2}(125°C) = 4.5$ Hz, so by use of Eq. (9.60) we find

$$\nu_E(125°C) = \frac{(9.6)^2 (3.14)}{(1.41)(4.5)} = 45.5 \text{ sec}^{-1}$$

Applying the same method to the 135°C singlet, we find (using $2\delta\nu_{1/2}$ (135°C) = 3.0 Hz)

$$\nu_E(135°C) = \frac{(9.6)^2 (3.14)}{(1.41)(3.0)} = 68.3 \text{ sec}^{-1}$$

Using these values of $\nu_E(T)$ we can construct the graph shown in Fig. 9.18. From the slope of this curve we calculate $E_a = 15$ kcal/mole, which should be the barrier to internal rotation about the C–N bond. This rela-

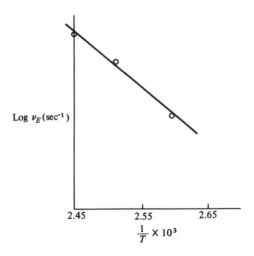

FIG. 9.18.   Exchange rate as a function of temperature for pure dimethylformamide.

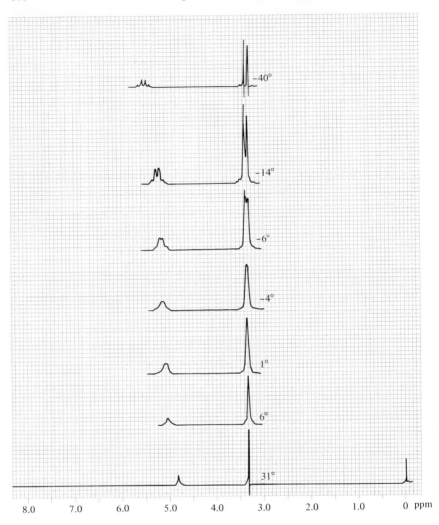

8.0   7.0   6.0   5.0   4.0   3.0   2.0   1.0   0 ppm

FIG. 9.19. Spectrum of pure methanol at different temperatures. 500 Hz sweep width. From John R. Dyer, *Applications of Absorption Spectroscopy of Organic Compounds*, © 1965. Reprinted by permission of Prentice-Hall, Inc., Englewood Cliffs, N. J.

tively large barrier is produced by the double bond character of the amide CN bond as shown by

In addition to the *intramolecular* exchange processes such as internal rotation or inversion, *intermolecular* ones are amenable to the same kind of treatments. We have mentioned several times the rapid exchange of protons in acidified alcohols. This proton transfer can be studied even in the neat (pure) alcohols if the temperature is lowered sufficiently to give the slow exchange case. This is illustrated for CH_3OH in Fig. 9.19, where we clearly see the methyl doublet go from fast to slow exchange as the temperature is lowered.

The theory which we have discussed prior to this example has involved a nucleus jumping between two *different* positions, with Eq. (9.55) showing how the resonant frequencies of the *chemically-shifted* nuclei become identical under fast exchange. In this case (CH_3OH) the exchange is between positions having the *same* chemical shift.

$$CH_3OH_a + CH_3OH_b \rightleftarrows CH_3OH_b + CH_3OH_a$$

Nevertheless we observe a behavior for the spin-spin multiplets that is similar to that described for chemically shifted nuclei. In fact, the previously described theory applies to the collapse of the spin-spin doublet since what happens is that the OH proton jumps between states of α and β spin in a random fashion as it is transferred from molecule to molecule. We note that at $-4°C$ the doublet is just coalesced, so by Eq. (9.56) the exchange frequency is

$$\nu_E = \frac{\pi(5.2)}{\sqrt{2}} = 12 \text{ sec}^{-1}$$

where 5.2 is the doublet separation at low temperatures.

Other interesting situations which have been studied involve the rotational isomers in $CH_2P–CQRS$ molecules where P, Q, R, S are nonmagnetic groups. If internal rotation is very slow, we should observe the NMR spectrum of each of the three(stable) isomers:

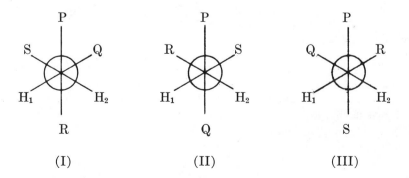

(I)   (II)   (III)

Thus each isomer should show an AB- (or AX-) type spectrum. The appearance of the spectrum of the mixture depends, of course, upon the relative amounts of each isomer. If internal rotation is fast, all three isomers become identical (that is, they are averaged by the rapid exchange) so a single AB spectrum arises. Note that the two protons are not equivalent even under rapid rotation, since in no isomer are they equivalent.

9.5 QUANTUM-MECHANICAL TREATMENT OF COMPLEX NMR SPECTRA IN LIQUIDS

In this section we return to a more detailed description of the theory needed to analyze NMR spectra exhibiting spin-spin hyperfine splitting. We show how the energy levels and transitions are calculated for complex spectra, and how the simple first-order rules arise. As in past sections our discussion does not pertain to solids but is appropriate for liquids and gases.

A. Hamiltonian, Wave Functions, and Matrix Elements

We have seen in previous sections of this chapter that the Hamiltonian is of the form

$$\widetilde{\mathfrak{IC}} = \widetilde{\mathfrak{IC}}^{(0)} + \widetilde{\mathfrak{IC}}^{(1)} \tag{9.61}$$

$\widetilde{\mathfrak{IC}}^{(0)}$ represents the large Zeeman term while $\widetilde{\mathfrak{IC}}^{(1)}$ represents the smaller spin-spin energy. For the purposes of this section we can write the Zeeman term generally as

$$\widetilde{\mathfrak{IC}}^{(0)} = -\sum_i \gamma_i \tilde{I}_z(i) H_i \tag{9.62a}$$

where the sum is over all magnetic nuclei and H_i is the *effective* magnetic field felt by nucleus i. It is convenient for our work here to convert energies in ergs directly to energies in $\sec^{-1}$. Since the angular momenta $(\tilde{I}, \tilde{I}_x, \text{etc})$ come in units of $\hbar$, we can pull this factor out of $\tilde{I}_z(i)$ immediately. Then dividing by h we get

$$\widetilde{\mathfrak{IC}}^{(0)} = -\frac{1}{2\pi} \sum_i \gamma_i \tilde{I}_z(i) H_i \sec^{-1} \tag{9.62b}$$

in which it must be remembered that the $\hbar$'s are to be suppressed when the matrix elements of $\tilde{I}_z(i)$ are evaluated.

The spin-spin term $\widetilde{\mathfrak{IC}}^{(1)}$ may be written

$$\widetilde{\mathfrak{IC}}^{(1)} = \sum_{i<j} J_{ij} \tilde{I}(i) \cdot \tilde{I}(j) \tag{9.63}$$

We suppress the $\hbar$'s completely for the matrix elements of $\tilde{I}$ (as above) and consider J_{ij} to have units of $\sec^{-1}$. Note that the sum is written such that all pairwise interactions are counted only once.

Our job is straightforward and by now well known in principle. We obtain the energy levels by forming the Hamiltonian matrix $\mathfrak{IC}$ and by bringing it to diagonal form by our established methods. In order to calculate the Hamiltonian matrix elements we need to choose a representation or basis set with which to work. The simplest and most logical choice is to use the spin-product functions of the form

$$| n \rangle = \phi(1)\phi(2)\phi(3)\ldots\phi(p) \tag{9.64}$$

where each $\phi(i)$ is one of the $2I(i) + 1$ eigenfunctions of $\tilde{I}^2(i)$ [and $\tilde{I}_Z(i)$] for nucleus i. This choice is appropriate since in the absence of $\tilde{\mathfrak{K}}^{(1)}$ these would be the correct eigenfunctions of $\tilde{\mathfrak{K}} = \tilde{\mathfrak{K}}^{(0)}$. As examples, if the system of interest contained two N^{14} nuclei only, the appropriate basis functions would be

$$\phi_1(1)\phi_1(2) \qquad \phi_1(1)\phi_0(2) \qquad \phi_1(1)\phi_{-1}(2)$$

$$\phi_0(1)\phi_1(2) \qquad \phi_0(1)\phi_0(2) \qquad \phi_0(1)\phi_{-1}(2)$$

$$\phi_{-1}(1)\phi_1(2) \qquad \phi_{-1}(1)\phi_0(2) \qquad \phi_{-1}(1)\phi_{-1}(2)$$

while a system consisting of two protons would utilize

$$\alpha(1)\alpha(2) \qquad \alpha(1)\beta(2)$$

$$\beta(1)\alpha(2) \qquad \beta(1)\beta(2)$$

In this section we treat almost exclusively systems of spin-$\frac{1}{2}$ nuclei, which include protons, F^{19}, C^{13}, etc. Thus for spin-$\frac{1}{2}$ systems there are always 2^p basis functions of the form of Eq. (9.64), and $\mathfrak{K}$ is of dimension 2^p also.

As we now show, it is a straightforward, but somewhat tedious task to evaluate the matrix elements of $\tilde{\mathfrak{K}}$ in the spin-product basis. For $I = \frac{1}{2}$ nuclei, we need only those angular momentum matrix elements given by the Pauli matrices, Eq. (4.96). For convenience we list them here with the $\hbar$ suppressed.

$$\mathbf{I}_X = \frac{1}{2}\begin{pmatrix} 0 & 1 \\ 1 & 0 \end{pmatrix} \qquad \mathbf{I}_Y = \frac{1}{2}\begin{pmatrix} 0 & -i \\ i & 0 \end{pmatrix} \qquad \mathbf{I}_Z = \frac{1}{2}\begin{pmatrix} 1 & 0 \\ 0 & -1 \end{pmatrix} \tag{9.65}$$

Matrix elements of $\tilde{\mathfrak{K}}^{(0)}$.

It is easy to show that $\tilde{\mathfrak{K}}^{(0)}$ has only diagonal elements by operating upon $| n \rangle$. We get

$$\tilde{\mathfrak{K}}^{(0)} | n \rangle = -\frac{1}{2\pi}\sum_i \gamma_i H_i \tilde{I}_Z(i)[\phi(1)\phi(2)\ldots\phi(p)]$$

$$= -\frac{1}{2\pi}\{\gamma_1 H_1 \tilde{I}_Z(1)[\phi(1)\phi(2)\ldots]$$

$$+ \gamma_2 H_2 \tilde{I}_Z(2)[\phi(1)\phi(2)\ldots] + \ldots\} \tag{9.66}$$

$$= -\frac{1}{2\pi}\{\gamma_1 H_1 M_I(1) | n \rangle + \gamma_2 H_2 M_I(2) | n \rangle + \ldots\}$$

$$= -\frac{1}{2\pi}\sum_i \gamma_i H_i M_I(i) | n \rangle$$

since $\tilde{I}_Z(i)$ operates only upon $\phi(i)$ to produce the eigenvalue $M_I(i)$. Thus it is obvious that

$$\langle n \mid \tilde{\mathfrak{IC}}^{(0)} \mid m \rangle = 0 \tag{9.67}$$

and that

$$\langle n \mid \tilde{\mathfrak{IC}}^{(0)} \mid n \rangle = -\frac{1}{2\pi} \sum_i \gamma_i H_i M_I(i) \tag{9.68a}$$

But by Eq. (9.17) we see that $(1/2\pi)\gamma_i H_i$ is the absorption frequency $\nu_0(i)$ considering only the Zeeman effect. We write, therefore,

$$\langle n \mid \tilde{\mathfrak{IC}}^{(0)} \mid n \rangle = -\sum_i \nu_0(i) M_I(i) \tag{9.68b}$$

Later we shall find that $\nu_0(i)$ is always the center of the ith nuclear resonance multiplet.

Diagonal matrix elements of $\tilde{\mathfrak{IC}}^{(1)}$.

The operator terms in $\tilde{\mathfrak{IC}}^{(1)}$ are of the form

$$\tilde{I}(m) \cdot \tilde{I}(n) = \tilde{I}_X(m)\tilde{I}_X(n) + \tilde{I}_Y(m)\tilde{I}_Y(n) + \tilde{I}_Z(m)\tilde{I}_Z(n) \tag{9.69}$$

so they operate only upon the $\phi(m)\phi(n)$ part of the function given by Eq. (9.64). For spin-$\frac{1}{2}$ nuclei there are only four possible product combinations of $\phi(m)\phi(n)$ that can occur, namely, $\alpha(m)\alpha(n)$, $\alpha(m)\beta(n)$, $\beta(m)\alpha(n)$, and $\beta(m)\beta(n)$. Considering only the $ij = mn$ term in the sum of Eq. (9.63), one of the possibilities is evaluated as follows:

$$\langle \phi(1)\ldots\alpha(m)\alpha(n)\ldots \mid J_{mn}\tilde{I}(m)\cdot\tilde{I}(n) \mid \phi(1)\ldots\alpha(m)\alpha(n)\ldots \rangle$$

$$= J_{mn}\{ \langle \alpha \mid \tilde{I}_X(m) \mid \alpha \rangle\langle \alpha \mid \tilde{I}_X(n) \mid \alpha \rangle + \langle \alpha \mid \tilde{I}_Y(m) \mid \alpha \rangle\langle \alpha \mid \tilde{I}_Y(n) \mid \alpha \rangle$$

$$+ \langle \alpha \mid \tilde{I}_Z(m) \mid \alpha \rangle\langle \alpha \mid \tilde{I}_Z(n) \mid \alpha \rangle\} \tag{9.70}$$

As we do often, we have dropped the nuclear labels on the α's when there is no cause for confusion. By use of Eqs. (9.65) we find that the matrix element has the value

$$\langle \ldots\alpha\alpha\ldots \mid J_{mn}\tilde{I}(m)\cdot\tilde{I}(n) \mid \ldots\alpha\alpha\ldots \rangle = \tfrac{1}{4}J_{mn} \tag{9.71a}$$

In an exactly similar manner the other possibilities lead to

$$\langle \ldots\alpha\alpha\ldots \mid \tilde{\mathfrak{IC}}^{(1)}{}_{mn} \mid \ldots\beta\beta\ldots \rangle = \tfrac{1}{4}J_{mn} \tag{9.71b}$$

$$\langle \ldots\alpha\beta\ldots \mid \tilde{\mathfrak{IC}}^{(1)}{}_{mn} \mid \ldots\alpha\beta\ldots \rangle = -\tfrac{1}{4}J_{mn} \tag{9.71c}$$

$$\langle \ldots\beta\alpha\ldots \mid \tilde{\mathfrak{IC}}^{(1)}{}_{mn} \mid \ldots\beta\alpha\ldots \rangle = -\tfrac{1}{4}J_{mn} \tag{9.71d}$$

The complete *diagonal* matrix element of $\tilde{\mathfrak{IC}}^{(1)}$ is therefore simply a sum of $\frac{1}{4}$ or $-\frac{1}{4}$ times J_{ij}. We can write this as

$$\langle n \mid \tilde{\mathfrak{IC}}^{(1)} \mid n \rangle = \tfrac{1}{4} \sum_{i<j} J_{ij}T_{ij} \tag{9.72}$$

where

$$T_{ij} = +1 \quad \text{if} \quad \phi(i) \text{ and } \phi(j) \text{ are identical}$$

$$T_{ij} = -1 \quad \text{if} \quad \phi(i) \text{ and } \phi(j) \text{ are different.}$$

As an example of the use of Eq. (9.72), we evaluate the diagonal element for $|n\rangle = \alpha\beta\alpha$. Thus $\langle\alpha\beta\alpha \mid \widetilde{\mathfrak{IC}}^{(1)} \mid \alpha\beta\alpha\rangle = \frac{1}{4}(-J_{12} - J_{23} + J_{13})$ is obtained by easy application of the formula.

Off-diagonal elements of $\widetilde{\mathfrak{IC}}^{(1)}$.

If we consider a general kl element of $\mathfrak{IC}^{(1)}$, we again find that the operator terms in $\widetilde{\mathfrak{IC}}^{(1)}$ of the form of Eq. (9.69) operate only upon the $\phi(m)\phi(n)$ part of $|l\rangle$. But we know then that the kl element must go to zero because of orthogonality unless the remaining parts of $\langle k|$ and $|l\rangle$ are identical. *We can state, therefore, that*

$$\langle k \mid \widetilde{\mathfrak{IC}}^{(1)} \mid l\rangle = 0$$

if more than two spin functions are changed in $\langle k|$ and $|l\rangle$. For example, for three interchanges,

$$\langle\alpha(1)\beta(2)\alpha(3) \mid \widetilde{\mathfrak{IC}}^{(1)} \mid \beta(1)\alpha(2)\beta(3)\rangle$$

$$= \langle\alpha(1)\beta(2) \mid \widetilde{\mathfrak{IC}}^{(1)}{}_{12} \mid \beta(1)\alpha(2)\rangle\langle\alpha(3) \mid \beta(3)\rangle$$

$$+ \langle\alpha(1)\alpha(3) \mid \widetilde{\mathfrak{IC}}^{(1)}{}_{13} \mid \beta(1)\beta(3)\rangle\cdot\langle\beta(2) \mid \alpha(2)\rangle$$

$$+ \langle\beta(2)\alpha(3) \mid \widetilde{\mathfrak{IC}}^{(1)}{}_{23} \mid \alpha(2)\beta(3)\rangle\langle\alpha(1) \mid \beta(1)\rangle = 0$$

because the overlap integrals go to zero.

Let us consider the case where two or less interchanges are made. One possibility is $\langle\cdot\cdot\alpha\alpha\cdot\cdot \mid \widetilde{\mathfrak{IC}}^{(1)} \mid \cdot\cdot\beta\beta\cdot\cdot\rangle$, which gives

$$\langle\cdot\cdot\alpha\alpha\cdot\cdot \mid \widetilde{\mathfrak{IC}}^{(1)} \mid \cdot\cdot\beta\beta\cdot\cdot\rangle = J_{mn}\{\langle\alpha \mid \tilde{I}_X(m) \mid \beta\rangle\langle\alpha \mid \tilde{I}_X(n) \mid \beta\rangle$$

$$+ \langle\alpha \mid \tilde{I}_Y(m) \mid \beta\rangle\langle\alpha \mid \tilde{I}_Y(n) \mid \beta\rangle + \langle\alpha \mid \tilde{I}_Z(m) \mid \beta\rangle\langle\alpha \mid \tilde{I}_Z(n) \mid \beta\rangle$$

$$\tag{9.73a}$$

Using Eqs. (9.65) we find easily

$$\langle\cdot\cdot\alpha\alpha\cdot\cdot \mid \widetilde{\mathfrak{IC}}^{(1)} \mid \cdot\cdot\beta\beta\cdot\cdot\rangle = 0 \tag{9.73b}$$

In a similar manner it is found that

$$\langle\cdot\cdot\alpha\alpha \mid \widetilde{\mathfrak{IC}}^{(1)} \mid \cdot\cdot\alpha\beta\cdot\cdot\rangle = \langle\cdot\cdot\alpha\alpha \mid \widetilde{\mathfrak{IC}}^{(1)} \mid \cdot\cdot\beta\alpha\cdot\cdot\rangle$$

$$= \langle\cdot\cdot\alpha\beta \mid \widetilde{\mathfrak{IC}}^{(1)} \mid \cdot\cdot\alpha\alpha\cdot\cdot\rangle = \langle\cdot\cdot\alpha\beta \mid \widetilde{\mathfrak{IC}}^{(1)} \mid \cdot\cdot\beta\beta\cdot\cdot\rangle$$

$$= \langle\cdot\cdot\beta\beta \mid \widetilde{\mathfrak{IC}}^{(1)} \mid \cdot\cdot\alpha\beta\cdot\cdot\rangle = \langle\cdot\cdot\beta\beta \mid \widetilde{\mathfrak{IC}}^{(1)} \mid \cdot\cdot\beta\alpha\cdot\cdot\rangle$$

$$= \langle\cdot\cdot\beta\alpha \mid \widetilde{\mathfrak{IC}}^{(1)} \mid \cdot\cdot\alpha\alpha\cdot\cdot\rangle = \langle\cdot\cdot\beta\alpha \mid \widetilde{\mathfrak{IC}}^{(1)} \mid \cdot\cdot\beta\beta\cdot\cdot\rangle = 0 \tag{9.73c}$$

The only two remaining possibilities yield

$$\langle \cdots \alpha\beta \cdots | \widetilde{\mathcal{3C}}^{(1)} | \cdots \beta\alpha \cdots \rangle = \langle \cdots \beta\alpha \cdots | \widetilde{\mathcal{3C}}^{(1)} | \cdots \alpha\beta \cdots \rangle = \tfrac{1}{2}J_{mn} \qquad (9.73d)$$

We see then that the off-diagonal elements are zero except when *one pair* of spin functions in $\langle k |$ is *interchanged* in $| l \rangle$. In general,

$$\langle k | \widetilde{\mathcal{3C}}^{(1)} | l \rangle = \tfrac{1}{2}UJ_{ij} \qquad (9.74)$$

where $U = 1$ if the spins of nuclei i and j are interchanged in $\langle k |$ and $| l \rangle$

$U = 0$ otherwise.

As examples we evaluate the following matrix elements using Eq. (9.74).

$$\langle \alpha\alpha\beta\beta | \widetilde{\mathcal{3C}}^{(1)} | \alpha\beta\alpha\beta \rangle = \tfrac{1}{2}J_{23}$$

$$\langle \alpha\beta\beta\beta | \widetilde{\mathcal{3C}}^{(1)} | \beta\beta\beta\alpha \rangle = \tfrac{1}{2}J_{14}$$

$$\langle \alpha\alpha\beta\beta | \widetilde{\mathcal{3C}}^{(1)} | \alpha\beta\beta\beta \rangle = 0$$

$$\langle \alpha\alpha\beta\beta | \widetilde{\mathcal{3C}}^{(1)} | \beta\beta\alpha\beta \rangle = 0$$

In each of the first two examples one pair of spin functions were interchanged. In the third there was just one change ($\tfrac{1}{2}$ pair) between $\langle k |$ and $| l \rangle$ while in the fourth there were three changes.

With these rules for determining matrix elements it is a straightforward but often tedious matter to determine the energy levels for any system of spin-$\tfrac{1}{2}$ nuclei. Later we find that some important simplifications accrue by utilizing group theory.

B. An Illustration Involving Equivalent Protons

We consider a system composed of two equivalent protons (or F^{19}, etc) such as exist in H_2O or H_2. The four spin-product functions are

$$\psi_1 = \alpha(1)\alpha(2) = \alpha\alpha$$

$$\psi_2 = \beta\beta \qquad\qquad\qquad\qquad (9.75)$$

$$\psi_3 = \alpha\beta$$

$$\psi_4 = \beta\alpha$$

The matrix elements of the 4×4 Hamiltonian are easily evaluated by use of Eqs. (9.68b), (9.72), and (9.74).

$H^{(0)}_{kk}$ *elements.*

$$H^{(0)}_{11} = -\{\tfrac{1}{2}\nu_0(1) + \tfrac{1}{2}\nu_0(2)\} = -\nu_0 \qquad (9.76a)$$

This result follows by using $M_I(1) = \tfrac{1}{2}$ and $M_I(2) = \tfrac{1}{2}$ in Eq. (9.68b) and remembering that since the two protons are equivalent, $\nu_0(1) = \nu_0(2) = \nu_0$.

Similarly we get

$$H^{(0)}_{22} = -\{-\tfrac{1}{2}\nu_0(1) - \tfrac{1}{2}\nu_0(2)\} = \nu_0 \qquad (9.76b)$$

$$H^{(0)}_{33} = H^0_{44} = 0 \qquad (9.76c)$$

$H^{(1)}_{kk}$ elements.

$$H^{(1)}_{11} = \langle \alpha\alpha \mid \widetilde{\mathcal{3C}}^{(1)} \mid \alpha\alpha \rangle = \tfrac{1}{4}J_{12}(+1)$$

$$= \tfrac{1}{4}J_{12} = \tfrac{1}{4}J \qquad (9.77a)$$

Here we used Eq. (9.72) with $T_{ij} = +1$, and since only one coupling constant exists we take $J_{12} = J$. We get also,

$$H^{(1)}_{22} = \tfrac{1}{4}J \qquad (9.77b)$$

$$H^{(1)}_{33} = H^{(1)}_{44} = -\tfrac{1}{4}J \qquad (9.77c)$$

$H^{(1)}_{kl}$ elements.

By Eq. (9.74) all elements are zero except $H^{(1)}_{34} = H^{(1)}_{43}$. Thus

$$H^{(1)}_{34} = H^{(1)}_{43} = \tfrac{1}{2}J \qquad (9.78a)$$

$$H^{(1)}_{kl} = 0 \quad \text{for } k, l \neq 3, 4 \text{ or } 4, 3 \qquad (9.78b)$$

The Hamiltonian matrix is then

$$\mathcal{3C} = \mathcal{3C}^{(0)} + \mathcal{3C}^{(1)} = \begin{pmatrix} -\nu_0 + \tfrac{1}{4}J & 0 & 0 & 0 \\ 0 & \nu_0 + \tfrac{1}{4}J & 0 & 0 \\ 0 & 0 & -\tfrac{1}{4}J & \tfrac{1}{2}J \\ 0 & 0 & \tfrac{1}{2}J & -\tfrac{1}{4}J \end{pmatrix} \qquad (9.79)$$

Two eigenvalues are immediately evident,

$$E_1 = -\nu_0 + \tfrac{1}{4}J \qquad (9.80a)$$

$$E_2 = \nu_0 + \tfrac{1}{4}J \qquad (9.80b)$$

and solution of the quadratic equation resulting from the 2 × 2 portion gives

$$E_3 = \tfrac{1}{4}J \qquad (9.80c)$$

$$E_4 = -\tfrac{3}{4}J \qquad (9.80d)$$

Note that ψ_1 and ψ_2 were eigenfunctions of $\widetilde{\mathcal{3C}}$, but that ψ_3 and ψ_4 were not. The eigenfunctions corresponding to E_3 and E_4 are perhaps obvious, but may be found by returning to the 2 × 2 secular equation. Thus for the root

$E_3 = \frac{1}{4}J$ we get the equations

$$(-\tfrac{1}{4}J - \tfrac{1}{4}J)a_1 + \tfrac{1}{2}Ja_2 = 0$$

and

$$\tfrac{1}{2}Ja_1 + (-\tfrac{1}{4}J - \tfrac{1}{4}J)a_2 = 0 \qquad (9.81)$$

which are compatible with $a_1 = a_2$. By normalization

$$a_1 = a_2 = 1/\sqrt{2} \qquad (9.82)$$

so the eigenfunction corresponding to E_3 is

$$\psi(E_3) = \frac{1}{\sqrt{2}}(\alpha\beta + \beta\alpha) \qquad (9.83a)$$

By the same approach

$$\psi(E_4) = \frac{1}{\sqrt{2}}(\alpha\beta - \beta\alpha) \qquad (9.83b)$$

The energy-level diagram for the system of two equivalent protons can be plotted from Eqs. (9.80). It is not possible to determine the *sign* of the coupling constant from the usual NMR measurements, but we assume it to be positive. Taking $J > 0$, we obtain the level diagram of Fig. 9.20.

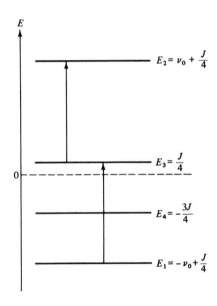

FIG. 9.20. Schematic energy-level diagram for two equivalent protons. Permitted transitions are shown by vertical arrows.

We can determine the permitted transitions by means of our previously stated selection rules. This is done most easily by listing the M_F values of each state and also the symmetry species. The latter may be listed as simply s (symmetric) or a (antisymmetric) since any two dimensional group (C_2, C_s, i) appropriate to two equivalent protons has only the two kinds of species. Thus we find the following properties of the wave functions:

$$\psi_1 = \alpha\alpha: \quad M_F = +1; s$$

$$\psi_2 = \beta\beta: \quad M_F = -1; s$$

$$\psi_3 = \frac{1}{\sqrt{2}}(\alpha\beta + \beta\alpha): \quad M_F = 0; s \tag{9.84}$$

$$\psi_4 = \frac{1}{\sqrt{2}}(\alpha\beta - \beta\alpha): \quad M_F = 0; a$$

Requiring that $\Delta M_F = \pm 1$ and $s \to s$ or $a \to a$, we find only two transitions, namely,

$$\psi_1 \to \psi_3$$
$$\tag{9.85}$$
$$\psi_3 \to \psi_2$$

These selection rules can be verified by actually evaluating the matrix elements given in Eq. (9.33b). For example,

$$(\psi_1 \mid \tilde{I}_X(1) + \tilde{I}_X(2) \mid \psi_4) \propto (\alpha\alpha \mid \tilde{I}_X(1) + \tilde{I}_X(2) \mid \alpha\beta - \beta\alpha)$$

$$= (\alpha \mid \tilde{I}_X(1) \mid \alpha)(\alpha \mid \beta) + (\alpha \mid \tilde{I}_X(2) \mid \beta)(\alpha \mid \alpha)$$

$$- (\alpha \mid \tilde{I}_X(1) \mid \beta)(\alpha \mid \alpha) - (\alpha \mid \tilde{I}_X(2) \mid \alpha)(\alpha \mid \beta)$$

$$= (\alpha \mid \tilde{I}_X(2) \mid \beta) - (\alpha \mid \tilde{I}_X(1) \mid \beta) = 0$$

Since the transition probability is proportional to the square of this matrix element we see that the transition $\psi_1 \to \psi_4$ does not occur, even though it satisfies the rule $\Delta M_F = \pm 1$. It fails, of course, to satisfy the symmetry requirement.

Finally we can calculate the frequencies of the allowed transitions by use of Eqs. (9.80). We find

$$\nu_{1 \to 3} = \tfrac{1}{4}J - (-\nu_0 + \tfrac{1}{4}J) = \nu_0$$
$$\tag{9.86}$$
$$\nu_{3 \to 2} = \nu_0 + \tfrac{1}{4}J - \tfrac{1}{4}J = \nu_0$$

Hence radiation is absorbed at only one frequency, and this is entirely independent of the magnitude of the coupling constant. This is an example of our earlier statement (Sec. 9.3 B) that the resonance of an equivalent set of nuclei is not split by the spin-spin interactions among those nuclei. This is always true, so in the remainder of the chapter we do not consider spin-spin coupling between equivalent nuclei since it produces no observable results.

EXAMPLE 9-7

Consider a system of two *non-equivalent* spin-$\frac{1}{2}$ nuclei. The $H^{(0)}{}_{kk}$ elements are now

$$H^{(0)}{}_{11} = -\tfrac{1}{2}[\nu_0(1) + \nu_0(2)]$$

$$H^{(0)}{}_{22} = \tfrac{1}{2}[\nu_0(1) + \nu_0(2)] \qquad (9.87)$$

$$H^{(0)}{}_{33} = \tfrac{1}{2}(\nu_0(1) - \nu_0(2)]$$

$$H^0{}_{44} = \tfrac{1}{2}[\nu_0(2) - \nu_0(1)]$$

where $\nu_0(1)$ and $\nu_0(2)$ are now different. The $\mathbf{H}^{(1)}$ elements are identical to those of Eqs. (9.77) and (9.78). In place of Eq. (9.79) we have then

$$\mathfrak{IC} = \frac{1}{2}\begin{pmatrix} -\nu_0(1) - \nu_0(2) \\ + \tfrac{1}{2}J & 0 & 0 & 0 \\[2mm] 0 & \begin{array}{c}\nu_0(1) + \nu_0(2) \\ + \tfrac{1}{2}J\end{array} & 0 & 0 \\[2mm] 0 & 0 & \begin{array}{c}\nu_0(1) - \nu_0(2) \\ - \tfrac{1}{2}J\end{array} & J \\[2mm] 0 & 0 & J & \begin{array}{c}\nu_0(2) - \nu_0(1) \\ - \tfrac{1}{2}J\end{array} \end{pmatrix}$$

$$(9.88)$$

which has eigenvalues

$$E_1 = -\tfrac{1}{2}[\nu_0(1) + \nu_0(2)] + \tfrac{1}{4}J$$

$$E_2 = \tfrac{1}{2}[\nu_0(1) + \nu_0(2)] + \tfrac{1}{4}J \qquad (9.89)$$

$$E_3 = -\tfrac{1}{4}J + \tfrac{1}{2}(\Delta^2 + J^2)^{1/2}$$

$$E_4 = -\tfrac{1}{4}J - \tfrac{1}{2}(\Delta^2 + J^2)^{1/2}$$

where $\Delta = \nu_0(1) - \nu_0(2)$.

For nonequivalent nuclei the s and a labels do not apply, that is the symmetry group is simply C_1 with one irreducible representation. Thus, all four $\Delta M_F = \pm 1$ transitions may occur, with the following frequencies:

$$\nu_{1\to 4} = -\tfrac{1}{2}J - \tfrac{1}{2}(\Delta^2 + J^2)^{1/2} + \bar{\nu}_0$$

$$\nu_{1\to 3} = -\tfrac{1}{2}J + \tfrac{1}{2}(\Delta^2 + J^2)^{1/2} + \bar{\nu}_0 \qquad (9.90)$$

$$\nu_{4\to 2} = \tfrac{1}{2}J + \tfrac{1}{2}(\Delta^2 + J^2)^{1/2} + \bar{\nu}_0$$

$$\nu_{3\to 2} = \tfrac{1}{2}J - \tfrac{1}{2}(\Delta^2 + J^2)^{1/2} + \bar{\nu}_0$$

where $\bar{\nu}_0 = \tfrac{1}{2}[\nu_0(1) + \nu_0(2)]$.

The four transitions of Eqs. (9.90) will not, in general, have the same intensities. We can obtain the relative intensities by obtaining the wave functions for the states $\psi(3)$ and $\psi(4)$ and then evaluating the square of the matrix element, $\langle m \mid \tilde{I}_x(1) + \tilde{I}_x(2) \mid n \rangle$. The easiest method of doing this is to define the angle θ by means of Eq. (2.14), which gives

$$\tan 2\theta = J/\Delta \qquad (9.91)$$

The matrix of Eq. (2.13) and its inverse then brings the 2×2 portion of Eq. (9.88) into diagonal form with the eigenvalues E_3 and E_4 as given in Eqs. (9.89). Now, as shown by Eq. (2.21), the desired eigenfunctions $\psi(3)$ and $\psi(4)$ are given by

$$\begin{pmatrix} \psi(3) \\ \psi(4) \end{pmatrix} = (\alpha\beta \ \ \beta\alpha) \begin{pmatrix} \cos\theta & -\sin\theta \\ \sin\theta & \cos\theta \end{pmatrix} \qquad (9.92a)$$

which gives

$$\psi(3) = \alpha\beta \cos\theta + \beta\alpha \sin\theta$$

$$\psi(4) = -\alpha\beta \sin\theta + \beta\alpha \cos\theta \qquad (9.92b)$$

The relative intensity of the transition $\psi_1 \to \psi(4)$ can be calculated as an example. We find

$$I \propto \mid \langle \alpha\alpha \mid \tilde{I}_x(1) + \tilde{I}_x(2) \mid -\alpha\beta \sin\theta + \beta\alpha \cos\theta \rangle \mid^2$$

$$= \mid -\langle \alpha \mid \tilde{I}_x(2) \mid \beta \rangle \sin\theta + \langle \alpha \mid \tilde{I}_x(1) \mid \beta \rangle \cos\theta \mid^2$$

$$= \tfrac{1}{4}(1 - 2\sin\theta \cos\theta) = \tfrac{1}{4}(1 - \sin 2\theta)$$

In a similar manner we can find the relative intensities for all the transitions. These have been summarized in Table 9.3, which also lists the expressions for transition frequencies.

We can now examine three cases.

Case A: Equivalent protons

Our formulas should reduce to the previous results if $\nu_0(1) = \nu_0(2)$. In this case $\Delta = 0$ and $\theta = 45°$. Thus the intensity of the $1 \to 4$ and $4 \to 2$

TABLE 9.3
Frequencies and relative intensities of transitions for an AB system of spin-$\frac{1}{2}$ nuclei.

Transition	Frequency[a]	Intensity[b]
$1 \rightarrow 4$	$-\frac{1}{2}[J + (\Delta^2 + J^2)^{1/2}] + \bar{\nu}_0$	$1 - \sin 2\theta$
$1 \rightarrow 3$	$-\frac{1}{2}[J - (\Delta^2 + J^2)^{1/2}] + \bar{\nu}_0$	$1 + \sin 2\theta$
$4 \rightarrow 2$	$\frac{1}{2}[J + (\Delta^2 + J^2)^{1/2}] + \bar{\nu}_0$	$1 - \sin 2\theta$
$3 \rightarrow 2$	$\frac{1}{2}[J - (\Delta^2 + J^2)^{1/2}] + \bar{\nu}_0$	$1 + \sin 2\theta$

   [a]   See text for definition of Δ and $\bar{\nu}_0$.
   [b]   See text for definition of θ.

transitions go to zero. The frequencies of the $1 \rightarrow 3$ and $3 \rightarrow 2$ transitions both become equal to $\nu_0(1) = \nu_0(2) = \nu_0$.

Case B: $\Delta \gg J$

In this case, $\theta \rightarrow 0$, and the intensities of all transitions become equal, assuming both nuclei have the same gyromagnetic ratio γ.[17] In case γ differs for the two nuclei we obtain two pairs of 1:1 doublets of differing intensities. The transition frequencies become

$$\nu_{1\rightarrow3} = \nu_0(1) - \tfrac{1}{2}J$$
$$\nu_{4\rightarrow2} = \nu_0(1) + \tfrac{1}{2}J \qquad\qquad (9.93)$$
$$\nu_{1\rightarrow4} = \nu_0(2) - \tfrac{1}{2}J$$
$$\nu_{3\rightarrow2} = \nu_0(2) + \tfrac{1}{2}J$$

The spectrum is shown schematically in Fig. 9.21(a). We have two doublets which is exactly the result predicted by the simple "first-order" rules. Note that the separation of each doublet is equal to J as expected. As practical examples of this AX-type spectrum, we might list HD, HF, the protons

$$\begin{array}{c} \text{O} \\ \parallel \end{array}$$
of HC—CHCl$_2$, and the protons of

$$\begin{array}{c} \text{NO}_2 \\ \diagdown \\ \text{C}=\text{CH}_2 \\ \diagup \\ \text{Cl} \end{array}$$

Case C: $\Delta \sim J$

This is what we call an AB spectrum and no special simplifications result. The appearance of the spectrum depends very strongly upon the ratio J/Δ. In Fig. 9.21(b) we have shown the spectrum for the case $\Delta = J$,

---

[17] It must be remembered that P_{mn} contains γ^2 [see Eqs. (9.33)].

while in Fig. 9.21(c) the spectrum is drawn for $\Delta = 3J$. Note that the separation between the outer pairs of lines is equal to the coupling constant J and that the separation of the inner two lines is $(\Delta^2 + J^2)^{1/2} - J$. Examples of AB systems include the protons of

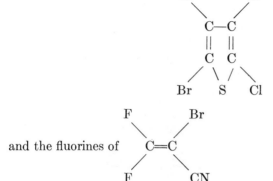

and the fluorines of

C. A Useful Commutation Relation

Although we gave no attention to the fact earlier, it is clear from Eqs. (9.79) and (9.88) that there were no matrix elements connecting states of different M_F, where M_F is the eigenvalue of $\tilde{F}_Z$, the total spin angular momentum along the Z axis. In fact this is a general result arising from the commutativity of $\tilde{\mathcal{3C}}$ and $\tilde{F}_Z$, that is

$$\tilde{\mathcal{3C}}\tilde{F}_Z - \tilde{F}_Z\tilde{\mathcal{3C}} = 0 \tag{9.94}$$

The relation Eq. (9.94) implies that $\tilde{\mathcal{3C}}$ and $\tilde{F}_Z$ have the same eigenfunctions, and furthermore that $\tilde{\mathcal{3C}}$ can have no elements off-diagonal in M_F.

To prove Eq. (9.94) we examine $\tilde{\mathcal{3C}}^{(0)}$ and $\tilde{\mathcal{3C}}^{(1)}$ separately. We form $\tilde{F}_Z\tilde{\mathcal{3C}}^{(0)}$ and substitute for each factor as in the following.

$$\tilde{F}_Z\tilde{\mathcal{3C}}^{(0)} = \sum_i \tilde{I}_Z(i)\{-\sum_i \nu_0(i)\tilde{I}_Z(i)\} \tag{9.95a}$$

We have used the definition of $\tilde{F}_Z$ and Eq. (9.62b). But now $\tilde{I}_Z(i)$ commutes with itself and with all $\tilde{I}_Z(j)$ so that

$$\tilde{F}_Z\tilde{\mathcal{3C}}^{(0)} = \{-\sum_i \nu_0(i)\tilde{I}_Z(i)\} \sum_i \tilde{I}_Z(i) = \tilde{\mathcal{3C}}^{(0)}\tilde{F}_Z \tag{9.95b}$$

Thus $\tilde{\mathcal{3C}}^{(0)}$ and $\tilde{F}_Z$ commute.

To prove that $\tilde{\mathcal{3C}}^{(1)}$ and $\tilde{F}_Z$ commute, we shall be nongeneral for simplicity. Consider just two nuclei, so

$$\tilde{\mathcal{3C}}^{(1)} = J\tilde{I}(1)\cdot\tilde{I}(2) \tag{9.96}$$

and

$$\tilde{F}_Z = \tilde{I}_Z(1) + \tilde{I}_Z(2) \tag{9.97}$$

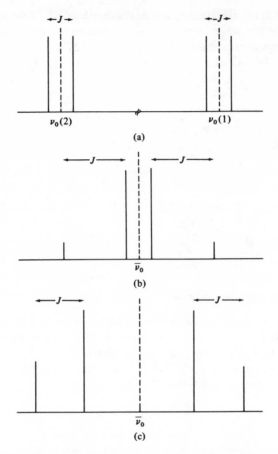

FIG. 9.21. Spectra for two-proton systems. (a) Spectrum for AX system. (b) Spectrum for AB system with $\Delta = J$. (c) Spectrum for AB system with $\Delta = 3J$.

Then

$$\tilde{F}_Z \tilde{\mathcal{K}}^{(1)} = J[\tilde{I}_Z(1) + \tilde{I}_Z(2)][\tilde{I}(1) \cdot \tilde{I}(2)]$$

$$= J[\tilde{I}_Z(1) + \tilde{I}_Z(2)][\tilde{I}_X(1)\tilde{I}_X(2) + \tilde{I}_Y(1)\tilde{I}_Y(2) + \tilde{I}_Z(1)\tilde{I}_Z(2)]$$

$$= J[\tilde{I}_X(2)\tilde{I}_Z(1)\tilde{I}_X(1) + \tilde{I}_Y(2)\tilde{I}_Z(1)\tilde{I}_Y(1) + \tilde{I}_Z(2)\tilde{I}_Z(1)\tilde{I}_Z(1)]$$

$$(9.98)$$

Using the well-known angular momentum commutation relations

$$\tilde{I}_Z\tilde{I}_X = -i\hbar\tilde{I}_Y + \tilde{I}_X\tilde{I}_Z$$

$$\tilde{I}_Z\tilde{I}_Y = i\hbar\tilde{I}_X + \tilde{I}_Y\tilde{I}_Z \qquad (9.99)$$

$$\tilde{I}_Z\tilde{I}_Z = \tilde{I}_Z\tilde{I}_Z$$

Eq. (9.98) becomes

$$\tilde{F}_Z\tilde{\mathcal{3C}}^{(1)} = J[\tilde{I}(1)\cdot\tilde{I}(2)][\tilde{I}_Z(1) + \tilde{I}_Z(2)] = \tilde{\mathcal{3C}}(1)\tilde{F}_Z \qquad (9.100)$$

A little reflection should convince the reader that $\tilde{\mathcal{3C}}^{(1)}$ and $\tilde{F}_Z$ commute for any number of nuclei, not just two as in this derivation.

Since $\tilde{\mathcal{3C}}^{(0)}$ and $\tilde{\mathcal{3C}}^{(1)}$ each commute with $\tilde{F}_Z$, we have proved Eq. (9.94) as desired. The great value of this is that it permits us to identify the possible nonzero matrix elements before actually doing the matrix element evaluation.

D. Several Sets of Equivalent Nuclei

We have considered a simple case consisting of two sets of nuclei, each set made up of one nucleus. This formalism can be generalized for any number and size of sets in the following way. Define first the Z component of total spin angular momentum for each chemically equivalent set, A, B, C, ...X...,

$$\tilde{F}_Z(A) = \sum_i^{A\ \text{nuclei}} \tilde{I}^{(A)}{}_Z(i), \text{ etc.} \qquad (9.101a)$$

so the *total* $\tilde{F}_{Z,T}$ is

$$\tilde{F}_{Z,T} = \sum_X^{\text{all sets}} \tilde{F}_Z(X) \qquad (9.101b)$$

The Hamiltonian terms $\tilde{\mathcal{3C}}^{(0)}$ and $\tilde{\mathcal{3C}}^{(1)}$ may now be written

$$\tilde{\mathcal{3C}}^{(0)} = - \sum_{X=A,B\cdots}^{\text{all sets}} \nu_0(X)\tilde{F}_Z(X) \qquad (9.102)$$

and

$$\tilde{\mathcal{3C}}^{(1)} = \sum_X^{\text{all sets}} \sum_{i,j<i}^{\text{Set } X} J_{XX}\tilde{I}_i(X)\cdot\tilde{I}_j(X) + \sum_i^{A'} \sum_j^{B'} J_{A'B'}\tilde{I}_i(A')\cdot\tilde{I}_j(B')$$

$$+ \sum_i^{A'} \sum_j^{C'} J_{A'C'}\tilde{I}_i(A')\cdot\tilde{I}_j(C') + \sum_i^{B'} \sum_j^{C'} J_{B'C'}\tilde{I}_i(B')\cdot\tilde{I}_j(C') + \cdots \qquad (9.103)$$

In Eq. (9.103) we use primes on the sets to indicate that the *magnetically* equivalent sets may be different from the chemically equivalent sets. In case the sets A, B, C... are the same as the sets A', B', C'..., it is most convenient to rewrite Eq. (9.103) in terms of the total angular momenta, $\tilde{F}(X) = \sum_i \tilde{I}_i(X)$, which gives

$$\tilde{\mathcal{3C}}^{(1)} = \sum_X \sum_{i,j<i} J_{XX}\tilde{I}_i(X)\cdot\tilde{I}_j(X) + J_{AB}\tilde{F}(A)\cdot\tilde{F}(B) + J_{AC}\tilde{F}(A)\cdot\tilde{F}(C)$$

$$+ J_{BC}\tilde{F}(B)\cdot\tilde{F}(C) + \cdots \qquad (9.104)$$

Since the couplings (J_{XX}) within the sets produce no observable spectral results, the first sum in 9.103 and 9.104 may be legitimately left out of any calculation. The energy levels which are calculated are not correct, of course, but transition frequencies are unaffected by the omission. This fact produces a considerable simplification in setting up the Hamiltonian matrix.

EXAMPLE 9-8

The hypothetical XY_2W_2 planar molecules of Fig. 9.22 provide useful examples for writing the correct Hamiltonian. We assume that X is a non-magnetic nucleus. Then there are two *chemically* equivalent sets of nuclei in each molecule. In the molecule of Fig. 9.22(a) the Y nuclei and the W nuclei each form *magnetically* equivalent sets also. Thus the Hamiltonian for the system of Fig. 9.22(a) is

$$\tilde{\mathfrak{IC}} = -\nu_0(Y)\tilde{F}_z(Y) - \nu_0(W)\tilde{F}_z(W) + J_{WY}[\tilde{I}(1)\cdot\tilde{I}(3) + \tilde{I}(2)\cdot\tilde{I}(3)$$
$$+ \tilde{I}(2)\cdot\tilde{I}(4) + \tilde{I}(1)\cdot\tilde{I}(4)]$$
$$= -\nu_0(Y)\tilde{F}_z(Y) - \nu_0(W)\tilde{F}_z(W) + J_{WY}\tilde{F}(W)\cdot\tilde{F}(Y) \qquad (9.105)$$

since there is simply one unique coupling constant J_{WY}.

For the molecule of Fig. 9.22(b) neither the Y nor W nuclei form magnetically equivalent sets. Thus there are two distinct coupling constants and the Hamiltonian is

$$\tilde{\mathfrak{IC}} = -\nu_0(Y)\tilde{F}_z(Y) - \nu_0(W)\tilde{F}_z(W) + J[\tilde{I}(1)\cdot\tilde{I}(4) + \tilde{I}(2)\cdot\tilde{I}(3)]$$
$$+ J'[\tilde{I}(1)\cdot\tilde{I}(3) + \tilde{I}(2)\cdot\tilde{I}(4)] \quad (9.106)$$

where $J = J_{14} = J_{23}$ and $J' = J_{13} = J_{24}$.

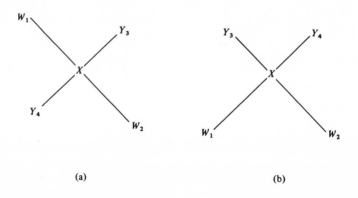

(a)   (b)

FIG. 9.22. Two isomeric forms of a planar XY_2W_2 molecule. The subscripts in the figure are merely labels for the nuclei.

The basis functions needed for treating a system having several sets of equivalent nuclei [and having the Hamiltonian of Eqs. (9.102) and (9.103)] are again simply the product functions like Eq. (9.64). It is usually most convenient to write all the permissible spin product functions for each set of nuclei and then to form all possible total products by combining the various sets. Thus for an AB_2 proton system we have for the A nucleus

$$\phi_1(A) = \alpha(1) \equiv \alpha$$
$$\phi_2(A) = \beta(1) \equiv \beta \tag{9.107a}$$

and for the B nuclei,

$$\phi_1(B) = \alpha(2)\alpha(3) \equiv \alpha\alpha$$
$$\phi_2(B) = \beta\beta \tag{9.107b}$$
$$\phi_3(B) = \alpha\beta$$
$$\phi_4(B) = \beta\alpha$$

The total wave functions are then

$$\phi_1 = \phi_1(A)\phi_1(B) = \alpha\alpha\alpha$$
$$\phi_2 = \alpha\beta\beta$$
$$\phi_3 = \alpha\alpha\beta$$
$$\phi_4 = \alpha\beta\alpha \tag{9.107c}$$
$$\phi_5 = \beta\alpha\alpha$$
$$\phi_6 = \beta\beta\beta$$
$$\phi_7 = \beta\alpha\beta$$
$$\phi_8 = \beta\beta\alpha$$

At this point it is useful to observe that much simplification of the Hamiltonian matrix occurs if symmetry functions are formed. In the A_2 case previously described, we found for example that $\alpha\beta$ and $\beta\alpha$ were not eigenfunctions but that $\alpha\beta + \beta\alpha$ and $\beta\alpha - \alpha\beta$ were. In fact we might have expected this since we know that all eigenfunctions of the Hamiltonian must have the symmetry of the group of Schrödinger's equation. If symmetry functions are utilized there are no matrix elements of $\tilde{\mathcal{K}}$ connecting states of different symmetry.

In place of Eqs. (9.107c) for the AB_2 system it is more satisfactory to use the symmetry functions summarized in Table 9.4. The general approach is to form symmetry functions for each set of nuclei and then form the over-all products. Table 9.4 lists, in addition to the symmetry functions, the $M_{F,T}$ values for each function. By means of the symmetries and $M_{F,T}$ values we can predict immediately that the Hamiltonian matrix has no nonzero elements other than those listed explicitly in Eq. (9.108).

$$
\mathfrak{K} = \begin{pmatrix}
H_{11} & 0 & 0 & 0 & 0 & 0 & 0 & 0 \\
0 & H_{22} & 0 & 0 & 0 & 0 & H_{27} & 0 \\
0 & 0 & H_{33} & 0 & H_{35} & 0 & 0 & 0 \\
0 & 0 & 0 & H_{44} & 0 & 0 & 0 & 0 \\
0 & 0 & H_{53} & 0 & H_{55} & 0 & 0 & 0 \\
0 & 0 & 0 & 0 & 0 & H_{66} & 0 & 0 \\
0 & H_{72} & 0 & 0 & 0 & 0 & H_{77} & 0 \\
0 & 0 & 0 & 0 & 0 & 0 & 0 & H_{88}
\end{pmatrix}
\tag{9.108}
$$

By rearrangement of rows and columns the above matrix reduces to two 2×2 blocks and four 1×1 blocks. Actual matrix element evaluation proceeds easily by using the equations of Sec. 9.5 A. It must, however, be remembered that these matrix element equations are for *simple* product functions. The symmetry-function matrix elements require that the appropriate linear combinations be taken.

It is easy to see now how the AB_2 system simplifies to the AX_2 type. The off-diagonal elements $H_{27} = H_{72}$, for example, are on the order of 10 Hz for protons, since they are proportional to J. If $H_{22} - H_{77}$ is large, as it is

TABLE 9.4

Symmetry functions and their $M_{F,T}$ values for an AB_2 system of protons.

Function	$M_{F,T}$	*Symmetry*
$\phi_1 = \alpha\alpha\alpha$	$\frac{3}{2}$	s
$\phi_2 = \alpha\beta\beta$	$-\frac{1}{2}$	s
$\phi_3 = \alpha(\alpha\beta + \beta\alpha)(2)^{-1/2}$	$\frac{1}{2}$	s
$\phi_4 = \alpha(\alpha\beta - \beta\alpha)(2)^{-1/2}$	$\frac{1}{2}$	a
$\phi_5 = \beta\alpha\alpha$	$\frac{1}{2}$	s
$\phi_6 = \beta\beta\beta$	$-\frac{3}{2}$	s
$\phi_7 = \beta(\alpha\beta + \beta\alpha)(2)^{-1/2}$	$-\frac{1}{2}$	s
$\phi_8 = \beta(\alpha\beta - \beta\alpha)(2)^{-1/2}$	$-\frac{1}{2}$	a

if A and B are different nuclei or if A and B are the same nuclei with large chemical shifts, the elements H_{27} and H_{72} may be neglected as a first approximation. The same applies to H_{35} and H_{53}. Thus in this AX_2 case the Hamiltonian matrix is entirely diagonal, so the stationary state energies are given directly by the diagonal elements $H^{(0)}{}_{ii} + H^{(1)}{}_{ii}$. Evaluation of these elements followed by application of the selection rules would lead to the AX_2 first-order spectrum described by the rules of Sec. 9.3 B. We show this more generally in the following section.

E.   Derivation of "first-order" rules.

Suppose we have several sets of chemically and magnetically equivalent nuclei, A_a, M_m, X_x, etc. Neglecting the spin-spin coupling within each set (since it produces no observable result), the Hamiltonian is [from Eq. (9.102) and (9.104)]

$$\mathcal{H} = \widetilde{\mathcal{H}}^{(0)}(A) + \widetilde{\mathcal{H}}^{(0)}(M) + \widetilde{\mathcal{H}}^{(0)}(X) + \cdots$$

$$+ \widetilde{\mathcal{H}}^{(1)}(AM) + \widetilde{\mathcal{H}}^{(1)}(AX) + \widetilde{\mathcal{H}}^{(1)}(MX) + \cdots \quad (9.109)$$

where

$$\widetilde{\mathcal{H}}^{(0)}(A) = -\nu_0(A)\tilde{F}_Z(A) \text{ etc.}$$

$$\widetilde{\mathcal{H}}^{(1)}(AM) = J_{AM}\tilde{F}(A) \cdot \tilde{F}(M) \text{ etc.}$$

We know that normally $\widetilde{\mathcal{H}}^{(0)} \gg \widetilde{\mathcal{H}}^{(1)}$, so it is natural to consider $\widetilde{\mathcal{H}}^{(1)}$ as a small perturbation upon $\widetilde{\mathcal{H}}^{(0)}$. The approximation is valid when the off-diagonal terms H_{ij} are much smaller in magnitude than $H_{ii} - H_{jj}$. In this case the zero-order energy is given by

$$E^{(0)} = \langle n \mid \widetilde{\mathcal{H}}^{(0)}(A) + \widetilde{\mathcal{H}}^{(0)}(M) + \cdots \mid n \rangle$$

$$= -\nu_0(A)M_F(A) - \nu_0(M)M_F(M) - \nu_0(X)M_F(X) - \cdots \quad (9.110)$$

The energy contribution from $\widetilde{\mathcal{H}}^{(1)}$ is, to first order,

$$E^{(1)} = \langle n \mid \widetilde{\mathcal{H}}^{(1)}(AM) + \widetilde{\mathcal{H}}^{(1)}(AX) + \cdots \mid n \rangle$$

$$= J_{AM}\langle n \mid \tilde{F}(A) \cdot \tilde{F}(M) \mid n \rangle$$

$$+ J_{AX}\langle n \mid \tilde{F}(A) \cdot \tilde{F}(X) \mid n \rangle + \cdots \quad (9.111)$$

Expanding a typical matrix element, we obtain the form

$$J_{AM}\langle n \mid \tilde{F}(A) \cdot \tilde{F}(M) \mid n \rangle$$

$$= J_{AM} \sum_i^A \sum_j^M \langle n \mid \tilde{I}_{iX}(A) \cdot \tilde{I}_{jX}(M) + \tilde{I}_{iY}(A) \cdot \tilde{I}_{jY}(M) + \tilde{I}_{iZ}(A) \cdot \tilde{I}_{jZ}(M) \mid n \rangle$$

$$(9.112a)$$

But $\tilde{I}_X$ and $\tilde{I}_Y$ have no diagonal elements so Eq. (9.112a) becomes

$$J_{AM}\langle n \mid \tilde{F}(A) \cdot \tilde{F}(M) \mid n \rangle = J_{AM} \sum_i \sum_j \langle n \mid \tilde{I}_{iZ}(A) \cdot \tilde{I}_{jZ}(M) \mid n \rangle$$

$$= J_{AM} \sum_i^A \sum_j^M M_I(i) M_I(j) = J_{AM} M_F(A) M_F(M) \quad (9.112b)$$

Thus Eq. (9.111) becomes

$$E^{(1)} = J_{AM} M_F(A) M_F(M) + J_{AX} M_F(A) M_F(X) + \cdots \quad (9.113)$$

and the total energy correct to first order is

$$E = E^{(0)} + E^{(1)} = -\nu_0(A) M_F(A) - \nu_0(M) M_F(M) - \cdots$$

$$+ J_{AM} M_F(A) M_F(M) + J_{AX} M_F(A) M_F(X) + \cdots \quad (9.114)$$

The transitions are easily obtained by application of the rules

$$\Delta M_F(A) = +1, \quad \Delta M_F(M) = +1, \text{ etc.,}$$

which gives

$$\nu(A) = \nu_0(A) + J_{AM} M_F(M) + J_{AX} M_F(X) + \cdots \quad (9.115)$$

$$\nu(M) = \nu_0(M) + J_{AM} M_F(A) + J_{MX} M_F(X) + \cdots \quad (9.116)$$

and so forth. This is the mathematical formalism of the previously stated "first-order" result, namely, that a resonance from a given set of nuclei is split into $2F + 1$ components by each set of nuclei to which it is coupled. Note that the results embodied in Eqs. (9.115–9.116) are based on the assumption of magnetic equivalence and that the chemical shifts are large relative to the coupling constants.

EXAMPLE 9-9

To illustrate these results we might consider methylamine in acid solution, in which case we are dealing with $CH_3—NH_3^+$. The proton resonance spectrum should show a $-CH_3$ and a $-NH_3$ resonance. We would predict for the methyl resonance

$$\nu(CH_3) = \nu_0(CH_3) + J_{HN} M_F(N) + J_{HH} M_F(NH_3)$$

$$= \nu_0(CH_3) + J_{HN} \left\{ \begin{array}{c} 1 \\ \\ 0 + J_{HH'} \\ \\ -1 \end{array} \right. \left\{ \begin{array}{c} \frac{3}{2} \\ \frac{1}{2} \\ -\frac{1}{2} \\ -\frac{3}{2} \end{array} \right. \quad (9.117)$$

Experimentally we find a 1:3:3:1 quartet at $\delta = 2.60$ ppm with spacing 7.4 Hz. Thus the N^{14} coupling is very small, so

$$J_{NH} \sim 0, \qquad J_{HH'} = 7.4 \text{ Hz}$$

For the NH_3 resonance

$$\nu(NH_3) = \nu_0(NH_3) + J_{H'N}M_F(N) + J_{HH'}M_F(CH_3)$$

$$= \nu_0(NH_3) + J_{H'N} \left\{ \begin{array}{c} 1 \\[1em] 0 + J_{HH'} \left\{ \begin{array}{c} \frac{3}{2} \\[0.5em] \frac{1}{2} \\[0.5em] -\frac{1}{2} \\[0.5em] -\frac{3}{2} \end{array} \right. \\[1em] -1 \end{array} \right. \qquad (9.118)$$

A 1:1:1 broad triplet is observed at $\delta = 7.3$ ppm with spacing of 52 Hz. The broad ($\sim$15 Hz) triplet members show on close inspection a poorly resolved 1:3:3:1 structure with spacing of 6–7 Hz. We may conclude that

$$J_{H'N} = 52 \text{ Hz}$$

and that the quartet structure is in semiquantitative agreement with $J_{HH'} = 7.4$ Hz as previously found. The N^{14} resonance of CH_3—NH_3^+ has been observed also.[18] It is a broad 1:3:3:1 quartet with spacings of 52 Hz, which again agrees with the previous findings.

9.6 THEORY OF THE CHEMICAL SHIFT AND THE SPIN-SPIN INTERACTION

The manner in which the chemical shift and the nuclear spin–spin interaction affect the magnetic energy levels has been described in some detail. In this section we wish to look briefly into the theoretical basis for these properties and their relationship to the electronic structure of molecules. The theory for each of these properties is rather complex so our view shall be somewhat qualitative.

A. Chemical Shift

Earlier we have seen that the fundamental quantity of interest is the screening constant σ which is the proportionality constant between the applied and induced magnetic field. Actually, the screening constant should be represented by a tensor just as the electric polarizability or moment of inertia. Liquid-phase measurements provide us with simply the average of the three principal components of the tensor because of the rapid tum-

[18] R. A. Ogg and J. D. Ray, *J. Chem. Phys.* **26**, 1339 (1957).

bling of the molecules, so we need consider only the scalar quantity σ in most of our discussion.

Using first- and second-order perturbation theory Ramsey[19] has derived an expression for the induced magnetic field at any nucleus in a molecule. If we imagine the z molecular axis to coincide with the Z space-fixed axis, the zz element of the tensor σ has the form

$$\sigma_{zz} = \frac{e^2}{2mc^2} \left\langle 0 \left| \sum_i \frac{x_i^2 + y_i^2}{r_i^3} \right| 0 \right\rangle - \frac{e^2}{2m^2c^2} \sum_{n \neq 0} (E_n - E_0)^{-1}$$

$$\cdot \left[\langle 0 | \sum_j \tilde{L}_{zj} | n \rangle \left\langle n \left| \sum_k \frac{\tilde{L}_{zk}}{r_k^3} \right| 0 \right\rangle + \langle 0 | \sum_k \tilde{L}_{zk} | n \rangle \right.$$

$$\left. \cdot \left\langle n \left| \sum_j \frac{\tilde{L}_{zj}}{r_j^3} \right| 0 \right\rangle \right] \quad (9.119)$$

where the matrix elements are evaluated for the electronic states of the molecule. $\tilde{L}_{zi}$ is the orbital angular momentum operator of the ith electron in the z direction, the sums over i, j, and k are over all electrons, and the sum over n includes all excited electronic states. The origin of the x, y, z axes is at the nucleus in question. Similar expressions apply for σ_{yy} and σ_{xx}, and the observed liquid- or gas-phase value is $\frac{1}{3}(\sigma_{xx} + \sigma_{yy} + \sigma_{zz})$.

Equation (9.119) has been applied to several rather simple molecules. Exact use of the equation is usually limited, however, by our knowledge of the excited-state energies and wave functions. The first term can be handled relatively easily since it involves only the ground-state matrix element. It should be pointed out also that for atoms, only the first term in (9.119) is nonzero. This first term is positive and is the origin of the dominant *diamagnetic* screening. On the other hand the second term of (9.119) is not negligible and is of opposite sign so appreciable cancellations may occur. The second term has been called the *paramagnetic* contribution to the diamagnetic screening.

The best calculations for H_2 predict a proton screening constant of about 26 ppm; of this total $+32$ ppm arise from the first term of (9.119) while -6 ppm are from the second. Since an absolute magnetic-field measurement to an accuracy of even 100 ppm is not possible, no experimental value for σ is available. It has been possible to check experimentally the paramagnetic contribution of -6 ppm, since this can be inferred from measurements of spin-rotation constants by microwave or molecular-beam measurements.[20] The experimental value of -5.9 ppm is, in fact, in excellent agreement.

[19] N. F. Ramsey, *Phys. Rev.* **78**, 699 (1950).
[20] N. F. Ramsey, Am. Scientist **49**, 509 (1961).

Since accurate values of σ cannot be measured, most theoretical investigations are aimed at predicting or correlating changes in σ for different molecules. These differences in σ values are precisely measurable and are in fact proportional to the chemical shifts δ defined by Eq. (9.24b). As we now show, many *qualitative* features of proton chemical shifts are rather easily understood, although their quantitative interpretation remains a formidable task.

In many cases the chemical shifts may be interpreted in terms of changes in the diamagnetic contributions from electrons in the region of the nucleus of interest. Thus the chemical shifts δ for the methyl halide proton resonances increase in the order $\delta(CH_3I) < \delta(CH_3Br) < \delta(CH_3Cl) < \delta(CH_3F)$. From Eq. (9.24b) we conclude that $\sigma(CH_3I) > \sigma(CH_3Br) > \sigma(CH_3Cl) > \sigma(CH_3F)$. This is just what is expected if one considers the changes in electron density around the proton that are produced by the inductive effect of the electronegative halogens. As the halogen electronegativity increases the protons become progressively less shielded by the local diamagnetic mechanism. Although some paramagnetic contribution might be expected from the halogen the effect is minor due to the relatively large separations of the protons and halogen.

A similar local diamagnetic screening mechanism is useful in describing the chemical shifts of the *ortho* protons in monosubstituted benzenes. The greater the withdrawing inductive effect of the substituent the greater is the diamagnetic deshielding of the *ortho* proton. Assuming all other shielding contributions to be constant, we predict, therefore, $\sigma(\phi NO_2) > \sigma(\phi Cl) > \sigma(\phi F) > \sigma(\phi OCH_3) > \sigma(\phi NH_2)$. This is the experimental finding, the total range in δ being about 1 ppm for the above compounds.

On the other hand it is easy to find many cases in which local diamagnetic shielding cannot be the complete explanation. For example, it is an experimental fact [See Fig. 9.6(b)] that $\sigma(SiH_4) < \sigma(CH_4)$ and $\sigma(PH_3) < \sigma(NH_3)$. From the previous arguments based on the local electron density about the protons, we would predict just the opposite result since the electronegativities of C and N are greater than those of Si and P, respectively. We must conclude that changes in paramagnetic screening contributions have taken over the dominant role. This is reasonable since from Eq. (9.119) it is clear that the paramagnetic terms become more important as the excited states become more abundant and lower in energy, which occurs for silicon and phosphorus atoms because of the presence of low-energy d states. Thus it may be concluded that the chemical shifts of CH_4 and SiH_4, and NH_3 and PH_3 are reasonably explained qualitatively by *neighboring* paramagnetic screening effects. Note that these effects must be present for the methyl halides also, but due to the greater distances involved the r^{-3} terms in Eq. (9.119) make them unimportant.

Dominant effects of this type from neighboring atoms are not uncommon. They must be invoked to explain the chemical shift of acetylene ($CH{\equiv}CH$), which from Fig. 9.6(a) is seen to be out of order with respect to ethylene ($CH_2{=}CH_2$) and ethane ($CH_3{-}CH_3$). The chemical shift of acetylene seems too small, that is, the acetylenic protons experience a diamagnetic shielding much greater than would be predicted due to any local shielding effects.

We can understand the large acetylene proton shielding if we consider the kinds of induced fields that occur in a linear $X–H$ molecule. In Fig. 9.23 we show in a qualitative way the induced fields at the atom X and their secondary effects upon the proton. Note that the primary induced fields may be either diamagnetic or paramagnetic in behavior. For a molecule with axial symmetry (such as a linear molecule) no paramagnetic contribution may occur when the field is along the symmetry axis, so we have not shown this case in Fig. 9.23.

We see that in general the proton screening is a function of the orientation of the molecule in the external field. As usual, the observed screening constant is an average,

$$\sigma = \tfrac{1}{3}(\sigma_{||} + \sigma_{\perp} + \sigma_{\perp}') \tag{9.120}$$

where $\sigma_{||}$ is the proton screening contribution when H_0 is along the axis and $\sigma_{\perp}$ and $\sigma_{\perp}'$ are the contributions when H_0 is perpendicular to the axis. This *neighbor-anisotropy* screening mechanism can lead to several possible results depending on the magnitudes of the various terms in (9.120). In the acetylene case the proton shielding is increased markedly since the contributions of (a) and (c) in Fig. 9.23 far outweigh that of (b).

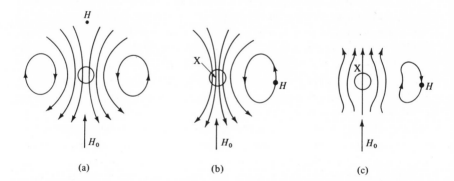

(a)   (b)   (c)

FIG. 9.23. Diamagnetic and paramagnetic screening of proton by a neighboring atom. In (b) the proton is deshielded, that is, $\sigma(H)$ is smaller, while in (a) and (c) the proton is shielded [$\sigma(H)$ is greater].

A similar mechanism probably applies to the hydrohalic acids, for which the experimental results are $\sigma(\text{HF}) < \sigma(\text{HCl}) < \sigma(\text{HBr}) < \sigma(\text{HI})$. Here the paramagnetic screening in Fig. 9.23(c) must predominate for the halogens other than fluorine because of the presence of low-energy states. The ordering is in agreement with that based on electronegativities also, but it seems that this cannot be the dominant term since all of these species are more screened than H_2 itself, that is $\sigma(H_2) < \sigma(\text{HF})$. Local diamagnetic effects based on electronegativities would surely cause $\sigma(H_2) > \sigma(\text{HI})$.

Finally we might inquire about the rather large chemical shift (compared to ethylene, say), of the benzene protons as shown in Fig. 9.6(a).

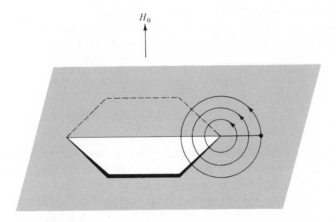

H_0

FIG. 9.24.   Field produced by ring currents in benzene. For simplicity only one proton is shown and only the induced fields in the plane containing the C–H bond are shown.

This large deshielding of the aromatic protons arises from "ring currents" produced by the six π electrons of the carbon skeleton. These currents flow around the ring when the external magnetic field is perpendicular to the ring, producing the magnetic field shown in Fig. 9.24. The result is a large paramagnetic contribution at the protons, that is they are deshielded. On the other hand when the external field is in the plane of the molecule no ring currents of this type are produced, and no additional proton shielding or deshielding occurs. Thus the screening tensor is very anisotropic and the net result is that the protons are greatly deshielded.

B.   Nuclear Spin–Spin Coupling

In Sec. 9.3 B we described in a qualitative manner the way in which two nuclei in a molecule influence one another energetically via the surrounding electrons. Theoretically this mechanism may occur in three ways:[21]

(1) Nuclear spin angular momentum-electron orbital angular momentum coupling.
(2) Nuclear magnetic moment-electron spin magnetic moment interaction.
(3) Nuclear spin angular momentum-electron spin angular momentum coupling.

Of these three mechanisms it seems that the last one, known as the "Fermi contact term" is the predominant contributor to the coupling constants between most nuclei. For proton-proton couplings the contact term probably represents some 90% or more of the total.

Using the contact mechanism Ramsey[22] has derived the following expression for the coupling constant between nuclei N and N':

$$J_{NN'} = -\frac{2}{3\hbar}\left(\frac{16\pi\beta_e\hbar}{3}\right)^2 \gamma_N\gamma_{N'}\frac{1}{\Delta E}$$

$$\cdot \langle 0 \mid \sum_k \sum_j \delta(\mathbf{r}_{kN})\delta(\mathbf{r}_{jN'})\tilde{S}_k\cdot\tilde{S}_j \mid 0\rangle \quad (9.121)$$

ΔE is the mean excitation energy for the excited electronic triplet states and the Dirac δ functions impose the condition that the integral has nonzero contributions only when the electrons are at the nuclei.

Unfortunately the computation of $J_{NN'}$ by Eq. (9.121) has proved to be a very difficult problem with many pitfalls. A good recent review of proton coupling constant calculations has been given by Davies.[23] We do not wish to go into these calculations except to mention that not only has there been difficulty computing the correct *magnitude* of J, in many cases the *sign* of J has been incorrectly calculated.

The reader should now wonder where or how the sign of J occurs experimentally. A quick look back at the theory of the experimental determination of J by ordinary NMR (Sec. 9.5) shows that the sign of J is indeterminate for the cases considered in detail. For simplicity in our previous discussions however, we have always taken J to be positive. It is possible to determine *relative* signs of coupling constants by conventional NMR spectroscopy for some nuclear spin systems such as AMX types. Recently nuclear magnetic double resonance has also proved very useful in determining relative signs of coupling constants.[24] *Absolute* signs of coupling constants may, in principle, be determined from molecular beam

---

[21] See footnote 2, chapter 8.

[22] N. F. Ramsey, *Phys. Rev.* **91,** 303 (1953).

[23] D. W. Davies, *The Theory of the Electric and Magnetic Properties of Molecules* (Wiley, New York, 1967), chapter 4, Sec. 3.2.

[24] J. D. Baldeschwieler and E. W. Randall, *Chem. Rev.* **63,** 81 (1963).

resonance experiments,[25] but this is difficult for most molecules of chemical interest. Recently, Bernheim and Lavery[26] have determined the absolute sign of $J_{HF} (J_{HF} > 0)$ for CH_3F by means of NMR measurements in a *nematic* liquid.

It is easy to see qualitatively that nuclear spin-spin coupling constants might not all be of the same sign. We showed in Fig. 9.8(a) the most stable coupling configuration of the nuclear and electron magnetic moments. This configuration of spins might be written ↑ ↓ ↑ ↓ for simplicity. The convention of Eq. (9.121) is such that the nuclear-spin configuration ↑ ↓ is defined to lead to a positive coupling constant between nuclei A and B. Suppose now a second atom C is covalently bonded to B. Then A and C are coupled through *two* bonds, the lowest energy configuration of magnetic dipoles being ↑ ↓ ↑ ↓ ↑ ↓ ↑. Thus the magnetic moments of nuclei A and C are in the configuration ↑ ↑. Since we defined opposite spin projections to lead to positive J we see that the A-C coupling through two bonds must lead to negative J. If the same through-bond argument is extended to longer chains it is seen that nuclei separated by an *odd* number of bonds lead to *positive* coupling constants while those separated by an *even* number of bonds lead to *negative* coupling constants.

It does appear that the coupling constants of directly-bonded atoms are always positive. However the coupling constants of nuclei separated by an even number of bonds are not always negative. For example the proton coupling constants in benzene or its derivatives are all positive. Likewise in

$$\begin{array}{ccc} H_1 & & H_2 \\ \diagdown & & \diagup \\ & C{=}C & \\ \diagup & & \diagdown \\ H & & H_3 \end{array}$$

J_{23} is positive. Most of these geminal coupling constants (such as J_{23}) do turn out to be negative, however; for example, $J_{23} = J_{gem} = -12.4$ in CH_4 and $J_{gem} = -1.4$ in

$$\begin{array}{ccc} H & & H \\ \diagdown & & \diagup \\ & C{=}C & \\ \diagup & & \diagdown \\ C(CH_3)_3 & & H \end{array}$$

While *ab initio* calculations of coupling constants still present great difficulties, a number of semi-empirical relations have been observed and derived. We simply mention a few of these.

[25] See for example, G. Graff, W. Paul, Ch. Schlier, *Z. Physik* **153**, 38 (1958).
[26] R. A. Bernheim and B. J. Lavery, *J. Am. Chem. Soc.* **89**, 1279 (1967).

For molecules of the type $H_2C \!\!=\!\! CHX$ the *cis* and *trans* proton coupling constants are related to the electronegativity difference ΔE between X and H.

$$J_{\text{trans}} = 19(1 - 0.17\ \Delta E) \tag{9.122a}$$

$$J_{\text{cis}} = 11.7(1 - 0.34\ \Delta E) \tag{9.122b}$$

It also seems generally true that the geminal coupling constants are proportional to the s character of the carbon-bonding orbitals.

A relationship between s character and coupling constant is not unexpected, since the contact mechanism [Eq. (9.121)] depends upon the electron density at the nucleus. One of the more interesting relationships is that relating the C^{13}—H coupling constant and the s-character of the carbon hybrid orbital. Muller and Pritchard[27] have proposed

$$J_{C^{13}H} = 5.0\rho_{CH} \tag{9.123}$$

where ρ_{CH} is the % s character in the carbon-bonding hybrid orbital. The relation is based upon very simple molecular-orbital arguments and seems to be in agreement with other chemical bonding properties.

Finally we mention two interesting relationships involving angular variables. The equation

$$J = 4.2 - 0.5 \cos \phi + 4.5 \cos 2\phi \tag{9.124}$$

predicts satisfactorily the *vicinal* coupling constants of substituted ethanes.

Some care must be exercised in using the relation since it takes no account of the substituents. Equation (9.124) predicts $J = 8.2$ Hz for *cis* interactions ($\phi = 0°$), $J = 9.2$ for *trans* interactions ($\phi = 180°$), and $J = -0.3$ for $\phi = 90°$C. The relation has proved especially useful for saturated cyclic molecules. Thus in substituted cyclohexanes the axial-axial dihedral angle[28] is approximately 180°, so J (axial-axial) should be about 8 Hz, in agreement

---

[27] N. Muller and D. E. Pritchard, *J. Chem. Phys.* **31**, 1471 (1959).

[28] Molecular models are useful in visualizing these angles.

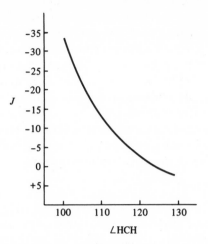

FIG. 9.25. Geminal proton coupling constants as function of ∠HCH.

with experiment. The axial-equatorial and equatorial-equatorial angles are about 60° so J(axial-equatorial) $\sim J$(equatorial-equatorial) is predicted to be about 2 Hz, again in agreement with experiment.

Gutowsky[29] et al. have computed a relationship between geminal proton-proton coupling constants and the CH_2 angle. The computed relation is shown in Fig. 9.25. Note that the magnitude of the coupling constant is predicted to change rapidly between 105° and 115°, and then to show relatively less variation in the region around 120°.

The interested reader can find useful reviews of investigations of nuclear spin-spin coupling constants in several of the issues of the Annual Review of Physical Chemistry. That by Grant[30] is particularly recommended although it is not the most recent.

9.7 OTHER NUCLEAR MAGNETIC RESONANCE TECHNIQUES

In addition to the common high-resolution NMR studies of liquids and gases a number of more specialized techniques have proved useful for certain applications. We simply mention some of these briefly in this section.

A. Nuclear Magnetic Double Resonance

This technique consists essentially of irradiating the sample of interest with a second radiofrequency field. The experiments have been performed

[29] H. S. Gutowsky, M. Karplus, and D. M. Grant, J. Chem. Phys. **31**, 1278 (1959).
[30] David M. Grant, Annual Review of Physical Chemistry (Annual Reviews, Palo Alto, 1964), Vol. 15, pp. 489–528.

several ways, for example, by holding H_0 and ν_1 constant while varying ν_2, or by holding ν_1 and ν_2 constant while varying H_0. The observed spectra differ somewhat depending upon the method. The essential feature is that if a resonance $\nu_0(A)$, which normally shows hyperfine splitting due to coupling with some other nucleus X, is observed while $\nu_0(X)$ is strongly irradiated, the normal splitting of the A resonance disappears. The theory and application of this *spin-decoupling* technique have been reviewed by Baldeschwieler and Randall.[31]

Nuclear magnetic double resonance has found its greatest application in the analysis of complex spectra, and it has also proved extremely useful for determining relative signs of coupling constants.

B.   Relaxation Techniques

The field here is very broad and growing rapidly. A great diversity of experiments have been performed, but the common characteristic is the observation of an NMR signal whose behavior is directly related to the relaxation (see Sec. 9.4 B) of the nuclear spin system. Most often pulse techniques are utilized. These measurements have led to much information concerning relaxation times and relaxation mechanisms. A good review has been given by Redfield[32] and a publication of a Chemical Society symposium[33] is also useful.

C.   Orientation Effects

We have mentioned earlier the direct interaction of two nuclear magnetic dipoles. A Hamiltonian term for this dipole-dipole interaction [as in Eq. (9.54)] is always present and should be added to those of Eqs. (9.62a) and (9.63). If the scalar products of Eq. (9.54) are expanded it may be written (taking the $\tilde{I}$'s as dimensionless)

$$\tilde{\mathcal{3C}}_D = \frac{g_n{}^2\beta_n{}^2}{r^5} \{\tilde{I}_{1x}\tilde{I}_{2x}(r^2 - 3x^2) + \tilde{I}_{1y}\tilde{I}_{2y}(r^2 - 3y^2) + \tilde{I}_{1z}\tilde{I}_{2z}(r^2 - 3z^2)$$

$$- (\tilde{I}_{1x}\tilde{I}_{2y} + \tilde{I}_{1y}\tilde{I}_{2x})3xy - (\tilde{I}_{1y}\tilde{I}_{2z} + \tilde{I}_{1z}\tilde{I}_{2y})3yz$$

$$- (\tilde{I}_{1z}\tilde{I}_{2x} + \tilde{I}_{1x}\tilde{I}_{2z})3zx\} \text{ ergs} \quad (9.125a)$$

or in matrix notation

$$\tilde{\mathcal{3C}}_D = g_n{}^2\beta_n{}^2\mathbf{I}_1\mathbf{DI}_2 \quad (9.125b)$$

where $\mathbf{D}$ is the matrix form of the dipolar tensor and $\mathbf{I}_1$ and $\mathbf{I}_2$ are row and

---

[31] See footnote 24.

[32] A. G. Redfield, *Advances in Magnetic Resonance* (Academic, New York, 1965), Vol. I, pp. 1–32.

[33] *Molecular Relaxation Processes* (Academic, New York, 1966), Chem. Soc. (London), Publ. *20*.

column matrices, respectively. Explicitly, $\mathbf{D}$ has the form

$$\mathbf{D} = \begin{pmatrix} (r^2 - 3x^2)/r^5 & -3xy/r^5 & -3xz/r^5 \\ -3xy/r^5 & (r^2 - 3y^2)/r^5 & -3yz/r^5 \\ -3xz/r^5 & -3yz/r^5 & (r^2 - 3z^2)/r^5 \end{pmatrix} \qquad (9.126)$$

It is seen easily that the trace of the tensor (matrix) is zero, and that the tensor is diagonal if $\mathbf{r}$ lies along any one of the coordinate axes.

For a randomly rotating molecule (in a gas or liquid) the tensor elements of $\mathbf{D}$ in the *space-fixed* axis system X, Y, Z fluctuate randomly. On the average then, the principal (diagonal) values of $\mathbf{D}$ in the space-fixed axis system have the value one-third the sum of the principal values of $\mathbf{D}$ in the molecule-fixed axis system, that is,

$$D_{XX} = D_{YY} = D_{ZZ} = 1/3 \operatorname{Tr} \mathbf{D}(x, y, z) \qquad (9.127)$$

But the trace of $\mathbf{D}$ is zero as shown in (9.126) so the dipole-dipole interaction vanishes. This is why it was not necessary to consider the dipole-dipole Hamiltonian term in the earlier discussions involving liquids, except with respect to linewidths.

For oriented molecules such as in solids, the situation is different. Let us consider two protons at fixed separation r in a magnetic field along the z direction. We treat $\tilde{\mathcal{K}}_D$ as a small perturbation to $\tilde{\mathcal{K}}^{(0)}$ and neglect the indirect spin–spin coupling term $\tilde{\mathcal{K}}^{(1)}$. If we wish to compute the energy correct to first order the basis functions of Eqs. (9.84) are appropriate. The energies (in Hz) correct to first order are

$$E_1 = -\nu_0 + \frac{1}{4\hbar} (r^2 - 3z^2) g_n^2 \beta_n^2 / r^5$$

$$= -\nu_0 + \frac{1}{4\hbar} g_n^2 \beta_n^2 (1 - 3\cos^2 \theta)/r^3 \qquad (9.127\mathrm{a})$$

$$E_2 = \nu_0 + \frac{1}{4\hbar} g_n^2 \beta_n^2 (1 - \cos^2 \theta)/r^3 \qquad (9.127\mathrm{b})$$

$$E_3 = -\frac{1}{2\hbar} g_n^2 \beta_n^2 (1 - 3\cos^2 \theta)/r^3 \qquad (9.127\mathrm{c})$$

$$E_4 = 0 \qquad (9.127\mathrm{d})$$

where $\cos \theta = z/r$.

The permitted transitions are the previous ones of Eq. (9.85), which lead to

$$\nu_{1\to3} = \nu_0 - \frac{3}{4\hbar} g_n{}^2 \beta_n{}^2 (1 - 3\cos^2\theta)/r^3 \qquad (9.128)$$

$$\nu_{3\to2} = \nu_0 + \frac{3}{4\hbar} g_n{}^2 \beta_n{}^2 (1 - 3\cos^2\theta)/r^3$$

A doublet is predicted therefore, with a separation of

$$\Delta\nu = \frac{\frac{3}{2}g_n{}^2\beta_n{}^2(1 - 3\cos^2\theta)}{\hbar r^3} \text{ Hz} \qquad (9.129)$$

Note that when $3\cos^2\theta = 1$ the doublet coalesces into a single line.

The important point is that a measurement of $\Delta\nu$ as a function of angle permits the *determination of the internuclear distance r*. This method has found some use for crystalline substances but unfortunately the linewidths are usually very large, approximately equal to $\Delta\nu$ itself as given by Eq. (9.129). Thus resolution of peaks is a major problem for solids. Some relief from this difficulty has been achieved by analyzing line widths with Van Vleck's second moment formula.[34]

More recently it has been found that solute molecules dissolved in nematic liquid crystals are highly oriented, yet give very sharp NMR spectral lines (a few Hz wide). This has stimulated much new work aimed at determining molecular geometries[35] by analysis of the dipole-dipole effects in NMR spectra. The results have been very encouraging and with proper calibration are in good agreement with known structural data.

SUPPLEMENTARY REFERENCES

A. Abragam, *The Principles of Nuclear Magnetism* (Clarendon Press, Oxford, England, 1961).

E. R. Andrew, *Nuclear Magnetic Resonance* (Cambridge U.P., Cambridge, 1955).

A. Carrington and A. D. McLachlan, *Introduction to Magnetic Resonance* (Harper and Row, New York, 1967).

D. W. Davies, *The Theory of the Electric and Magnetic Properties of Molecules* (Wiley, London, 1967).

[34] J. H. Van Vleck, *Phys. Rev.* **74,** 1168 (1948).

[35] See, for example, G. Englert and A. Saupe, *Mol. Crystals* **1,** 503 (1966); L. C. Snyder and S. Meiboom, *J. Chem. Phys.* **47,** 1480 (1967).

J. R. Dyer, *Applications of Absorption Spectroscopy of Organic Compounds* (Prentice-Hall, Englewood Cliffs, N.J., 1965).

J. A. Pople, W. G. Schneider, and H. J. Bernstein, *High-Resolution Nuclear Magnetic Resonance* (McGraw-Hill, New York, 1959).

N. F. Ramsey, *Nuclear Moments* (Wiley, New York, 1953).

R. M. Silverstein and G. C. Bassler, *Spectrometric Identification of Organic Compounds* (Wiley, New York, 1967), 2nd ed.

C. P. Slichter, *Principles of Magnetic Resonance* (Harper and Row, New York, 1963).

CHAPTER 10

ELECTRON PARAMAGNETIC RESONANCE

10.1 GENERAL CONSIDERATIONS

The first observation of direct transitions between the Zeeman energy states of a paramagnetic substance was performed in 1945 using very small magnetic fields and radiofrequency radiation.[1] It became clear immediately that great improvements in signal observation could be achieved by utilizing microwave radiation and larger magnetic field strengths. Consequently the many years of fruitful research in electron paramagnetic resonance since 1945 have been carried out in the microwave spectral region.

What sort of substances exhibit electron paramagnetic resonance (EPR) spectra? We might answer this by stating that in general EPR spectra are exhibited by those substances which are classed as *paramagnetic*.[2] These substances, whose magnetic susceptibility follows the Curie law,

$$\chi \propto \frac{1}{T} \tag{10.1}$$

are those having magnetic moments produced by electron-spin and orbital-

---

[1] E. Zavoisky, *J. Phys. USSR* **9**, 245 (1945).

[2] See, for example, C. Kittel, *Introduction to Solid State Physics* (Wiley, New York, 1956), chapter 9.

electronic angular momentum.[3] Thus we can state somewhat more concisely that EPR spectra are expected for substances having nonzero electron-spin angular momentum S, except in cases for which cancellation of spin and orbital contributions occur or when certain other experimental difficulties arise.

The classes of substances which have been studied by EPR are numerous. First we might mention atoms with nonsinglet ($S \neq 0$) ground states, such as hydrogen, oxygen, nitrogen, fluorine, etc., but not helium or neon. Atomic EPR spectra have been observed in both the gas phase (using continuous production methods) and in the solid phase (by trapping at liquid-helium temperatures).

A relatively small number of stable molecules have an odd number of electrons or unpaired spins. Included in this small class of more or less stable species are NO, NO_2, NF_2, ClO_2, O_2, and diphenylpicryl hydrazil (DPPH). Of this class oxygen is the only triplet species and DPPH is one of the few stable organic molecules with an odd number of electrons. All of these substances and the previously mentioned atoms are commonly termed *free radicals*, although the term is a rather imprecise one. We take it to mean a molecule or molecular fragment having an odd number of electrons or unpaired electrons in its valence or bonding orbitals.

Numerous *unstable* free radicals have been studied by EPR in either solid, liquid or gas phases. These include inorganic species such as OH, BrO, CO_2^-, CH_3, and NS. Note that CO_2^- is the first ionic radical that we have mentioned. A much larger class of unstable free radicals are of the organic variety such as the negative ions of naphthalene, benzene, p-benzo-semiquinone, and nitrobenzene; the positive ions of p-phenylenediamine and anthracene; and the neutral radicals ethyl, isopropyl, cyclopentadienyl, benzyl, and allyl. Investigations of these organic radicals are generally performed only in the solid or liquid phases, and it is seen later that the EPR spectra are rather different for the two cases. Certainly the study of organic radical EPR spectra is one of the most important and active areas in the entire field.

The remaining large class of substances amenable to EPR study consists of the transition metal and rare-earth ions and their complexes. These ions, by virtue of their incomplete d or f electronic shells, have unpaired electrons and are therefore generally paramagnetic. The observed spectra are often strongly influenced by the ionic environment (the ligand or crystal field) and therefore provide information concerning the environment as well as the metal ion itself.

From this brief summary it should be clear that a great variety of substances have been studied by electron paramagnetic resonance spec-

---

[3] Nuclei with $I \neq 0$ produce paramagnetism also but the usual static methods of measuring magnetic susceptibilities cannot detect this property since the nuclear moments are ca. 1000 times smaller than the electronic moments.

troscopy. Indeed, the list is not yet complete and could be extended much further. Certainly the crystal defects such as F and V centers[4] should be mentioned, since they yield EPR spectra. These damage centers in crystals are typically radiation induced, and are simply sites in the lattice having trapped electrons or trapped electron deficiencies (that is, positive "holes") in the case of F and V centers, respectively. In any case these paramagnetic centers do yield spectra of the same general nature as those of the previously described substances.

Slightly farther away from the interests of the present chapter are magnetic resonance studies of metal conduction electrons, semiconductors,[5] and ferromagnetic materials.[6] The interested reader should consult the references for information about these and related studies.

The experimental apparatus used for EPR experiments has much in common with the microwave (rotational) spectrometer described in Sec. 7.1. This may be seen from Fig. 10.1 which shows a block diagram of a simple EPR spectrometer. The klystron is the typical source of radiation and signal detection is performed by silicon crystal diodes. Amplification and phase sensitive detection is performed as described in Sec. 7.1 also.

An EPR spectrometer differs from a microwave spectrometer in two major ways. The first is, of course, the use of a magnet in the resonance

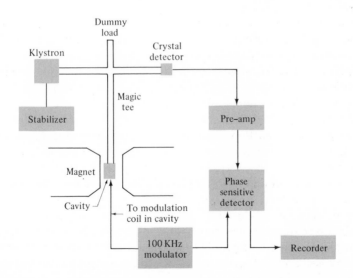

FIG. 10.1.   Simple block-diagram of balanced bridge EPR spectrometer.

---

[4] See footnote 2, chapter 17.

[5] G. Feher, *Phys. Rev.* **114**, 1219 (1959).

[6] W. A. Yager and R. M. Bozorth, *Phys. Rev.* **72**, 80 (1947).

experiments. Here the requirements as to field homogeneity are similar to those for NMR (Sec. 9.1). However, the required magnetic fields for an X band (10 GHz) spectrometer are only about 3000 G, so the magnet power and cooling requirements are considerably less.

The second distinguishing feature is the use of a microwave cavity in a "balanced bridge" *reflection* system in place of the *transmission* cell used in microwave spectroscopy. In this method microwaves flow into the "magic-T" and then into each of the two arms containing the cavity and the dummy load. If these two arms were perfectly matched no power would flow down the fourth arm of the magic-T to the crystal. Only when an unbalance exists does power flow to the crystal. Thus if balance exists initially, no signal appears at the detector until power is absorbed in the cavity by sweeping the magnetic field through the resonance condition of the sample in the cavity. While this description is a slight oversimplification of the actual operation of the cavity reflection system the general ideas are satisfactory for our purposes.

The cavities used in EPR spectrometers are typically rectangular or cylindrical in shape and are constructed of a good conducting metal. A given cavity stores incident microwaves efficiently only at a specific frequency determined by the precise geometry of the cavity. Simple discussions of cavity design have been given by Ingram[7] and Townes and Schawlow[8] and are not considered here.

Samples in liquid or powder form are usually inserted into the cavity in thin-walled glass or quartz tubes through an opening in the cavity wall. The position of the opening depends on the cell design but is normally located such that the sample is in the position of maximum microwave magnetic-field strength. Solid (single crystal) samples are placed in the cavity on the ends of nonlossy rods or holders.

As shown in Fig. 10.1, the entire cavity is positioned between the magnet pole faces. EPR systems are naturally operated at fixed frequency because of the fixed-frequency nature of cavities. Thus it is common to utilize some method of klystron stabilization as shown in the figure. Spectral scans are then obtained by varying the magnetic field via the current through the magnet. Amplitude modulation of the absorption signals is achieved by small modulation coils introduced into the cavity. Typical modulation frequencies are in the range of 5-100 KHz with magnetic field amplitudes that are usually less than one gauss for narrow resonance lines. In the typical mode of operation EPR signals appear as derivatives of the absorption line since the modulation amplitude is chosen to be several times smaller than the linewidth.

[7] D. J. E. Ingram, *Free Radicals as Studied by Electron Spin Resonance* (Butterworths, London, 1958).

[8] C. H. Townes and A. L. Schawlow, *Microwave Spectroscopy* (McGraw-Hill, New York, 1955).

Considerations concerning the sensitivity of an EPR spectrometer are rather complex and are not treated here. Typical spectrometers have the ability to detect a minimum of about 10^{11} or 10^{12} radicals or unpaired spins. Translated to moles this represents 10^{-13} to 10^{-12} moles. Thus the sensitivity of electron paramagnetic resonance spectroscopy is very good and has made possible extensive studies of many unstable free radical species.

Books and articles supplementing this cursory discussion of the experimental aspects of EPR are listed at the end of the chapter.

10.2 ENERGY LEVELS AND SPECTRUM OF THE HYDROGEN ATOM

A. Energy Levels

As has been known since the pioneering experiments of Stern and Gerlach,[9] an atom or molecule may exhibit a magnetic moment attributable to the electron *spin* angular momentum of the species, that is

$$\boldsymbol{\mu}_S = -g_e\beta_e\mathbf{S} \tag{10.2}$$

where β_e, the Bohr magneton, has the value 0.9273×10^{-20} erg G^{-1}. Simple theories or even the relativistic Dirac theory predict that $g_e = 2.000$ for a single free electron. More sophisticated quantum corrections lead to a free-spin g_e value of 2.00232. Later we find it useful to take g to be an experimentally determined quantity for most systems. Nevertheless, contrary to the situation with nuclei, electron g factors are always positive and usually have values close to g_e.[10]

In addition to the intrinsic spin magnetic moment, an electron having nonzero *orbital* angular momentum has a magnetic moment contribution of

$$\boldsymbol{\mu}_L = -\beta_e\mathbf{L} \tag{10.3}$$

Thus the net magnetic moment of a system of electrons is

$$\boldsymbol{\mu} = -(g_e\beta_e\mathbf{S} + \beta_e\mathbf{L}) \tag{10.4}$$

For a hydrogen atom in its ground state ($L = 0$) we need not consider $\boldsymbol{\mu}_L$. The potential energy or Hamiltonian term produced by interaction of $\boldsymbol{\mu}_S$ with a static magnetic field is then

$$\mathfrak{IC}_{\text{Zeeman}} = -\boldsymbol{\mu}_S\cdot\mathbf{H} \tag{10.5a}$$

or

$$\widetilde{\mathfrak{IC}}_0 = \widetilde{\mathfrak{IC}}_{\text{Zeeman}} = g_e\beta_e\widetilde{S}\cdot\tilde{H} \tag{10.5b}$$

[9] O. Stern and W. Gerlach, *Z. Physik* **8**, 110; **9**, 349 (1922).

[10] If S has the usual units of erg sec, the right-hand side of Eq. (10.2) should be divided by $\hbar$ as in Eq. (9.4b). The matrix elements of $\widetilde{S}$, $\widetilde{S}_x$, $\widetilde{S}_y$, and $\widetilde{S}_z$ are in units of $\hbar$ as usual, so the $\hbar$ in the denominator always cancels out. For simplicity in this chapter we take the matrix elements of angular momentum to be dimensionless and consequently suppress the $\hbar$ in equations such as (10.2) in this chapter.

With the customary choice of H along the Z axis, we obtain

$$\widetilde{\mathcal{K}}_0 = g_e\beta_e\widetilde{S}_Z H_Z \tag{10.5c}$$

While $\widetilde{\mathcal{K}}_0$ here is about 2000 times that for the nuclear Zeeman term (chapter 9) it still represents a very small energy compared to the electronic energy (that is, the kinetic and coulombic energies). We also find when we speak of molecules that the electron Zeeman energy is small compared to vibrational energies. In these cases, as outlined in chapter 3, we consider implicitly that the g values are averaged over electronic and vibrational wave functions. For small molecules in the gas phase the rotational energies are comparable to the Zeeman term and because the rotational and spin angular momenta become coupled it is not possible in general to separate rotational and spin parts of the Hamiltonian.

We should now add to $\widetilde{\mathcal{K}}_0$ of Eq. (10.5c) the nuclear Zeeman term described in the last chapter, namely $-g_n\beta_n\widetilde{I}_Z H_Z$, where we do not worry about distinguishing H_Z from $H_{\text{effective}}$ since the screening is very small as previously described.

There is yet one more term which needs to be added to $\widetilde{\mathcal{K}}_0$. This is the Fermi contact term which has been mentioned in connection with the nuclear spin-spin interaction in chapter 9. For our purposes here we may write this as $A\widetilde{S}\cdot\widetilde{I}$ where A is the coupling constant.[11] The total Hamiltonian is therefore

$$\widetilde{\mathcal{K}} = g_e\beta_e\widetilde{S}_Z H_Z - g_n\beta_n\widetilde{I}_Z H_Z + A\widetilde{S}\cdot\widetilde{I} \tag{10.6}$$

In the absence of the coupling term it is clear that a wave function constructed as the product of an electron spin and a nuclear spin wave function would be an exact eigenfunction. Since the coupling term is small these same product functions are satisfactory as zero-order functions for a perturbation calculation. Since $S = \frac{1}{2}$ and $I = \frac{1}{2}$ for the hydrogen atom, the zero-order functions are

$$\psi_1 = \alpha_e\alpha_n$$
$$\psi_2 = \alpha_e\beta_n$$
$$\psi_3 = \beta_e\alpha_n \tag{10.7}$$
$$\psi_4 = \beta_e\beta_n$$

where we use α and β as eigenfunctions corresponding to spin eigenvalues of $+\frac{1}{2}$ or $-\frac{1}{2}$, respectively. The subscripts indicate either the electron or the proton.

The energies of the four states are easily calculated to first order. $\widetilde{S}_Z$ and $\widetilde{I}_Z$ have entirely diagonal matrices, of course, while $\widetilde{S}\cdot\widetilde{I} = \widetilde{S}_X\widetilde{I}_X + \widetilde{S}_Y\widetilde{I}_Y + \widetilde{S}_Z\widetilde{I}_Z$ has both diagonal and off-diagonal elements. We obtain the

[11] E. Fermi, *Z. Physik* **60**, 320 (1930).

energies to first order as (writing $H_Z = H$)

$$E_1 = \tfrac{1}{2}g_e\beta_e H - \tfrac{1}{2}g_n\beta_n H + \tfrac{1}{4}A$$
$$E_2 = \tfrac{1}{2}g_e\beta_e H + \tfrac{1}{2}g_n\beta_n H - \tfrac{1}{4}A$$
$$E_3 = -\tfrac{1}{2}g_e\beta_e H - \tfrac{1}{2}g_n\beta_n H - \tfrac{1}{4}A \qquad (10.8)$$
$$E_4 = -\tfrac{1}{2}g_e\beta_e H + \tfrac{1}{2}g_n\beta_n H + \tfrac{1}{4}A$$

For example, E_1 is obtained as follows.

$$\begin{aligned}
E_1 &= \langle \alpha_e\alpha_n \mid \tilde{\mathcal{H}} \mid \alpha_e\alpha_n \rangle \\
&= g_e\beta_e H \langle \alpha_e \mid \tilde{S}_Z \mid \alpha_e \rangle - g_n\beta_n H \langle \alpha_n \mid \tilde{I}_Z \mid \alpha_n \rangle \\
&\quad + A \langle \alpha_e\alpha_n \mid \tilde{S}_X\tilde{I}_X + \tilde{S}_Y\tilde{I}_Y + \tilde{S}_Z\tilde{I}_Z \mid \alpha_e\alpha_n \rangle \\
&= g_e\beta_e H(\tfrac{1}{2}) - g_n\beta_n H(\tfrac{1}{2}) \\
&\quad + A \langle \alpha_e \mid \tilde{S}_Z \mid \alpha_e \rangle \langle \alpha_n \mid \tilde{I}_Z \mid \alpha_n \rangle \\
&= \tfrac{1}{2}g_e\beta_e H - \tfrac{1}{2}g_n\beta_n H + \tfrac{1}{4}A \qquad (10.9)
\end{aligned}$$

Note that $\tilde{S}_X\tilde{I}_X$ and $\tilde{S}_Y\tilde{I}_Y$ have no diagonal elements. Also note that the terms involving H are naturally considered to be the zero-order energies, while the $\pm(\tfrac{1}{4})A$ terms are the first-order contributions of the $A\tilde{S}\cdot\tilde{I}$ term.

We could write the complete 4×4 matrix $\mathcal{H}$ and find the eigenvalues by direct diagonalization. Instead we continue with perturbation theory to find the second-order contributions. Only the terms in X and Y are involved here since only these terms have off-diagonal elements. This procedure leads to

$$E_1{}^{(2)} = 0$$
$$E_2{}^{(2)} = \tfrac{1}{4}[A^2/(g_e\beta_e H + g_n\beta_n H)]$$
$$E_3{}^{(2)} = -\tfrac{1}{4}[A^2/(g_e\beta_e H + g_n\beta_n H)] \qquad (10.10)$$
$$E_4{}^{(2)} = 0$$

As an example, $E_2{}^{(2)}$ is evaluated as follows.

$$E_2{}^{(2)} = -\sum_{i \neq 2} \frac{\mid \langle \psi_i \mid \tilde{S}_X\tilde{I}_X + \tilde{S}_Y\tilde{I}_Y \mid \psi_2 \rangle \mid^2 A^2}{E_i{}^{(0)} - E_2{}^{(0)}} \qquad (10.11a)$$

Now $\psi_2 = \mid \alpha_e\beta_n \rangle$ so $\tilde{S}_X\tilde{I}_X$ and $\tilde{S}_Y\tilde{I}_Y$ may have nonzero elements connecting only with $\psi_i = \beta_e\alpha_n$. Thus we reduce $E_2{}^{(2)}$ to

$$E_2{}^{(2)} = \frac{-A^2 \mid \langle \beta_e\alpha_n \mid \tilde{S}_X\tilde{I}_X \mid \alpha_e\beta_n \rangle + \langle \beta_e\alpha_n \mid \tilde{S}_Y\tilde{I}_Y \mid \alpha_e\beta_n \rangle \mid^2}{E_3{}^{(0)} - E_2{}^{(0)}} \qquad (10.11b)$$

The matrix elements are easily evaluated from our well-known spin-$\tfrac{1}{2}$ matrices and the denominator contains the zero-order parts of Eqs. (10.8).

We find

$$E_2^{(2)} = \frac{\mid \frac{1}{2}(\frac{1}{2}) + (-\frac{1}{2}i)(\frac{1}{2}i) \mid^2}{g_e\beta_e H + g_n\beta_n H} A^2$$

$$= \tfrac{1}{4}[A^2/(g_e\beta_e H + g_n\beta_n H)] \tag{10.11c}$$

Finally, then, the energies correct to second order are

$$E_1 = \frac{1}{2} H(g_e\beta_e - g_n\beta_n) + \frac{1}{4} A$$

$$E_2 = \frac{1}{2} H(g_e\beta_e + g_n\beta_n) - \frac{1}{4}\left(A - \frac{A^2}{g_e\beta_e H + g_n\beta_n H}\right)$$

$$E_3 = \frac{1}{2} H(-g_e\beta_e - g_n\beta_n) - \frac{1}{4}\left(A + \frac{A^2}{g_e\beta_e H + g_n\beta_n H}\right) \tag{10.12}$$

$$E_4 = \frac{1}{2} H(-g_e\beta_e + g_n\beta_n) + \frac{1}{4} A$$

B. Selection Rules and Spectrum

The transition probabilities in the presence of electromagnetic radiation are easily determined by use of our basic formula, Eq. (5.29). In this case the time-dependent perturbation is

$$\mathcal{H}'(t) = -\boldsymbol{\mu}_S \cdot \mathbf{H}(t)$$

$$= -2\boldsymbol{\mu}_S \cdot \mathbf{H}_1 \cos 2\pi\nu t \tag{10.13}$$

where ν is the frequency of the electromagnetic radiation and $\mathbf{H}_1$ the amplitude of the magnetic vector. But by use of Eq. (10.2), we get

$$\mathcal{H}'(t) = 2g_e\beta_e \mathbf{S} \cdot \mathbf{H}_1 \cos 2\pi\nu t \tag{10.14}$$

In EPR experiments the cavity is normally oriented such that $\mathbf{H}_1$ is perpendicular to the static magnetic field; that is, it is entirely in the X (or Y) direction. Thus we find

$$\widetilde{\mathcal{H}}'(t) = 2g_e\beta_e \widetilde{S}_X H_1 \cos 2\pi\nu t \tag{10.15}$$

Thus the matrix element of Eq. (5.29) takes the form

$$F_{mn} = g_e\beta_e H_1 \langle m \mid \widetilde{S}_X \mid n \rangle \tag{10.16}$$

so the transition probability is

$$P_{mn} \propto \mid \langle m \mid \widetilde{S}_X \mid n \rangle \mid^2 \tag{10.17}$$

for the EPR experiment.

Since the eigenstates to first order are merely products of electron and nuclear functions, ϕ_e and ϕ_n, respectively, the matrix element may be written as

$$\langle \phi_e \phi_n \mid \tilde{S}_X \mid \phi_e' \phi_n' \rangle$$

for two states ψ and ψ'. But $\tilde{S}_X$ does not operate upon the nuclear function so the scalar product factors as

$$\langle \phi_e \phi_n \mid \tilde{S}_X \mid \phi_e' \phi_n' \rangle = \langle \phi_n \mid \phi_n' \rangle \langle \phi_e \mid \tilde{S}_X \mid \phi_e' \rangle \tag{10.18}$$

Thus the transition probability is zero unless $\phi_n = \phi_n'$, which means

$$\Delta M_I = 0 \tag{10.19a}$$

is required. The M_I are the eigenvalues of I_Z which serve to label the different nuclear spin states.

The remaining factor in (10.18) contains matrix elements of $\tilde{S}_X$. The well-known results are that this factor is zero unless

$$\Delta M_S = \pm 1 \tag{10.19b}$$

Equations (10.19a) and (10.19b) are the first-order selection rules for the hydrogen atom. For the states listed in Eqs. (10.7), we find two permitted transitions,

$$\psi_3 \rightarrow \psi_1, \qquad \psi_4 \rightarrow \psi_2 \tag{10.20}$$

with transition frequencies of (to first order)

$$\nu_{3 \rightarrow 1} = \frac{E_1 - E_3}{h} = \frac{g_e \beta_e H + \frac{1}{2} A}{h}$$

$$\nu_{4 \rightarrow 2} = \frac{E_2 - E_4}{h} = \frac{g_e \beta_e H - \frac{1}{2} A}{h} \tag{10.21}$$

Thus a doublet is predicted with spacing of A, the hyperfine coupling constant.

This was the result first observed by Beringer and Rawson[12] in 1952. A more accurate measurement of A comes from molecular beam measurements,[13] however, the value being 1420.405 MHz. The value of g_e obtained from the EPR study of the hydrogen atom is not exactly the "free-electron" value of 2.002322 but is about 18 ppm smaller because of its binding to the proton. Note that at a microwave frequency of 10 000 MHz, the two tran-

[12] R. Beringer and E. B. Rawson, *Phys. Rev.* **87**, 228 (1952).
[13] P. Kusch, *Phys. Rev.* **100**, 1188 (1955).

sitions of Eq. (10.21) occur at magnetic fields of

$$H_{3 \to 1} = \frac{(10000 - 1420)(10^6)(6.625 \times 10^{-27})}{2.0023(9.273 \times 10^{-21})} = 3061 \text{ G}$$

and

$$H_{4 \to 2} = \frac{(10000 + 1420)(10^6)(6.625 \times 10^{-27})}{2.0023(9.273 \times 10^{-21})} = 4075 \text{ G}$$

Since the $\tilde{S}_X \tilde{I}_X$ and $\tilde{S}_Y \tilde{I}_Y$ terms in the Hamiltonian have off-diagonal elements coupling the ψ_2 and ψ_3 states, these are not really eigenstates of the complete Hamiltonian. This means that ψ_2 and ψ_3 become mixed, that is the states $\alpha_e \beta_n$ and $\beta_e \alpha_n$ become $\alpha_e \beta_n + \delta \beta_e \alpha_n$ and $\beta_e \alpha_n + \delta \alpha_e \beta_n$, respectively, where δ is the mixing coefficient. It is easily seen then, that the transitions $\psi_2 \to \psi_1$ and $\psi_4 \to \psi_3$ become weakly allowed, the intensity being proportional to δ^2. For magnetic fields of 5000 G or greater, these "forbidden" transitions have intensities that are less than 0.1% of those of the allowed ones, and consequently are not easily observed.

Finally it is interesting to observe that the hyperfine term $A\tilde{S} \cdot \tilde{I}$ persists when $H = 0$, in which case the Zeeman terms disappear. This term produces a "zero-field" splitting of the hydrogen-atom ground state into two states separated by an energy of $A = 1420$ MHz. This is most easily seen by observing that in the absence of the external magnetic field I and S are coupled to produce the total angular momentum F, that is

$$\mathbf{F} = \mathbf{I} + \mathbf{S} \tag{10.22}$$

Then we see that

$$\tilde{I} \cdot \tilde{S} = \frac{\tilde{F}^2 - \tilde{I}^2 - \tilde{S}^2}{2} \tag{10.23}$$

follows directly. F may have the values 1 or 0 according to (10.22) so the possible eigenvalues of $A\tilde{I} \cdot \tilde{S}$ are

$$\langle F | A\tilde{I} \cdot \tilde{S} | F \rangle = \frac{F(F + 1) - I(I + 1) - S(S + 1)}{2} \tag{10.24}$$

This equation leads to values of $+\frac{1}{4}A$ and $-\frac{3}{4}A$.

The transition between these two states has been observed in molecular beam experiments and leads to the accurate value of A mentioned earlier. It is interesting also that certain stellar sources emit radiation of this frequency, indicating the existence of a population inversion in the hydrogen-atom source. It has also been possible to build a *maser* operating at this frequency.

10.3 SPECTRA OF ORGANIC RADICALS IN LIQUID SOLUTIONS

A. Energies of Radicals Having One Unpaired Spin ($S = \frac{1}{2}$)

We are not particularly interested in the EPR spectra of atoms. However, the theory presented for the hydrogen atom in the gas phase may be taken over with no fundamental change to describe the spectra of organic radicals in solution. The major difference, in fact, is that the Fermi contact coupling constant is normally much smaller than for hydrogen atom. Thus the hyperfine splittings of the EPR absorption lines of organic radicals are considerably smaller, usually in the range of several gauss for proton couplings.

The g factor for organic radicals is not expected to have the precise free-spin value g_e, but due to the absence of any marked orbital magnetic moment contributions its value is expected to differ little from 2.002. It might be thought that this electron g factor might play a role for electrons similar to that of the screening constant σ or chemical shift δ for nuclei. However, it seems that no generally useful simple correlations exist for organic radical g factors, and since the variations are so small they are commonly measured with rather low precision.

In the last chapter we discussed the direct spin-spin interaction of two nuclei and showed that in liquids or gases the random motions of the molecules would cause this Hamiltonian term to average to zero. For the present case a $\mathbf{\mu}_S \cdot \mathbf{\mu}_I$ term also exists, and again averages to zero for rapidly tumbling radicals in the liquid (or gas) phase. This is why no such term was included for the hydrogen atom and why it need not be included in the present case. The direct dipolar electron spin-nuclear spin coupling does play a role in the relaxation processes of the radicals in solution, and is a necessary Hamiltonian term for radicals in the solid state.

We can now write the appropriate Hamiltonian using Eq. (10.6) as a basis. Since the selection rules require $\Delta M_I = 0$, the nuclear Zeeman term makes no contribution to the observed frequencies as was shown in Eqs. (10.21). Thus we omit this term from the Hamiltonian in the present discussion. Also, since the coupling constants are much smaller than for the hydrogen atom case, the second-order terms arising from $\widetilde{S}_X \cdot \widetilde{I}_X$ and $\widetilde{S}_Y \cdot \widetilde{I}_Y$ can be neglected. For a free radical having $S = \frac{1}{2}$ and *one* interacting nucleus we write simply

$$\widetilde{\mathcal{K}} = g\beta_e \widetilde{S}_Z H + A\,\widetilde{S}_Z \widetilde{I}_Z \tag{10.25}$$

If there are n sets of equivalent nuclei,[14] each set having m_i members

---

[14] An equivalent set is one whose member nuclei have the same value of the coupling constant A.

of nuclear spin I_i and coupling constant A_i, the Hamiltonian is written

$$\tilde{\mathcal{H}} = g\beta_e \tilde{S}_Z H + \sum_{i=1}^{n} A_i \tilde{S}_Z m_i \tilde{I}_{Zi} \tag{10.26}$$

Since $m_i \tilde{I}_{Zi}$ is the total nuclear spin angular momentum along the Z axis, we may rewrite (10.26) as

$$\tilde{\mathcal{H}} = g\beta_e \tilde{S}_Z H + \sum_{i=1}^{n} A_i \tilde{S}_Z \tilde{F}_{Zi} \tag{10.27}$$

The energy levels of systems described by the first-order Hamiltonian (10.27) are then the diagonal elements using basis functions labeled by their M_S and M_F values, that is,

$$E = \langle M_S M_F \mid \tilde{\mathcal{H}} \mid M_S M_F \rangle = g\beta_e H M_S + \sum_i A_i M_S M_{Fi} \tag{10.28}$$

For $S = \frac{1}{2}$ systems M_S may have only the values $\pm\frac{1}{2}$. The M_{Fi} values are determined by the number and spin of the nuclei in each set.

B.   Selection Rules and Spectra for $S = \frac{1}{2}$ Systems

The selection rules are identical to those described for the hydrogen atom in Sec. 9.2. Since M_{Fi} is the sum of M_{Ii} for any set, the selection rule $\Delta M_I = 0$ is modified to read $\Delta M_F = 0$ for this case. Thus the observed transitions obey (to first order)

$$\Delta M_S = \pm 1$$
$$\tag{10.29}$$
$$\Delta M_{Fi} = 0$$

This leads to the following expression for the frequencies of the allowed transitions:

$$\nu = g\beta_e H/h + \sum_i A_i M_{Fi}/h \tag{10.29a}$$

Note that according to Eq. (10.28) A_i has the units of ergs so A_i/h is the coupling constant in Hz. Since most spectra are obtained by varying the magnetic field at constant ν, it is common to express coupling constants in units of gauss. Coupling constants in units of Hz may be converted to units of gauss by application of the factor $h/g\beta_e = 2.80 \text{ MHz/G}$ at $g = 2.00$. For simplicity in our discussion we omit the h in the second term of (10.29a) and write the first term as ν_0. We get therefore

$$\nu = \nu_0 + \sum_i A_i M_{Fi} \tag{10.29b}$$

The application of Eq. (10.29b) is straightforward. Consider the methyl radical ($\cdot C^{12}H_3$) in solution. There is just one set of nuclei having nonzero spin. The three protons lead to an F value of $\frac{3}{2}$ and M_F values of $\frac{3}{2}, \frac{1}{2}, -\frac{1}{2}$ and $-\frac{3}{2}$. By Eq. (10.29b) there are four equally spaced transitions centered about ν_0, namely

$$\nu = \nu_0 + A_H \cdot \begin{cases} \frac{3}{2} \\[2mm] \frac{1}{2} \\[2mm] -\frac{1}{2} \\[2mm] -\frac{3}{2} \end{cases} \qquad (10.30)$$

where A_H is the proton coupling constant.

The methyl-radical quartet should have a symmetrical 1:3:3:1 intensity distribution since there are three states having $M_F = \frac{1}{2}$ or $-\frac{1}{2}$ and only one having $M_F = \frac{3}{2}$ or $-\frac{3}{2}$. Note that the arguments here look very much like those used to describe the hyperfine splitting of nuclear magnetic resonance lines by the indirect spin-spin interaction.

Indeed, the results for proton (or any $I = \frac{1}{2}$ nucleus) coupling in EPR liquid-phase spectra may be stated in the same way as for coupling in NMR spectra, namely, each equivalent set of n protons splits the EPR resonance into $n + 1$ components with intensities given by the binomial coefficients (see Sec. 9.3 B). The more general rule, of course, is that any set of magnetic nuclei produce $2M_{Fi} + 1$ components.

It must be remembered that the total number of components when several sets of equivalent nuclei exist is determined by *multiplying* the splittings produced by each set. Thus for $C^{13}H_3$ we expect a maximum of $4(2) = 8$ components since the C^{13} produces a doublet and the H_3 a quartet. Explicitly, Eq. (10.29b) shows for this case

$$\nu = \nu_0 + A_H \cdot \begin{cases} \frac{3}{2} \\[2mm] \frac{1}{2} \\[2mm] -\frac{1}{2} \\[2mm] -\frac{3}{2} \end{cases} + A_C \cdot \begin{cases} \frac{1}{2} \\[2mm] -\frac{1}{2} \end{cases} \qquad (10.31)$$

It should be mentioned here also that the algebraic sign of the coupling constants cannot be determined from these simple spectra. In this section we imagine them to be positive for the sake of simplicity. Thus for the $\cdot C^{13}H_3$ radical, $A_H = 23$ G and $A_C = 41$ G are the experimentally determined coupling constants.

Finally, before continuing to the analysis of some slightly more complicated examples, we show the spectrum of the duroquinone anion radical

$$
\left[
\begin{array}{c}
\text{O} \\
\text{CH}_3 \;\; || \;\; \text{CH}_3 \\
\\
\text{CH}_3 \;\; || \;\; \text{CH}_3 \\
\text{O}
\end{array}
\right]^{-}
$$

The twelve protons are all equivalent and therefore lead to a 13-line multiplet centered at $g = 2.0$ as shown in Fig. 10.2. The equally spaced lines yield a coupling constant of 1.76 G.

EXAMPLE 10-1

We mentioned that the coupling constants for methyl radical were $A_H = 23$ and $A_C = 41$ G. Let us reconstruct the experimental spectrum assuming a C^{13} enrichment to 16.7 mole %. Equation (10.29b) gives for the $\cdot C^{13}H_3$ species

$$
\nu = \nu_0 + 23 \cdot
\begin{cases}
\frac{3}{2} \\[4pt]
\frac{1}{2} \\[4pt]
-\frac{1}{2} \\[4pt]
-\frac{3}{2}
\end{cases}
+ 41 \cdot
\begin{cases}
\frac{1}{2} \\[4pt]
-\frac{1}{2}
\end{cases}
\tag{10.32a}
$$

or

$$
\nu - \nu_0 = 55.0, \ 32.0, \ 14.0, \ 9.0, \ -9.0, \ -14.0, \ -32.0, \ -55.0 \ \text{G} \tag{10.32b}
$$

For the $\cdot C^{12}H_3$ species we get

$$
\nu - \nu_0 = 34.5, \ 11.5, \ -11.5, \ -34.5 \ \text{G} \tag{10.33}
$$

The intensities must be rather carefully considered. Consider first the simplest spectrum, that of $\cdot C^{12}H_3$. Here the quartet has relative intensities of 1:3:3:1 by our previous results.

The $\cdot C^{13}H_3$ spectrum may be considered to be a 1:3:3:1 quartet, each line split further into a 1:1 doublet. Because of crossovers near the center of the multiplet, the eight components of Eq. (8.32b) show relative intensities of 1:3:1:3:3:1:3:1 reading from high to low or vice versa. These intensities are obtained simply by keeping track of the pair of components into which each member of the 1:3:3:1 quartet is split.

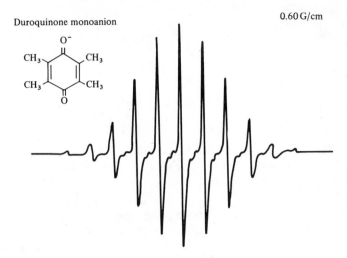

Duroquinone monoanion

0.60 G/cm

FIG. 10.2. EPR spectrum of duroquinone anion radical.

Now the sum of the intensities of the $\cdot C^{12}H_3$ species must be 5.0 times (83.3/16.7) that of the $\cdot C^{13}H_3$ species. Thus if the $\cdot C^{13}H_3$ species *in the mixture* produces resonance lines with intensities of $1:3:1:3:3:1:3:1$, the intensities of the $\cdot C^{12}H_3$ species *in the mixture* must be $10:30:30:10$. When the spectrum is sketched as a stick diagram, using the numerical resonance fields of Eqs. (10.32) and the intensities just deduced, it has the appearance shown in Fig. 10.3.

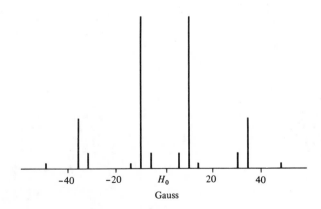

FIG. 10.3. Stick spectrum of methyl radicals having C^{13} abundance of 16.6 mole percent.

EXAMPLE 10-2

We consider now the spectrum of the dianion of p-nitrobenzoate,

$$\begin{bmatrix} O & & & & O \\ \diagdown & & & & \diagup \\ & N-\bigcirc-C & \\ \diagup & & & & \diagdown \\ O & & & & O \end{bmatrix}^{-2}$$

shown in Fig. 10.4. The analysis of such spectra to obtain coupling constants is, of course, the desired goal for studies of organic radicals. We note first that there are three possible sets of equivalent nuclei which may couple with the electron spin. These include two pairs of hydrogens (on the NO_2 and CO_2 ends of the molecule) and the lone N^{14}.

The N^{14} should produce a triplet since $I = 1$ and $2I + 1 = 3$. The intensities of the triplet components should be $1:1:1$ since there is just one function of each M_I value $(1, 0, -1)$. Each line of the triplet should then be split into a $1:2:1$ triplet by one of the pairs of protons, and each of the resulting lines further splits into a second $1:2:1$ triplet, which produces a total of $3(3)(3) = 27$ lines. This is the number appearing on the spectrum of Fig. 10.4.

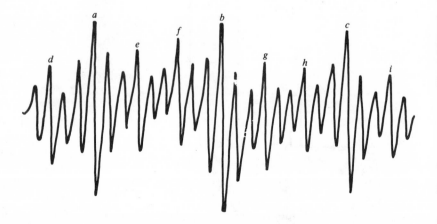

p-Nitrobenzoate dianion (Et$_4$N$^+$ salt)

0.43 G/cm

FIG. 10.4.   EPR spectrum of p-nitrobenzoate dianion.

The three nearly equally intense components marked a, b, and c stand out clearly. Their equal spacing and symmetrical arrangement leads one to assign the spacing as A_N. The average of the two spacings is

$$A_N = 9.0 \text{ G}$$

Note now the 1:2:1 patterns about each of the strongest lines. For example, (d, a, e), (f, b, g), and (h, c, i) form identical 1:2:1 triplets. Thus the spacings within each of these triplets give the value of one of the proton coupling constants, $A_{H(1)}$. The average value obtained is

$$A_{H(1)} = 3.1 \text{ G}$$

Finally each of the lettered lines is the central component of a 1:2:1 triplet. The spacings within these give the second proton coupling constant, the experimental value being

$$A_{H(2)} = 1.0 \text{ G}$$

As a check the stick diagram of the spectrum may be rather rapidly constructed by the method used in Fig. 9.10. This is shown in Fig. 10.5, and the agreement would be seen to be very good if the result were superimposed on Fig. 10.4 with the same scales.

There is no way to identify the experimental values of $A_{H(1)}$ and $A_{H(2)}$ with the proper environment in the molecule on the basis of this spectrum

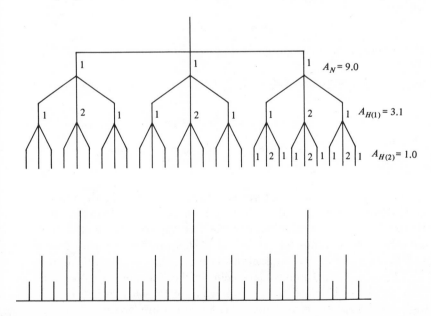

FIG. 10.5.   Construction of stick spectrum of p-nitrobenzoate dianion.

alone. One experimental technique for deciding would be to substitute deuterium atoms for *one* pair of equivalent protons. Then this pair of deuterons $I(D) = 1$ would produce $2F + 1 = 2(2) + 1 = 5$ lines instead of the three with protons. The other proton triplet would remain and hence a clear experimental identification would have been achieved.

It should be obvious from this last example that the analysis of an EPR spectrum can run into serious problems, particularly when lines overlap badly. The example of Fig. 10.4 was a barely resolved spectrum with only 27 lines, and it is not uncommon to have considerably more complex spectra. The common approach to the analysis is to use a digital computer to simulate the spectrum including the complete lineshapes.

We do not attempt any interpretation of the coupling constants at this time, but return to this topic in a later section.

10.4   ORGANIC RADICALS WITH TRIPLET STATES

It may be wondered why we limited ourselves to $S = \frac{1}{2}$ systems in the last section. The additional problems that occur when $S > \frac{1}{2}$ may be understood by considering a system containing two unpaired electrons ($S = 1$). Now the key point is that the two electron spins may interact with each other.

Two interaction mechanisms are possible, a Fermi contact-type interaction

$$\tilde{\mathcal{3C}}_{\text{Fermi}} \propto \tilde{S}_1 \cdot \tilde{S}_2 \qquad (10.34)$$

or a dipole-dipole interaction

$$\tilde{\mathcal{3C}}_{\text{dipole}} = g_e{}^2\beta_e{}^2\mathbf{S_1 D S_2} = \frac{g_e{}^2\beta_e{}^2}{r^5} \{ \tilde{S}_{1x}\tilde{S}_{2x}(r^2 - 3x^2) + \tilde{S}_{1y}\tilde{S}_{2y}(r^2 - 3y^2)$$

$$+ \tilde{S}_{1z}\tilde{S}_{2z}(r^2 - 3z^2) - (\tilde{S}_{1x}\tilde{S}_{2y} + \tilde{S}_{1y}\tilde{S}_{2x})3xy$$

$$- (S_{1y}S_{2z} + S_{1z}S_{2y})3yz - (S_{1z}S_{2x} + S_{1x}S_{2z})3zx$$

$$(10.35)$$

In the latter equations the form is entirely analogous to the nuclear magnetic dipole interactions described by Eqs. (9.125). Note in the above that $\mathbf{S_1}$ and $\mathbf{S_2}$ are row and column vectors. It may be shown that the Fermi contact term vanishes for triplet ($S = 1$) states so we are left with Eq. (10.35) to describe the interactions between the two unpaired spins.

The complete Hamiltonian includes the Zeeman terms for each electron and in addition, if magnetic nuclei are present, the nuclear Zeeman and nuclear spin-electron spin $(A\tilde{I} \cdot \tilde{S})$ terms are present. For simplicity we neglect the latter two terms in order to describe the primary new effects.

Thus the Hamiltonian may be written

$$\tilde{\mathfrak{R}} = g\beta_e H \cdot (\tilde{S}_1 + \tilde{S}_2) + \tilde{\mathfrak{R}}_D \qquad (10.36)$$

where the first term includes the Zeeman effect for each electron and $\tilde{\mathfrak{R}}_D$ is given by Eq. (10.35).

Let us consider a further simplification to begin our discussion. Suppose the unpaired spins are separated by a relatively large distance so that $\tilde{\mathfrak{R}}_D \ll g\beta_e H \cdot (\tilde{S}_1 + \tilde{S}_2)$. Then we need consider only the first term which we now write as

$$\tilde{\mathfrak{R}} = g\beta_e H \cdot \tilde{S} \qquad (10.37a)$$

or

$$\tilde{\mathfrak{R}} = g\beta_e H_z \tilde{S}_z \qquad (10.37b)$$

where $\tilde{S} = \tilde{S}_1 + \tilde{S}_2$ and we assume that H is along the Z axis.

The energies are now obtained directly using the eigenvalues for the $S = 1$ case. They are simply

$$E = g\beta_e H M_S \qquad (10.38)$$

where $M_S = 1, 0, -1$. Notice that we have used g, not g_e, in expectation that the spins are not entirely "free." Nevertheless we expect g to be close to 2.00 as usual. With the usual $\Delta M_S = \pm 1$ selection rules we find the two possible transitions to have the frequency

$$\nu = \frac{g\beta_e H}{h} \qquad (10.39)$$

which is the same result as for the spin-$\frac{1}{2}$ case. For an equivalent number of radicals the absorption line intensity would be twice as great in this case however, because of the two coincident transitions.

Numerous spectra of this type have been observed in liquid solution, an example being that from

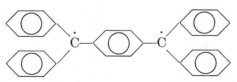

Radicals of this type have commonly been called *biradicals* due to the isolation of the two unpaired electrons. This isolation is a requirement, in fact, since otherwise the dipole-dipole terms would not be negligible. Because of this fact, biradicals are often thought of as having two entirely independent unpaired electrons. Hyperfine structure is normally observed as described in Sec. 9.3, with each electron showing coupling only with its own immediate environment. Thus in the above molecule the electron on the left-hand side has vanishing probability of appearing or coupling with nuclei on the right-hand side and vice versa.

In summary, then, biradicals are a curious type of triplet molecule showing no effects that distinguish them particularly from ordinary spin-$\frac{1}{2}$ radicals. In particular they show no splitting of the triply degenerate $(2S + 1)$ state in the absence of a field, as would be expected for a ^3P atom which would show a zero-field splitting into $^3P_2$, $^3P_1$, and $^3P_0$ states.

The situation becomes much different when the coupling of the electrons becomes appreciable relative to the Zeeman energy. To see what happens we first perform some manipulations upon $\tilde{\mathfrak{K}}_D$ in order to obtain a simpler Hamiltonian. We might begin by noting that there exists a molecular axis system for which the off-diagonal elements of the tensor **D** vanish. In fact this axis system would be that shown for a D_{2h} molecule such as naphthalene in Fig. 10.6.

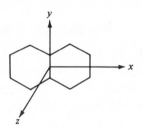

FIG. 10.6.   Principal-axis system for dipole tensor of triplet naphthalene.

The vanishing of the off-diagonal elements is easily verified by use of our group-theory principles. For example, the xz element is the following average over the ground electronic wave function of the molecule.

$$g^2\beta_e^2 \int \psi^* \left(\frac{3xz}{r^5}\right) \psi \, d\tau = g^2\beta_e^2 D_{xz} \tag{10.40}$$

Thus the integral vanishes unless

$$\Gamma^{(\psi)} \times \Gamma^{(\psi)} \times \Gamma^{(xz)} \times \Gamma^{(r^{-5})} \to \Gamma^{(A_g)}$$

The D_{2h} group has only one-dimensional representations, so regardless of the symmetry of ψ itself, $\Gamma^{(\psi)} \times \Gamma^{(\psi)} = A_g$ must hold. From the group character table, $\Gamma^{(xz)} = B_{2g}$ and $\Gamma^{(r^{-5})} = A_g$. Inserting these results we find

$$\Gamma^{(\psi)} \times \Gamma^{(\psi)} \times \Gamma^{(xz)} \times \Gamma^{(r^{-5})} = A_g \times B_{2g} \times A_g = B_{2g} \tag{10.41}$$

so the integral in Eq. (10.40) must vanish. Similar arguments hold for the D_{xy} and D_{yz} elements. Hence we immediately simplify $\tilde{\mathcal{H}}_D$ by dropping the off-diagonal terms.

If we now utilize the identity

$$\tilde{S}_1 \cdot \tilde{S}_2 = \tilde{S}_{1x}\tilde{S}_{2x} + \tilde{S}_{1y}\tilde{S}_{2y} + \tilde{S}_{1z}\tilde{S}_{2z}$$

the simplified form of $\tilde{\mathcal{H}}_D$ may be rearranged to

$$\tilde{\mathcal{H}}_D = \frac{3}{2}\frac{(x^2 - y^2)}{r^5}(\tilde{S}_{1x}\tilde{S}_{2x} - \tilde{S}_{1y}\tilde{S}_{2y})g^2\beta_e^2$$

$$+ \frac{1}{2}\frac{(3z^2 - r^2)}{r^5}(3\tilde{S}_{1z}\tilde{S}_{2z} - \tilde{S}_1 \cdot \tilde{S}_2)g^2\beta_e^2$$

$$= e(\tilde{S}_{1x}\tilde{S}_{2x} - \tilde{S}_{1y}\tilde{S}_{2y}) + d(3\tilde{S}_{1z}\tilde{S}_{2z} - \tilde{S}_1 \cdot \tilde{S}_2) \qquad (10.42)$$

where e and d are convenient temporary definitions. Now we observe the following identities involving *total* spin angular momenta,

$$\tfrac{1}{2}(\tilde{S}_x^2 - \tilde{S}_y^2) = \tfrac{1}{2}\{(\tilde{S}_{1x} + \tilde{S}_{2x})^2 - (\tilde{S}_{1y} + \tilde{S}_{2y})^2$$

$$= \tfrac{1}{2}\{\tilde{S}_{1x}^2 + 2\tilde{S}_{1x}\tilde{S}_{2x} + \tilde{S}_{2x}^2 - \tilde{S}_{1y}^2 - 2\tilde{S}_{1y}\tilde{S}_{2y} - \tilde{S}_{2y}^2$$

$$= (\tilde{S}_{1x}\tilde{S}_{2x} - \tilde{S}_{1y}\tilde{S}_{2y}) + \tfrac{1}{2}(\tilde{S}_{1x}^2 + \tilde{S}_{2x}^2 - \tilde{S}_{1y}^2 - \tilde{S}_{2y}^2) \qquad (10.43a)$$

and

$$\tfrac{3}{2}\tilde{S}_z^2 = \tfrac{3}{2}(\tilde{S}_{1z} + \tilde{S}_{2z})^2 = \tfrac{3}{2}(\tilde{S}_{1z}^2 + \tilde{S}_{2z}^2) + 3\tilde{S}_{1z}\tilde{S}_{2z} \qquad (10.43b)$$

Substituting Eqs. (10.43) into Eq. (10.42) we obtain

$$\tilde{\mathcal{H}}_D = e\{\tfrac{1}{2}(\tilde{S}_x^2 - \tilde{S}_y^2)\} + d\{\tfrac{3}{2}\tilde{S}_z^2\} + e\{\tfrac{1}{2}(\tilde{S}_{1y}^2 + \tilde{S}_{2y}^2 - \tilde{S}_{1x}^2 - \tilde{S}_{2x}^2)\}$$

$$+ d\{\tfrac{3}{2}(-\tilde{S}_{1z}^2 - \tilde{S}_{2z}^2 - \tilde{S}_{1x}\tilde{S}_{2x} - \tilde{S}_{1y}\tilde{S}_{2y} - \tilde{S}_{1z}\tilde{S}_{2z})\} \qquad (10.44)$$

Note now that the first terms involve only *total* angular momenta while the remaining ones contain terms for each electron separately.

We shall be interested shortly in evaluating matrix elements of $\tilde{\mathcal{H}}_D$ using spin-product triplet wave functions. If the last two terms are carried along in the calculations it is found that they provide no contribution to the *separation* of the energy levels. Thus we drop them at this time and utilize the simpler Hamiltonian,

$$\tilde{\mathcal{H}}_D = D\tilde{S}_z^2 + E(\tilde{S}_x^2 - \tilde{S}_y^2) \qquad (10.45)$$

where

$$D = \tfrac{3}{4}g^2\beta_e{}^2 \frac{3z^2 - r^2}{r^5}$$

$$E = \tfrac{3}{4}g^2\beta_e{}^2 \frac{x^2 - y^2}{r^5}$$

The complete Hamiltonian is then

$$\widetilde{\mathfrak{K}} = g\beta_e \widetilde{S}\cdot\widetilde{H} + D\widetilde{S}_z{}^2 + E(\widetilde{S}_x{}^2 - \widetilde{S}_y{}^2) \tag{10.46}$$

and is often referred to as the "spin Hamiltonian."

We have seen in Sec. 9.5 that two spin-$\tfrac{1}{2}$ particles lead to four spin-product states, three symmetric and one antisymmetric. In fact the three symmetric states are eigenfunctions of $\widetilde{S}^2$ and $\widetilde{S}_z$ with eigenvalues given by

$$\widetilde{S}^2 \begin{cases} \alpha\alpha \\[2mm] \dfrac{1}{\sqrt{2}}(\alpha\beta + \beta\alpha) \\[2mm] \beta\beta \end{cases} = 2 \begin{cases} \alpha\alpha \\[2mm] \dfrac{1}{\sqrt{2}}(\alpha\beta + \beta\alpha) \\[2mm] \beta\beta \end{cases} \tag{10.47a}$$

$$\widetilde{S}_z \begin{cases} \alpha\alpha \\[2mm] \dfrac{1}{\sqrt{2}}(\alpha\beta + \beta\alpha) \\[2mm] \beta\beta \end{cases} = \begin{cases} \alpha\alpha \\[2mm] 0 \\[2mm] -\beta\beta \end{cases} \tag{10.47b}$$

Thus the functions

$$|T_1\rangle = \alpha\alpha$$

$$|T_0\rangle = \frac{1}{\sqrt{2}}(\alpha\beta + \beta\alpha) \tag{10.48}$$

$$|T_{-1}\rangle = \beta\beta$$

are satisfactory functions for describing a *triplet* state. The singlet function $(1/\sqrt{2})(\alpha\beta - \beta\alpha)$ does not enter into our considerations here since we wish to describe the energy levels of the triplet state, and of course, the singlet state is nonmagnetic ($S = 0$) anyway.

If we wished to use basis functions which diagonalized the zero-field terms of the Hamiltonian (10.46), the functions of Eq. (10.48) would not be appropriate. For this case functions which are eigenfunctions of $\tilde{S}_x{}^2$, $\tilde{S}_y{}^2$, and $\tilde{S}_z{}^2$ would be needed. The reader can verify that

$$| T_x \rangle = \frac{1}{\sqrt{2}} (\alpha\alpha - \beta\beta)$$

$$| T_y \rangle = \frac{1}{\sqrt{2}} (\alpha\alpha + \beta\beta) \qquad (10.49)$$

$$| T_z \rangle = \frac{1}{\sqrt{2}} (\alpha\beta + \beta\alpha)$$

have this property, that is,

$$\tilde{S}_x{}^2 \begin{cases} | T_x \rangle & 0 \\ | T_y \rangle = | T_y \rangle \\ | T_z \rangle & | T_z \rangle \end{cases}$$

$$\tilde{S}_y{}^2 \begin{cases} | T_x \rangle & | T_x \rangle \\ | T_y \rangle = & 0 \\ | T_z \rangle & | T_z \rangle \end{cases} \qquad (10.50)$$

$$\tilde{S}_z{}^2 \begin{cases} | T_x \rangle & | T_x \rangle \\ | T_y \rangle = | T_y \rangle \\ | T_z \rangle & 0 \end{cases}$$

Later we see by explicit calculation that these functions are correct when $H = 0$.

A problem which arises now is that the Zeeman term tends to quantize **S** along the field (Z axis) direction while the zero-field terms tend to quantize **S** along the *molecular* axes. Since the dipolar tensor elements are most meaningful when expressed in the molecular axes we have chosen to work entirely in this system. Thus as the molecule's orientation in the magnetic field changes, the matrix elements of the Zeeman term vary markedly while those of the zero-field terms remain constant.

Let us set up the Hamiltonian matrix for the Hamiltonian of (10.46) using the functions of (10.48). The necessary matrix elements with this basis are directly obtained by using the results of Table 4.1(c) with $S = 1$, $M_S = 1, 0, -1$. Thus

$$S_x = \begin{pmatrix} 0 & \sqrt{2}/2 & 0 \\ \sqrt{2}/2 & 0 & \sqrt{2}/2 \\ 0 & \sqrt{2}/2 & 0 \end{pmatrix}$$

$$S_y = \begin{pmatrix} 0 & -i\sqrt{2}/2 & 0 \\ i\sqrt{2}/2 & 0 & -i\sqrt{2}/2 \\ 0 & i\sqrt{2}/2 & 0 \end{pmatrix} \qquad (10.51a)$$

$$S_z = \begin{pmatrix} 1 & 0 & 0 \\ 0 & 0 & 0 \\ 0 & 0 & -1 \end{pmatrix}$$

and by squaring these

$$S_x{}^2 = \begin{pmatrix} \tfrac{1}{2} & 0 & \tfrac{1}{2} \\ 0 & 1 & 0 \\ \tfrac{1}{2} & 0 & \tfrac{1}{2} \end{pmatrix}$$

$$S_y{}^2 = \begin{pmatrix} \tfrac{1}{2} & 0 & -\tfrac{1}{2} \\ 0 & 1 & 0 \\ -\tfrac{1}{2} & 0 & \tfrac{1}{2} \end{pmatrix} \qquad (10.51b)$$

$$S_z{}^2 = \begin{pmatrix} 1 & 0 & 0 \\ 0 & 0 & 0 \\ 0 & 0 & 1 \end{pmatrix}$$

The Hamiltonian becomes by use of Eqs. (10.51)

$$\mathfrak{IC} = \begin{pmatrix} g\beta_e nH + D & \tfrac{1}{2}\sqrt{2}\, g\beta_e H(1 - im) & E \\ \tfrac{1}{2}\sqrt{2}\, g\beta_e H(1 + im) & 0 & \tfrac{1}{2}\sqrt{2}\, g\beta_e H(1 + im) \\ E & \tfrac{1}{2}\sqrt{2}\, g\beta_e H(1 - im) & -g\beta_e nH + D \end{pmatrix}$$

(10.52)

where l, m, and n are the cosines of the angles between the magnetic field direction and the x, y, and z axes, respectively. When E and D are zero, the eigenvalues of Eq. (10.52) are found to be $E = \pm g\beta_e H$ and 0.[15] This is just the result expected and is that given already by Eq. (10.38).

On the other hand, when $H = 0$, the eigenvalues of Eq. (10.52) are found to be

$$W = 0, D + E, D - E \qquad (10.53a)$$

and the eigenfunctions are

$$\frac{1}{\sqrt{2}}\,(\alpha\beta + \beta\alpha)$$

$$\frac{1}{\sqrt{2}}\,(\alpha\alpha - \beta\beta) \qquad (10.53b)$$

$$\frac{1}{\sqrt{2}}\,(\alpha\alpha + \beta\beta)$$

respectively. The exact appearance of the zero-field energy levels given by Eq. (10.53a) depends strongly upon the magnitudes and signs of the quantities D and E.

At conditions intermediate to these two extreme cases the energy levels have a more complicated form, and in particular depend upon the orientation (that is, upon l, m, and n). As an example when the magnetic field lies along the z axis the Hamiltonian matrix becomes

$$\mathfrak{IC} = \begin{pmatrix} g\beta_e H + D & 0 & E \\ 0 & 0 & 0 \\ E & 0 & -g\beta_e H + D \end{pmatrix}$$

(10.54)

[15] The relation $l^2 + m^2 + n^2 = 1$ must be utilized in obtaining these results.

which has eigenvalues

$$W = 0, D \pm [(g\beta_e H)^2 + E^2]^{1/2}. \tag{10.55}$$

On the other hand, if the magnetic field is in the x direction, the eigenvalues of $\mathcal{3C}$ are

$$W = D - E, \frac{D + E \pm [(D + E)^2 + 4(g\beta_e H)^2]^{1/2}}{2} \tag{10.56}$$

If we consider the possible transitions which may occur between the three levels we immediately see a difficulty with the liquid-phase spectrum of a triplet molecule. If M_S were a good quantum number the two allowed transitions would be between the states having energies approximated by 0 and $\pm g\beta_e H$. Thus when the molecular z axis is lined up with the field the transition frequencies would be [from (10.55)]

$$h\nu = [(g\beta_e H)^2 + E^2]^{1/2} \pm D \tag{10.57a}$$

while when the x axis lined up with H we would find

$$h\nu = \frac{[(D + E)^2 + 4(g\beta_e H)^2]^{1/2} \pm (3E - D)}{2} \tag{10.57b}$$

from (10.56). Clearly the transition frequencies are strongly dependent upon angular orientation. Thus a collection of triplet free radicals in the liquid phase shows a complete range of all possible absorption frequencies because of the random orientation in the liquid phase. The result is that the resonance becomes very broad and the peak intensity very small. Consequently these strongly allowed transitions have not been observed in the liquid phase. In this regard it is useful to note that the magnitudes of D and E are such that the resonance frequencies may vary easily over several thousand gauss.

The problem is actually somewhat more involved. It might have been guessed that the dipolar Hamiltonian would have averaged to zero for the freely tumbling molecules in the liquid as was the case for the nuclear dipole-dipole interaction described in the previous chapter. If this had occurred only the Zeeman term would remain and a single sharp resonance (at $g\beta_e H/h$) would be expected. Since this is not observed experimentally we know that the averaging process is not effective. The reason for this is the rapid relaxation of the spin system produced by the highly anisotropic dipolar interaction. Qualitatively, the transitions and subsequent relaxation occur in a time scale that is short compared to the averaging time due to molecular motion.

EPR spectra of triplets have been observed in the solid phase using dilute solutions of oriented molecules. The first such reported study was

TABLE 10.1

Representative parameters for triplet radicals.[a]

	$\mid D \mid$, cm^{-1}	$\mid E \mid$, cm^{-1}
(1) [phenyl–C with H structure]	0.518	
(2) [naphthalene structure]	0.101	0.014
(3) [benzene fused with N=N ring structure]	0.101	0.018
(4) [diphenyl methylene structure, C··]	0.405	0.019
(5) [benzene ring with φ substituents]$^{-2}$	0.042	

ᵃ   Taken in part from M. Bersohn and J. C. Baird, *An Introduction to Electron Paramagnetic Resonance*, Benjamin, New York (1966), Table 8-1.

performed on a dilute solid solution of naphthalene in durene crystals.[16] The values of the parameters D and E for this system were found to be 3010 MHz and -411 MHz, respectively. Table 10.1 lists the values found for several other systems also.

EXAMPLE 10-3

In order to see the highly anisotropic nature of the triplet naphthalene system it is interesting to calculate the magnetic fields for which resonance would be obtained (in an oriented solid) at a microwave frequency of 10 000

¹⁶ C. A. Hutchison and B. W. Mangum, *J. Chem. Phys.* **29**, 952 (1958); **34**, 908 (1961).

MHz. When the field is in the z direction Eq. (10.57a) gives

$$H = \frac{1}{g\beta_e} [(h\nu \pm D)^2 - E^2]^{1/2} \tag{10.58}$$

where D and E have units of ergs. We have from the known values

$$D = 3010 \times 10^6 \times h \text{ erg}$$

$$E = -411 \times 10^6 \times h \text{ erg}$$

so
$$H = \frac{h}{g\beta_e} \sqrt{(1.00 \times 10^{10} \pm 3.01 \times 10^9)^2 - (0.411 \times 10^9)^2}$$

$$= \frac{10^9 h}{g\beta_e} \sqrt{(10.0 \pm 3.01)^2 - (0.411)^2}$$

The term under the square-root sign has the values 13.00 and 6.98, while the factor in front has the value

$$\frac{10^9 (6.62 \times 10^{-27})}{(2.00)(0.927 \times 10^{-20})} = 3.57 \times 10^2$$

using $g = 2.00$ (the experimental result). We get for the two magnetic fields

$$H = 4641, 2492 \text{ G}$$

In this example the resonant fields were rather insensitive to the magnitude of E. At lower frequencies, $\nu = 3500$ MHz for example, the low field line becomes much more sensitive to this parameter.

10.5 ELECTRON PARAMAGNETIC RESONANCE OF TRANSITION METAL COMPLEXES

A great amount of EPR work has been performed with transition-metal complexes in the solid and liquid states. Ions whose complexes have been extensively studied include Cu^{++}, V^{+++}, Mn^{++}, and Cr^{+++}. As remarked earlier the paramagnetism arises because of the unshared electrons of the incompletely filled d orbitals of the metal ion. However, the data obtained from the analysis of the spectra provide rather detailed knowledge about the environment of the ion, namely the complex in which the ion resides. In order to understand the nature of the spectra it is necessary for us to look briefly into the theory of ionic energy states, both for free and complexed ions. Our discussion is necessarily brief at this point, but we return to a more detailed discussion of the energy levels and spectra of transition-metal complexes in chapter 11.

A.   Energy Levels of Transition-Metal Ions in Crystal Fields

We wish eventually to describe the energy states of an ion in the environment of its ligands or "crystal field."[17] To begin with we recall the important energy terms for a free ion or atom,

$$\tilde{\mathcal{K}} = \sum_i \left(-\frac{\hbar^2}{2m} \nabla i^2 - Ze^2/r_i \right) + \sum_{i>j} \frac{e^2}{r_{ij}} + \tilde{\mathcal{K}}_{so} \qquad (10.59)$$

The first term represents the kinetic energy of the electrons and the Coulombic interaction of electrons with the nucleus of charge Z, while the second term represents the interelectronic interaction. The third term is the spin-orbit interaction and for our purposes may be written as

$$\tilde{\mathcal{K}}_{so} = \zeta \tilde{L} \cdot \tilde{S} \qquad (10.60)$$

where ζ is the spin-orbit coupling constant which has a magnitude in the range 50–1000 cm^{-1} for transition-metal ions.

The energy levels of the free ion may be calculated through such approaches as the self-consistent field method[18] which yield the energies of the various electron *configurations*. Thus the ground-state configuration of Ti^{+3} is $1s^2 2s^2 2p^6 3s^2 3p^6 3d^1$, while an excited-state configuration would be $1s^2 2s^2 2p^6 3s^2 3p^5 3d^2$, for example. Because of the interelectronic interactions a given configuration consists of one or more *terms* which differ in energy by an amount on the order of 15 000 cm^{-1} and which are specified by the term symbol

$$^{2S+1}L$$

where S is the total electron spin angular momentum and L is the total orbital angular momentum. The ground-state configuration of Ti^{+3} consists of only one term, the ^2D. The ground-state configuration of V^{+3} is $1s^2 \ldots 3d^2$, which leads to five terms: ^1G, ^3F, ^1D, ^3P, ^1S. Of these latter terms the ^3F is the ground state.

Finally the spin-orbit term given by Eq. (10.60) splits each term (except S state terms) into $2J + 1$ sublevels where $\tilde{J} = \tilde{L} + \tilde{S}$ is the total angular momentum of the atom or ion apart from any nuclear spin angular momentum which we neglect here. The Ti^{+3} ^2D term therefore is split by spin-orbit interaction into two sublevels, the $^2D_{5/2}$ and $^2D_{3/2}$, where the subscripts are the J values. Experimentally the $^2D_{3/2}$ state lies lowest in energy in agreement with theoretical expectations. The *ground-state* level of any

---

[17] In the older literature the environment's effect upon the ion has been called the "crystal field." More recently chemists have preferred the term "ligand field" for essentially the same phenomenon. We use the terms more or less interchangeably.

[18] D. R. Hartree, *The Calculation of Atomic Structures* (Wiley, New York, 1957).

free atom or ionic term can always be predicted by Hund's rules:

(1) The lowest term is that having maximum spin S.

(2) For terms having the same spin multiplicity the lowest is that of maximum L.

(3) When only one atomic shell is incompletely filled, the lowest term sublevel is (a) that having the minimum value of J for less than half-filled shells, (b) that having the maximum value of J for more than half-filled shells.

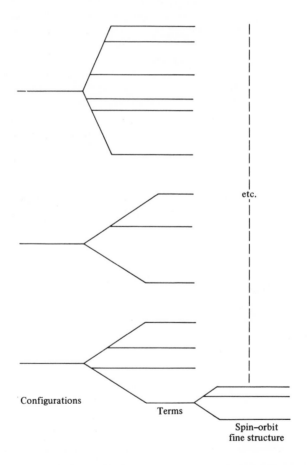

FIG. 10.7.   Qualitative behavior of atomic or free-ion energy levels. Number of terms of a given configuration depends upon the precise electron configuration. Number of fine structure components depends upon S and L values of each term.

In Fig. 10.7 we show the qualitative behavior of the atomic (or ionic) energy levels as just described. The reader should consult the literature[19] for a more thorough discussion of the energy states.

For later discussions it is useful to point out here that an electron in a d orbital of a free transition-metal ion (or atom) has a precise value of orbital angular momentum along the Z axis. Thus a d electron must have a Z-component angular momentum of either ± 2, ± 1, or 0 $\hbar$.

We are interested in the ionic energy levels in a crystal or ligand environment. In this case the spherical symmetry of free space normally does not exist and the energy-level pattern of the free ion is modified. Formally we need to add to the Hamiltonian of the free ion [Eq. (10.59)] a term $V_{cf}(r_i)$ which describes the crystal field potential felt by the ion. We do not consider the explicit mathematical form of the interaction but merely look at the situation qualitatively. The interested reader may consult the general references for more detailed discussions.

The qualitative behavior of the ^2D term for the $3d^1$ configuration is

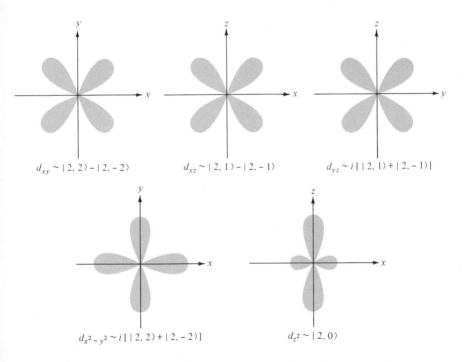

FIG. 10.8.   Planar projections of the real forms of the d orbitals.

[19] G. Herzberg, *Atomic Spectra and Atomic Structure* (Dover, New York, 1944); E. U. Condon and G. H. Shortley, *The Theory of Atomic Spectra* (Cambridge U.P., London, 1935).

best seen by considering the real forms of the d-orbital free ion eigenfunctions shown in Fig. 10.8. These are conventionally designated as d_{z^2}, $d_{x^2-y^2}$, d_{xy}, d_{xz}, and d_{yz}. In the presence of an octahedral distribution of ligands as shown in Fig. 10.9(a), it is clear that the ligand-ion interactions break the degeneracy of the five d states. The resulting structure of the energy levels depends strongly upon the relative magnitudes of the term energies, spin-orbit coupling, and crystal field. Three distinct situations can be recognized.

(1) Weak-field case: Here the crystal field term is smaller than the spin-orbit term. The crystal field term then splits each J state into a maximum of $2J + 1$ levels. Group-theoretical considerations[20] may be used to predict the level splittings. For an octahedral (cubic) field one finds that $D_{1/2}$ and $D_{3/2}$ states remain

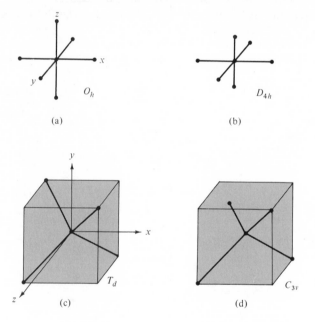

(a)   (b)

(c)   (d)

FIG. 10.9.   Symmetric ligand configurations. The tetragonal configuration (b) arises from the octahedral (a) by compressing the ligands in the z direction. The threefold configuration (d) arises from the tetrahedral case (c) by moving one ligand toward the origin.

[20] M. Tinkham, *Group Theory and Quantum Mechanics* (McGraw-Hill, New York, 1964), p. 75.

unsplit, while a $D_{5/2}$ splits into two states with degeneracy 4 and 6. Note that the total degeneracy of a given J state is $2J + 1$. We do not treat this case in any detail but simply point out that it is applicable to the rare-earth ions whose $4f$ electrons are relatively shielded from the crystal field by the $5s$ and $5p$ electron shells.

(2) Medium-field case: Here the crystal field V_{cf} is intermediate in magnitude to the term spacings and the sublevel spacings produced by spin-orbit interactions. In this case it is usually most appropriate to see how the crystal field splits the various terms, and then to consider the further splitting produced by spin-orbit coupling as a perturbation. The first and second transition-metal ions fall commonly into this medium-field case.

(3) Strong-field case: An extension of case (2), this situation is specified by $V_{cf} \gg \mathcal{3C}_{so}$, and, in fact, V_{cf} may get as large as the term splittings. It occurs when strongly interacting ligands such as cyanide (CN⁻) are present, and is more common for the $4d$

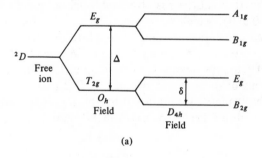

(a)

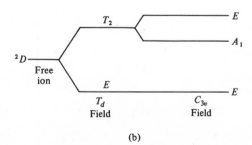

(b)

FIG. 10.10.   Crystal field splitting of a ^2D state in various crystal-field configurations.

and $5d$ transition metals than for the first series. In this strong-field case the terms of a given configuration may not be identifiable and indeed the best qualitative way to describe the states of a given configuration is to neglect the r_{ij}^{-1} terms in (10.59) for the valence electrons to the first approximation. The splitting of the configuration produced by V_{cf} may then be determined. Finally, the further effects of the r_{ij}^{-1} terms may be included.

For the purposes of this chapter we consider only the medium-field case. Let us return now to a consideration of the $3d^1$ (or nd^1) ion in an octahedral field. From Figs. 10.8 and 10.9 it is seen that the d_{xy}, d_{xz}, and d_{yz} orbitals do not point directly at the ligands as do the $d_{x^2-y^2}$ and d_{z^2} orbitals. The result is that the former three orbital states are stabilized relative to the latter two because of the smaller electrostatic and/or covalent bonding interactions. In fact calculations or group theory show that the ^2D term gives a doubly degenerate and a triply degenerate set of levels in the octahedral field as shown in Fig. 10.10(a). The symbols used in the diagram are the symmetry labels for the O_h group, and the splitting Δ is known as the crystal-field splitting.

If now the ligand-field is reduced to D_{4h} symmetry (known as tetragonal), a further splitting is observed. Such a ligand arrangement is shown in Fig. 10.9(b), which by comparison to Fig. 10.8 again shows that the d_{z^2} and $d_{x^2-y^2}$ orbitals are now affected differently. Similarly, the d_{xy} state differs from the d_{xz}, d_{yz} pair which remain degenerate. For the case shown in Fig. 10.9(b), the ligands have been moved in along the z axis so we expect the ordering of the states shown in Fig. 10.10(a). δ is the crystal-field splitting of the triplet state produced by the D_{4h} ligand arrangement.

Many other types of crystal-field symmetry may occur, of course.

TABLE 10.2

Representative spin-orbit coupling constants and crystal-field splitting parameters.

Free ion	$\zeta\,(cm^{-1})$	Covalently bonded atom	$\zeta\,(cm^{-1})$	Complex	$\Delta\,(cm^{-1})$
Ti^{+3}	150	C	30	$Ti(H_2O)_6^{+3}$	20 400
V^{+3}	210	N	70	$V(H_2O)_6^{+3}$	17 000
Cr^{+3}	270	F	270	CrF_6^{-3}	15 200
Mn^{+2}	350	P	230	$Mn(H_2O)_6^{+2}$	25 000
Fe^{+2}	410	Cl	590	$FeCl_4^-$	18 800
Ni_{+2}^{+2}	650	I	4060	$Ni(NH_3)_6^{+2}$	10 800
Cu^{+2}	830				

Our purpose is not to delve deeply into this very interesting problem at this time, but we show one more example in Fig. 10.10(b). In this case we have a tetrahedral field with an additional perturbation which lowers the symmetry to only C_{3v}. Figs. 10.9(c) and 10.9(d) show the ligand arrangements for this situation. We conclude this brief section by listing in Table 10.2 a sampling of crystal-field splitting parameters Δ, and spin-orbit constants ζ.

B. Origin of the Anisotropy of the EPR Spectra

The EPR spectra of transition metal ions show invariably moderate to large anisotropic effects; that is, the spectrum is highly dependent upon the crystal orientation in the magnetic field. This is reminiscent of the anisotropic dipolar interaction in organic triplets. However, for transition metals the anisotropy persists even for ions with only one d electron, so one must search further for an explanation.

The answer to our dilemma is rather easily found if we return to Eq. (10.4), which showed that electron orbital angular momentum is expected to contribute to the total magnetic moment and hence to the Zeeman energy. Up to this point we have assumed that the average value of orbital angular momentum along the Z axis was zero, that is, $\tilde{L}_Z$ was "quenched." For these cases the g value was always very close to 2.0023 and no great anisotropy was evident. It can be shown that the proper inclusion of $\tilde{L}$ in the present discussion leads to the explanation of the anisotropic spectral results.

First, let us obtain a better picture of the phenomenon of quenching. We have observed already for a free ion that the orbital angular momentum $\tilde{L}^2$ and the Z component $\tilde{L}_Z$ have definite values, in general not zero. What happens when the crystal-field splitting is present?

To answer this, consider the situation pictured in Fig. 10.10(a). It is easily shown by simple quantum-mechanical or group-theoretical arguments that the zero-order wave functions of the five states existing in the D_{4h} field are simply linear combinations of the free-ion d-state functions having definite values of $\tilde{L}^2$ and $\tilde{L}_Z$. Thus in the order of increasing energy the state functions are

$$| B_{2g} \rangle = \frac{1}{\sqrt{2}} (| 2, 2 \rangle - | 2, -2 \rangle) \qquad (10.61a)$$

$$| E_g^{\pm} \rangle = \frac{1}{\sqrt{2}} (| 2, 1 \rangle \pm | 2, -1 \rangle) \qquad (10.61b)$$

$$| B_{1g} \rangle = \frac{1}{\sqrt{2}} (| 2, 2 \rangle + | 2, -2 \rangle) \qquad (10.61c)$$

$$| A_{1g} \rangle = | 2, 0 \rangle \qquad (10.61d)$$

where the $| LM_L \rangle$ are the free-ion d-state wave functions. Note that these functions have been identified with the common d_{z^2}, etc., notation in Fig. 10.8. Actually the functions $| B_{2g} \rangle$ and $| E_g^- \rangle$ are pure imaginary as written in Eqs. (10.61). They need to be multiplied by i to obtain the real functions.

Note that all the states of Eqs. (10.61) have a value of zero for the average Z component of angular momentum. For $| A_{1g} \rangle$ the zero value is explicit. For the ground state we find

$$\langle B_{2g} | \tilde{L}_Z | B_{2g} \rangle = \tfrac{1}{2} \{ \langle 2, 2 | \tilde{L}_Z | 2, 2 \rangle + \langle 2, -2 | \tilde{L}_Z | 2, -2 \rangle$$

$$- \langle 2, 2 | \tilde{L}_Z | 2, -2 \rangle - \langle 2, -2 | \tilde{L}_Z | 2, 2 \rangle \}$$

$$= \tfrac{1}{2} \{ 2 - 2 - 0 - 0 \} = 0$$

also, as we do for all the states. Thus $\langle L_Z \rangle$ is zero; that is, there is no angular momentum along the preferred crystal axis. This is what is meant by quenching. A similar phenomenon operates for organic radicals, so our neglect of L was justified.

The curious result now is that we are again back to a quenched system which should lead to isotropic spectral behavior. In the medium-field case the only remaining Hamiltonian term is $\tilde{\mathcal{K}}_{so}$, and we can show quantitatively that this $\tilde{L} \cdot \tilde{S}$ term mixes some angular momentum back into the electron spin states of the system.

In the absence of $\tilde{\mathcal{K}}_{so}$, the ground-state is doubly degenerate due to the two possible spin functions, α or β. Thus the ground-state functions are

$$\alpha | B_{2g} \rangle = \frac{1}{\sqrt{2}} \left(| 2, 2, \tfrac{1}{2} \rangle - | 2, -2, \tfrac{1}{2} \rangle \right) \tag{10.62a}$$

$$\beta | B_{2g} \rangle = \frac{1}{\sqrt{2}} \left(| 2, 2, -\tfrac{1}{2} \rangle - | 2, -2, -\tfrac{1}{2} \rangle \right) \tag{10.62b}$$

We now use first-order perturbation theory to determine the appropriate ground-state wave functions in the presence of $\tilde{\mathcal{K}}_{so}$.

In the standard fashion,

$$| + \rangle = \alpha | B_{2g} \rangle - \sum_k{}' \frac{| \langle k | \zeta \tilde{L} \cdot \tilde{S} | \alpha B_{2g} \rangle |^2}{\Delta E} | k \rangle \tag{10.63}$$

where ΔE is the energy of the state $| k \rangle$ minus that of the state $\alpha | B_{2g} \rangle$, and $\zeta \tilde{L} \cdot \tilde{S}$ is the spin-orbit perturbation. We assume that this term is small compared to the terms Δ and δ in Fig. 10.10(b). As usual the sum is over all states exclusive of $| \alpha B_{2g} \rangle \equiv \alpha | B_{2g} \rangle$. The matrix elements may be evaluated if we recall that

$$\tilde{L} \cdot \tilde{S} = \tilde{L}_x \tilde{S}_x + \tilde{L}_y \tilde{S}_y + \tilde{L}_z \tilde{S}_z \tag{10.64}$$

in which case nonzero matrix elements exist only for states $|k\rangle$ which have M_L and M_S values which differ from those of $\alpha\,|\,B_{2g}\rangle$ by 0 or ± 1. Using Eqs. 10.62 and Eqs. 10.61b, (10.61c) and (10.61d) with inclusion of spin α and β, Eq. (10.63) becomes

$$|+\rangle = \alpha\,|\,B_{2g}\rangle - \tfrac{1}{2}\zeta\{\langle 2, 2, \tfrac{1}{2}\,|\,\tilde{L}_z\tilde{S}_z\,|\,2, 2, \tfrac{1}{2}\rangle$$

$$- \langle 2, -2, \tfrac{1}{2}\,|\,\tilde{L}_z\tilde{S}_z\,|\,2, -2, \tfrac{1}{2}\rangle\}\alpha\,|\,B_{1g}\rangle(1/\Delta)$$

$$- \tfrac{1}{2}\zeta\{\langle 2, 1, -\tfrac{1}{2}\,|\,\tilde{L}_x\tilde{S}_x + \tilde{L}_y\tilde{S}_y\,|\,2, 2, \tfrac{1}{2}\rangle$$

$$- \langle 2, -1, -\tfrac{1}{2}\,|\,\tilde{L}_x\tilde{S}_x + \tilde{L}_y\tilde{S}_y\,|\,2, -2, \tfrac{1}{2}\rangle\}\beta\,|\,E_g{}^+\rangle(1/\delta)$$

$$- \tfrac{1}{2}\zeta\{\langle 2, -1, -\tfrac{1}{2}\,|\,\tilde{L}_x\tilde{S}_x + \tilde{L}_y\tilde{S}_y\,|\,2, -2, \tfrac{1}{2}\rangle$$

$$+ \langle 2, 1, -\tfrac{1}{2}\,|\,\tilde{L}_x\tilde{S}_x + \tilde{L}_y\tilde{S}_y\,|\,2, 2, \tfrac{1}{2}\rangle\}\beta\,|\,E_g{}^-\rangle(1/\delta) \qquad (10.65)$$

We can evaluate all these matrix elements by referring to the general results for L and S given in Table 4.1. Somewhat tedious algebra leads to

$$|+\rangle = \frac{1}{\sqrt{2}}\,(|\,2, 2\rangle - |\,2, -2\rangle)\alpha - \frac{\zeta}{\Delta}\frac{1}{\sqrt{2}}\,(|\,2, 2\rangle + |\,2, -2\rangle)\alpha$$

$$+ \frac{\zeta}{\Delta}\frac{1}{\sqrt{2}}\,|\,2, -1\rangle\beta \qquad (10.66a)$$

In an entirely similar manner we find that

$$|-\rangle = \frac{1}{\sqrt{2}}\,(|\,2, 2\rangle - |\,2, -2\rangle)\beta + \frac{\zeta}{\Delta}\frac{1}{\sqrt{2}}\,(|\,2, 2\rangle + |\,2, -2\rangle)\beta$$

$$- \frac{\zeta}{\Delta}\frac{1}{\sqrt{2}}\,|\,2, -1\rangle\alpha \qquad (10.66b)$$

Thus we see that the $|+\rangle$ and $|-\rangle$ states (correct to first order) are not "pure" in either spin or orbital variables as were the states $\alpha\,|\,B_{2g}\rangle$ and $\beta\,|\,B_{2g}\rangle$. Indeed the reader can verify that the average value of $\tilde{L}_z$ using the functions of (10.66a) or (10.66b) is not zero. Thus the spin-orbit interaction brings orbital angular momentum back into the picture.

We can find the energy levels and transitions if we utilize Eq. 10.4 and write the Zeeman energy as

$$\tilde{\mathcal{3C}}_{\text{Zeeman}} = -\boldsymbol{\mu}\cdot\mathbf{H} = \beta_e(g_e\mathbf{S}\cdot\mathbf{H} + \mathbf{L}\cdot\mathbf{H}) \qquad (10.67)$$

When H lies along the z axis of the radical, we find

$$\tilde{\mathcal{3C}}_{\text{Zeeman}} = \beta_e(g_e\tilde{S}_Z H_Z + \tilde{L}_Z H_Z) \qquad (10.68)$$

Evaluating the diagonal elements of $\tilde{\mathcal{3C}}$ using the $|+\rangle$ and $|-\rangle$ functions we find

$$\langle + \mid \widetilde{\mathcal{K}} \mid + \rangle = \beta_e H_Z \left\{ g_e \left[\frac{1}{2} + \frac{\zeta^2}{\Delta^2} \left(\frac{1}{2} \right) - \frac{\zeta^2}{\delta^2} \left(\frac{1}{4} \right) \right] \right.$$

$$\left. + \frac{1}{2} (2 - 2) - \frac{\zeta^2}{\Delta^2} \left(\frac{1}{2} \right) (2 - 2) + \frac{\zeta^2}{\delta^2} \left(\frac{1}{2} \right) (-1) - \frac{1}{2} (2) \frac{\zeta}{\Delta} (2 + 2) \right\}$$

$$= \beta_e H_Z \left(\frac{1}{2} g_e - 4 \frac{\zeta}{\Delta} \right) = \frac{1}{2} \beta_e H_Z \left(g_e - 8 \frac{\zeta}{\Delta} \right) \quad (10.69a)$$

where we have neglected terms involving ζ^2 in the last line. Similarly we find

$$\langle - \mid \widetilde{\mathcal{K}} \mid - \rangle = \frac{1}{2} \beta_e H_Z \left(-g_e + 8 \frac{\zeta}{\Delta} \right) \quad (10.69b)$$

Thus the transition frequency for $\mid - \rangle \rightarrow \mid + \rangle$ is

$$\nu = \left(g_e - 8 \frac{\zeta}{\Delta} \right) \beta_e H_Z / h \quad (10.70)$$

Now if we replace the Z's with X's in Eq. (10.68) we can find the energy states when the field is in the x direction. In this case there are no diagonal elements of $\widetilde{\mathcal{K}}_{\text{Zeeman}}$ so a 2×2 matrix must be diagonalized. When this is done we find

$$E_1 = \tfrac{1}{2} \beta_e H_X \left(-g_e + 2 \frac{\zeta}{\delta} \right) \quad (10.71)$$

$$E_2 = \tfrac{1}{2} \beta_e H_X \left(g_e - 2 \frac{\zeta}{\delta} \right) \quad (10.72)$$

which lead to a transition frequency of

$$\nu = \left(g_e - 2 \frac{\zeta}{\delta} \right) \beta_e H_X / h \quad (10.72)$$

By symmetry we know that the results for the y axis are identical to those for x.

Comparing Eqs. (10.70) and (10.72) we see clearly that anisotropy exists. Since ζ / Δ may be ~ 0.1, the frequencies predicted by these equations are significantly different for the same value of magnetic field. Thus the reestablishment of orbital angular momentum via the spin-orbit interaction provides a suitable explanation of the anisotropic nature of the EPR spectra of transition metal complexes.

It should be noted right now that the precise form of the transition formulas depends upon the symmetry of the crystal field and the term state of the ion. Thus our detailed example above applies only to a ^2D state (such as arises from a d^1 configuration) ion in a tetragonally distorted octahedral environment. Detailed theoretical considerations or simple physical intuition lead to the conclusion that the crystal field splitting for a d^9 ^2D term is just like that for a d^1 ^2D term except the levels are inverted. Thus Figs. 10.10 may be applied directly to Ti^{+3} ($3d^1$) complexes. If they are inverted they apply to Cu^{+2} ($3d^9$) complexes. The reason for this inverse behavior is simply that a d^9 configuration of electrons acts like a d^1 configuration of positive "holes."

Treatments similar to that given here for the d^1 (or d^9) octahedral-tetragonal case may be worked out for any d-electron terms in any type of crystal field. The possibilities are too numerous for us to consider here so the reader is urged to consult the literature for further details.

C. Effective Spin Hamiltonian

The results of the previous section lead us to a particularly simple way of empirically fitting transition metal EPR spectra. For the case treated, the transition frequencies of Eqs. (10.70) and (10.72) may be written

$$\nu = g_{||}\beta_e H_{||}/h$$

$$\nu = g_{\perp}\beta_e H_{\perp}/h$$

(10.73)

for the transitions with H parallel and perpendicular to the z axis, respectively. The g factors are, according to our theory,

$$g_{||} = g_e - 8\frac{\zeta}{\Delta} = g_z$$

(10.74)

$$g_{\perp} = g_e - 2\frac{\zeta}{\delta} = g_x = g_y$$

This form for the transition frequencies suggests the following form for the Hamiltonian in place of Eq. (10.67):

$$\tilde{\mathcal{3C}}_{\text{Zeeman}} = \beta_e[g_{\perp}(\tilde{S}_x'H_x + \tilde{S}_y'H_y) + g_{||}\tilde{S}_z'H_z]$$

(10.75)

The primes on $\tilde{S}_x$, $\tilde{S}_y$, and $\tilde{S}_z$ indicate that these do not represent the real spin angular momenta. In fact we define these quantities by requiring that the functions $|+\rangle$ and $|-\rangle$ of Eqs. (10.66) be eigenfunctions of $\tilde{S}_z'$ with eigenvalues $\pm\frac{1}{2}\hbar$. Thus we have (suppressing the $\hbar$)

$$\tilde{S}_z'|+\rangle = \frac{1}{2}|+\rangle$$

$$\tilde{S}_z'|-\rangle = -\frac{1}{2}|-\rangle$$

(10.76)

and also

$$\widetilde{S}_x' \,|\,+\rangle = \tfrac{1}{2}\,|\,-\rangle$$

$$\widetilde{S}_y' \,|\,+\rangle = -\tfrac{1}{2}i\,|\,-\rangle \tag{10.77}$$

Note that we are simply defining a fictitious spin having the same eigenvalues and matrix elements as the true spin.

Now if we evaluate the Hamiltonian matrix of Eq. (10.75) using matrix elements of the fictitious spin, we get for $H_x = H_y = 0$,

$$\langle \pm \,|\, \widetilde{\mathcal{3C}} \,|\, \pm \rangle = \pm\tfrac{1}{2}g_{\|}\beta_e H_{\|} \tag{10.78}$$

which gives a transition frequency as given already by Eqs. (10.73). Similarly, if H lies along either the x or y axis we obtain precisely the previous results again. Thus the "effective spin" Hamiltonian is a satisfactory means of empirically describing the spectrum.

These arguments lead us to the concept of the g tensor,

$$\mathbf{g} = \begin{pmatrix} g_{xx} & g_{xy} & g_{xz} \\ g_{xy} & g_{yy} & g_{yz} \\ g_{xz} & g_{yz} & g_{zz} \end{pmatrix} \tag{10.79}$$

which may be regarded as the proportionality constant relating $\boldsymbol{\mu}$ to $\mathbf{S}'$. Of course a principal axis system always exists so a proper choice of axes requires only diagonal elements. In general then, the effective spin Hamiltonian is

$$\widetilde{\mathcal{3C}}_{\text{Zeeman}} = \beta_e(g_{xx}\widetilde{S}_x'H_x + g_{yy}\widetilde{S}_y'H_y + g_{zz}\widetilde{S}_z'H_z) \tag{10.80}$$

For axially symmetric g tensors, such as in the tetragonal symmetry case,

$$g_{xx} = g_{yy} = g_\perp$$

and

$$g_{zz} = g_{\|}$$

which is the case we treated in detail.

Still more generally we may now include other terms in the spin Hamiltonian. These additional terms would include the spin–spin interaction $\widetilde{I}\cdot\widetilde{S}$ described in Sec. 10.2 and the dipole–dipole terms described in Sec. 10.4. Hence in the principal axis system a rather general spin Hamiltonian form is

$$\widetilde{\mathcal{3C}} = \beta_e(g_xH_x\widetilde{S}_x + g_yH_y\widetilde{S}_y + g_zH_z\widetilde{S}_z) + D\widetilde{S}_z^2 + E(\widetilde{S}_x^2 + \widetilde{S}_y^2)$$
$$+ A_x\widetilde{I}_x\widetilde{S}_x + A_y\widetilde{I}_y\widetilde{S}_y + A_z\widetilde{I}_z\widetilde{S}_z \tag{10.81}$$

where we have suppressed the primes on the $\widetilde{S}_x$, $\widetilde{S}_y$, and $\widetilde{S}_z$ for simplicity. The last three terms of Eq. (10.81) are included under the assumption

TABLE 10.3

Representative g values for transition-metal complexes.

Complex	g value			Reference
$CsTi(III)(SO_4)_2 \cdot 12 H_2O$	$g_{\parallel} = 1.25$	$g_{\perp} = 1.14$		a
Co^{+2} in $ZnSO_4 \cdot 7 H_2O$	$g_x = 2.30$	$g_y = 3.30$	$g_z = 6.90$	b
$K_2Cu(II)(SO_4)_2 \cdot 6 H_2O$	$g_x = 2.14$	$g_y = 2.04$	$g_z = 2.36$	c
Copper(II) phthalocyanine	$g_{\parallel} = 2.165$	$g_{\perp} = 2.045$		d
Copper(II) citrate	$g_{\parallel} = 2.349$	$g_{\perp} = 2.074$		e
VCl_4	$g_{\parallel} = 1.920$	$g_{\perp} = 1.899$		f

[a] B. Bleaney, G. S. Bogle, A. H. Cooke, R. J. Duffus, M. C. M. O'Brien, and K. W. H. Stevens, *Proc. Phys. Soc.* **68**, 57 (1955).
[b] B. Bleaney and D. J. E. Ingram, *Proc. Roy. Soc.* **205**, 336 (1951); **208**, 143 (1952).
[c] B. Bleaney, R. P. Penrose, and B. I. Plumpton, *Proc. Roy. Soc.* **198**, 406 (1949).
[d] D. Kivelson and R. Neiman, *J. Chem. Phys.* **35**, 149 (1961).
[e] B. G. Malmstrom and T. Vanngard, *J. Mol. Biol.* **2**, 118 (1960).
[f] R. B. Johannesen, G. A. Candela, and T. Tsang, *J. Chem. Phys.* **48**, 5544 (1968).

that only one nucleus has a nuclear spin; if several are present the terms should be reproduced for each nucleus as previously described.

Table 10.3 lists the experimental values found for the g tensor elements in a variety of transition-metal complexes.

EXAMPLE 10-4

Until very recently the EPR spectrum VCl_4 had escaped detection. While it is not a very representative example, the spectrum observed by Johannesen, Candela, and Tsang[21] exhibits in a rather compact way the spin Hamiltonian and g-tensor anisotropy concepts that we have been describing. The experimentally observed spectrum is shown in Fig. 10.11, and was obtained from a dilute solution of VCl_4 in $TiCl_4$ at 4°K. The spectrum of this *glassy* solution was essentially orientation independent, even though the g tensor and nuclear spin-electron spin tensor are anisotropic. The point is that all orientations of VCl_4 molecules are frozen into the glass, so the orientation of the macroscopic glassy sample is unimportant.

According to our earlier discussion we would expect the randomly oriented VCl_4 species (with one unpaired d electron) to absorb at all frequencies permitted by the continuous range of orientations of H relative to the x-, y-, and z-molecular axes. Indeed this does occur, but the transition probabilities and density of absorption lines is such that the spectrum is

[21] R. B. Johannesen, G. A. Candela, and T. Tsang, *J. Chem. Phys.* **48**, 5544 (1968).

relatively sharp and distinct only for those species oriented either parallel or perpendicular to the direction of H.

The spectrum is therefore described in the following way. We take the spin Hamiltonian of Eq. (10.81) for the axially symmetric case

$$g_x = g_y = g_\perp$$

$$g_z = g_{||}$$

$$A_x = A_y = A_\perp$$

$$A_z = A_{||}$$

and neglect D and E since there are no electron dipolar interactions for this ^2D species. The abundant isotope of vanadium has spin $\frac{7}{2}$ so an 8-line multiplet $(2I + 1)$ is expected for both the $||$ and $\perp$ orientations. While the multiplet of Fig. 10.11 is somewhat overlapped, it may be decomposed into two sets of equally spaced 8-line multiplets. The stronger series is ascribed to the $\perp$ group, the weaker series to the $||$ group, and the best values of the coupling constants obtained by Johannesen $et\ al.$ are

$$A_{||} = 72 \text{ G}$$

$$A_\perp = 120 \text{ G}$$

The magnetic field value at the center of each series of lines along with the

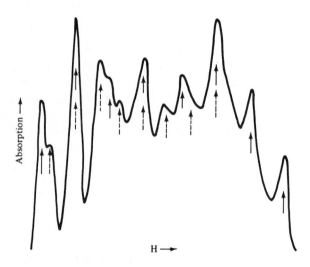

FIG. 10.11.   EPR spectrum of VCl$_4$ in TiCl$_4$ at 4°K. From R. B. Johannesen, G. A. Candela, and T. Tsang, $J.\ Chem.$ $Phys.$ **48**, 5546 (1968). With permission.

fixed value of the microwave frequency permitted the g values to be found as follows:

$$g_{||} = 1.920 \qquad g_{\perp} = 1.899$$

Johannesen *et al.* have discussed these results in terms of the spin-orbit/crystal field parameter ratio, ζ/Δ. The interested reader is urged to consult the original work for this and other theoretical interpretations of the EPR parameters.

In leaving this example, we want to emphasize that if oriented single crystal studies were possible, the two 8-line multiplets would not be observed simultaneously. Instead, only a single multiplet would be observed at each orientation of the crystal relative to the magnetic field (assuming only one orientation of VCl_4 molecules per unit cell).

10.6   TIME-DEPENDENT PHENOMENA

We shall not expend any further effort in this chapter describing transition probabilities or selection rules. The results are simple as we have shown and the theory parallels very closely that described for NMR in chapter 9. Indeed, much of the discussion of Sec. 9.4 is appropriate to the electron resonance case.

A.   Relaxation and Line Shapes

The natural line shapes of EPR transitions are the Lorentzian functions of Eq. (9.53). Again we stress here that the spin-lattice relaxation time T_1 is important in determining the saturation properties of the spin system, while T_2 is important in determining the linewidth. Spin-spin relaxation times are somewhat variable but are often in the range of $T_2 \sim 10^{-7} - 10^{-9}$ sec, compared to values of, say, 1 sec for nuclear relaxation in liquids. Spin-lattice relaxation times vary over a broad range and are often strongly temperature dependent. Typical values may vary from about $10^{-10} - 1$ sec, the latter value being appropriate for the lower range of temperatures (say 4°K) or for systems whose orbital angular momentum is strongly quenched. For rapidly tumbling radicals in solution it is often found that $T_1 \sim T_2$ as with the nuclear case.

For liquid-phase spectra of spin-$\frac{1}{2}$ radicals, linewidths on the order of 100 mG are often obtained. This corresponds to about 0.3 MHz at $g = 2.0$, which gives from Eq. (9.52) a T_2 value of about 5×10^{-7} sec^{-1}. On the other hand some transition-metal complexes with strong relaxation mechanisms have such short relaxation times ($\sim 10^{-10}$ sec) that their resonances are broadened enough to be unobservable at room temperature.

The relaxation mechanisms are many, and as with the nuclear case their theoretical description is not easy. In general, relaxation contributions are produced by any time dependent interaction which involves Hamiltonian matrix elements connecting the states between which magnetic resonance transitions occur. For the EPR case, these interactions include

(1) Electron dipole–dipole
(2) Spin exchange ($S_1 \cdot S_2$)
(3) Spin-orbit coupling

The first of these is expected to produce very broad lines in samples having relatively high radical concentrations, since in this case the electron magnetic dipoles are relatively close together. As shown by Eq. (9.54) the dipolar interaction falls off quite rapidly with increasing separation. Line broadening from these dipole-dipole interactions is circumvented by performing resonance studies in dilute solutions.

Strangely enough, it was observed very early that linewidths were often not nearly as large as expected from a dipole-dipole model. The reason for this was that the spin exchange relaxation, (2) above, often is very effective in averaging out the dipole interactions. The spin-exchange interaction is strictly quantum mechanical and arises because of the overlap of wave functions on one radical with those on a neighboring radical.[22] Physically, one often speaks of this as being a real exchange of the unpaired electrons between molecules. If this exchange becomes rapid enough, the dipole interactions become inefficient in producing relaxation and the EPR line is said to be "exchange narrowed." Under conditions where dipolar broadening is not great, however, such as dilute solutions of radicals in the liquid phase, the spin exchange mechanism may lead to broadening of EPR multiplets in much the same way that proton exchange leads to broadening of NMR hyperfine multiplets (Sec. 9.4 E).

Finally, we mention briefly the third relaxation mechanism above. It is particularly important when the spin-orbit coupling constant ζ is large relative to the crystal-field splitting. Effectively this produces very strong relaxation via the crystal lattice, and hence very short relaxation times (T_1) occur. Ti^{+3} and V^{+4} complexes are examples of species having such strong spin-lattice relaxation effects. It is usually necessary to study these ions at very low temperatures where the relaxation times become much longer, since otherwise the resonance lines are broadened so much that they are unobservable.

These latter results may seem somewhat puzzling since we stressed earlier that T_2 was primarily responsible for the width of the resonance lines. If T_1 becomes very small, however, we know that the Uncertainty Principle puts a limitation on the precision of the resonance frequency as described

[22] J. H. Van Vleck, *Electric and Magnetic Susceptibilities* (Oxford U.P., New York, 1932), p. 316.

in Sec. 5.2, that is, the linewidth is given by

$$\delta \nu \sim \frac{1}{T_1}$$

In a semiquantitative way the contributions of T_1 and T_2 to the total linewidth can be expressed by

$$\delta \nu = \frac{1}{2\pi} \left(\frac{1}{T_1} + \frac{1}{T_2} \right) \tag{10.82}$$

Thus in the previously described case where T_1 was very small ($T_1 \ll T_2$) at room temperature due to strong spin-lattice relaxation, $\delta \nu = (1/2\pi)(1/T_1)$ and the linewidth is very broad. As the temperature is lowered T_1 becomes large due to the ineffective energy transfer and the "normal" linewidth $\delta \nu = (1/2\pi)(1/T_2)$ is established.

B.   Exchange Processes

Exchange processes affect EPR spectra in much the same way as they do NMR spectra. Thus if an electron is able to jump (or exchange) from one site to another with a rate comparable to the frequency separation of resonance lines for the two sites, distinct spectral effects are observed. The time scale for the EPR exchange processes is much different, however, from that for NMR. For example, hyperfine splittings ($A\tilde{I}\cdot\tilde{S}$) in EPR spectra may be on the order of 100 MHz, so exchange processes may be observed if they have rate constants on the order of 10^{+8} sec^{-1}.

We do not describe any of the processes in detail since they are not nearly so common or systematic as for the NMR case. We simply mention that it has been possible to investigate effects ascribable to inversion-type motions, internal rotation, and electron transfer between two different species or between different portions of the same radical species. Considerable detail about these processes can be found by consulting the general references at the end of the chapter.

10.7   THEORETICAL INTERPRETATION OF EPR PARAMETERS

A.   Fermi Contact Term

The theory of the isotropic $A\tilde{S}\cdot\tilde{I}$ term, which is so prominent in the interpretation of the spectra of organic radicals in solution, was first worked out by Fermi.[23] For a single unpaired electron this term has the form

$$A\tilde{S}\cdot\tilde{I} = \frac{8\pi}{3} g_e \beta_e g_n \beta_n \delta(\mathbf{r}) \tilde{S}\cdot\tilde{I} \tag{10.83}$$

[23] See footnote 11.

where $\delta(\mathbf{r})$ is the Dirac δ function operator and $\mathbf{r}$ is the electron position vector relative to the nucleus. This operator has the following properties:

$$\delta(\mathbf{r})\psi(\mathbf{r}) = \psi(\mathbf{r}) \quad \text{when} \quad r = 0 \qquad (10.84a)$$

$$\delta(\mathbf{r})\psi(\mathbf{r}) = 0 \qquad \text{when} \quad r \neq 0 \qquad (10.84b)$$

where $\psi(\mathbf{r})$ is an electronic wave function, that is, a function satisfying Eq. (3.22). Recalling our earlier discussions, we know that the experimentally measured value of A is an average (usually) in the ground electronic state. Thus the measured value of A is

$$\langle \psi \mid A \mid \psi \rangle = \frac{8\pi}{3} g_e\beta_e g_n\beta_n \langle \psi \mid \delta(\mathbf{r}) \mid \psi \rangle$$

$$= \frac{8\pi}{3} g_e\beta_e g_n\beta_n \mid \psi(0) \mid^2 \qquad (10.85)$$

where $\psi(0)$ is the value of the electronic wave function at $r = 0$.

For the hydrogen atom the value of A is 1420 MHz. Actually the units of A as expressed in (10.85) are in ergs, so it is necessary to divide by h to obtain a comparison with the experimental result. The ground-state wave function for the hydrogen atom is well known to everyone, namely

$$\psi = \left(\frac{1}{\pi a_0{}^3}\right)^{1/2} \exp(-r/a_0) \qquad (10.86)$$

where $a_0 = 0.529$ Å. Substitution of Eq. (10.86) (with $r = 0$) into (10.85) and using the well-known values of g_e, g_n, β_e, and β_n leads to a theoretical value of $A = 1422$ MHz. The agreement is quite excellent and the small discrepancy need not concern us here.

What does concern us now is that the Fermi mechanism given in Eq. (10.85) predicts hyperfine splitting only for s states, since it will be recalled that p, d, f, etc. states have wave functions which vanish at the nucleus. For the molecular case similar conclusions exist, namely that the coupling constant should differ from zero only if the unpaired electron has some σ-type character to its distribution. This is puzzling at first since our first hunch would be that the unpaired electron in benzene negative ion, for example, would be in a π-molecular orbital. Consequently this electron should produce no Fermi contact coupling with the protons (or C^{13} ring carbons) since the π-electron wave functions have nodes in the plane of the ring. Yet the spectrum shows equivalent coupling to all protons with a value of 3.75 G (or 10.5 MHz).

Of course the point is that the "perfect-pairing" approximation of chemical bonding is an oversimplification. σ and π electrons are not really independent as this simple approximation assumes. Indeed, interelectronic interactions between the electrons produces a breakdown in the σ, π concept. In a very simple way we can see how an odd electron in a π-orbital induces some excess electron distribution in the σ orbital of a proton on an aromatic ring. Suppose the odd electron has spin function α, while the σ electrons in a C–H bond are α and β. In the absence of interelectronic interactions the sigma bond would be equally well described by having the carbon sp^2 electron α and the hydrogen $1s$ electron β, or vice versa. But due to interactions between the σ and π electrons, the pairing scheme having the sp^2 electron α is slightly preferred when the π electron is α. The result is that the carbon atom is relatively richer in α electron density while the hydrogen is relatively richer in β electron density. Consequently the π electron has caused some excess σ electron density to be present at the proton and the Fermi term is nonzero.

This spin polarization problem has been treated by several workers using molecular orbital theory.[24,25] Here the argument involves the concept of *configuration interaction*. In MO Theory a three electron system consisting of a C–H bond with two σ electrons and a carbon $2p_z\pi$ electron would be described by the ground-state Slater determinant

$$\psi_G = \frac{1}{\sqrt{6}} \, || \, \sigma(1)\alpha(1)\sigma(2)\beta(2)\pi(3)\alpha(3) \, || \qquad (10.87)$$

where only the diagonal elements of the determinant have been recorded. Now a variety of excited state configurations exist, for example,

$$\psi_E = \frac{1}{\sqrt{6}} \, || \, \sigma(1)\alpha(1)\sigma^*(2)\beta(1)\pi(3)\alpha(3) \, || \qquad (10.88)$$

where σ^* is the first antibonding orbital. It is known that a better approximation than (10.87) to the ground state is obtained by including configurations such as that of (10.88), that is

$$\psi = \psi_G + \sum_i \lambda_i \psi_E^{(i)} \qquad (10.89)$$

where λ_i is a small coefficient prescribing the contribution of the $\psi_E^{(i)}$th excited state configuration.

Now for the three-electron system the value of the coupling constant at the proton would be (for the z components of $\tilde{S} \cdot \tilde{I}$ only)

$$A_H \tilde{I}_Z \sum_i \tilde{S}_Z(i) = \frac{8\pi}{3} \, g_e \beta_e g_n \beta_n \langle \psi \, | \, \delta_1 \mathcal{P}(1) + \delta_2 \mathcal{P}(2) + \delta_3 \mathcal{P}(3) \, | \, \psi \rangle \tilde{S}_Z \tilde{I}_Z$$

$$(10.90)$$

[24] S. J. Weissman, *J. Chem. Phys.* **25**, 890 (1956).
[25] H. M. McConnell, *J. Chem. Phys.* **24**, 764 (1956).

by an extension of Eq. (10.83) to the three electron case. $\mathcal{P}(i)$ is an operator which has the value ± 1 depending on whether $\langle \tilde{S}_Z(i) \rangle$ is α or β. $\tilde{S}_Z$ is the net spin, which for the case of Eqs. (10.87) to (10.89) has the expectation value of $+\frac{1}{2}$, but of course may equally well be $-\frac{1}{2}$.

When (10.89) is substituted into Eq. (10.90) we find three types of matrix elements. First we have terms like $\langle \psi_G \,||\, \psi_G \rangle$ which leads to $+ \,|\, \sigma(0) \,|^2 - |\, \sigma(0) \,|^2 + |\, \pi(0) \,|^2$. The first two terms cancel and the third term is zero since π orbitals vanish in the molecular plane. This verifies our earlier conclusion that a perfect pairing approximation would lead to $A = 0$ if the odd electron were of π type. We also get terms of the form $\langle \sum \lambda_i \psi_E{}^{(i)} \,||\, \sum \lambda_i \psi_E{}^{(i)} \rangle$ which are of order $\lambda_i{}^2$ and are consequently very small. Finally we find the cross terms which are linear in λ and do not vanish. Thus one concludes that A_H should be proportional to λ, the mixing coefficient.

The theoretical evaluation of λ has been carried through by means of first-order perturbation theory. This treatment shows that $\lambda \propto \rho$, where ρ is the unpaired spin density of the π electron at the carbon nucleus of interest.[26] Thus, the result is that one predicts

$$A_H \propto \lambda \propto \rho \qquad \text{or} \qquad A_H = Q\rho \qquad (10.91)$$

where the theoretical value for the proportionality constant Q is about -28 G in the most useful units.

It turns out that Eq. (10.91) is in relatively good agreement with experimental values of A_H and theoretical values of ρ as determined by Hückel molecular orbital calculations, but a Q value of -22.5 G gives better agreement and is generally used in preference to the larger value. The validity of this treatment may be verified easily. The measured coupling constant for $C_6H_6{}^-$ is -3.75 G.[27] Because of the symmetry of the radical we know that $\rho = \frac{1}{6}$ at each carbon. Thus $Q = -3.75(6) = -22.5$ G. Similarly, the measured value of A_H for the methyl radical $\cdot CH_3$ is -23.0 G. Since the odd electron occupies a π orbital centered at the carbon, we expect $\rho = 1.0$ so $Q = -23.0$ is predicted. On the other hand, if we predict A_H for the tropylium radical (C_7H_7) using $Q = -22.5$, we get $A_H = 1/7(-22.5) = -3.22$ G. The experimental value of -3.91 G is therefore in fair but not perfect agreement.

We list in Table 10.4 the experimentally determined coupling constants for several representative organic radicals. Most of the data refer to proton coupling constants but some results for N^{14} and C^{13} couplings are also listed.

[26] See footnote 24.
[27] Actually, the signs are often not experimentally known.

TABLE 10.4

Experimental values of the isotropic Fermi contact coupling constant

Radical	A_H (gauss)	A_X (gauss)
$\cdot CH_3$	23.04	
$C_6H_6^-$	5.98	
$\cdot C_7H_7$	3.91	
(naphthalene anion, positions a, b)	$A_a = 4.90, \quad A_b = 1.83$	
$\cdot C(CH_3)_3$	22.72	
(anthracene anion, positions c, b, a)	$A_a = 5.56, \quad A_b = 2.74,$ $A_c = 1.57$	
(pyrazine anion)	2.66	$A_N = 7.22$
(4,4'-bipyridine anion)	2.35, 0.43	$A_N = 3.64$
$\cdot CPh_3$	$A_0 = 2.53, \quad A_m = 1.11,$ $A_p = 2.77$	$A_{C^{13}} = 26$
$O - \langle \bigcirc \rangle - O^-$	2.37	

EXAMPLE 10-5

As a more stringent test of the relation (10.91) we consider the naphthalene negative ion

(naphthalene structure with positions 1–10 labeled)

which has two sets of equivalent protons, (1, 4, 5, 8) and (2, 3, 6, 7). The 25-line spectrum of the radical in solution leads to $A_1 = 4.90$ and $A_2 = 1.83$ G, the assignment being confirmed by deuteration experiments.

According to the very simplest π-electron Hückel molecular-orbital theory the odd electron occupies the following molecular orbital

$$\psi = 0.425(\phi_1 + \phi_4 + \phi_5 + \phi_8) - 0.263(\phi_2 + \phi_3 + \phi_6 + \phi_7)$$

Thus the spin densities at carbons 1 and 2 are

$$\rho_1 = (0.425)^2 = 0.181$$

$$\rho_2 = (0.263)^2 = 0.0691 \tag{10.92}$$

Using Eq. (10.91) with $Q = 22.5$, we obtain

$$A_1 = 22.5(0.181) = 4.07 \text{ G}$$

$$A_2 = 22.5(0.0691) = 1.55 \text{ G} \tag{10.93}$$

These results certainly can not be said to be in remarkable agreement with the measured values. Yet A_1/A_2 has the experimental value of 2.68 while the theoretical ratio is 2.62. Indeed, this merely reflects again the variability in the value of Q from molecule to molecule. In the present case a value of $Q \sim 27$ G would give a much better fit. In any case it does usually occur that the relative values of proton coupling constants are predicted quite well by Hückel molecular orbital theory, so one begins to believe that there is really merit to the procedure.

We shall not pursue further the topic of proton coupling constants and their understanding and interpretation via valence bond or molecular-orbital theory, but simply mention that this very interesting subject may be investigated further by consulting the reference material.

As has been illustrated in earlier examples and in Table 10.4, C^{13} and N^{14} nuclei show interesting and valuable Fermi contact coupling constants. In contrast to the situation occurring in NMR spectra, the N^{14} quadrupole moment produces no appreciable broadening so the N^{14} EPR hyperfine lines are very sharp. This difference occurs simply because the nuclear electric quadrupole interactions are much smaller than the dominant hyperfine energies of EPR so their relaxation effects are negligible. N^{14} coupling constants in aromatic radicals have been successfully related to the unpaired spin density in the nitrogen $2p_z$ orbital, the proportionality constant being on the same order as that for proton couplings in aromatic radicals.

B. g Values and the Dipole–Dipole Constants, D and E

In Sec. 10.5 we have seen the general principles involved in the theoretical interpretation of transition-metal anisotropic g values in terms of spin-orbit coupling constants and crystal-field parameters. These latter parameters are useful in obtaining a complete description of the electric and magnetic properties of the complexes, and provide knowledge of the chemical bonding. Further information may be obtained from texts dealing with ligand field theory.[28]

In liquid solution we have seen that any possible g-tensor anisotropy is averaged out so that only an average g value is observed, usually very close to $g = g_e = 2.0023$ for spin-$\frac{1}{2}$ radicals. While some theoretical interpretations of the small deviations from 2.0023 can be made they are of less applicability than for the corresponding changes in the chemical shift δ in NMR spectroscopy. Table 10.5 provides a brief list of some g values for organic radicals in solution. It can be noted that the g-values increase as atoms with larger spin-orbit coupling (such as O or Cl) are incorporated in the radical. This qualitative dependence upon ζ is not unexpected according to our earlier discussions. It indicates a slight removal of total quenching of orbital angular momentum. Since this reinstatement of orbital angular momentum is in some way proportional to the magnitude of ζ, we may expect deviations from $g = g_e$ as ζ increases.

The dipole-dipole constants D and E have been defined by Eq. (10.45) and some representative values have been presented in Table 10.1. A qualitative or semiquantitative understanding of these parameters may be obtained by evaluating D and E for a case involving simply two negative point charges. Let us imagine that these charges are situated with the (x, y, z) coordinates: $(0, 0, 1)$ and $(0, 0, -1)$ in units of Angstroms. Then from Eq. (10.45) we find

$$E = 0 \tag{10.94a}$$

and

$$D = \frac{3}{4} g^2 \beta_e^2 \left(\frac{2}{r^3} \right)$$

$$= \frac{3}{2} g^2 \beta_e^2 \left(\frac{1}{r^3} \right)$$

$$= \frac{3}{2} g^2 \beta_e^2 \left(\frac{1}{2 \times 10^{-8}} \right)^3 \text{erg}$$

$$= 0.32 \text{ cm}^{-1} \tag{10.94b}$$

From this simple calculation we see that a small or zero value for E is expected when the unpaired electrons have a high probability of being located along the z principal axis. Furthermore a 2 Å charge separation in the z direction leads to a value of D similar to all those of Table 10.1 except for the last entry. Of course, D and E must really be evaluated quan-

---

[28] See, for example, C. Ballhausen, *Introduction to Ligand Field Theory* (McGraw-Hill, New York, 1962).

TABLE 10.5

g values of organic radicals in liquid solution [a]

Radical	g
Vinyl	2.00220
Anthracene[+]	2.00249
Ethyl	2.00260
Naphthalene[-]	2.00263
Benzene[-]	2.00276
Benzophenone[-]	2.00359
1,4 Benzosemiquinone	2.00468
2 Chlorobenzosemiquinone anion	2.00486

[a] Taken in part from M. Bersohn and J. C. Baird, *An Introduction to Electron Paramagnetic Resonance*, Benjamin, New York (1966), Table 5.4.

tum mechanically by averaging over the triplet electronic state of the radical, rather than using a point-charge model. Nevertheless the qualitative features of the point-charge model are useful. In particular, as the separation of the unpaired electrons increases we expect D to decrease.

The qualitative interpretations of the D values in Table 10.1 may thus be given as follows. Radicals 1 and 4 have both electrons localized on the carbon atom as shown and are delocalized relatively little into the aromatic rings. Radicals 2, 3, and 5 have the unpaired electrons relatively delocalized over the aromatic rings, particularly in the last case.

Much more sophisticated and quantitative interpretations of the dipolar parameters may be made,[29] but we leave the topic at this very elementary stage.

10.8 OTHER EPR STUDIES

There is yet one rather large class of EPR studies of great importance to chemistry. This involves small inorganic and organic spin-$\frac{1}{2}$ radicals trapped in the solid phase or in glasses. Species of this type, such as CO_2^- and $HCOOH^-$, are typically generated by ultraviolet or nuclear irradiation. The EPR spectra of these species illustrate rich hyperfine effects including those from anisotropic g tensors and anisotropic nuclear spin-electron spin interactions. Analysis of these spectra provides much complementary information about the electronic structure of the radicals, and due to the relative simplicity of the radicals it is often possible to carry out reasonably

[29] For example, see S. A. Boorstein and M. A. Gouterman, *J. Chem. Phys.* **39**, 2443 (1963).

sophisticated quantum chemical calculations. Further information about these systems can be obtained by consulting the general reference material.

SUPPLEMENTARY REFERENCES

M. Bersohn and J. C. Baird, *An Introduction to Electron Paramagnetic Resonance* (Benjamin, New York, 1966).

A. Carrington and A. D. McLachlan, *Introduction to Magnetic Resonance* (Harper and Row, New York, 1967).

J. S. Griffith, *The Theory of Transition Metal Ions* (Cambridge U.P., Cambridge, England, 1961).

D. J. E. Ingram, *Free Radicals as Studied by Electron Spin Resonance* (Butterworths, London, 1958).

G. E. Pake, *Paramagnetic Resonance* (Benjamin, New York, 1962).

C. P. Slichter, *Principles of Magnetic Resonance* (Harper and Row, New York, 1963).

CHAPTER 11

ELECTRONIC SPECTROSCOPY

11.1 INTRODUCTION

We have left for last the description of the spectra arising from transitions between the electronic states of molecules. This choice is entirely arbitrary but was made with a predetermined prejudice—namely, that it is best to study first the spectroscopic methods whose theoretical interpretation and understanding are the soundest. It has been shown in the previous chapters that very precise quantum-mechanical treatments (although generally not exact) of rotational, vibrational, and nuclear and electron Zeeman energy levels were possible. In the present case, however, the quantum-mechanical problem is so difficult that even for the very simplest molecules it is not generally possible to specify the absolute or relative values of the energies of the electronic states with any precision.

On the other hand, this comparison is not entirely fair, since the previous four spectroscopic descriptions were by no means *ab initio*, but proceeded with various natural parametrizations. For example, the rotational energy states were specified in terms of moments of inertia, the vibrational states in terms of force constants, the nuclear Zeeman states in terms of chemical shifts and spin-spin coupling constants, and the electron Zeeman states in terms of g tensors and coupling constants. No suitably precise parametrizations are generally possible for the electronic energy levels although a variety of approximate ones are used. As we shall see, the quantum-mechanical problem here is simply that of the energies of electrons moving in the Coulombic fields of nuclei and other electrons; a problem easy to set down but difficult to solve.

These notions are by no means meant to detract from the importance or significance of the field. Indeed, if the electronic energy of only the ground state could be correctly calculated as a function of all internuclear separations R_α, nearly all the results of rotational, vibrational, ESR and NMR spectroscopy could be predicted. Fortunately for experimental chemists and spectroscopists, the immensity of the computational problems make it unlikely that the quantum theorists will have all the answers in any foreseeable future. Still the progress in this direction has been very impressive and will unquestionably continue.

In this section we depart from the format of the last several chapters by providing for the reader no discussion of experimental methods of electronic spectroscopy. It probably has been presumptuous to describe the experimental methods in such a cursory manner as was done in the previous four chapters. It would be even more presumptuous to present a simple block diagram and attending discussion for electronic spectroscopy, the oldest form of molecular spectroscopy. We justify this decision by the fact that the techniques and instrumentation are really too numerous to mention or to choose from, and also upon the supposition that the reader has already a pretty good feeling about the general experimental methods. After all, electronic spectroscopy is performed normally with visible or ultraviolet radiation. Even the novice scientist knows about sources of visible radiation (incandescent lamps, etc.), and about detection devices (the eye and photographic film, for example). Thus we limit our experimental comments to the recommendation that the reader review chapter 1 for general background, and consult the reference literature for experimental details.

A theoretical description of the molecular electronic energy states proceeds via the Born–Oppenheimer approximation as detailed in Sec. 3.2. In this approximation the electronic stationary states are obtained by solution of Eq. (3.22),

$$\tilde{\mathfrak{K}}_e\psi_e(r_i, R_\alpha) = E_e(R_\alpha)\psi_e(r_i, R_\alpha) \tag{3.22}$$

where the eigenfunctions ψ_e and energies E_e are determined as functions of the internuclear separations R_α. For the electronic Hamiltonian we have as in Eq. (3.21)

$$\tilde{\mathfrak{K}}_e = -\frac{\hbar^2}{2m_e}\sum_i \nabla_i^2 + e^2 \sum_{\alpha>\beta} \frac{Q_\alpha Q_\beta}{R_{\alpha\beta}} + e^2 \sum_{i>j} \frac{1}{r_{ij}} - e^2 \sum_{i,\alpha} \frac{Q_\alpha}{r_{i\alpha}} \tag{3.21}$$

For even a simple molecular system such as hydrogen (H_2), an exact solution of Eq. (3.22) is impossible. However, many approximate means of solving Eq. (3.22) exist for molecular systems. Furthermore, it is possible to describe qualitatively the electronic states which may exist even though precise calculations are not possible. Specifically, it is always possible to describe the symmetry properties of the electronic states, and to some extent to describe some of the angular momentum properties of the molecular energy states.

Earlier it has been mentioned that under appropriate conditions (approximations) the total molecular energy may be expressed as a sum of electronic, vibrational, and rotational terms plus smaller types such as Zeeman or Stark terms [see Eq. (3.92), for example]. Thus, just as vibrational transitions occur as a band of transitions between the various rotational states, electronic transitions occur as bands consisting of vibrational transitions which themselves are further split into a rotational fine structure. For this reason electronic spectra contain, in principle, an incredible amount of information. The problem is that except for very small molecules it is not possible to resolve the rotational lines; indeed, for large molecules even the vibrational band structure becomes impossible to resolve. We show in Fig. 11.1 a portion of the electronic spectrum of dimethylsulfide in the vapor phase.

In the following sections we investigate in a preliminary fashion the analysis and interpretation of molecular electronic spectra, and the quantitative and/or qualitative descriptions of the electronic states. Little con-

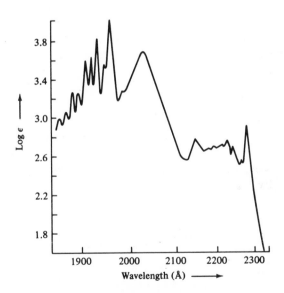

FIG. 11.1. A portion of the electronic spectrum of dimethylsulfide in the vapor phase at 23°C. From S. D. Thompson, D. G. Carroll, F. Watson, M. O'Donnell, and S. P. McGlynn, *J. Chem. Phys.* **45**, 1367 (1966). With permission.

centration is placed upon the study of high-resolution spectra of small molecules, since other means of determining the vibrational and rotational energy states have been previously described. We begin our discussion, however, by looking into the spectra of diatomic molecules in some detail.

11.2 ELECTRONIC SPECTRA OF DIATOMIC MOLECULES

A. General Form of the Energy Levels

Neglecting the smaller energy terms the manifold of states of a diatomic molecule have energies given by the sum of electronic, vibrational, and rotational energies. Using Herzberg's notation,[1] the wavenumber energies are

$$T = T_e + G(v) + F(J) \tag{11.1}$$

where T_e is the electronic energy,

$$G(v) = \omega_e(v + \tfrac{1}{2}) - \omega_e x_e(v + \tfrac{1}{2})^2 + \cdots \tag{11.2}$$

and

$$F(J) = B_v J(J + 1) - D_v J^2(J + 1)^2 + \cdots \tag{11.3}$$

The latter two equations are by now very familiar. The electronic energy states T_e are measured with respect to the minimum in the potential surface (see Fig. 3.3) of the lowest state. We leave until later the investigation of the theory of the electronic energy, merely considering it here to be an experimental quantity. Representing the upper electronic state by *single* primes and the lower by *double* primes, the frequency of any electronic transition may be written

$$\omega = T_e' - T_e'' + G(v') - G(v'') + F(J') - F(J'') \tag{11.4}$$

$T_e' - T_e''$ is logically known as the electronic transition frequency ω_E. Note that Eq. (11.4) and the previous results restated in Eqs. (11.2) and (11.3) assume separability of electronic and nuclear motions (that is, the Born–Oppenheimer approximation), but rotation-vibration interaction is accounted for by B_v in the rotational energies.

B. Selection Rules

As always, the electric dipole selection rules, or the transition probabilities, are given by

$$P_{mn} \propto \left| \int \psi_m{}^* \tilde{\mu} \psi_n d\tau \right|^2 \tag{11.5}$$

[1] G. Herzberg, *Spectra of Diatomic Molecules* (Van Nostrand, New York, 1950), p. 149.

where $\tilde{\mu}$ is the molecular electric dipole moment operator as given by Eq. (7.37). The appropriate wave functions are the products

$$\psi_e(r_i, R_\alpha)\psi_v(R_\alpha)\psi_r(\Omega)$$

where r_i represents electron coordinates, R_α represents the nuclear vibrational position coordinates, and Ω represents the angular coordinates of the rotational wave function. An electron spin wave function should be included also but we defer this for the present.

As in Sec. 7.3 B the dipole moment components along space-fixed axes may be expressed in terms of the dipole moment along the molecular z axis by means of the direction cosines. Since the rotational wave function contains no coordinates in common with ψ_e, its contribution to the transition probability is just the same as derived in chapter 7. Thus the integral always vanishes unless

$$\Delta J = 0, \pm 1 \tag{11.6}$$

We have only then to investigate the selection rules for the electronic vibrational states $\psi_e\psi_v$, which are often termed *vibronic* states. The transition probability between two such states $| ev \rangle$ and $| e'v' \rangle$ is proportional to the square of the matrix element

$$\langle ev \mid \tilde{\mu} \mid e'v' \rangle = \int \psi_e{}^*\psi_v{}^*\tilde{\mu}\psi_e{}'\psi_v{}'dr \tag{11.7a}$$

Since the dipole moment includes contributions from both nuclei and electrons, we may write

$$\tilde{\mu}(r_i, R_\alpha) = \tilde{\mu}_e(r_i) + \tilde{\mu}_n(R_\alpha) \tag{11.8}$$

Then Eq. (11.7a) becomes

$$\langle ev \mid \tilde{\mu} \mid e'v' \rangle = \int \psi_v{}^*\psi_v{}' \left[\int \psi_e{}^*\tilde{\mu}_e\psi_e{}'dr_i \right] dR_\alpha$$

$$+ \int \left[\int \psi_e{}^*\psi_e{}'dr_i \right] \psi_v{}^*\tilde{\mu}_n\psi_v{}'dR_\alpha \tag{11.7b}$$

where the volume element $d\tau$ is written symbolically as $dR_\alpha dr_i$.

Consider the second term in Eq. (11.7b). The integration over electronic coordinates may be carried out first; and since the eigenfunctions belonging to different energy states are orthogonal this term vanishes except when $\psi_e = \psi_e{}'$. This case would correspond to pure vibrational-rotational spectra and has been treated in earlier chapters. Thus for the electronic transitions we are interested only in the first term in Eq. (11.7b).

Within the framework of the Born-Oppenheimer approximation, therefore, electronic transitions occur only if the electronic matrix element

$$\langle \tilde{\mu}_e \rangle = \int \psi_e^* \tilde{\mu}_e \psi_e' dr_i \tag{11.9}$$

does not vanish. It should be remembered that ψ_e depends upon R_α as a parameter, so $\langle \tilde{\mu}_e \rangle$ is a slowly varying function of R_α. For this reason the integration over dr_i cannot strictly be carried out independently. However, since the electronic charge distribution changes only very slightly as R_α varies, that is, the dipole moment is only slightly dependent upon the vibrational state v, it is usually a good approximation to replace $\langle \tilde{\mu}_e \rangle$ in Eq. (11.7b) by an average over R_α, which we designate as $\bar{\mu}_e$. Then Eq. (11.7b) becomes for the possible vibronic transitions,

$$\langle ev \mid \tilde{\mu} \mid e'v' \rangle = \bar{\mu}_e \int \psi_v^* \psi_v' dr_i \tag{11.7c}$$

We discuss further the remaining integral in the above equation at a later time, but we note here that it is in general nonzero, since the vibrational eigenfunctions in *different* electronic states are not orthogonal.

We can determine easily when the electronic transition moment of Eq. (11.9) does not vanish by resorting to group theory. The integral is nonzero only when one of the triple products contains the totally symmetrical representation,

$$\Gamma^{(e)} \times \Gamma^{(\mu_g)} \times \Gamma^{(e')} \to \Gamma^{(A)} \tag{11.10a}$$

where $g = x, y, z$. In fact, as we have done previously, we replace μ_g by g since they are linearly related. Multiplying both sides of Eq. (11.10a) by $\Gamma^{(g)}$ we have then

$$\Gamma^{(e)} \times \Gamma^{(e')} \to \Gamma^{(g)} \tag{11.10b}$$

Thus an electronic transition occurs only if the direct product of the representations of the two states contains the representations of x, y, or z.

There are two classes of diatomic molecules, homonuclear and heteronuclear, which have $D_{\infty h}$ and $C_{\infty v}$ symmetry, respectively. The electronic eigenfunctions must therefore form bases for the irreducible representations of the appropriate point group, that is, the electronic states are labeled by the symmetry species of the $D_{\infty h}$ or $C_{\infty v}$ point group.

As examples, consider states of symmetry Σ^+ and Σ^- for a heteronuclear molecule. From the $C_{\infty v}$ character table in Appendix V, we find the characters of the direct product to be

$$\chi(E) = (1)(1) = 1$$

$$\chi(C_\phi) = (1)(1) = 1$$

$$\chi(\sigma_v) = (1)(-1) = -1$$

which gives the Σ^- representation. Since this representation does not contain x, y, or z as a basis, the transition $\Sigma^+ \leftrightarrow \Sigma^-$ is not allowed. On the other hand, $\Sigma^+ \leftrightarrow \Pi$ is allowed since

$$\chi(E) = (1)(2)$$

$$\chi(C_\phi) = (1)2\cos\phi$$

$$\chi(\sigma_v) = (1)(0)$$

This gives the Π representation which has x and y as bases. In this manner all the allowed transitions for diatomic molecules may be obtained. We list in Table 11.1 some of the allowed transitions.

Although a molecule such as O^{16}–O^{18} is strictly heteronuclear, the selection rules for the homonuclear case apply for this molecule to a high approximation. This is because the electronic eigenfunctions are independent of the nuclear mass to the accuracy of the Born–Oppenheimer approximation.

From the results of Table 11.1 we can write down some general selection rules for diatomic molecules, namely

$$+ \leftrightarrow + \qquad - \leftrightarrow - \qquad + \nleftrightarrow - \qquad (11.11)$$

and

$$\Delta\Lambda = 0, \pm 1 \qquad (11.12)$$

where Λ has values of $0, 1, 2, 3\ldots$ for Σ, Π, Δ, $\Phi\ldots$ states. We see shortly that $\Lambda\hbar$ is the magnitude of the electronic orbital angular momentum along

TABLE 11.1

Allowed transitions for diatomic molecules.

Homonuclear	Heteronuclear
$\Sigma_g^+ \leftrightarrow \Sigma_u^+$	$\Sigma^+ \leftrightarrow \Sigma^+$
$\Sigma_g^- \leftrightarrow \Sigma_u^-$	$\Sigma^- \leftrightarrow \Sigma^-$
$\Pi_g \leftrightarrow \Sigma_u^+$	$\Pi \leftrightarrow \Sigma^+$
$\Pi_u \leftrightarrow \Sigma_g^+$	$\Pi \leftrightarrow \Sigma^-$
$\Pi_g \leftrightarrow \Sigma_u^-$	$\Pi \leftrightarrow \Pi$
$\Pi_u \leftrightarrow \Sigma_g^-$	$\Pi \leftrightarrow \Delta$
$\Pi_g \leftrightarrow \Pi_u$	$\Delta \leftrightarrow \Delta$
$\Pi_g \leftrightarrow \Delta_u$	$\Delta \leftrightarrow \Phi$
$\Pi_u \leftrightarrow \Delta_g$	$\Phi \leftrightarrow \Phi$
$\Delta_g \leftrightarrow \Delta_u$	etc.
etc.	

the molecule axis. Note from the character tables that the $+$, $-$ symmetry refers to symmetry with respect to reflection in a plane containing the molecule. In addition to the rules of Eqs. (11.11) and (11.12), for homonuclear molecules we have also

$$g \leftrightarrow u \qquad g \nleftrightarrow g \qquad u \nleftrightarrow u \qquad (11.13)$$

Here g and u refer to symmetry with respect to inversion about the center of symmetry.

From these results it is clear that the possible electronic transitions of diatomic molecules are rather easily classified. The assignment of observed transitions of a given molecule to the various possible states is not always an easy task, however. Generally, the qualitative ordering of the lower energy states is known from theoretical considerations to be described later, so this provides one aid to spectral assignments. In addition, the rotational fine structure often shows characteristic intensity and splitting patterns which aid in the assignment of the electronic quantum states involved in the transitions. We do not go into this in any detail but do describe some of the vibrational structure effects in Sec. 11.2 C. Excellent descriptions of the rotational structure of electronic bands are given by Herzberg[2] in the first of his monumental series on molecular spectra.

Before leaving this general discussion of the selection rules we should comment briefly upon the role of electron spin. The point is that the total wave function must properly include spin, which, as is more completely described later, means that the electronic wave function must include spin orbitals in such a way that the total wave function is antisymmetric (Pauli principle) with respect to interchange of any pair of electrons. While the spin wave function cannot generally be factored from the orbital functions, the correct symmetry properties are obtained from this approximation. Thus we need to multiply $\psi_e(r_i)$ by $\phi_e(\sigma_i)$ where σ_i is the symbolic spin variable.

Consequently the electric dipole matrix element in Eq. (11.9) becomes

$$\langle \tilde{\mu}_e \rangle = \int \psi_e^* \tilde{\mu}_e \psi_e' dr_i \int \phi_e^* \phi_e' d\sigma \qquad (11.14)$$

The spin functions form an orthonormal set of functions in spin space and thus the matrix element vanishes unless $\phi_e = \phi_e'$. Classifying the spin states by the total spin S as usual, we can state the following selection rule:

$$\Delta S = 0 \qquad (11.15)$$

Very often one classifies electronic states by their multiplicities, $2S + 1$, which leads to singlet, doublet, triplet, quartet, etc., states for $S = 0$, $\frac{1}{2}$,

[2] See footnote 1, chapter V.

TABLE 11.2
Electronic spectral data[a,b] for O_2^{16}

State	T_e	ω_e	$\omega_e x_e$	B_e	α_e	Transition
$B\ ^3\Sigma_u^-$	49 802	700.4	8.002	0.819	0.011	$B \leftrightarrow X$
$A\ ^3\Sigma_u^+$	36 096	819	22.5	...	...	$A \leftrightarrow X$
$b\ ^1\Sigma_g^+$	13 195	1432.7	13.95	1.4004	0.182	$b \leftrightarrow X$
$a\ ^1\Delta_g$	7 918	1509	12.9	1.4264	0.0171	$a \leftarrow X$
$X\ ^3\Sigma_g^-$	0	1580.4	12.07	1.4457	0.0158	

    [a]   A short version of Table 39 from G. Herzberg, *Spectra of Diatomic Molecules* (Van Nostrand, Princeton, N.J., 1950).
    [b]   All values in cm^{-1}.

1, $\frac{3}{2}$, etc., respectively. Hence we see that singlet $\leftrightarrow$ triplet transitions are not permitted.

All of the selection rules that we have derived apply for the electric dipole case. Of course, if magnetic dipole and electric quadrupole matrix elements are considered the results would differ. These latter selection rules usually lead to very weak transitions, so we are not much concerned with them.

A famous violation of the rules is the weak but clear occurrence of singlet-triplet transitions. This arises because the spin and orbital states become mixed (as seen in Sec. 10.5) so that the simple product wave function is not correct. The states are therefore no longer pure singlet and pure triplet so even in the electric dipole approximation singlet-triplet transitions do occur.

We close this section by showing in Table 11.2 some of the observed electronic states and energies of the oxygen (O_2^{16}) molecule. The table lists also some vibrational and rotational constants obtained by analysis of the electronic bands, and the observed transitions.[3] Note that in addition to the group theory state symbol, electronic states are further labeled by a left superscript which specifies the spin multiplicity, and a small or large alphabetic symbol, X being reserved usually for the ground state. It should be noted that the listed transitions range from the near infrared to the ultraviolet.

C. Vibrational Structure of Electronic Bands

Electronic transitions of diatomic molecules are normally accompanied by a vibrational band structure, and in the gas phase at high-resolution by

---

[3] Standard notation, which we have not always followed, shows the *upper* state first, followed by the *lower* state of the transition. The arrow then indicates clearly whether the transition was observed in emission, absorption, or both.

a further rotational fine structure within each vibrational band head. As an example Figure 11.2 shows a rather low-resolution absorption spectrum of iodine in the gas phase. Neglecting the rotational energies, Eqs. (11.2) and (11.4) show the transition frequencies to be given by

$$\omega = \omega_E + \omega_e'(v' + \tfrac{1}{2}) - \omega_e'x_e'(v' + \tfrac{1}{2})^2 - \omega_e''(v'' + \tfrac{1}{2})$$
$$+ \omega_e''x_e''(v'' + \tfrac{1}{2})^2 + \cdots \quad (11.16)$$

Selection rules for the vibrational states are determined by the vibrational wave-function overlap integral given in Eq. (11.7c), which in general has nonzero values. Nevertheless the magnitude of the integral may vary greatly, so that all transitions $v'' \to v'$ are not equally intense. At room temperatures, the predominant transitions in an absorption spectrum originate from the lowest vibrational state $v'' = 0$, since the Boltzmann factor causes the population of the $v'' > 0$ states to fall off rapidly. Thus a typical electronic transition consists of a series of vibrational transitions as shown in Fig. 11.3. Such a series of transitions is known as a *progression*, and it is seen from Eq. (11.16) that the $v'' = 0$ progression has frequencies given by

$$\omega = \omega_E + \omega_e'(v' + \tfrac{1}{2}) - \omega_e'x_e'(v' + \tfrac{1}{2})^2 - \tfrac{1}{2}\omega_e'' + \tfrac{1}{4}\omega_e''x_e'' \quad (11.17)$$

For the vibrational transitions as pictured in Fig. 11.3 and given by Eq. (11.17), it is seen that as higher v' values are attained the vibrational band separation becomes less; that is, the bands converge to a limit. This

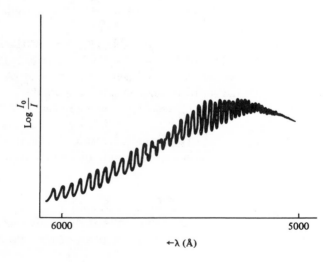

FIG. 11.2.   Low-resolution spectrum of iodine (I_2) in gas phase at room temperature.

convergence limit, designated as the energy E^* in Fig. 11.3, cannot usually be reached, although transitions terminating in v' states near to the convergence limit are often observed. The spectrum of iodine shows this characteristic convergence of bandheads near 5000 Å. Recall now that this convergence limit corresponds to the dissociation of the iodine molecule into two atoms, since it represents the onset of the vibrational continuum at which point the bound molecule no longer exists. Note that the atoms are in excited states, having total energy $E^*(X)$ relative to the energy of the atoms produced at the dissociation limit of the ground electronic state.

The intensity distribution in a progression of bands in an electronic transition is determined entirely by the integral of Eq. (11.7c). Thus the intensity of the vibrational transitions is greatest between states whose wave functions have strong overlap. We show two typical cases in Fig. 11.4. In (a), the two stable electronic states have their minima at essentially the same value of R, that is, R_e is similar in the two states. The strongest transition in the $v'' = 0$ progression is therefore the $v' = 0 \leftarrow v'' = 0$ or $(0, 0)$ band, since each of these states have their maximum values of ψ_v at $\sim R_e$, and consequently the overlap is large. For the next transition, the $(1, 0)$ band, the wave-function overlap is considerably less due to the cancellations from the positive and negative lobes of the $v' = 1$ wave function. Similarly, each higher band becomes increasingly weak and the spec-

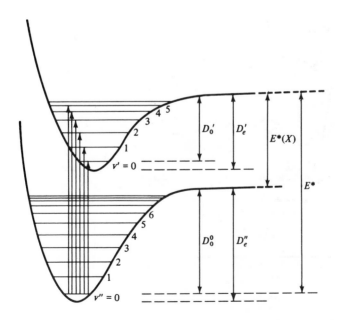

FIG. 11.3. $v'' = 0$ progression in electronic transition of a diatomic molecule.

trum has the appearance shown below the potential curves. In Fig. 11.4b
we have shown a case where the upper state has a larger value of R_e than
the ground state. In the case illustrated the most probable transition is
the (4, 0) since the $v' = 4$ state has its maximum directly above R_e for the
ground state. Transitions to states of both smaller and larger values of v'
lead to lower intensities and a band structure as shown below the potential
curves of Fig. 11.4(b). This is the type of intensity distribution shown by
the iodine spectrum in Fig. 11.2, in which case the maximum occurs at a
rather large value of v' so that transitions near the convergence limit are
observable.

These considerations lead one to the Franck-Condon principle, which
may be stated as follows: *an electronic transition occurs so rapidly compared
to the vibrational motions that the internuclear distance is relatively unchanged
immediately afterwards,* that is, the most probable transitions are "vertical."
In view of the fact that the largest maxima in the vibrational wave functions
occur (a) in the center of the potential curve ($\sim R_e$) for $v = 0$, or (b) at or
near the point where the total energy equals the potential energy (that is,
at point where energy level intersects potential curve) for $v > 0$, the most

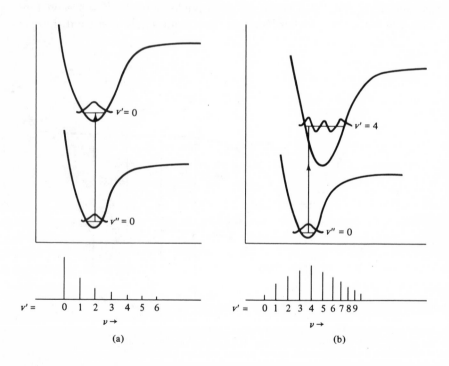

FIG. 11.4.   Potential energy curves and resulting $v'' = 0$ progressions.

probable transitions are those which have their termini in the middle of $v = 0$ states or at either end of higher vibrational states to a first approximation.

This latter interpretation is most useful for transitions originating in other than $v = 0$ states. Thus in Fig. 11.5(a) the most probable transitions from the $v'' = 3$ state are those shown. The principle is similarly useful for the emission case in Fig. 11.5(b).

It has been mentioned previously that transitions which approach the convergence limit excite molecules to the onset of dissociation into atoms. If the upper potential energy curve is displaced far enough to the right, it may occur that the most probable transition arrives (vertically)

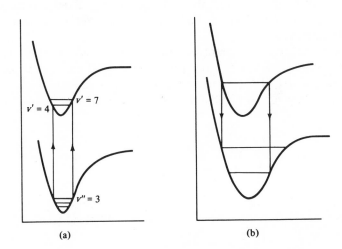

(a)   (b)

FIG. 11.5.   Franck-Condon transitions not involving $v = 0$ states. In each figure only a sampling of vibrational states is shown, and in (b) all displayed states are for $v \gg 0$.

at a point which is above the dissociation limit of the excited state. In this case *continuous* absorption occurs for frequencies down to the dissociation limit, at which point some discrete vibrational bands begins appearing as the frequency is lowered. For transitions into the continuum the excess energy imparted to the atoms appears as kinetic energy.

By reference to Fig. 11.3 it should be clear that if one measures the frequency at which the continuum begins, information concerning dissociation energies can be derived if other appropriate data are available. Typically the energy E^* (see Fig. 11.3) is found spectroscopically by observing the low frequency limit of the continuum, or in other cases it can be found by extrapolating the converging bands to the convergence limit.

When E^* is thus known, the ground-state dissociation energy, $D_0{}^0$, is given by

$$D_0{}^0 = E^* - E^*(X) \qquad (11.18)$$

where $E^*(X)$ is the excitation energy of the atoms. While the energies of excited states of atoms are very well known from atomic spectra, the difficulty here is knowing the precise state of the atoms which occur. Usually it is possible to make sensible guesses, or else one resorts to rough thermochemical data (say $\Delta H^0{}_{\text{diss}}$) to obtain a rough value of $E^*(X)$. Then finding from tables of atomic energy levels the atomic state having a similar value, the exact $E^*(X)$ is used to calculate $D_0{}^0$. A good description of such analyses has been given by Gaydon.[4] We have previously listed several of these spectroscopic values of $D_0{}^0$ in Table 8.2, and Herzberg[5] gives a complete list in an appendix.

Before considering a practical example, we should mention two other common types of electronic transitions which lead to dissociation in different ways. First, if the transition is to an excited state that is entirely unstable, such as $E^{(4)}$ in Fig. 3.3, the entire spectrum consists of continuous

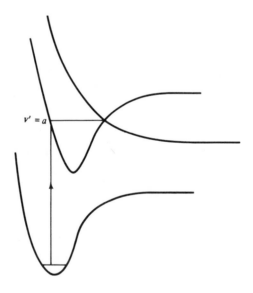

FIG. 11.6. Simple illustration of a predissociation mechanism via an intersecting unstable state.

[4] A. G. Gaydon, *Dissociation Energies and Spectra of Diatomic Molecules* (Dover Publications, New York, 1950).

[5] Footnote 1, Table 39.

absorption at frequencies above the limiting energy of the state. In this case all transitions lead to immediate dissociation.

Dissociation can occur by an unusual mechanism if the upper electronic state to which a transition has occurred is intersected by an unstable electronic state as shown in Fig. 11.6. In this case transitions to vibrational states below and above $v' = a$ are normal and no dissociation occurs. Transitions to $v' = a$ and states very close to this lead to *radiationless* transitions into the unstable state, thereby producing atoms. This nonradiative transition occurs within a few periods of vibration and has a relatively high probability because the wave-function overlap for the two states is rather large at the point of intersection. This dissociation phenomenon, which produces atoms at lower frequencies than would occur if the unstable state were absent, is known as *predissociation*. It shows up in the absorption spectrum by producing broadened, diffuse bands near $v' = a$. If high resolution spectra are taken, the rotational structure is washed out, since the cross-over occurs in a time short compared to the rotational periods of molecules. Thus the discreteness of the rotational transitions is lost.

EXAMPLE 11-1

The spectrum of gaseous iodine shown in Fig. 11.2 is one that can be obtained on many medium-resolution recording visible spectrophotometers. We can illustrate the analysis of such a band spectrum by considering the typical data given in Table 11.3. This spectrum arises primarily from the $v'' = 0$ progression of the $X\ ^1\Sigma_g^+$ to $B\ ^3\Pi_u$.[6] It should be noted that we have here a case of the breakdown of the selection rule $\Delta S = 0$ due to large spin-orbit coupling. The entries in the table are the measured wavelengths of the bandheads and their assignments to the upper states v'. These vibrational assignments cannot be deduced easily from this spectrum, so the original work has been consulted.[7]

When $v'' = 0$, Eq. (11.17) represents the transition frequencies through terms in $(v + \frac{1}{2})^2$. From this it is clear that the band *separations* $\Delta\omega$ are predicted to be

$$\Delta\omega = \omega(v' + 1) - \omega(v') = \omega_e' - 2\omega_e'x_e'(v' + 1) \qquad (11.19)$$

Thus a plot of $\Delta\omega$ versus v' is predicted to be linear. Of course, deviations from linearity are not unexpected for $v' \gg 1$ since Eq. (11.17) requires more terms in $(v + \frac{1}{2})^n$. This graph, known as a Birge-Sponer plot, has been drawn in Fig. 11.7 from the data of Table 11.3. In this case the data are adequately described by a straight line, although there is considerable scatter due to the estimated ± 2 cm^{-1} uncertainties in the measured fre-

---

[6] See footnote 1, p. 541; also R. Mecke, *Ann. Physik* **71**, 104 (1923).
[7] See footnote 6.

TABLE 11.3

Measured bandheads in the $v'' = 0$ progression in the $X \rightarrow B$ electronic transition of molecular iodine.

Upper state quantum number	Wavelength (Å)	Upper state quantum number	Wavelength (Å)
15	5775.5	37	5242.2
16	5741.7	38	5226.4
17	5710.2	39	5210.7
18	5678.7	40	5196.4
19	5647.2	41	5182.2
20	5617.2	42	5168.7
21	5587.2	43	5156.7
22	5559.5	44	5144.7
23	5533.2	45	5133.4
24	5507.7	46	5124.4
25	5482.2	47	5113.2
26	5458.2	48	5103.4
27	5435.7	49	5093.7
28	5412.4	50	5085.4
29	5390.7	51	5077.2
30	5369.7	52	5069.7
31	5349.4	53	5062.2
32	5329.9	54	5056.2
33	5311.2	55	5050.2
34	5293.2	56	5044.9
35	5275.2		
36	5257.9		

quencies. From Eq. (11.19) we see that the slope of the line gives $-2\omega_e'x_e'$ while the extrapolated intercept at $v' = 0$ gives $\omega_e' - 2\omega_e'x_e'$.

Referring to Fig. 11.3, it is seen that E^* is given by

$$E^* = \omega(v' = 0) + \sum \Delta\omega \qquad (11.20a)$$

where the sum is from $v' = 0$ to the apparent convergence value v_c', which from the Birge-Sponer plot is 65.8. Since the $(0, 0)$ band was not observed, Eq. (11.20a) may be usefully replaced by

$$E^* = \omega(v' = 15) + \sum \Delta\omega \qquad (11.20b)$$

where the sum begins with the lowest observed band, the $(15, 0)$. Since the straight line represents $\Delta\omega$ as a continuous function of v', the sum in Eqs. (11.20) may be replaced by an integral to a good approximation.

$$\sum \Delta\omega = \int_{v'=15}^{v'_c} \Delta\omega \, dv' \qquad (11.21)$$

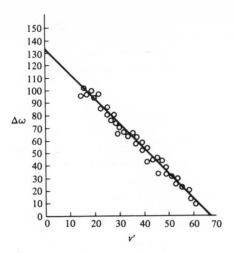

FIG. 11.7.   Birge–Sponer plot from data of Table 11.3.

Thus the summation is merely the area under the Birge–Sponer curve from $v' = 15$ to $v_c' = 65.8$.

From the Birge–Sponer plot we find by these procedures

$$\omega_e' = 133.9 \text{ cm}^{-1}$$

$$\omega_e' x_e' = 1.002 \text{ cm}^{-1}$$

$$\sum \Delta\omega = 2591 \text{ cm}^{-1}$$

By Eq. (11.18) the ground-state dissociation energy D_0^0 can be obtained if $E^*(X)$ is known. It turns out that the products of the dissociation from the $B\ ^3\Pi$ state are a normal $^2P_{3/2}$ atom and an excited one, namely $^2P_{1/2}$. Thus $E^*(X)$ is the energy of the $^2P_{1/2}$ state relative to the $^2P_{3/2}$ state, which has the value[8] 7603 cm^{-1}. Consequently we obtain by use of Eq. (11.18),

$$D_0^0 = 12302 \text{ cm}^{-1}$$

$$= 1.5252 \text{ eV}$$

$$= 35.173 \text{ kcal/mole}$$

If ω_e'' and $\omega_e'' x_e''$ were known it would also be possible to determine

[8] Charlotte E. Moore, *Atomic Energy Levels*, N.B.S. Circular No. 467 (U.S. Government Printing Office, Washington, D.C., 1949).

the electronic transition frequency ω_E by use of Eq. (11.17). On Fig. 11.3, this is the separation of the minima of the two potential curves. The former ground-state parameters could be gotten if the $v'' = 1$ progression were observed. This is possible if the temperature is raised to increase the population of the $v'' = 1$ state. Using the accepted results,[9] namely, $\omega_e'' = 215$ cm^{-1} and $\omega_e'' x_e'' = 0.61$ cm^{-1}, we find

$$\omega_E = 15587 \text{ cm}^{-1}$$

Note that the term $\frac{1}{4}\omega_e'' x_e''$ is practically negligible, and that even a rough value of ω_e'' permits an accurate determination of ω_E.

The data could be pursued further to obtain D_0', the dissociation energy of the excited state. We leave this analysis to the reader. For comparison purposes, the accepted values of the parameters we have calculated are:

$$\omega_e' = 128.0 \text{ cm}^{-1}$$

$$\omega_e' x_e' = 0.83 \text{ cm}^{-1}$$

$$D_0^0 = 1.542 \text{ eV}$$

$$\omega_E = 15642 \text{ cm}^{-1}$$

11.3   ELECTRONIC STATES OF DIATOMIC MOLECULES

A.   One-Electron Molecular Orbitals

For the simplest molecule, the H_2^+ ion, the wave equation can be written in terms of a set of variables which permit the partial differential equation to be separated into three ordinary differential equations whose solutions may be obtained to any desired accuracy. In Cartesian coordinates the Hamiltonian is

$$\tilde{\mathcal{H}}_e = -\frac{\hbar^2}{2m_e} \nabla^2 - \frac{e^2}{r_a} - \frac{e^2}{r_b} + \frac{e^2}{R} \tag{11.22}$$

where r_a, r_b, and R are the separations of the electron and nucleus a, the electron and nucleus b, and the two nuclei, respectively. If the z axis lies along the internuclear line, and if φ is the angle about the z axis measured from the xz plane, the following equations define a transformation to el-

---

[9] See footnote 1, p. 541.

liptical coordinates:

$$x = \tfrac{1}{2}R[(\mu^2 - 1)(1 - \nu^2)]^{1/2}\cos\varphi$$

$$y = \tfrac{1}{2}R[(\mu^2 - 1)(1 - \nu^2)]^{1/2}\sin\varphi$$

$$z = \tfrac{1}{2}R\mu\nu$$

$$\mu = \frac{r_a + r_b}{R} \tag{11.23}$$

$$\nu = \frac{r_a - r_b}{R}$$

After transforming the Hamiltonian (11.22), Schrödinger's equation separates into three ordinary differential equations if we write[10]

$$\psi_e = M(\mu)N(\nu)\Phi(\varphi) \tag{11.24}$$

The equation involving Φ is

$$\frac{d^2\Phi}{d\varphi^2} = -\lambda^2\varphi \tag{11.25}$$

which has the trivial solution

$$\Phi = \frac{1}{(2\pi)^{1/2}}\exp(-i\lambda\varphi) \tag{11.26}$$

where $\lambda = 0, \pm 1, \pm 2, \pm 3, \ldots$

The equations in $M(\mu)$ and $N(\nu)$ are rather complicated but may be solved to any degree of accuracy desired in order to find the total energy of the H_2^+ molecule ion.[11] The results of these calculations for the lowest two states of the molecule ion are given in Fig. 11.8, from which it is seen that the lower state is stable with respect to dissociation to $H + H^+$. It should be noted here that the equations in $M(\mu)$ and $N(\nu)$ contain a dependence upon λ only as λ^2, so the energy states are independent of the sign of λ.

As the reader may have suspected, the quantum number λ is associated with the angular momentum of the electron. In fact, the angular momentum

---

[10] See, for example, L. Pauling and E. B. Wilson, Jr., *Introduction to Quantum Mechanics* (McGraw-Hill, New York, 1935), p. 333.

[11] See for example, E. A. Hylleraas, *Z. Physik* **71**, 739 (1931); E. Teller, *Z. Physik* **61**, 458 (1930); G. Jaffe, *Z. Physik* **87**, 535 (1934).

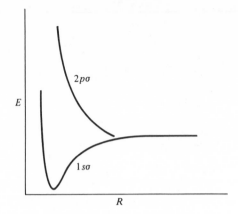

FIG. 11.8.   Lowest two energy states of H_2^+. For $1s\sigma$ state, $R_e = 1.06$
Å and $D_e = 2.79$ eV.

in the z direction is given by the operator

$$\tilde{\lambda} = \frac{\hbar}{i}\frac{d}{d\varphi} \tag{11.27}$$

so that

$$\tilde{\lambda}\psi_e = \lambda\hbar\psi_e \tag{11.28}$$

since $M(\mu)$ and $N(\nu)$ are independent of the variable φ. Thus λ gives the angular momentum of the electron along the internuclear axis in units of $\hbar$. In analogy to the nomenclature for atomic orbitals, the molecular orbitals having $|\lambda| = 0, 1, 2, 3, \ldots$ are designated as $\sigma, \pi, \delta, \ldots$.

Since the total energy is calculated as a function of R, it is possible to correlate the molecular orbital energy states with those of the *united atom*, that is, the imaginary atom existing when $R = 0$, which in this case would be He^+. The quantum number λ is a good quantum number at all internuclear distances, and therefore correlates with m_l for the united atom. Thus the molecular orbitals may be labeled by the united atom state nl values (or symbols) along with the value of λ (or m_l). This leads to the states $1s\sigma, 2s\sigma, 2p\sigma, 2p\pi, 3s\sigma, 3p\sigma, 3p\pi, 3d\sigma, 3d\pi, 3d\delta$, etc. For small values of R, the order of the states is expected to be in this order, although crossovers may occur for larger values of R.

An approximate means of calculating the energies of the molecular orbitals of H_2^+ might utilize the linear-variation method. In this method one needs to choose the linear-variation basis functions. A first approximation, and also one of the simplest, is to choose the functions as two hydrogen atom functions, one centered at each nucleus. This seems like a physically reasonable choice, since as $R \to \infty$ a hydrogen atom plus a bare proton results. For the lowest molecular orbital the proper choices are $1s$ orbitals centered at each nucleus.

In the usual fashion the energies result from the solution of a secular determinant, in this case a 2×2 of the form

$$\begin{vmatrix} H_{aa} - E & H_{ab} - SE \\ H_{ab} - SE & H_{bb} - E \end{vmatrix} = 0 \tag{11.29}$$

where

$$H_{ii} = \int \phi_i \tilde{\mathcal{3C}} \phi_i d\tau$$

$$H_{ij} = \int \phi_i \tilde{\mathcal{3C}} \phi_j d\tau \tag{11.30}$$

$$S = \int \phi_i \phi_j d\tau$$

$\tilde{\mathcal{3C}}$ is given by Eq. (11.22), i and j equal a or b and the ϕ_i are hydrogen $1s$ orbitals. The resulting energies are

$$E_1 = \frac{H_{aa} + H_{ab}}{1 + S} \tag{11.31a}$$

$$E_2 = \frac{H_{aa} - H_{ab}}{1 - S} \tag{11.31b}$$

and the corresponding wave functions are

$$\psi_1 = \frac{1}{(2 + 2S)^{1/2}} (\phi_a + \phi_b) \tag{11.32a}$$

$$\psi_2 = \frac{1}{(2 - 2S)^{1/2}} (\phi_a - \phi_b) \tag{11.32b}$$

The integrals appearing in Eqs. (11.29) can be evaluated in straightforward but somewhat tedious fashion.[12] The results lead to potential curves in qualitative agreement with the accurate ones of Fig. 11.8. E_1 is the state corresponding to $1s\sigma$ and E_2 corresponds to $2p\sigma$. For comparison to the accurate values for the stable $1s\sigma$ state, the variation calculation leads to a dissociation energy of 1.76 eV and an internuclear distance of 1.32 Å.

[12] H. Eyring, J. Walter, and G. E. Kimball, *Quantum Chemistry* (Wiley, New York, 1944), pp. 196–197.

A reason for the stability of the state represented by Eq. (11.32a) may be seen from the plot of $(\phi_a + \phi_b)^2$ in Fig. 11.9. It is seen that an appreciable probability exists for the electron to be found in the region between the two nuclei. On the other hand the state ψ_2, represented except for the multiplying factor by $\phi_a - \phi_b$, has a node midway between the nuclei and hence has a point of zero probability as shown in Fig. 11.9.

This variation method based upon atomic orbitals situated on each of the separated atoms leads not only to a useful computational method for the molecular orbital energies, but also provides another means of labeling the molecular orbitals. Thus the molecular orbitals must correlate with the atomic states of the *separated atoms*.

To see how this correlation is made we note that for a homonuclear molecule such as H_2^+, all the one-electron molecular orbitals are formed as the sum and difference of identical atomic orbitals, just as in the case considered. Therefore we may again label the states by the quantum numbers n and l of the separated atoms, plus the quantum number λ, which again correlates with m_l. We note further that each pair of atomic orbitals leads to a symmetric and an antisymmetric orbital with respect to inversion through the center of symmetry of the molecule. That is, if R is the inversion operator, we find for the case of $1s$ atomic orbitals

$$\tilde{R}(\phi_a + \phi_b) = +(\phi_a + \phi_b) \quad g \qquad (11.33a)$$

$$\tilde{R}(\phi_a - \phi_b) = -(\phi_a - \phi_b) \quad u \qquad (11.33b)$$

where g and u are the symbols we use to represent symmetric and antisymmetric, respectively. The validity of Eqs. (11.33) is clearly shown by interpreting the molecular orbitals in Fig. 11.10 (to be described below) geometrically. It must not be assumed that the sum of atomic orbitals is always g and the difference u. For example, if ϕ_a and ϕ_b are $2p_x$ orbitals (x being $\perp$ to the molecular z axis), the sum represents u symmetry and the difference g.

The separated atom notation is further distinguished from that of the united atom by writing the symbol for λ first. Thus the two states repre-

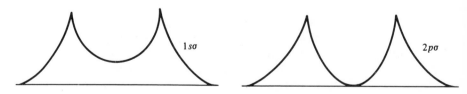

FIG. 11.9.   Probability distributions of $1s\sigma$ and $2p\sigma$ states of H_2^+ from simple MO method.

sented by Eqs. (11.32a) and (11.32b) are designated as $\sigma_g 1s$ and $\sigma_u 1s$.
Similarly, $2s$ atomic orbitals lead to $\sigma_g 2s$ and $\sigma_u 2s$ and $2p$ orbitals lead to
$\sigma_g 2p$, $\sigma_u 2p$, $\pi_g 2p$ and $\pi_u 2p$. Figure 11.10 shows the well-known qualitative
sketches of the molecular orbitals which arise from this scheme. These
diagrams present qualitative contours of the angular distribution of the
functions, and the signs in the various closed loops represent the relative
algebraic sign of the *radial* functions in these regions. Nodes are easily
recognized as regions between closed contours of opposite sign. Note that
σ states are cylindrically symmetric about the z axis, while π states have
their maximum density in either the xz or yz planes.

The ordering of the separated atom states for small R follows the
scheme $\sigma_g 1s < \sigma_u 1s < \sigma_g 2s < \sigma_u 2s < \sigma_g 2p < \pi_u 2p < \pi_g 2p < \sigma_u 2p < \sigma_g 3s \ldots$ etc. For σ orbitals the g state is always lowest, while for π states
the u is lowest. We call the lowest energy orbital of each g, u pair a *bonding
orbital*, since it leads to a minimum in the potential curve, while the higher

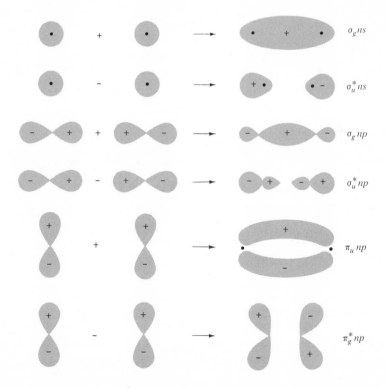

FIG. 11.10. Qualitative angular distributions of the molecular
orbitals of homonuclear diatomic molecules.

energy member is known as an *antibonding* orbital. Note that the antibonding $\sigma_u 2p$ orbital is more antibonding than the $\pi_g 2p$ orbital. As an additional aid in remembering the bonding-antibonding character of the molecular orbitals we often affix an asterisk to the antibonding orbital symbols. Thus our previous list becomes $\sigma_g 1s < \sigma_u * 1s < \sigma_g 2s < \sigma_u * 2s < \sigma_g 2p < \pi_u 2p < \pi_g * 2p < \sigma_u * 2p \ldots$ etc.

The final link now is to join up the united atom and separated atom concepts, since they each specify the same molecular orbitals. This link is provided by a *correlation* diagram as shown in Fig. 11.11. In order to construct this diagram we note first that the quantitative solutions of the differential equations in the variables μ and ν provide an exact numerical correlation from $R = 0$ to $R = \infty$ for the H_2^+ case. We can, however, perform the correlation without resort to the quantitative numerical results. First we note that λ is a good quantum number for any value of R and hence it must correlate in the two extreme cases. Next we note that the g, u symmetry must be conserved, even though our united atom states were not

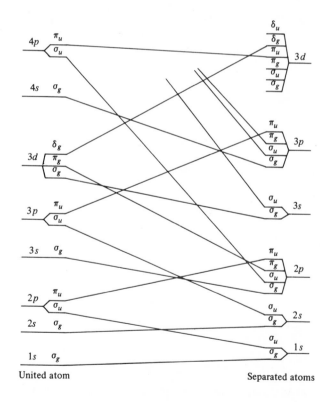

FIG. 11.11. Correlation diagram for homonuclear diatomic molecule.

explicitly labeled in this fashion. For the united atom the σ, π, δ... states have g, u, g, u... symmetry as is easily verified from the well-known form of the atomic orbitals. With these two rules we can construct the entire correlation diagram by beginning at the bottom and working up, choosing always the lowest states of the correct λ and g, u symmetry. It should be clearly noted that, in general, the nl labels do not correlate for the united and separated atoms.

A useful additional concept is the no-crossing rule,[13] which says that electronic energy curves (or more simply, electronic states) of the same symmetry may not cross. Thus two σ_g states may not intersect, but a σ_g may intersect with a σ_u or a π_g or π_u, etc. It can be seen that this rule is satisfied in the correlation diagram of Fig. 11.11.

A simple proof of this rule can be given as follows. Suppose two states are specified by $\psi_1^0(R)$ and $\psi_2^0(R)$ and the corresponding energies as a function of internuclear distance, $E_1^0(R)$ and $E_2^0(R)$. We imagine that these states have been calculated to some low order of approximation and that at some value of R the two energy curves cross. In a higher approximation (more accurate calculation) we can now see the circumstances under which the crossing may or may not really occur.

Suppose some higher-order perturbation is included in the calculation, such as spin-orbit interaction. The energies of the two states near the intersection in the presence of the perturbation $\widetilde{\mathcal{3C}}^{(1)}$ are the solutions of the secular determinant,

$$\begin{vmatrix} H_{11} - E & H_{12}^{(1)} \\ H_{21}^{(1)} & H_{22} - E \end{vmatrix} = 0 \tag{11.34}$$

where

$$H_{ij}^{(1)} = \int \psi_i^0 \widetilde{\mathcal{3C}}^{(1)} \psi_j^0 d\tau \tag{11.35a}$$

$$H_{ii} = E_i^0 + \int \psi_i^0 \widetilde{\mathcal{3C}}^{(1)} \psi_i^0 d\tau \tag{11.35b}$$

The general solution of Eq. (11.34) is given by

$$E = \tfrac{1}{2}(H_{11} + H_{22}) \pm \tfrac{1}{2}(4 \mid H_{12}^{(1)} \mid^2 + \Delta^2)^{1/2} \tag{11.36}$$

where $\Delta = H_{22} - H_{11}$.

Two cases now arise. If ψ_1^0 and ψ_2^0 have different symmetry, we know by our general matrix element theorem of chapter 6 that $H_{12}^{(1)}$ must vanish, since $\widetilde{\mathcal{3C}}^{(1)}$ must belong to the totally symmetric representation. That is,

[13] J. von Neumann and E. P. Wigner, *Z. Physik* **30**, 467 (1929).

$H_{12}^{(1)}$ vanishes if

$$\Gamma^{(\psi_1^0)} \neq \Gamma^{(\psi_2^0)}$$

In this case, Eq. (11.36) shows that $E = H_{11}$ and H_{22}, and it is possible for these two energies to become equal at some value of R, that is

$$H_{11}(R = R') = H_{22}(R = R').$$

Hence crossing is possible in this case.

On the other hand, if the two states have identical symmetry, $H_{12}^{(1)}$ is not zero. In this case it is impossible for the two energies to be equal at any value of R and consequently crossing is impossible. This is true, because equality of the two energies would require

$$(H_{11} + H_{22}) + (4 \mid H_{12}^{(1)} \mid^2 + \Delta^2)^{1/2} = (H_{11} + H_{22}) - (4 \mid H_{12}^{(1)} \mid^2 + \Delta^2)^{1/2}$$

$$(11.37)$$

which is not possible when $H_{12}^{(1)} \neq 0$.

B. Electron Configurations and Term Symbols of Diatomic Molecules

Just as complex atoms may be built up by filling one-electron atomic orbitals with electrons according to the Pauli principle, the configurations of molecules may be built up by placing electrons in the molecular orbitals. Of course, we cannot distinguish one electron from another, and that is the point of the general Pauli principle which requires that an electronic wave function representing any configuration be antisymmetric with respect to interchange of any pair of electrons. This antisymmetry is guaranteed if the electronic wave function is written as a Slater determinant.[14]

As a simple example we consider the ground-state configuration of He_2, which according to the molecular orbital (MO) scheme of Fig. 11.11 is $(\sigma_g 1s)^2 (\sigma_u^* 1s)^2$. The appropriate Slater determinant is therefore

$$\psi = \frac{1}{\sqrt{4!}} \begin{vmatrix} \phi_1(1)\alpha(1) & \phi_1(1)\beta(1) & \phi_2(1)\alpha(1) & \phi_2(1)\beta(1) \\ \phi_1(2)\alpha(2) & \phi_1(2)\beta(2) & \phi_2(2)\alpha(2) & \phi_2(2)\beta(2) \\ \phi_1(3)\alpha(3) & \phi_1(3)\beta(3) & \phi_2(3)\alpha(3) & \phi_2(3)\beta(3) \\ \phi_1(4)\alpha(4) & \phi_1(4)\beta(4) & \phi_2(4)\alpha(4) & \phi_2(4)\beta(4) \end{vmatrix} \quad (11.38)$$

where we have included spin functions and have taken

$$\phi_1 = \sigma_g 1s$$

$$\phi_2 = \sigma_u^* 1s$$

$$(11.39)$$

[14] J. C. Slater, *Phys. Rev.* **34**, 1293 (1929).

Computations show this state to be unstable, in agreement with experiment. Qualitatively this is reasonable since this configuration consists of two bonding and two antibonding electrons. One finds more or less generally that the antibonding contributions outweigh the bonding ones when there are the same number of each type of electrons.

In Sec. 11.3 C we look further into the theoretical calculations. At this point we wish to introduce the molecular *term symbol*, the analog of the atomic-term symbol. For the molecular case, we define the net angular momentum along the z axis, Λ as

$$\Lambda = \sum_i \lambda_i \qquad (11.40)$$

where the sum is over all electrons. Note that this quantum-mechanical vector addition merely sums up parallel or antiparallel vectors. In a similar way we define the total electron spin angular momentum

$$S = \sum_i s_i \qquad (11.41)$$

The basic molecular-term symbol is now written as Σ, Π, Δ, ... depending upon whether $|\Lambda| = 0, 1, 2, \ldots$, plus a left superscript giving $2S + 1$. In addition, a right superscript $+$ or $-$ sign is added to indicate symmetry with respect to reflection in a plane containing the molecule, and a right subscript g or u is added to indicate symmetry with respect to inversion about the center of symmetry.

It is possible, therefore, to write down all the possible terms for a given molecular orbital configuration. Hund's rules for atoms (see Sec. 10.5 A) apply for molecules also so the lowest state may always be chosen. Furthermore, we can now see some physical significance to the group-theory symmetry labels described in Sec. 11.2. Specifically, the symbols Σ, Π, Δ, etc., correspond to $|\Lambda| = 0, 1, 2, \ldots$ according to the scheme described here. It should be mentioned that other coupling schemes are possible, such as the (jj) coupling scheme in atoms.[15] In this case the group theoretical labels are of course still valid, but the correlation of the label with $|\Lambda|$ may no longer be made.

EXAMPLE 11-2

For nonequivalent electron configurations, that is, those whose electrons occupy different one-electron molecular orbitals, all combinations of S and Λ values are permitted. We consider two examples to illustrate this.

[15] See footnote 1, p. 337.

$(\sigma)^1(\sigma)^1$:

Two values of S are possible, given by

$$S = s_1 + s_2 \cdots \mid s_1 - s_2 \mid \qquad (11.42)$$

or $S = 0$ and 1. The possible Λ values are given by

$$\Lambda = \lambda_1 + \lambda_2 \quad \text{and} \quad \lambda_1 - \lambda_2 \qquad (11.43)$$

which in this case gives $\Lambda = 0$. Thus the possible terms are $^3\Sigma$ and $^1\Sigma$. The g, u symmetry is determined by the g, u symmetry of the particular orbitals. Thus if both are g or both are u, the symmetry of the terms is g, while if one of the orbitals is g and the other is u, the symmetry of the terms is u. For configurations arising from σ orbitals only, the $+$, $-$ reflection symmetry of the terms is always $+$, since the orbitals are always symmetric with respect to reflection in a plane containing the z axis.

As a concrete example, a $(\sigma_g 1s)^1(\sigma_u{}^* 1s)^1$ configuration leads to $^3\Sigma_u{}^+$ and $^1\Sigma_u{}^+$ terms, while the $(\sigma_g 1s)^1(\sigma_g 2s)^1$ leads to $^3\Sigma_g{}^+$ and $^1\Sigma_g{}^+$. In the former case, one orbital is g and the other u, while in the latter they are both g.

$(\pi)^1(\pi)^1$:

Again $S = 0$ or 1. By Eq. (11.43) the Λ values are

$$1 + 1 = 2$$
$$-1 - 1 = -2$$
$$1 - 1 = 0$$
$$-1 + 1 = 0$$

That is, we have two ways of forming each of $\mid \Lambda \mid = 0$ and 2. This double degeneracy of the $\pm \mid \Lambda \mid$ states is always present when $\Lambda \geqq 1$, unless the interaction of electronic and rotational motions is added to the previously described electronic Hamiltonian. From the above values of S and $\mid \Lambda \mid$ we see that the possible terms are: one $^3\Delta$, one $^1\Delta$, two $^3\Sigma$, and two $^1\Sigma$. Note that the two ways of forming $\Lambda = 0$ lead to terms of different energy since the two electrons are nonequivalent.

For any particular case the g, u symmetry is obtained as described above. The vector model method cannot predict the correct $+$, $-$ symmetry. This is best done by group theory, noting that the possible states which may arise from two singly occupied π orbitals are given by the direct product $\pi \times \pi$. Using the $C_{\infty v}$ character table,

$$\pi \times \pi \to \Sigma^+ + \Sigma^- + \Delta$$

Thus the possible terms are

$$^1\Sigma^+, \, ^3\Sigma^+, \, ^1\Sigma^-, \, ^3\Sigma^-, \, ^1\Delta \quad \text{and} \quad ^3\Delta.$$

EXAMPLE 11-3

For *equivalent* electrons the terms that are permitted must satisfy the Pauli principle. Again we consider two examples.

$(\sigma)^2$:

As shown in Example 11-2, $^3\Sigma$ and $^1\Sigma$ are compatible with the vector-model coupling scheme. But now only the $^1\Sigma$ is permitted since two electrons may occupy the same space orbital only if their spin functions or quantum numbers are different. As previously stated the Σ state must have $+$ reflection symmetry and since the two electrons occupy the same space orbital the inversion symmetry must be g. Thus any $(\sigma)^2$ configuration leads to a single $^1\Sigma_g{}^+$ term.

As with atoms it is found that all *closed* or filled subshells lead to $^1\Sigma_g{}^+$ terms. Thus $(\sigma_g 1s)^2$, $(\sigma_g 1s)^2(\sigma_u{}^*1s)^2$, $(\sigma_g 1s)^2(\sigma_u{}^*1s)^2(\sigma_g 2s)^2$, etc. lead to $^1\Sigma_g{}^+$ terms.

$(\pi)^2$:

From Example 11-2 again, the terms permitted by the vector model were $^3\Delta$, $^1\Delta$, $^3\Sigma$, and $^1\Sigma$. The $^3\Delta$ state is immediately ruled out since it would violate Pauli's principle by requiring the electrons to have identical quantum numbers. Both $^3\Sigma$ and $^1\Sigma$ states are possible, however, since the Σ terms arise from space orbitals having different values of λ (namely, ± 1). In this case, since the electrons are equivalent there is now only one $^3\Sigma$ and one $^1\Sigma$ in contrast to the case for inequivalent electrons. The remaining question is what is the correct $+$, $-$ reflection symmetry.

We can answer this question by writing down the four Σ wave functions. We take Π_+ and Π_- to be the spatial molecular orbitals of the form of Eq. (11.24) corresponding to $\lambda = 1$ and -1, respectively. The four possible wave functions are then

$$\psi_S{}^0 = \tfrac{1}{2}[\Pi_+(1)\Pi_-(2) + \Pi_+(2)\Pi_-(1)][\alpha(1)\beta(2) - \beta(1)\alpha(2)] \quad (11.44a)$$

$$\psi_T{}^1 = \frac{1}{\sqrt{2}}[\Pi_+(1)\Pi_-(2) - \Pi_+(2)\Pi_-(1)][\alpha(1)\alpha(2)] \quad (11.44b)$$

$$\psi_T{}^0 = \tfrac{1}{2}[\Pi_+(1)\Pi_-(2) - \Pi_+(2)\Pi_-(1)][\alpha(1)\beta(2) + \beta(1)\alpha(2)] \quad (11.44c)$$

$$\psi_T{}^{-1} = \frac{1}{\sqrt{2}}[\Pi_+(1)\Pi_-(2) - \Pi_+(2)\Pi_-(1)][\beta(1)\beta(2)] \quad (11.44d)$$

in which S and T stand for singlet and triplet, respectively, and the superscripts on the ψ's give the M_S values of the states. Note that $\psi_T{}^0$ and $\psi_S{}^0$ are formed as the sum and difference, respectively, of two Slater determinants having $\alpha\beta$ and $\beta\alpha$ spin configurations.

The reflection symmetry is easily determined now by noting from Eq. (11.26) that $\Pi_\pm$ goes into $\Pi_\mp$ upon reflection ($\varphi \rightarrow -\varphi$) in a molecular plane. Thus the spatial (orbital) part of ψ_S^0 is symmetric $(+)$ upon reflection, while the spatial part of ψ_T^0 is antisymmetric $(-)$. The possible terms of the $(\Pi)^2$ configuration are therefore $^3\Sigma^-$, $^1\Sigma^+$ and $^1\Delta$, and Hund's rules predict $^3\Sigma^-$ lies lowest.

With this classification scheme the configurations and terms for all the simple homonuclear diatomic molecules may be written down easily. The lowest energy configuration may be picked out by means of the order described in Sec. 11.3 A and shown on the right-hand side of the correlation diagram of Fig. 11.11. Note from this diagram, however, that some uncertainty in ordering is expected in regions of orbital crossovers, such as occurs for σ_u^*2s, σ_g2p, and π_u2p. In Table 11.4 we have listed the lowest configuration and term for the simple homonuclear molecules, and the experimental values of D_0^0. As a simple rule we see that $\sigma_g2p < \pi_u2p$ for molecules heavier than N_2.

TABLE 11.4

Configurations and terms of some homonuclear diatomic molecules.[a]

Molecule	Configuration	Term	D_0^0
H_2^+	$(\sigma_g1s)^1$	$^2\Sigma_g^+$	2.65 eV
H_2	$(\sigma_g1s)^2$	$^1\Sigma_g^+$	4.48
He_2	$(\sigma_g1s)^2(\sigma_u^*1s)^2$	$^1\Sigma_g^+$	$\ldots$
Li_2	$(\sigma_g1s)^2(\sigma_u^*1s)^2(\sigma_g2s)^2$	$^1\Sigma_g^+$	1.03
Be_2	$(\sigma_g1s)^2(\sigma_u^*1s)^2(\sigma_g2s)^2(\sigma_u^*2s)^2$	$^1\Sigma_g^+$	$\ldots$
B_2	$(\sigma_g1s)^2(\sigma_u^*1s)^2(\sigma_g2s)^2(\sigma_u^*2s)^2(\pi_u2p)^2$	$^3\Sigma_g^-$	3.0
C_2	$(\sigma_g1s)^2(\sigma_u^*1s)^2(\sigma_g2s)^2(\sigma_u^*2s)^2(\pi_u2p)^4$	$^1\Sigma_g^+$	6.2
N_2^+	$(\sigma_g1s)^2(\sigma_u^*1s)^2(\sigma_g2s)^2(\sigma_u^*2s)^2(\pi_u2p)^4(\sigma_g2p)^1$	$^2\Sigma_g^+$	8.73
N_2	$(\sigma_g1s)^2(\sigma_u^*1s)^2(\sigma_g2s)^2(\sigma_u^*2s)^2(\pi_u2p)^4(\sigma_g2p)^2$	$^1\Sigma_g^+$	9.76
O_2^+	$(\sigma_g1s)^2(\sigma_u^*1s)^2(\sigma_g2s)^2(\sigma_u^*2s)^2(\sigma_g2p)^2(\pi_u2p)^4$ $(\pi_g^*2p)^1$	$^2\Pi_g$	6.48
O_2	$(\sigma_g1s)^2(\sigma_u^*1s)^2(\sigma_g2s)^2(\sigma_u^*2s)^2(\sigma_g2p)^2(\pi_u2p)^4$ $(\pi_g^*2p)^2$	$^3\Sigma_g^-$	5.08
F_2	$\ldots (\pi_u2p)^4(\pi_g^*2p)^4$	$^1\Sigma_g^+$	1.64
Ne_2	$\ldots (\pi_u2p)^4(\pi_g^*2p)^4(\sigma_u^*2p)^2$	$^1\Sigma_g^+$	$\ldots$

[a] Abstracted from Ref. 1 and JANAF Thermochemical Tables, U.S. Dept. of Commerce, Washington, D.C. (1968).

A simple example is Li_2 with six electrons. The lowest configuration is $(\sigma_g 1s)^2(\sigma_u*1s)^2(\sigma_g 2s)^2$. The only possible term is $^1\Sigma_g^+$ and the state is stable experimentally. This is predictable since there are 4 bonding and 2 anti-bonding electrons. If we define the bond order (BO) as

$$BO = \frac{N - N_*}{2} \tag{11.45}$$

where N is the number of bonding electrons and N_* the number of anti-bonding (starred) electrons, the BO of Li_2 is 1. It is clear that the bond-order concept is equivalent qualitatively to the Lewis model of covalent bonding (single, double, triple bonds, etc.).

O_2 represents a more interesting example. The lowest configuration is $(\sigma_g 1s)^2(\sigma_u*1s)^2(\sigma_g 2s)^2(\sigma_u*2s)^2(\sigma_g 2p)^2(\pi_u 2p)^4(\pi_g*2p)^2$ with 16 electrons. As shown in Example 11-3 the electrons outside closed shells lead to $^3\Sigma_g^-$ as the ground state. Thus O_2 is a triplet-state molecule and is paramagnetic, as experimentally observed (see chapter 10). Furthermore, BO = 2, which agrees with the Lewis double-bond picture, and also with the relatively large D_0^0. The two excited-state terms, $^1\Delta_g$ and $^1\Sigma_g^+$, of the ground configuration have been observed experimentally also, as shown by the spectral data of Table 11.2.

A few of the states of the first excited configuration, $\ldots (\sigma_g 2p)^2 (\pi_u 2p)^3(\pi_g*2p)^3$, have been observed also, namely the $^3\Sigma_u^+$ and $^3\Sigma_u^-$ states of Table 11.2. Hence we see that the first electronic transition leading to an excited configuration is $X \rightarrow A$ at 36 000 cm^{-1} or about 2800 Å (the near uv). Other possible terms of this configuration such as $^1\Delta$, $^3\Delta$, and $^1\Sigma$ have not been observed.

The O_2^+ molecule ion is easily described by simply removing one electron from the O_2 configuration, which leads to $\ldots (\sigma_g 2p)^2(\pi_u 2p)^4(\pi_g*2p)^1$. The only possible term is $^2\Pi_g$ since $|\Lambda| = 1$ and $S = \frac{1}{2}$. A bond order of $2\frac{1}{2}$ is predicted, which agrees with the experimental observation that D_0^0 is larger for O_2^+ than for O_2.

The electronic states of heteronuclear diatomic molecules may be described in a fashion similar to that for homonuclear molecules. A major change is that the molecular orbitals no longer have g or u symmetry since the nuclear charges and consequently the electron distribution differs at the two ends of the molecules. For nuclei which are not too different, a correlation diagram such as Fig. 11.11 may be constructed. Note here that even for completely separated atoms $(R = \infty)$ the orbitals (states) of nuclei a and b differ. A diagram of this type is most useful for molecules such as CO, CN, NO, etc., and the separated atom symbolism for the molecular orbitals is appropriate. For molecules such as CH, OH, HF, etc., the united atom symbolism is most appropriate since the electronic energy is determined largely by the heavy atom. Thus CH with seven electrons would have a $(1s\sigma)^2(2s\sigma)^2(2p\sigma)^2(2p\pi)^1$ ground-state configuration with a $^2\Pi$ term.

When the corresponding states of the separated atoms are much different, such as in HF or PN, the separated atom molecular-orbital notation of Fig. 11.11 may be grossly inaccurate, since the lowest separated atom states (the most tightly bound electrons) are those of the heavy nucleus. Rather than redrawing the diagram for each case, it is usual to denote the effective molecular orbitals by $z\sigma$, $y\sigma$, $x\sigma$, $w\pi$, $v\pi$.... Thus for PN only the valence electrons contribute greatly to the molecular orbitals. The ground configuration might then be written $[(1s_P)^2(2s_P)^2(2p_P)^6(1s_N)^2]$ $(z\sigma)^2(y\sigma)^2(x\sigma)^2(w\pi)^4$, where the inner shell electrons are shown in brackets.

11.4   THEORY OF THE ELECTRONIC ENERGIES OF DIATOMIC MOLECULES

A.   Molecular-Orbital Method

Since exact solutions of the electronic wave equation (3.22) are impossible for any molecule other than H_2^+, some approximate means must be utilized in any attempt at calculations of electronic energies. The theoretical and computational approaches which have been utilized are many and varied.[16] Nearly all computations have relied upon the quantum-mechanical variation method, which assures that an upper bound is obtained. Thus the final accuracy (agreement with experiment) is limited only by the size and speed of available computers and the theorist's ingenuity in choosing the variation parameters.

The most common method, and that which seems most pleasing intrinsically, is the molecular-orbital method. We have described this qualitatively in Sec. 11.3. The method follows the approach common to atomic-energy calculations, namely, it utilizes one-electron orbitals combined in Slater determinants such as shown for He_2 in Eq. (11.38).

If such a one-electron molecular-orbital (MO) scheme is settled upon, the remaining choice is the form of the molecular orbital. One might like to use the H_2^+ orbitals, but they are nonanalytic and also do not permit sufficient flexibility for variational purposes. The most common choice of MO is a linear combination of orbitals, each orbital centered at one of the nuclear centers. That is, we write a molecular orbital ψ_i as

$$\psi_i = \sum_j c_{ji}\phi_j \tag{11.46}$$

where the ϕ_j are the basis orbitals. These may be hydrogen-like atomic orbitals, Slater atomic orbitals,[17] Gaussian functions or even the Hartree-

---

[16] See for example, J. C. Slater, *Quantum Theory of Molecules and Solids* (McGraw-Hill, New York, 1963), Vol. 1.

[17] J. C. Slater, *Phys. Rev.* **36,** 57 (1930).

Fock atomic orbitals resulting from atomic self-consistent field theory.[18] The energy is then determined by the variational procedure

$$E = \frac{\int \Psi^* \widetilde{\mathfrak{K}} \Psi d\tau}{\int \Psi^* \Psi d\tau} \tag{11.47}$$

where Ψ is the Slater determinant formed from the molecular orbitals ψ_i, and $\widetilde{\mathfrak{K}}$ is given by Eq. (3.21). Note that the determinantal wave function Ψ for an n electron closed-shell molecule requires a minimum of $\frac{1}{2}n$ molecular orbitals since at most two electrons may be placed in a given MO. This method, which utilizes a linear combination of atomic orbitals in the molecular orbital formation, is known as the LCAO-MO technique.

Note that the basis set ϕ_1, ϕ_2, ... must contain a minimum of $\frac{1}{2}n$ functions in order to generate the minimum required number of molecular orbitals. The basis set used is nearly always larger than $\frac{1}{2}n$ in any practical computation. When the basis set is chosen to include only those atomic orbitals involved in the ground-state configuration of the atoms, the basis set is said to be *minimal*. If a greater number of basis functions are included a corresponding increase in molecular-orbitals results, and the basis set is known as *extended*. Usually the extended basis sets produce a better molecular orbital representation but the size of the basis is largely limited by computational difficulties.

For closed shells, the best single-determinant variational calculations are the self-consistent-field (LCAO-MO-SCF) variety described by Roothaan.[19] The energy of Eq. (11.47) is required to be minimized, with the expansion coefficients of Eq. (11.46) used as the variational parameters. This self-consistent-field approach generates the best one-electron molecular orbitals consistent with the assumed LCAO basis set and the variational principle. These calculations of diatomic molecules other than H_2 are much too complex for us to go into in any detail, but we give some results for the F_2 molecule in Table 11.5. The 28 basis orbitals used in this computation were atomic Hartree-Fock results. The calculated dissociation energy D_e is not at all good, but it must be remembered that this energy represents only about 1% of the total and is obtained by subtracting the total molecular electronic energy from the energy of the dissociated atoms.

Several other molecular properties of F_2 have been evaluated from the molecular wave function and from the energy curve $E(R)$ in the region of the minimum. The results here are generally rather encouraging.

Exactly similar procedures may, of course, be used to treat any excited configuration. Because of the great amount of computational effort involved, however, little SCF work has been performed on molecular excited states.

[18] E. Clementi, *J. Chem. Phys.* **38,** 996, 1001 (1962).
[19] C. C. J. Roothaan, *Rev. Mod. Phys.* **23,** 69 (1951).

TABLE 11.5
LCAO-MO-SCF results for the fluorine molecule.[a]

Orbital	$\sigma_g 1s$	$\sigma_u{}^*1s$	$\sigma_g 2s$	$\sigma_u{}^*2s$	$\pi_u 2p$	$\sigma_g 2p$	$\pi_g{}^*2p$
Energy[b]	-26.42	-26.42	-1.76	-1.49	-0.80	-0.75	-0.66

Property	Computed	Experimental
R_e	1.33 Å	1.42 Å
E_{total}	-198.768 hartrees	-199.670
D_e	-1.370 eV	1.68
ω_e	1257 cm^{-1}	919
B_e	1.003 cm^{-1}	0.890

    [a]   A. C. Wahl, *J. Chem. Phys.* **41**, 2600 (1964).
    [b]   Energy in hartrees. 1 hartree = 27.2 eV.

To obtain results more accurate than those from the single-determinant LCAO-MO-SCF method, and to permit treatment of open shell systems, the *configuration interaction* technique is often used. Here it is recognized that any linear combination of Slater determinantal wave functions, each of which is a solution of the variational expression, is also a satisfactory wave function. Thus if Ψ_1 and Ψ_2 are such single determinant wave functions we might take

$$\Psi = a_1\Psi_1 + a_2\Psi_2 \qquad (11.48)$$

as the next approximation. This would lead to a 2×2 secular equation in the common variational procedure. Of course, if the matrix element H_{12} is zero no mixing or interaction occurs and the energies remain unchanged. We know from group theory that H_{12} is nonzero only if Ψ_1 and Ψ_2 have identical symmetry, that is, belong to the same irreducible group representation.

As a concrete example, the ground-state configuration of fluorine is $(\sigma_g 1s)^2(\sigma_u{}^*1s)^2(\sigma_g 2s)^2(\sigma_u{}^*2s)^2(\sigma_g 2p)^2(\pi_u 2p)^4(\pi_g{}^*2p)^4$ with symmetry $^1\Sigma_g{}^+$. Thus only configurations of the same symmetry need be considered if we wish to carry the ground-state calculation to a higher order of approximation. An appropriate configuration might be $\ldots (\sigma_g 2p)^2(\pi_u 2p)^4(\pi_g{}^*2p)^2$ $(\sigma_u{}^*2p)^2$ which has a $^1\Sigma_g{}^+$ term. The problem is a little complicated since the $^1\Sigma_g{}^+$ wave function would have to be written as a linear combination of the six possible Slater determinants for this excited π^2 configuration (see Sec. 11.3 B).

We shall not go further with detailed considerations of general diatomic molecules. Research in this area is being actively pursued and the reader is urged to consult the reference material for more details. In the following example we provide some further detail for the simple hydrogen molecule.

EXAMPLE 11-4

The electronic Hamiltonian for H_2 is

$$\tilde{\mathcal{H}}_e = -\frac{\hbar^2}{2m_e}(\nabla_1^2 + \nabla_2^2) - e^2\left(\frac{1}{R_{a1}} + \frac{1}{R_{a2}} + \frac{1}{R_{b1}} + \frac{1}{R_{b2}}\right)$$

$$+ e^2\left(\frac{1}{R_{ab}} + \frac{1}{R_{12}}\right) \quad (11.49)$$

where the electrons are labeled by the numeral subscripts and the protons by the alphabetic subscripts. The ground-state configuration is $(\sigma_g 1s)^2$ so the simplest determinantal function is

$$\Psi_1 = \frac{1}{\sqrt{2}}\begin{vmatrix} \psi_g(1)\alpha(1) & \psi_g(1)\beta(1) \\ \psi_g(2)\alpha(2) & \psi_g(2)\beta(2) \end{vmatrix}$$

$$\quad (11.50)$$

$$= \frac{1}{\sqrt{2}}\psi_g(1)\psi_g(2)[\alpha(1)\beta(2) - \beta(1)\alpha(2)]$$

where $\psi_g(i)$ is a σ_g $1s$ MO occupied by electron i. From our H_2^+ experience we write ψ_g as a simple LCAO form,

$$\psi_g(i) = \frac{1}{[2(1 + S)]^{1/2}}[\phi_a(i) + \phi_b(i)] \quad (11.51)$$

in which ϕ_a and ϕ_b are hydrogen atom $1s$ orbitals centered at nuclei a and b, respectively, and $S = \int \phi_a(i)\phi_b(i)d\tau$ is the atomic orbital overlap integral. The energy is then gotten as usual from

$$E = \frac{\int \Psi_1^* \tilde{\mathcal{H}}_e \Psi_1 d\tau}{\int \Psi_1^* \Psi_1 d\tau} \quad (11.52)$$

In this simplest of calculations we have no variational parameter.

The calculation outlined above was performed by Hellmann,[20] and led to the prediction of a stable state with $R_e = 1.6a_0$ and $D_e = 2.65$ eV, compared to the experimental $1.40a_0$ and 4.70 eV, respectively.[21] The major failure of this simple calculation is that it is totally incorrect at $R = \infty$, where the energy must be twice the energy of a hydrogen atom. The D_e value that we have quoted is based upon the correct energy at $R = \infty$.

[20] H. Hellmann, *Einführung in die Quantenchemie* (Franz Deuticke, Leipzig, 1933), p. 133.

[21] a_0, the radius of the first Bohr orbit of the hydrogen atom, has the value 0.529 Å.

The result is not spectacular, yet it is remarkable that so simple a molecular wave function has some validity.

Further improvement is expected if we permit configuration interaction. Let us examine the possible low-lying configurations for the hydrogen molecule. We recall from our earlier considerations that the MO lying immediately above the $\sigma_g 1s$ is the $\sigma_u{}^*1s$, which in the simplest LCAO form is

$$\psi_u(i) = \frac{1}{[2(1-S)]^{1/2}} [\phi_a(i) - \phi_b(i)] \tag{11.53}$$

From these two MO's the only permitted configurations are $(\sigma_g 1s)^2$, $(\sigma_g 1s)^1$, $(\sigma_u{}^*1s)^1$, and $(\sigma_u{}^*1s)^2$. The first and third of these are single-term configurations of $^1\Sigma_g{}^+$ symmetry. The second configuration, treated in Example 11-2, yields $^3\Sigma_u{}^+$ and $^1\Sigma_u{}^+$ terms. Thus the only excited configuration having the same symmetry as the ground configuration is $(\sigma_u{}^*1s)^2$, whose determinantal wave function is

$$\Psi_2 = \frac{1}{\sqrt{2}} \begin{vmatrix} \psi_u(1)\alpha(1) & \psi_u(1)\beta(1) \\ \psi_u(2)\alpha(2) & \psi_u(2)\beta(2) \end{vmatrix}$$

$$= \frac{1}{\sqrt{2}} \psi_u(1)\psi_u(2)[\alpha(1)\beta(2) - \beta(1)\alpha(2)] \tag{11.54}$$

Using the wave function in the form of Eq. (11.48) the secular equation may be set up and solved. Note that we have elements H_{ij}, where

$$H_{ij} = \int \Psi_i{}^*\widetilde{\mathcal{3C}}\Psi_j d\tau \tag{11.55}$$

H_{11} has already been obtained in the simpler preceding calculation so only H_{22} and H_{12} are needed. Slater[22] presents the numerical results of this calculation. Although the total electronic energy of the lowest state is only slightly lower than Hellmann's simple result in the region of the minimum, the behavior at $R = \infty$ is now correct, so the method represents an inherently better result. The second state arising from this computation is unstable. We have plotted these results in Fig. 11.12, the stable state being labeled as $^1\Sigma_{g1}$, the unstable state as $^1\Sigma_{g2}$.

The energies of the $^3\Sigma_u{}^+$ and $^1\Sigma_u{}^+$ terms could of course be determined in a similar fashion. It is found (See Fig. 11.12) that the $^3\Sigma_u{}^+$ and $^1\Sigma_u{}^+$ states fall between the two $^1\Sigma_g{}^+$ states. The $^1\Sigma_u{}^+$ state is very slightly bonding and lies higher than the $^3\Sigma_u{}^+$ term, in agreement with Hund's rules and experiment.

[22] See footnote 16, p. 69.

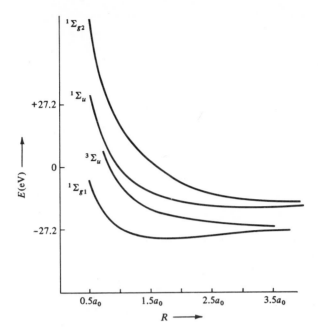

FIG. 11.12.   Electronic states of H_2 resulting from simple LCAO-MO-CI computation.

Still further improvement may be obtained from the MO calculation if an additional appropriate variational parameter is included in the wave function. The most sensible choice is a variable factor in the exponential of the basis functions, that is, a variable effective charge Z. Hence $\phi_a(i)$ is chosen to be of the form

$$\phi_a(i) \propto \exp(-ZR_{ai})$$

Then the configuration-interaction calculation[23] leads to $R_e = 1.41a_0$, $D_e = 4.02$ eV with an optimum Z value of 1.19.

Further improvements are possible but we pursue this example no further. The reader may consult the literature for further refinements.

B.   Valence-Bond Method

While the MO method has been stressed up to now, the first calculation on H_2 in 1927 used a somewhat different approach. Heitler and London[24] noted that for two noninteracting H atoms $[R_{ab}^{-1}, R_{12}, R_{a2}^{-1}, R_{b1}^{-1}$ each

23 S. Weinbaum, *J. Chem. Phys.* **1**, 593 (1933).

24 W. Heitler and F. London, *Z. Physik* **44**, 455 (1927).

equal to zero in Eq. (11.49)] an appropriate wave function would be $\phi_a(1)\phi_b(2)$ where the ϕ's are $1s$ hydrogen-atom orbitals as previously used. But clearly $\phi_a(2)\phi_b(1)$ is just as satisfactory because of the indistinguishability of the electrons. Thus it was reasonable to assume a linear variation function of the form

$$\psi_{HL} = c_1\phi_a(1)\phi_b(2) + c_2\phi_a(2)\phi_b(1) \tag{11.56}$$

The secular equation may now be set up and solved using the correct Hamiltonian (11.49). All integrals may be evaluated precisely as in the MO method. The wave functions for the two states are conveniently labeled ψ_S and ψ_T and in normalized form are

$$\psi_S = \frac{1}{\sqrt{2}(1+S)^{1/2}}[\phi_a(1)\phi_b(2) + \phi_a(2)\phi_b(1)] \tag{11.57a}$$

$$\psi_T = \frac{1}{\sqrt{2}(1-S)^{1/2}}[\phi_a(1)\phi_b(2) - \phi_a(2)\phi_b(1)] \tag{11.57b}$$

When the secular equation is solved it is found that the state E_S is stable with $R_e = 1.5a_0$, $D_e = 3.20$ eV. The E_T state is found to be entirely repulsive.

It is clear that the state ψ_S corresponds to the ground $^1\Sigma_g^+$ state of hydrogen, and hence the Heitler–London (HL) wavefunction is a better first approximation than the simplest LCAO-MO function described in Example 11-4. Since ψ_S is symmetric to interchange of electrons, a spin function which is antisymmetric must be attached. This gives

$$\Psi_S = \psi_S[\alpha(1)\beta(2) - \beta(1)\alpha(2)] \tag{11.58a}$$

On the other hand, the state ψ_T is antisymmetric and thus requires a symmetric spin function. Three of these are possible, leading to a triplet state.

$$\Psi_T = \psi_T \begin{cases} \alpha(1)\alpha(2) \\ \alpha(1)\beta(2) + \beta(1)\alpha(2) \\ \beta(1)\beta(2) \end{cases} \tag{11.58b}$$

Thus the state ψ_T represents the lowest triplet $^3\Sigma_g^+$ of H_2. Note that the HL method, also known as the *valence-bond* (VB) method, requires that the spin functions be added in a rather *ad hoc* fashion.

It is possible to improve on the simple HL wave function in several ways. For example, if the variable charge Z is introduced (as in Example 11-4) the value of D_e improves to 3.78 eV.[25]

In order to assess the relative merits of the VB and MO methods let

[25] S. Wang, *Phys. Rev.* **31,** 579 (1928).

us expand the bonding MO wave function of Eq. (11.50). Neglecting the constant factors and the spin part we find

$$\Psi_{MO} \propto \phi_a(1)\phi_a(2) + \phi_b(1)\phi_b(2) + \phi_a(1)\phi_b(2) + \phi_a(2)\phi_b(1) \quad (11.59)$$

We see that the third and fourth terms in Eq. (11.59) are exactly the entire VB function ψ_S. The first and second terms place both electrons on either nucleus a or nucleus b, and hence correspond to ionic structures, H^-H^+ and H^+H^-, respectively. The terms in the VB function ψ_S are of the covalent type H-H with essentially equal sharing of electrons.

Thus, in a very simple way, we conclude that the simple MO function is poorer than the simple VB function because the MO function places entirely too much weight on the intuitively unlikely ionic structures. Note that the simple HL valence bond function is very close to the chemist's concept of the covalent bond, and it is the quantitative success of this function which provides the concept with validity.

On the other hand, the VB function places all its emphasis upon keeping the electrons localized on different nuclei. One suspects, then, that the VB function might be improved if some allowance is provided for the ionic terms. If we use the variation function

$$\psi_S = c_1[\phi_a(1)\phi_b(2) + \phi_a(2)\phi_b(1)] + c_2[\phi_a(1)\phi_a(2) + \phi_b(1)\phi_b(2)] \quad (11.60)$$

we can find the optimum mixture of ionic and covalent terms. This calculation has been carried out, and in fact, the reader can verify that the function above is of identical form to that used for the MO configuration interaction calculation in Example 11-4, namely

$$\Psi = a_1\Psi_1 + a_2\Psi_2$$

It turns out that the optimum admixture of ionic terms is about 7%; that is, $c_2/c_1 = 0.26$ or $(c_2/c_1)^2 \sim 0.07$. This shows in another way why the VB function is better than the MO function as a first approximation, since $c_2/c_1 = 0$ for the VB case, while $c_2/c_1 = 1$ for the MO case.

The valence-bond method may be applied to other diatomic molecules and it might be expected that the results would be better than those of the MO method in its simple form. It turns out to be more difficult to handle computationally, however, and since the MO configuration interaction (CI) technique leads to proper electron correlation (that is, ionic term considerations) the VB method is used rather less frequently. Therefore we do not go further into the necessary methods.

11.5 ELECTRONIC SPECTRA OF POLYATOMIC MOLECULES

A. General Considerations

In a general way all the methods and techniques used to describe the electronic states and spectra of diatomic molecules are also applicable to polyatomic molecules. Thus the states must be specified by the symmetry species or irreducible representations of the point group to which the equilibrium molecule conforms. For example, the electronic states of a pyramidal XY_3 molecule such as ammonia are specified by the irreducible representations of C_{3v}, namely, A_1, A_2, and E. The states of a linear Y-X-X-Y molecule such as acetylene must conform to $D_{\infty h}$, and are therefore Σ_g^+, Σ_g^-, Π_g, Δ_g, etc. Note that the general linear molecule (just like the diatomic special case) has a distinct value of orbital angular momentum Λ along the internuclear axis. In nonlinear molecules electronic orbital angular momentum is usually quenched (see Sec. 10.5b) and consequently the symmetry species labels provide less information in these cases.

One special feature of polyatomic molecules is that the equilibrium conformation in excited states may differ from that of the ground state. Consequently it may be necessary to use different point groups to specify the symmetry of the possible states of ground- and excited-state molecules. Examples of this are CS_2, which is linear in its ground state $(D_{\infty h})$ but is bent in both its first and second excited states (C_{2v});[26] and $H_2C=O$, which is planar in its ground state (C_{2v}) but is nonplanar in several of its excited states (C_s).[27]

As with diatomics the electron spin must be accounted for also, and this is done by affixing the multiplicity $2S + 1$ to the symmetry label as a left superscript.

Once the possible states are delineated it is an easy matter to specify the possible transitions by group-theoretical techniques. Thus Eq. (11.9) applies for polyatomic molecules and Eqs. (11.10a) and (11.10b) therefore provide the group-theory restrictions. Once again the selection rule on S, $\Delta S = 0$, is expected to apply as a good first approximation. Many violations of this rule occur, however, due to appreciable spin-orbit coupling.

EXAMPLE 11-5

Let us consider the possible transitions of a symmetrical XY_3 molecule such as ammonia (NH_3). We imagine that both the planar (D_{3h}) and pyramidal (C_{3v}) forms are possible. We consider first the possible transitions between states of the planar molecule. Only those transitions are possible which satisfy Eq. (11.10b), which means the direct products must lead to E' or A_2''. We need to form all direct products. For example,

$$A_1' \times A_2' = A_2'$$

or less trivially,

$$\chi(A_2' \times A_1'') = 1, 1, -1, -1, -1, 1$$

or

$$A_2' \times A_1'' = A_2''.$$

[26] G. Herzberg, *Electronic Spectra and Electronic Structure of Polyatomic Molecules* (Van Nostrand, Princeton, N.J., 1966), p. 601.

[27] See footnote 26, p. 612.

Hence we have found that $A_2' \leftrightarrow A_1''$ is permitted but $A_1' \leftrightarrow A_2'$ is forbidden.

In this fashion the complete set of allowed transitions is found to be

$$A_1' \leftrightarrow A_2''$$

$$A_1' \leftrightarrow E'$$

$$A_2' \leftrightarrow A_1''$$

$$A_2' \leftrightarrow E' \tag{11.61}$$

$$A_1'' \leftrightarrow E''$$

$$A_2'' \leftrightarrow E''$$

$$E' \leftrightarrow A_2''$$

$$E'' \leftrightarrow A_2'$$

For the pyramidal (C_{3v}) molecule the direct products must lead to A_1 or E, which leads to

$$A_1 \leftrightarrow A_1$$

$$A_1 \leftrightarrow E$$

$$A_2 \leftrightarrow A_2 \tag{11.62}$$

$$A_2 \leftrightarrow E$$

$$E \leftrightarrow E$$

If we wish to consider possible transitions between states belonging to different molecular symmetry, in this case $D_{3h} \leftrightarrow C_{3v}$, we must correlate the symmetry species of the two groups.[28] In the present case we note that the classes of elements in C_{3v} are all included in D_{3h}. Consequently the species of the two groups are correlated by comparing the characters of the *common* classes only. We find that A_1 of C_{3v} correlates with A_1' and A_2'' of D_{3h} or

$$A_1(C_{3v}) \rightarrow A_1', A_2''(D_{3h})$$

Similarly

$$A_2(C_{3h}) \rightarrow A_1'', A_2'(D_{3h})$$

$$E(C_{3v}) \rightarrow E', E''(D_{3h})$$

Then to find the permitted transitions between C_{3v} and D_{3h} molecules we form the direct products only between the states of the lowest molecular symmetry, namely C_{3v}. Hence the results of Eqs. (11.62) give the appro-

[28] See, for example, E. B. Wilson, J. C. Decius, and P. C. Cross, *Molecular Vibrations* (McGraw-Hill, New York, 1955), Table X-14 for a tabulation of group correlations.

priate selection rules if we use the above correlations. This gives

$$A_1 \leftrightarrow A_1'$$

$$A_1 \leftrightarrow A_2''$$

$$A_1 \leftrightarrow E'$$

$$A_1 \leftrightarrow E''$$

$$A_2 \leftrightarrow A_1'' \tag{11.63}$$

$$A_2 \leftrightarrow A_2'$$

$$A_2 \leftrightarrow E'$$

$$A_2 \leftrightarrow E''$$

$$E \leftrightarrow E'$$

$$E \leftrightarrow E''$$

Thus a multitude of transitions are permitted. Of course, in any real case one needs to consider also the spin multiplicity of each state, and spin interconversions ($\Delta S > 0$) would be forbidden. For ammonia all observed excited states have D_{3h} symmetry and are singlets. Table 11.6 lists the observed transitions for ammonia. Note that the alphabetic labels used for diatomic molecules are used here also, except a tilde ($\sim$) is placed over the symbols. Note also that the ammonia electronic bands lie entirely in the ultraviolet region, for example, 50 000 cm^{-1} = 2000 Å.

In the gas phase the general fine-structure characteristics of the electronic bands are similar to those of diatomic molecules. Thus vibrational and rotational fine structure appears, and the band structure is of course

TABLE 11.6

Observed electronic states and transitions of ammonia.

State		Energy (cm^{-1})	Transition
$\tilde{E}$	$^1A_2''$	75 205	$\tilde{E} \leftarrow \tilde{X}$
$\tilde{D}$	$^1A_2''$	69 731	$\tilde{D} \leftarrow \tilde{X}$
$\tilde{C}$	$^1A_1'$	63 771	$\tilde{C} \leftarrow \tilde{X}$
$\tilde{B}$	$^1E''$	59 225	$\tilde{B} \leftarrow \tilde{X}$
$\tilde{A}$	$^1A_2''$	46 136	$\tilde{A} \leftarrow \tilde{X}$
$\tilde{X}$	$^1A_1$	0	

considerably more complicated. Only for the relatively small, simple polyatomic molecules has it been possible to resolve and analyze the vibrational and rotational structure. Perhaps the most interesting class of such studies has been that carried on by the flash photolysis technique introduced by Porter.[29] This method has permitted the spectra of numerous unstable species (particularly triatomic molecules) to be studied under high-resolution conditions. BH_2 and CH_2 are two such species whose electronic band spectra analyses have led to extensive quantitative structural information.

We shall not go into the details of the rotational and vibrational analysis for polyatomic molecules but refer the reader to the excellent treatise of Herzberg.[30]

B. Electron Configurations and the Aufbau Principle

Just as with atoms and diatomic molecules, the electronic structure or configuration of a polyatomic molecule may be specified by placing electrons into one-electron orbitals. Each of these molecular orbitals must have the symmetry of the irreducible representations of the molecular point group and may be so labeled. For consistency with atoms and diatomics the molecular-orbital symmetry labels are given in lower case. A prefix numerical index is used to enumerate the different orbitals of the same symmetry. Thus a C_{3v} molecule may have $1a_1, 2a_1, 3a_1, \ldots, 1a_2, 2a_2, 3a_2, \ldots,$ $1e, 2e, 3e, \ldots$ molecular orbitals. It is possible also to correlate the molecular orbitals with the united or separated atom states. Thus the CH_2 orbitals may be correlated with the united atom (O) and the separated atoms (2H and C). In many cases it is useful to correlate the molecular orbitals with group orbitals. For example, ethane (C_2H_6) molecular orbitals might be correlated with those of two methyl radicals (CH_3).[31]

Quantitative and/or semiquantitative calculations are needed, of course, to determine the order of the molecular orbital configurations. We do not go into this in great detail but present several typical examples.

Consider first the triatomic nonlinear symmetrical hydride molecules, XH_2, where X is a first row atom. The possible orbital symmetries are a_1, a_2, b_1, and b_2 (C_{2v} symmetry). The lowest two MO's correlate with the $1s$ and $2s$ atomic orbitals of the X atom and are consequently totally symmetrical in the molecular C_{2v} environment. Thus the lowest two MO's are $1a_1$ and $2a_1$. The next highest orbital depends upon the relative energies of the H $1s$ and X $2p$ atomic orbitals. Assuming the H $1s$ orbitals to lie lowest, these two atomic orbitals may lead to a_1 and b_2 molecular orbitals. That is,

[29] G. Porter, *Proc. Roy. Soc. (London)* **200A,** 284 (1950).
[30] See footnote 26.
[31] See footnote 26 for numerous correlation diagrams.

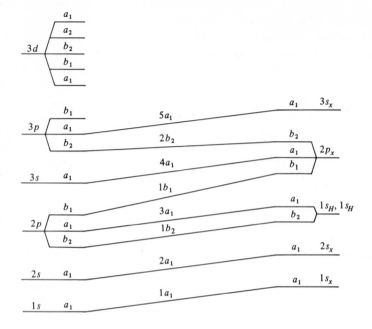

FIG. 11.13. Correlation diagram for nonlinear symmetrical XH_2 molecule.

the H $1s$ orbitals may be combined as $1s_A(1) + 1s_B(1)$ which is of a_1 symmetry; or as $1s_A(1) - 1s_B(1)$ which is of b_2 symmetry. Note that in the true molecular orbital there are contributions from the X atom orbitals also, and as we see later this lowers the energy of the b_2 orbital relative to the a_1 orbital. We therefore label these two molecular orbitals as $1b_2$ and $3a_1$ with $1b_2 < 3a_1$.

Finally we have three molecular orbitals which correlate with the $2p$ atomic orbital of X. These have symmetries of a_1, b_1, and b_2 which gives $1b_1$, $4a_1$, and $2b_2$ orbitals. Energetically it may be shown that $1b_1 < 4a_1 < 2b_2$. The MO scheme is therefore $1a_1 < 2a_1 < 1b_2 < 3a_1 < 1b_1 < 4a_1 < 2b_2$. Figure 11.13 presents a correlation diagram for this case.

Using this scheme the ground state configurations of BH_2, CH_2, NH_2, and OH_2 may be written as in Table 11.7. The term symbols including multiplicity follow easily by inspection. In general, the possible terms are obtained by decomposing the direct product representation formed by the molecular orbital representations. For example, the terms of H_2O are given by

$$(a_1 \times a_1) \times (a_1 \times a_1) \times (b_2 \times b_2) \times (a_1 \times a_1) \times (b_1 \times b_1)$$

which gives trivially A_1. Of course, Pauli's principle must be satisfied as usual.

Excited state configurations are similarly treated. The first H_2O excited configuration would be $(1a_1)^2(2a_1)^2(1b_2)^2(3a_1)^2(1b_1)^1(4a_1)^1$ which leads to $^1B_1$ and $^3B_1$ terms, since

$$b_1 \times a_1 = b_1.$$

As another example we consider briefly an XH_4 tetrahedral molecule such as methane. Using the T_d symmetry species to label the MO's, the lowest orbitals are $1a_1 < 2a_1 < 1t_2 < 3a_1$. . . . The lowest two a_1 MO's are essentially C $1s$ and $2s$ atomic orbitals while the triply degenerate state and following a_1 state are largely composed of H $1s$ and C $2p$. The $3a_1$ state is strongly antibonding. With ten electrons the ground configuration is $(1a_1)^2(2a_1)^2(1t_2)^6$, which gives the single term $^1A_1$. The first excited configuration would be $(1a_1)^2(2a_1)^2(1t_2)^5(3a_1)^1$. The easiest method of determining the terms for this configuration is to recall from atomic theory that a shell (orbital) which is one electron from being filled may be considered as a shell with *one* electron. Thus the term symmetries that are possible are given by

$$t_2 \times a_1 = t_2,$$

which gives $^3T_2$ and $^1T_2$ terms.

With this brief introduction we move on now to several more specific examples. It should be clear that if the molecular-orbital scheme is known for a given general form of molecule, the possible electron configurations and terms may be determined. In general the correct molecular orbital scheme must be arrived at by quantitative predictions and by comparison with experimental results. Of course the reader is probably aware by now that we are really speaking about "chemical bonding" in an indirect fashion, and consequently all the well-known concepts provide useful background.

TABLE 11.7

Ground-state configurations of some non-linear XH_2 molecules.

Molecule	Configuration	Term
BH_2	$(1a_1)^2(2a_1)^2(1b_2)^2(3a_1)^1$	$^2A_1$
CH_2[a]	$(1a_1)^2(2a_1)^2(1b_2)^2(3a_1)^2$	$^1A_1$
NH_2	$(1a_1)^2(2a_1)^2(1b_2)^2(3a_1)^2(1b_1)^1$	$^2B_1$
OH_2	$(1a_1)^2(2a_1)^2(1b_2)^2(3a_1)^2(1b_1)^2$	$^1A_1$

[a]   Ground state is most likely linear $^3\Sigma_g^-$. See G. Herzberg, *Proc. Roy. Soc.* **262A,** 291 (1961).

11.6 MOLECULAR ORBITALS AND SPECTRA OF POLYATOMIC MOLECULES

A. *Water Molecule*

A minimal basis-set molecular-orbital calculation for H_2O may be performed using the following atomic-orbital basis set:

$$\text{Hydrogen orbitals—} 1s_a, \ 1s_b$$

$$\text{Oxygen orbitals—} 1s_0, \ 2s_0, \ 2p_x, \ 2p_y, \ 2p_z$$

Here we choose z along the C_2 axis with x perpendicular to the plane as shown in Fig. 11.14. The oxygen atomic orbitals may be classified as to their symmetry under C_{2v}. This gives

$$\phi_1{}^{a_1} = 1s_0$$

$$\phi_2{}^{a_1} = 2s_0$$

$$\phi_3{}^{a_1} = 2p_z \tag{11.64a}$$

$$\phi_4{}^{b_1} = 2p_x$$

$$\phi_5{}^{b_2} = 2p_y$$

For example, $2p_x$ has the following characters under the C_{2v} operations: $\chi(E) = 1$, $\chi(C_2) = -1$, $\chi(\sigma_v) = 1$, $\chi(\sigma_v{}') = -1$, which corresponds to the B_1 species.

The hydrogen orbitals do not have symmetry under C_{2v} but the sum and difference lead to orbitals of A_1 and B_2 symmetry, respectively. Thus

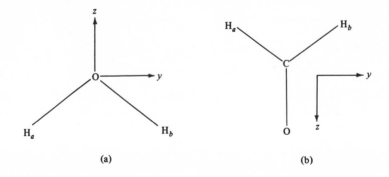

(a)   (b)

FIG. 11.14.   (a) Coordinate system for oxygen orbitals in H_2O, and (b) oxygen and carbon orbitals in formaldehyde. x axis is $\perp$ to plane.

we have the hydrogen symmetry orbitals

$$\phi_6{}^{a_1} = 1s_a + 1s_b$$

$$\phi_7{}^{b_2} = 1s_a - 1s_b$$

(11.64b)

With these seven basis functions the LCAO-MO-SCF treatment leads to the energies and analytic forms of the seven one-electron molecular orbitals in terms of the seven symmetry basis functions. Since there can be no Hamiltonian matrix elements between states of different symmetry the final MO's may contain terms of only the same symmetry. Thus the form of the MO's must be

$$\psi_1 = a_{11}\phi_1 + a_{12}\phi_2 + a_{13}\phi_3 + a_{16}\phi_6$$

$$\psi_2 = a_{21}\phi_1 + a_{22}\phi_2 + a_{23}\phi_3 + a_{26}\phi_6$$

$$\psi_3 = a_{31}\phi_1 + a_{32}\phi_2 + a_{33}\phi_3 + a_{36}\phi_6$$

$$\psi_4 = a_{41}\phi_1 + a_{42}\phi_2 + a_{43}\phi_3 + a_{46}\phi_6 \qquad (11.65)$$

$$\psi_5 = a_{55}\phi_5 + a_{57}\phi_7$$

$$\psi_6 = a_{65}\phi_5 + a_{67}\phi_7$$

$$\psi_7 = \phi_4$$

where we have omitted the superscript labels on the ϕ_i for simplicity.

Pitzer and Merrifield[32] carried out the above calculation using optimized Slater-type atomic orbitals and found the following energies[33] for the occupied molecular orbitals:

$$E(1a_1) = -559.3 \text{ eV}$$

$$E(2a_1) = -35.0 \text{ eV}$$

$$E(1b_2) = -17.0 \text{ eV} \qquad (11.66)$$

$$E(3a_1) = -12.7 \text{ eV}$$

$$E(1b_1) = -11.0 \text{ eV}$$

No values of orbital energy were stated for $4a_1$ and $2b_2$, but they are known from other computations to be about 18 and 22 eV more positive, respectively, than $1b_1$. This is the ordering which we previously described and it seems in good agreement with experimental observations.

[32] See S. Aung, R. M. Pitzer and S. I. Chan, *J. Chem. Phys.* **49**, 2071 (1968).

[33] These values do not include the nuclear repulsion contribution, which is of course a constant for fixed geometry.

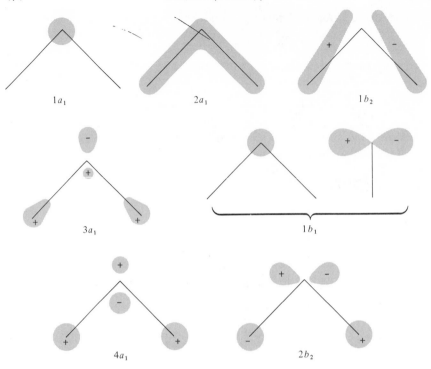

FIG. 11.15.   Qualitative forms of the molecular orbitals of H_2O.

The coefficients a_{ij} in 11.65 were determined also. Rather than list them we merely show in Fig. 11.15 the rough qualitative form of the resulting orbitals. The forms shown indicate in a very qualitative way that the $1a_1$ orbital is primarily $1s_O$, while $1b_1$ is nonbonding (lone-pair) $2p_x$. The bonding orbitals are $2a_1$ and $1b_2$ while $4a_1$ and $2b_2$ are strongly antibonding. The $3a_1$ orbital shows some bonding character but is largely a nonbonding $2p_z$-$2s_O$ hybrid.

B. Formaldehyde and Group Absorptions

Formaldehyde ($H_2C=O$) is an interesting simple example of the carbonyl class of molecules. As such it is the prototype of the entire class and its understanding is pertinent to larger carbonyl species. The simplest minimal basis set of atomic orbitals for a LCAO-MO treatment would consist of hydrogen $1s_a$ and $1s_b$ orbitals, oxygen $1s_O$, $2s_O$, $2p_{xO}$, $2p_{yO}$, $2p_{zO}$, and carbon $1s_c$, $2s_c$, $2p_{xc}$, $2p_{yc}$, and $2p_{zc}$. A calculation using such a basis set was carried out by Foster and Boys[34] using Gaussian orbitals to represent the

[34] J. M. Foster and S. F. Boys, *Rev. Mod. Phys.* **32**, 303 (1960).

Slater atomic orbitals. Rather than consider this treatment in detail we look at the problem in a more qualitative fashion.

Formaldehyde, as shown in Fig. 11.14, has C_{2v} symmetry. The lowest energy molecular orbitals are the inner shell orbitals of oxygen and carbon of A_1 symmetry. These are essentially unperturbed atomic orbitals as a first approximation.

$$\phi_1 = 1a_1 = 1s_O$$

$$\phi_2 = 2a_1 = 1s_O \qquad (11.67a)$$

$$\phi_3 = 3a_1 = 2s_O$$

The first bonding orbitals are those involving the CH bonds and have the form

$$\phi_4 = 4a_1 = 2s_c + 1s_a + 1s_b$$

$$\qquad (11.67b)$$

$$\phi_5 = 1b_2 = 2p_{yc} - 1s_a + 1s_b$$

Following these CH bonding orbitals are the CO σ bond MO and the CO π bond MO

$$\phi_6 = 5a_1 = 2p_{zO} + 2p_{zc}$$

$$\qquad (11.67c)$$

$$\phi_7 = 1b_1 = 2p_{xO} + 2p_{xc}$$

where we choose the p_z orbitals to be oppositely oriented. The eighth MO is largely nonbonding oxygen $2p_y$.

$$\phi_8 = 2b_2 = 2p_{yO} \qquad (11.67d)$$

This exhausts the bonding and nonbonding orbitals. The next orbital is undoubtedly the antibonding π^* orbital

$$\phi_9 = 2b_1 = 2p_{xO} - 2p_{xc} \qquad (11.67e)$$

while the last which we list is probably the antibonding σ^* orbital

$$\phi_{10} = 6a_1 = 2p_{zO} - 2p_{zc} \qquad (11.67f)$$

With this orbital scheme the 16 electrons of formaldehyde would lead to the lowest configuration

$$(1a_1)^2(2a_1)^2(3a_1)^2(4a_1)^2(1b_2)^2(5a_1)^2(1b_1)^2(2b_2)^2$$

which is a $^1A_1$ term. Recalling that the last three orbitals are σ, π, and nonbonding n, we may write more simply

$$\psi_1: [\phi_1 \ldots \phi_5]^{10}\sigma^2\pi^2n^2 \qquad {}^1A_1 \qquad (11.68)$$

for the ground configuration. Excited state configurations may also be written, the most likely being

$$\psi_2: \; [\phi_1 \ldots \phi_5]^{10}\sigma^2\pi^2 n^1\pi^{*1} \quad A_2$$

$$\psi_3: \; [\phi_1 \ldots \phi_5]^{10}\sigma^2\pi^1 n^2\pi^{*1} \quad A_1 \qquad (11.69)$$

$$\psi_4: \; [\phi_1 \ldots \phi_5]^{10}\sigma^2\pi^2 n^1\sigma^{*1} \quad B_2$$

where both singlet and triplet terms are possible. The term symbol is obtained as usual, for example, that for ψ_2 is given by $\Gamma^{(\phi_8)} \times \Gamma^{(\phi_9)} = B_2 \times B_1 = A_2$.

Since the states represented in (11.69) are expected to be the lowest excited states it is valuable to consider the possible electric dipole selection rules for transitions connecting these states to the ground state. For the C_{2v} molecule, electric dipole transitions are permitted only if $\Gamma^{(\psi_i)} \times \Gamma^{(\psi_j)} \to A_1$, B_1, or B_2. Table 11.8 summarizes the results of this analysis and also gives the selection rules based on a slightly nonplanar (C_s) excited-state structure. This is done by correlating the C_{2v} representations with those of the C_s group. Note that in this case all transitions turn out to be allowed. Finally we have listed the selection rules based on a molecular *magnetic moment* operator. This operator transforms like the infinitesimal rotations, R_x, R_y, and R_z, consequently the direct product must yield A_2, B_1, or B_2. Note that these rules are all based on *singlet* excited states. No transitions to triplet states are predicted as a first approximation.

The observed spectrum of formaldehyde in the near and far UV may be correlated reasonably well with the predictions in Table 11.8. The most intense band occurs at about 1600 Å and is assignable to the $\pi \to \pi^*$ transition. Somewhat higher, at about 1700 Å, the $n \to \sigma^*$ transition occurs. These are each relatively intense transitions, a result which agrees with the selection rules. A broad weak band appears at about 2800 Å and has been assigned to the $n \to \pi^*$ transition. This is not permitted by electric dipole selection rules for the planar C_{2v} molecule. It is, however, permitted for a nonplanar excited state molecule, which is the experimentally observed conformation. This band is also allowed via the magnetic dipole selection rules and recent studies indicate that this is probably the im-

TABLE 11.8

Selection rules for some low-lying states of formaldehyde

Transition	Electric dipole allowed?		Magnetic dipole allowed?
	C_{2v}	C_s	C_{2v}
$\psi_1 \to \psi_2 \; (n \to \pi^*)$	no	yes	yes
$\psi_1 \to \psi_3 \; (\pi \to \pi^*)$	yes	yes	no
$\psi_1 \to \psi_4 \; (n \to \sigma^*)$	yes	yes	yes

portant mechanism.[35] Note finally that the excitation of the nonbonding oxygen $2p_y$ electron to the antibonding CO $2p_x$ orbital accounts adequately for the nonplanar excited state, since electron density is leaving the plane of the molecule for the out-of-plane direction.

There is still one more very weak band that has been observed at about 3800 Å in emission. This band is the $n \rightarrow \pi^*$ transition with the upper state a triplet. In group-theory notation the transition is $^3A_2 \rightarrow {}^1A_1$.

The spectral results for formaldehyde are of rather great generality for systems with isolated carbonyl groups. Thus the relative energies of the carbonyl molecular orbitals are similar in acetaldehyde, acetone, etc., as are the more tightly bound states which did not enter into the previous spectral description. In Table 11.9 we have listed some observed wavelengths of maximum absorption (λ_{max}) and the corresponding molar extinction coefficients ϵ.

While a relatively reasonable constancy exists, it is seen that the carbonyl electronic bands are sensitive to the other molecular substituents. Thus, if atoms with nonbonded electrons are attached to the carbonyl group, such as in acetic acid or acetyl chloride, the n-π^* wavelength decreases. This is caused by a lowering of the energy of the nonbonding (n) molecular orbital relative to the π molecular orbital. If the carbonyl is conjugated with an ethylene linkage, as in methyl vinyl ketone, the n-π^* transition wavelength increases markedly. This conjugative effect is effective in lowering the energy of π orbitals due to electron delocalization, consequently decreasing the n-π^* energy difference. Note that the π-π^* transition is shifted to an even greater extent.

What we have described is an example of a *characteristic* electronic absorption wavelength (frequency). The concept here is very similar to that described for infrared spectra, which showed characteristic group vibrations. Just as with the latter case, characteristic electronic bands are

TABLE 11.9

Carbonyl absorption frequencies and intensities.

	n-π^*		n-σ^*		π-π^*	
	λ_{max}	ϵ	λ_{max}	ϵ	λ_{max}	ϵ
acetaldehyde	290[a]	17	180	10^4	160	2×10^4
acetone	280	15	190	10^3	165	2×10^4
cyclopentanone	300	20				
acetic acid	204	41				
acetyl chloride	235	53				
methyl vinyl ketone	300	27			210	0.6×10^4

[a]   Units are nanometers or millimicrons.

[35] J. H. Calloman and K. K. Innes, *J. Mol. Spec.* **10,** 166 (1963).

TABLE 11.10

Characteristic electronic transitions of simple structural units

Group	λ_{max}	ϵ
$\diagdown\diagup$ C=C $\diagup\diagdown$	1750[a]	15 000
—C≡C—	1750	10 000
—COOH	2050	100
—C≡N	1600	. . .
—N=N—	3450	10
—NO$_2$	2700	20

[a] Units are angstroms.

affected by the entire molecular (electronic) environment, and therefore they may serve as a more or less sensitive probe of the molecular structure. We shall not go into this in further detail but we present in Table 11.10 a short list of characteristic group absorptions in the easily accessible wavelength range. It should be mentioned that the absorption wavelengths are somewhat medium dependent; that is, they vary somewhat in gas and liquid phase, and also vary with solvent. The literature references should be consulted for more detailed discussions of these points.

C. Conjugated Hydrocarbons

The longest wavelength absorptions of aromatic and conjugated polyene molecules can be ascribed generally to π-π^* transitions. Consequently most theoretical descriptions of such molecular spectra have been carried out by neglecting the σ electrons, or by considering them in some kind of average way. While certain theoretical justifications of such treatments are possible,[36] they are best justified as semiempirical attempts to describe very complicated molecules.

Recalling the vast bulk of chemical data which points toward electron delocalization in π-conjugated molecules, it is useful to look first into a free electron molecular orbital (FEMO) approach. Here we neglect all σ electrons and write the Hamiltonian as a sum of one-electron π terms,

$$\widetilde{\mathcal{3C}} = \sum_i^\pi \widetilde{\mathcal{3C}}_i \tag{11.70}$$

The potential energy of each electron is taken to be the one-dimensional particle-in-a-box type, that is, $V_i = 0$ in a certain region, $0 \leqq x \leqq L$ and

---

[36] See, for example, R. G. Parr, *Quantum Theory of Molecular Electronic Structure* (Benjamin, New York, 1963).

$V_i = \infty$ everywhere else. The remainder of the $\tilde{\mathcal{H}}_i$ is simply the electronic kinetic energy. Then the solution of

$$\tilde{\mathcal{H}}\psi = E\psi \tag{11.71}$$

is clearly

$$\psi = \phi_1\phi_2\phi_3\ldots \tag{11.72}$$

and

$$E = \frac{h^2}{8mL^2} \sum_i^\pi n_i^2 \tag{11.73}$$

where the one-electron FEMO functions are the well-known particle-in-a box functions and n_i is the principal quantum number.

Taking the Pauli principle into account, two electrons may go into any FEMO. Thus for N π-electrons the lowest $\frac{1}{2}N$ MO's are filled in the ground-state configuration,

$$(\phi_1)^2(\phi_2)^2\ldots(\phi_{N/2})^2 \tag{11.74}$$

and the energy is given by

$$E = \frac{2h^2}{8mL^2}\left[(1)^2 + (2)^2 + \ldots(\tfrac{1}{2}N)^2\right] \tag{11.75}$$

Finally we predict the transition frequency to the first excited configuration to be

$$\nu = \frac{\Delta E}{h} = \frac{h}{8mL^2}(N + 1) \tag{11.76}$$

In order to test the usefulness of (11.76) we consider first the hydrocarbon polyenes and choose L to be the zigzag length of the molecule. For N carbon atoms there are N π-electrons and $(N - 1)$ CC linkages as shown for hexatriene below.

$$C{=}C{-}C{=}C{-}C{=}C$$

We draw the structure in a straight line since its correct conformation does not affect our treatment. For simplicity we now choose the average CC-bond distance as 1.40 Å, so that $L = (N - 1)(1.4)$. Using this result and Eq. (11.76), transition wavelengths in Å are given by

$$\lambda = 645\frac{(N - 1)^2}{(N + 1)} \tag{11.77}$$

For butadiene, hexatriene, and decapentaene we predict

$$\lambda_4 = 1160\text{ Å}$$

$$\lambda_6 = 2310\text{ Å}$$

$$\lambda_{10} = 4750\text{ Å}$$

respectively, while the experimental values are 2170, 2680, and 3340 Å, respectively.

We see that the agreement is quantitatively poor, but the theory does predict the proper qualitative dependence of λ upon molecular chain length. The computations can be improved in a variety of ways, and can be modified to apply to aromatic molecules. An excellent collection of articles on the subject by University of Chicago theorists has been published.[37]

The most well-known method for treating π-electron systems is the Hückel molecular orbital method (HMO). As with the FEMO method the Hamiltonian is approximated by a sum of one-electron terms for each of the π electrons. No explicit form of $\widetilde{\mathfrak{IC}}_i$ is specified. The problem resolves itself into the determination of the one electron HMO's and their energies with the aid of a certain set of assumptions.

To do this the LCAO approximation is used, with one $2p_z$ atomic orbital from each carbon making up the basis set. The linear-variation method then leads to the secular determinant

$$| \, \mathbf{H} - \lambda_i \mathbf{S} \, | = 0 \qquad (11.78)$$

where λ_i are the one-electron energies and

$$H_{jk} = \langle \, j \, | \, \widetilde{\mathfrak{IC}}_i \, | \, k \, \rangle \qquad (11.79a)$$

$$S_{jk} = \langle \, j \, | \, k \, \rangle \qquad (11.79b)$$

The functions used to evaluate the matrix elements of Eqs. (11.79) may be the atomic-orbital basis set, but we know that symmetrized functions lead to a simpler Hamiltonian matrix.

In the simplest HMO theory, the following assumptions are made concerning the elements of $\mathbf{H}$ and $\mathbf{S}$. If the carbon $2p_z$ basis orbitals are ϕ_1, ϕ_2, ϕ_3, ..., ϕ_N for the N π-electron system, we require:

$$S_{ij} = 0 \qquad (11.80a)$$

$$S_{ii} = 1 \qquad (11.80b)$$

$$H_{jj} = \int \phi_j^* \widetilde{\mathfrak{IC}}_i \phi_j d\tau = \alpha \qquad (11.80c)$$

$$H_{jk} = \int \phi_j^* \widetilde{\mathfrak{IC}}_i \phi_k d\tau \qquad (11.80d)$$

$$= \beta \text{ for adjacent atoms}$$

$$= 0 \text{ for nonadjacent atoms}$$

α is commonly known as the *Coulomb* integral and β the *resonance* integral.

[37] J. R. Platt, K. Ruedenberg, C. W. Scherr, N. S. Ham, H. Labhart, W. Lichten, *Free-Electron Theory of Conjugated Molecules* (Wiley, New York, 1964).

These quantities are usually chosen to agree with experiment in some way. Note that (11.80a) is not generally true, but the assumption is made for simplicity. Note that (11.80b) follows automatically by using normalized atomic orbitals.

We proceed now with an important example—benzene. Using the numbering scheme for the atoms shown in Fig. 11.16, we can immediately set up the secular determinant with no reliance on symmetry. This gives

$$\begin{vmatrix} \alpha - \lambda & \beta & 0 & 0 & 0 & \beta \\ \beta & \alpha - \lambda & \beta & 0 & 0 & 0 \\ 0 & \beta & \alpha - \lambda & \beta & 0 & 0 \\ 0 & 0 & \beta & \alpha - \lambda & \beta & 0 \\ 0 & 0 & 0 & \beta & \alpha - \lambda & \beta \\ \beta & 0 & 0 & 0 & \beta & \alpha - \lambda \end{vmatrix} = 0 \quad (11.81)$$

This determinant could be expanded and the resulting polynomial in λ solved. Rather than do this, we reformulate the problem using symmetry orbitals.

Since the benzene molecule conforms to D_{6h} symmetry we know that the HMO functions must transform like the irreducible representations of

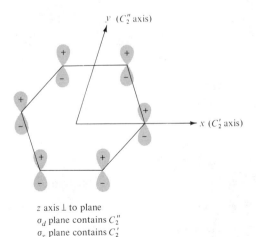

z axis $\perp$ to plane
σ_d plane contains C_2''
σ_v plane contains C_2'

FIG. 11.16.   Benzene basis orbitals and identification of axes and planes.

the group. By the methods of chapter 6 the six basis orbitals must form a reducible representation of the group. This gives the following characters.

$$\chi(E) = 6 \qquad \chi(C_6) = 0 \qquad \chi(C_3) = 0$$
$$\chi(C_2) = 0 \qquad \chi(C_2') = -2 \qquad \chi(C_2'') = 0 \qquad (11.82)$$
$$\chi(i) = 0 \qquad \chi(S_3) = 0 \qquad \chi(S_6) = 0$$
$$\chi(\sigma_h) = -6 \qquad \chi(\sigma_d) = 0 \qquad \chi(\sigma_v) = 2$$

Using the standard reduction formula we find

$$\Gamma = A_{2u} + B_{2g} + E_{1g} + E_{2u} \qquad (11.83)$$

Hence there are two nondegenerate and two doubly degenerate HMO's. Thus, although we do not yet know the precise form and energies of the resulting one-electron states, we do know that there are only four distinct orbital states which may arise. Furthermore we know that if the Hamiltonian matrix were evaluated using functions having the symmetries of the eigenfunctions [i.e., as given by (11.83)], it would have no matrix elements connecting states of different symmetry. Indeed it should be possible to choose the degenerate pairs of functions so that no off-diagonal matrix elements exist in the degenerate state blocks. We conclude, therefore, that the use of symmetry functions leads to a completely diagonal matrix $\mathfrak{IC}$ for the problem at hand. It must be possible to generate functions having the symmetries of Eq. (11.83). To do this we use the method of Sec. 6.4 A, which requires as a first step that we find $\tilde{R}\phi_i$ for all $\tilde{R}$ in the group and for an appropriate ϕ_i (or ϕ_i's). We choose ϕ_1 as a beginning, and find the following results.

$\tilde{R}$	E	C_6	C_3	C_2	C_2'	C_2''	i
$\tilde{R}\phi_1$	ϕ_1	ϕ_2, ϕ_6	ϕ_3, ϕ_5	ϕ_4	$-\phi_5, -\phi_1, -\phi_3$	$-\phi_2, -\phi_6, -\phi_4$	$-\phi_4$

$\tilde{R}$	S_3	S_6	σ_h	σ_d	σ_v
$\tilde{R}\phi_1$	$-\phi_3, -\phi_5$	$-\phi_2, -\phi_6$	$-\phi_1$	ϕ_2, ϕ_6, ϕ_4	ϕ_5, ϕ_1, ϕ_3

The function of A_{2u} symmetry is now easily found by applying Eq. (6.68).

$$
\begin{aligned}
\psi_1(A_{2u}) &= \mathfrak{R}\{\phi_1 + (\phi_2 + \phi_6) + (\phi_3 + \phi_5) + \phi_4 \\
&\quad - (-\phi_5 - \phi_1 - \phi_3) - (-\phi_2 - \phi_6 - \phi_4) - (-\phi_4) \\
&\quad - (-\phi_3 - \phi_5) - (-\phi_2 - \phi_6) - (-\phi_1) + (\phi_2 + \phi_6 + \phi_4) \\
&\quad + (\phi_5 + \phi_1 + \phi_3) \\
&= \mathfrak{R}\{4\phi_1 + 4\phi_2 + 4\phi_3 + 4\phi_4 + 4\phi_5 + 4\phi_6\} \\
&= \frac{1}{\sqrt{6}} (\phi_1 + \phi_2 + \phi_3 + \phi_4 + \phi_5 + \phi_6) \qquad (11.84)
\end{aligned}
$$

We have used the Hückel theory assumption that $S_{ij} = 0$, which, of course, is not true. In a similar fashion we find the function of B_{2g} type to be

$$\psi_2(B_{2g}) = \frac{1}{\sqrt{6}} (\phi_1 - \phi_2 + \phi_3 - \phi_4 + \phi_5 - \phi_6) \qquad (11.85)$$

One member of each of the degenerate sets is similarly found to be

$$\psi_3'(E_{1g}) = \frac{1}{\sqrt{12}} (2\phi_1 + \phi_2 - \phi_3 - 2\phi_4 - \phi_5 + \phi_6) \qquad (11.86a)$$

$$\psi_5'(E_{2u}) = \frac{1}{\sqrt{12}} (2\phi_1 - \phi_2 - \phi_3 + 2\phi_4 - \phi_5 - \phi_6) \qquad (11.87a)$$

If we had generated the E-type functions using $\tilde{R}\phi_2$ rather than $\tilde{R}\phi_1$ we would have obtained

$$\psi_4'(E_{1g}) = \frac{1}{\sqrt{12}} (\phi_1 + 2\phi_2 + \phi_3 - \phi_4 - 2\phi_5 - \phi_6) \qquad (11.86b)$$

$$\psi_6'(E_{2u}) = \frac{1}{\sqrt{12}} (\phi_1 - 2\phi_2 + \phi_3 + \phi_4 - 2\phi_5 + \phi_6) \qquad (11.87b)$$

which are just as satisfactory from the point of view of symmetry. However, ψ_3' and ψ_4' are not orthogonal, and the same is true for ψ_5' and ψ_6'. The pairs of degenerate functions may be orthogonalized however, by forming the sum and difference. Doing this we obtain

$$\psi_3(E_{1g}) = \tfrac{1}{2}(\phi_1 + \phi_2 - \phi_4 - \phi_5) \qquad (11.88)$$

$$\psi_4(E_{1g}) = \frac{1}{\sqrt{12}} (\phi_1 - \phi_2 - 2\phi_3 - \phi_4 + \phi_5 + 2\phi_6) \qquad (11.89)$$

$$\psi_5(E_{2u}) = \tfrac{1}{2}(\phi_1 - \phi_2 + \phi_4 - \phi_5) \qquad (11.90)$$

$$\psi_6(E_{2u}) = \frac{1}{\sqrt{12}} (\phi_1 + \phi_2 - 2\phi_3 + \phi_4 + \phi_5 - 2\phi_6) \qquad (11.91)$$

The functions (11.84), (11.85), and (11.88)–(11.91) are a complete set of orthonormal symmetry functions. There can be no off-diagonal elements of $\mathcal{3C}$ since all functions belong to different irreducible representations or different rows of degenerate representations. We may evaluate the diagonal elements, for example,

$$\langle A_{2u} | \tilde{\mathcal{3C}} | A_{2u} \rangle = E(A_{2u}) = \frac{1}{6} \{6\alpha + 12\beta\} = \alpha + 2\beta \qquad (11.92a)$$

Similarly

$$E(B_{2g}) = \alpha - 2\beta \qquad (11.92b)$$

$$E(E_{1g}) = \alpha + \beta \qquad (11.92c)$$

$$E(E_{2u}) = \alpha - \beta \qquad (11.92d)$$

Note that we have used the simple approximations listed in Eqs. (11.80). Since β as defined in Eq. (11.80d) is negative, the ordering of the one-electron MO's is $a_{2u} < e_{1g} < e_{2u} < b_{2g}$, as shown in Fig. 11.17. The simple qualitative orbital pictures in Fig. 11.17 show clearly the bonding-antibonding nature of the MO's.

While there are, of course, numerous properties of chemical interest which may be elucidated by HMO theory,[38] we are interested here only in

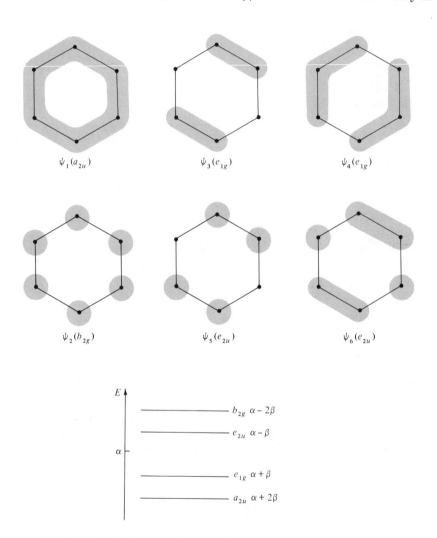

FIG. 11.17. Qualitative shapes of benzene π orbitals and Hückel energy levels.

[38] See, for example, A. Streitweiser, *Molecular Orbital Theory for Organic Chemists* (Wiley, New York, 1961).

pursuing some spectral features. The ground-state configuration of the benzene π electrons is

$$(a_{2u})^2(e_{1g})^4 \quad {}^1A_{1g} \tag{11.93}$$

while the lowest excited configuration is

$$(a_{2u})^2(e_{1g})^3(e_{2u})^1 \tag{11.94}$$

with the following six terms:

$${}^1B_{1u}, \; {}^1B_{2u}, \; {}^1E_{1u}$$
$$\tag{11.95}$$
$${}^3B_{1u}, \; {}^3B_{2u}, \; {}^3E_{1u}$$

In the simple HMO approximation these six terms are all degenerate, but we know that this is incorrect due to the total neglect of interelectronic repulsions. Nevertheless it is reasonable to assume that these states are probably the lowest-lying excited states.

Selection rules based on the orbital symmetries are now obtained by the well-known methods. We find (for transitions from the populated ground state)

$$A_{1g} \not\leftrightarrow B_{1u}$$

$$A_{1g} \not\leftrightarrow B_{2u} \tag{11.96}$$

$$A_{1g} \leftrightarrow E_{1u}$$

Thus only one transition is predicted to be both orbitally and spin allowed, the ${}^1A_{1g} \rightarrow {}^1E_{1u}$. Unfortunately the electronic spectrum of benzene at wavelengths above 1600 Å is quite dense all the way up to about 3500 Å, with at least four distinct bands in the region. The most intense absorption occurs at a λ_{max} of about 1800 Å, and is most likely the strongly allowed transition, $A_{1g} \rightarrow E_{1u}$.

The most extensively studied band in benzene occurs at about 2500 Å with an intensity about $\frac{1}{10}$ that of the 1800 Å band. Experimental and theoretical studies[39] have shown unambiguously that this transition is ${}^1A_{1g} \rightarrow {}^1B_{2u}$, which becomes allowed due to vibrational-electronic interactions.

At about 3400 Å a very weak band has been observed in both absorption and emission. The long phosphorescent lifetime of the excited state and the dependence of the absorption band intensity upon paramagnetic impurities (or additives) such as NO and O_2 have led to the conclusion that the transition is a triplet-singlet type. Theoretical considerations lead to the assignment ${}^1A_{1g} \leftrightarrow {}^3B_{1u}$.

Finally, a rather prominent band occurs just above the strongly al-

[39] R. G. Parr, D. P. Craig, and I. G. Ross, *J. Chem. Phys.* **18**, 1561 (1950).

lowed band, at about 2000 Å. Since the $^3B_{1u}$ state has been observed and assigned, it is not unreasonable to assign this band to $^1A_{1g} \rightarrow {}^1B_{1u}$. There are, however, some theoretical grounds for believing that this observed transition has a $^1E_{2g}$ upper state, which could arise from the

$$(a_{2u})^2(e_{1g})^3(b_{2g})^1$$

configuration with only $^1E_{2g}$ and $^3E_{2g}$ terms.

Thus the reader should be aware that the theoretical assignment and analysis of even rather simple π-electron spectra is a difficult task. The electronic-vibrational (vibronic) interactions are particularly important in causing breakdowns of the simple electronic selection rules. Herzberg[40] gives good descriptions of these situations.

Actually, some quantitative sense can be shown to arise even from the simple HMO method. For example, a $|\beta|$ value of 28 000 cm^{-1} (3.48 eV) has been shown to fit rather well the HMO predictions for the frequencies of the transitions which Platt[41] terms ^1L_a type. This transition in simple HMO theory is that from the ground-state configuration to the first excited configuration; that is, the transition is to the first (lowest) unfilled orbital. For benzene this is the $^1A_{1g} \rightarrow {}^1B_{1u}$ transition at 2000 Å. For many other benzenoid-type molecules the ^1L_a band is the longest wavelength band in absorption, and usually has a molar extinction coefficient of about 10 000. Thus according to HMO theory, for benzene this band occurs at an energy (or frequency) of 2β, which would be 56 000 cm^{-1} using $|\beta| = 28\ 000$ cm^{-1}. The observed value is about 50 000 cm^{-1} (2000 Å).

Numerous other calculations of π-electron spectra of aromatic molecules have been performed. These include more sophisticated HMO methods[42] (evaluate overlap integrals, empirically estimate certain integrals, etc.) and the previously mentioned free-electron model. Similar π-electron LCAO-MO treatments have been carried out for the nonaromatic conjugated molecules such as butadiene, etc., and the results have been of comparable accuracy to those for aromatic molecules. The reader should consult the abundant literature for the newer developments which are being spawned in the wake of the digital computer development.

11.7   SPECTRA OF TRANSITION METAL COMPLEXES

Transition-metal complexes provide an interesting type of spectra that is rather unlike that which we have previously described. The visible absorption spectra of complexes such as $Cu(NH_3)_4^{+2}$, $V(H_2O)_6^{+3}$, or $Cr(C_2O_4)_3^{-3}$

[40] See footnote 26, chapter II.

[41] J. R. Platt, *J. Chem. Phys.* **17**, 484 (1949).

[42] See, for example, M. Goeppart-Mayer and A. L. Sklar, *J. Chem. Phys.* **6**, 645 (1938); R. Pariser, *J. Chem. Phys.* **24**, 250 (1956); C. C. J. Roothaan and R. S. Mulliken, *J. Chem. Phys.* **16**, 118 (1948).

may be ascribed primarily to transitions between the various d-orbital states of the transition-metal ion. We have seen already in chapter 10 how the crystal- or ligand-field environment splits the degenerate free-ion states, and that this splitting is typically in the range of 20 000 cm^{-1}. Thus it is clear that transitions between such states would lead to absorption of radiation in or near the visible region of the spectrum. We shall not delve into this interesting subject in great detail but look at some of the simple, general principles involved.

First, it is perhaps useful to consider the group-theoretical properties of the states of an atom or ion perturbed by a crystal or ligand environment. We recall that the angular dependence of the atomic-orbital states is given by the spherical harmonics Y_{lm} which may be written[43]

$$Y_{lm}(\vartheta, \varphi) = P_{lm}(\vartheta) \exp(im\varphi) \qquad (11.97)$$

These wave functions must form bases for the irreducible representations of the group of Schrödinger's equation, which in this case includes the group of all rotations. That is, the Hamiltonian for the free ion must be invariant to a rotation of the axes (or a rotation of the system) used in defining the system. Since the axes are totally arbitrary in isotropic space, we may simply investigate how Y_{lm} behaves under the rotation by α about Z, which changes φ to $\varphi + \alpha$; the character of the transformation matrix then applies to *any other rotation by α about any other axis* since all rotations by α must belong to the same class. Performing this rotation $\tilde{R}_\alpha$ we find

$$\tilde{R}_\alpha Y_{lm} = P_{lm}(\vartheta) \exp[im(\varphi + \alpha)]$$
$$= P_{lm}(\vartheta) \exp(im\varphi) \exp(im\alpha)$$
$$= Y_{lm} \exp(im\alpha) \qquad (11.98)$$

Thus for the $2l + 1$ spherical harmonics of a particular l value, the transformation matrix is given by

$$\Gamma^{(1)}(\alpha) = \begin{pmatrix} e^{il\alpha} & & & & & \\ & e^{i(l-1)\alpha} & & & & 0 \\ & & \cdot & & & \\ & & & \cdot & & \\ & & & & e^{i(0)\alpha} & \\ & & & & \cdot & \\ 0 & & & & & \cdot \\ & & & & & & e^{-il\alpha} \end{pmatrix} \qquad (11.99)$$

since m takes the values $l, l - 1, \ldots, -l$.

---

[43] P_{lm} contains all the normalization factors in addition to the ϑ dependence.

The character of the representation formed by these $2l + 1$ functions is

$$\chi^{(l)}(\alpha) = \exp(il\alpha) + \exp[i(l-1)\alpha] + \cdots \exp(-il\alpha)$$

$$= \exp(-il\alpha)\{\exp(i2l\alpha) + \exp[i(2l-1)\alpha]$$

$$+ \cdots \exp[i(2l-2l)\alpha]\}$$

$$= [\exp(-il\alpha)]\frac{\exp[i(2l+1)\alpha]}{\exp(i\alpha)-1} \qquad (11.100)$$

where the last equality is easily verified by division. Further algebraic manipulation upon Eq. (11.100) leads to

$$\chi^{(l)}(\alpha) = \frac{\sin(l+\frac{1}{2})\alpha}{\sin\frac{1}{2}\alpha} \qquad (11.101)$$

Equation (11.101) gives the character of the $2l + 1$ odd-dimensional representations of the full-rotation group. For a free ion or atom with no spin-orbit coupling (an imaginary situation, of course) the $2l + 1$ states of a given l are degenerate. However, in a field having anything less than full-rotational (spherical) symmetry, the $2l + 1$ degeneracy is broken, in general. Our treatment so far has been couched implicitly in a one-electron terminology, hence the use of the symbols l and m. For a many electron system, we would substitute L and M_L for l and m, respectively.

Thus we may easily show how $S, P, D, F \ldots$ atomic (or ionic) terms behave in the medium crystal field case described in Sec. 10.5 A. For a crystalline field of a given point-group symmetry, the behavior of the $2L + 1$ states of given L is determined simply by reducing the representations given by Eq. (11.101) under the largest rotational subgroup of the group of interest. Thus for an octahedral (O_h) field, we use the point group O. In Table 11.11 we have listed the characters of the reducible representations for the $S, P, D,$ and F states using Eq. (11.101). Note that $\chi(E) = 2L + 1$, which is the result obtained if $\alpha = 0$.

TABLE 11.11

Characters of representations of S, P, D, F states in an octahedral field.

O	E	$8C_3$	$3C_2$	$6C_2$	$6C_4$
$\Gamma(S)$	1	1	1	1	1
$\Gamma(P)$	3	0	-1	-1	1
$\Gamma(D)$	5	-1	1	1	-1
$\Gamma(F)$	7	1	-1	-1	-1

Reducing these representations we find

$$\Gamma(S) = A_1$$
$$\Gamma(P) = T_1 \tag{11.102}$$
$$\Gamma(D) = E + T_2$$
$$\Gamma(F) = A_2 + T_1 + T_2$$

We see that S and P states remain unsplit in an octahedral field, while the D and F states split up as shown.

The same results apply to the behavior of one-electron states or orbitals. Thus the free-ion d orbitals lead to e and t_2 orbitals in the octahedral field, where we use lower case letters according to our previous conventions for one-electron orbitals. Note that the symmetry designations of O may be correlated with those of O_h and the result agrees with our notation in Fig. 10.10(a). The energy-level diagram in that figure thus embodies the group theory results, but in addition it shows the correct ordering of the one-electron states.

An identical treatment for the case of the tetrahedral (T_d) field would lead to corresponding results. For $\Gamma(D)$ the fivefold degeneracy is broken to yield the T_2 and E states shown in Fig. 10.10(b). Rather than treat this case in any detail we consider the behavior of the atomic states in a field of D_{4h} symmetry as a final example. Table 11.12 contains the characters for this case. Reducing these representations yields

$$\Gamma(S) = A_1$$
$$\Gamma(P) = A_2 + E$$
$$\Gamma(D) = A_1 + B_1 + B_2 + E \tag{11.103}$$
$$\Gamma(F) = A_2 + B_1 + B_2 + 2E$$

In this case we observe that P-state terms may be split in this rather low symmetry field. Note also that the result for the D state agrees with the previous results in Fig. 10.10(a), if we correlate D_4 and D_{4h}.

TABLE 11.12

Characters of the representations of S, P, D, F states in a tetragonal (D_{4h}) field.

D_4	E	$2C_4$	C_2	$2C_2'$	$2C_2''$
$\Gamma(S)$	1	1	1	1	1
$\Gamma(P)$	3	1	-1	-1	-1
$\Gamma(D)$	5	-1	1	1	1
$\Gamma(F)$	7	-1	-1	-1	-1

EXAMPLE 11-6

It is instructive to consider the group-theoretical result of applying a D_{4h} perturbation to a D-state system in an octahedral field. In this case the representations under the group O must be reduced to find the possible irreducible representations under the group D_4. This is most easily done by rewriting the character table for the group O such that the elements not common to D_4 are left out. This has been done in Table 11.13, in which we also reproduce the D_4 table for convenience. In producing these entries in the table it is necessary to note that the C_4^2 operation of the group O correlates with both C_4^2 and $2C_2'$ of D_4; and that the $6C_2$ of O correlates with $2C_2''$ of D_4. Reducing E and T_2 of O in the usual manner we find

$$E(O) = A_1 + B_1$$
$$\text{(11.104)}$$
$$T_2(O) = B_2 + E$$

These are, of course, the same representations given by reducing $\Gamma(D)$ directly for the free ion as shown in Eqs. (11.103). But now we have the additional knowledge as to which states of the octahedral system undergo a further splitting. The results are in agreement again with the previous diagram in Fig. 10.10(b).

We can now see the possible electronic transitions involved with transition metal-ion systems. For an octahedral d^1 ion such as $Ti(H_2O)_6^{+3}$, the lowest molecular orbital originating from the metal d orbitals is t_{2g} according to our earlier reasoning. Hence we may write the electron configuration of the ground state as $(t_{2g})^1$, which would have the term symbol $^2T_{2g}$. The lowest excited state is ^2E_g which arises from the $(e_g)^1$ configuration. Simple theoretical calculations[44] indicate that the separation of these two states

TABLE 11.13
Tables for reducing representations of O under the group D_4.

		E	$2C_4$	C_4^2	$2C_2'$	$2C_2''$
Group O	E	2	0	2	2	0
	T_2	3	-1	-1	-1	1
Group D_4	A_1	1	1	1	1	1
	A_2	1	1	1	-1	-1
	B_1	1	-1	1	1	-1
	B_2	1	-1	1	-1	1
	E	2	0	-2	0	0

[44] C. Ballhausen, *Introduction to Ligand Field Theory* (McGraw-Hill, New York, 1962).

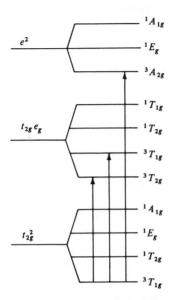

FIG. 11.18. Energy states of a d^2 ion in an octahedral field.

(identified usually as Δ) is about 20 000 cm^{-1}, which would lead to absorption at about 5000 Å.

However, if we investigate the transition-moment integral we find that the electronic transition probability vanishes since the direct product

$$T_{2g} \times E_g = T_{1g} + T_{2g}$$

does not contain $\Gamma^{(x)}$, $\Gamma^{(y)}$, or $\Gamma^{(z)}$, which is T_{1u} in the present case. Actually, this result can be seen more easily by noting simply that μ_x, μ_y, and μ_z are antisymmetric (u) with respect to inversion, while the states of a complex with inversion symmetry are all symmetric (g) with respect to inversion.

This result notwithstanding, the transition $^2T_{2g} \rightarrow {}^2E_g$ does occur rather weakly (molar extinction coefficient ~ 1) at 4900 Å. The mechanism which causes the selection rule breakdown here is that previously mentioned, electronic-vibrational coupling.

Considering a less trivial example, we treat now a d^2 case, such as $V(OC(NH_2)_2)_6{}^{+3}$. Three configurations are possible, $(t_{2g})^2$, $(t_{2g})^1(e_g)^1$, $(e_g)^2$, which have the following terms:

$$(t_{2g})^2: {}^3T_{1g},\ {}^1T_{2g},\ {}^1E_g,\ {}^1A_1$$

$$(t_{2g})^1(e_g)^1: {}^3T_{2g},\ {}^3T_{1g},\ {}^1T_{2g},\ {}^1T_{1g} \tag{11.105}$$

$$(e_g)^2: {}^3A_{2g},\ {}^1E_g,\ {}^1A_{1g}$$

Here the orbital term symmetries are obtained by reducing the direct products as usual. The multiplicities are either singlet or triplet, the precise choices being made so as to satisfy the Pauli principle. The reader is urged to consult the literature[45] for the systematic means of determining the correct multiplicities. These states have been sketched qualitatively in Fig. 11.18 in the expected order.

Since the ground state is $^3T_{1g}$ the only spin-allowed transitions would be those to other triplet states. Again, there are no orbitally permitted transitions so any observed transitions are expected to be rather weak. In Fig. 11.18 we have shown the possible spin-permitted transitions by vertical arrows. In the spectrum of $V(OC(NH_2)_2)_6^{+3}$ the lowest two of these bands have been identified at 16 200 cm^{-1} (6170 Å) and 24 200 cm^{-1} (4130 Å).

Entirely similar methods may be applied to other transition and rare-earth metal complexes. We do not pursue this further, but simply note in passing that spectra of this type have been of great value in elucidating the characteristics of coordination compounds and coordination chemistry.

SUPPLEMENTARY REFERENCES

R. P. Bauman, *Absorption Spectroscopy* (Wiley, New York, 1962).

W. R. Brode, *Chemical Spectroscopy* (Wiley, New York, 1943).

H. Eyring, J. Walter, and G. E. Kimball, *Quantum Chemistry* (Wiley, New York, 1944).

A. G. Gaydon, *Dissociation Energies and Spectra of Diatomic Molecules* (Dover, New York, 1950).

G. Herzberg, *Electronic Spectra and Electronic Structure of Polyatomic Molecules* (Van Nostrand, Princeton, N.J., 1966).

G. Herzberg, *Spectra of Diatomic Molecules* (Van Nostrand, Princeton, N.J., 1950).

C. Sandorfy, *Electronic Spectra and Quantum Chemistry* (Prentice-Hall, Englewood Cliffs, N.J., 1964).

R. M. Silverstein and G. C. Bassler, *Spectrometric Identification of Organic Compounds* (Wiley, New York, 1967), 2nd ed.

J. C. Slater, *Quantum Theory of Molecules and Solids* (McGraw-Hill, New York, 1963), Vol. 1.

[45] E. Wigner and E. E. Witmer, *Z. Physik* **51**, 859 (1928).

APPENDIX A

CLASSICAL MECHANICS

In classical mechanics the force on a particle of mass m is given by Newton's law

$$\mathbf{f} = m\ddot{\mathbf{r}} \qquad (A.1)$$

where $\ddot{\mathbf{r}}$ is the second time derivative of the position vector. For n particles of mass m_i, the components of the force may be expressed as

$$
\begin{aligned}
f_i{}^{(x)} &= m_i\ddot{x}_i \\
f_i{}^{(y)} &= m_i\ddot{y}_i \qquad i = 1\ldots n \\
f_i{}^{(z)} &= m_i\ddot{z}_i
\end{aligned}
\qquad (A.2)
$$

The kinetic energy for the system of n particles is written

$$
\begin{aligned}
T &= \frac{1}{2}\sum_i m_i\dot{r}_i{}^2 \\
&= \frac{1}{2}\sum_i m_i(\dot{x}_i{}^2 + \dot{y}_i{}^2 + \dot{z}_i{}^2)
\end{aligned}
\qquad (A.3)
$$

For most systems of physical interest the particles exist in a potential field, such as gravitational, electrostatic, mechanical, etc. If the potential field is such that the potential energy is a function of position coordinates only,

$$V = V\,(x_i, y_i, z_i) \qquad (A.4)$$

the system is known as *conservative*. In this case the force on the particles is given by

$$f_i^{(x)} = -\frac{\partial V}{\partial x_i}$$

$$f_i^{(y)} = -\frac{\partial V}{\partial y_i} \qquad i = 1 \dots n \qquad (A.5)$$

$$f_i^{(z)} = -\frac{\partial V}{\partial z_i}$$

While Eqs. (A.1)–(A.5) are suitable for solving many problems of classical mechanics there are other useful forms into which they may be cast. For example, using Eqs. (A.2), (A.3), and (A.5) it is easily shown that

$$\frac{d}{dt}\frac{\partial T}{\partial \dot{x}_i} + \frac{\partial V}{\partial x_i} = 0$$

$$\frac{d}{dt}\frac{\partial T}{\partial \dot{y}_i} + \frac{\partial V}{\partial y_i} = 0 \qquad i = 1 \dots n \qquad (A.6)$$

$$\frac{d}{dt}\frac{\partial T}{\partial \dot{z}_i} + \frac{\partial V}{\partial z_i} = 0$$

Remembering now that $T = T(\dot{x}_i, \dot{y}_i, \dot{z}_i)$ and $V = V(x_i, y_i, z_i)$, Eqs. (A.6) may be rewritten

$$\frac{d}{dt}\frac{\partial(T-V)}{\partial \dot{x}_i} - \frac{\partial(T-V)}{\partial x_i} = 0 \qquad \text{etc.} \qquad (A.7)$$

Defining the Lagrangian function,

$$L = T - V \qquad (A.8)$$

Eqs. (A.7) becomes

$$\frac{d}{dt}\frac{\partial L}{\partial \dot{x}_i} - \frac{\partial L}{\partial x_i} = 0 \quad \text{etc.} \qquad (A.9)$$

Note that L is a function of coordinates and velocities only. The most important feature of writing Newton's equations in the form of (A.9) is that the form of the equations is invariant to the coordinate system. Thus for any set of $3n$ orthogonal coordinates $q_1, q_2, q_3, \dots, q_{3n}$, Newton's equations in Lagrangian form are written

$$\frac{d}{dt}\frac{\partial L}{\partial \dot{q}_i} - \frac{\partial L}{\partial q_i} = 0 \qquad i = 1 \ldots 3n \qquad (A.10)$$

Note that L is a function of q_i and $\dot{q}_i$ only and that there are $3q_i$'s for each of the n particles. Note that this result is significant, since the Eqs. (A.1)–(A.5) have other forms for different (non-Cartesian) coordinate systems. We call the coordinates q_1, q_2, ..., q_{3n} *generalized* coordinates.

It is useful now to define a *generalized* momentum by the equation

$$p_k = \frac{\partial L}{\partial \dot{q}_k} \qquad (A.11)$$

where p_k is known as the momentum *conjugate* to the coordinate q_k. In Cartesian coordinates the Lagrangian function is (in one dimension)

$$L = \sum_i m_i \dot{x}_i^2 - V(x_i)$$

so that

$$p_k = m_k \dot{x}_k$$

which is the well-known definition of linear momentum in Cartesian coordinates. The definition (A.11) is found to be consistent with other common definitions of momentum also, such as that for angular momentum.

While Newton's equations in Cartesian or Lagrangian form are useful for many classical problems, there is yet one other extremely useful form. To obtain this we define

$$\mathfrak{IC} = \sum_k (p_k \dot{q}_k) - L(q_k, \dot{q}_k) \qquad (A.12)$$

and we choose the independent variables of this function $\mathfrak{IC}$ to be the p_i and q_i. Thus it is possible to write

$$d\mathfrak{IC} = \sum_k \left(\frac{\partial \mathfrak{IC}}{\partial p_k} dp_k + \frac{\partial \mathfrak{IC}}{\partial q_k} dq_k \right) \qquad (A.13)$$

But from Eq. (A.12) the total derivative is

$$d\mathfrak{IC} = \sum_k (p_k d\dot{q}_k + \dot{q}_k dp_k - \dot{p}_k dq_k - p_k d\dot{q}_k) \qquad (A.14)$$

where the fourth term on the right-hand side utilizes Eq. (A.11) and the third term uses the identity

$$\frac{\partial L}{\partial \dot{q}_k} = p_k \qquad (A.15)$$

which arises by substituting Eq. (A.11) into (A.10). Comparing Eqs. (A.13) and (A.14) we find

$$\dot{q}_k = \frac{\partial \mathcal{3C}}{\partial p_k} \tag{A.16}$$

$$\dot{p}_k = -\frac{\partial \mathcal{3C}}{\partial q_k} \tag{A.17}$$

$$k = 1 \ldots 3n$$

These are known as Hamilton's equations of motion, and the function $\mathcal{3C}(p, q)$ is known as the Hamiltonian function.

The Hamiltonian has two important properties. First, since $\mathcal{3C}$ is a function of p_k and q_k,

$$\frac{d\mathcal{3C}}{dt} = \sum_k \left(\frac{\partial \mathcal{3C}}{\partial q_k} \frac{dq_k}{dt} + \frac{\partial \mathcal{3C}}{\partial p_k} \frac{dp_k}{dt} \right) \tag{A.18}$$

and using Eqs. (A.16) and (A.17), we obtain

$$\frac{d\mathcal{3C}}{dt} = \sum_k \left(-\dot{p}_k \dot{q}_k + \dot{q}_k \dot{p}_k \right) = 0 \tag{A.19}$$

Thus for conservative systems the Hamiltonian is a constant of the motion, that is, it is independent of time.

Finally we can show the most important meaning of $\mathcal{3C}$. We begin by rewriting Eq. (A.12) as

$$\mathcal{3C} = \sum_k \frac{\partial T}{\partial \dot{q}_k} \dot{q}_k - T + V \tag{A.20}$$

In Cartesian coordinates the kinetic energy is given by Eq. (A.3). Since the generalized coordinates q are related to the Cartesian coordinates by an expression of the form

$$x_i = f(q_1, q_2, \ldots, q_{3n}) \tag{A.21}$$

the velocities may be expressed as

$$\dot{x}_i = \sum_j \frac{\partial x_i}{\partial q_j} \frac{dq_j}{dt} = \sum_j c_{ij} \dot{q}_j \tag{A.22}$$

Consequently the kinetic energy in terms of generalized velocities is of the form

$$T = \sum_{i,j} a_{ij} \dot{q}_i \dot{q}_j \tag{A.23}$$

and $\partial T/\partial \dot{q}_k$ is given by

$$\frac{\partial T}{\partial \dot{q}_k} = \sum_i a_{ik}\dot{q}_i + \sum_j a_{kj}\dot{q}_j \qquad (A.24)$$

Thus Eq. (A.20) becomes

$$\mathcal{3C} = \sum_{i,k} a_{ik}\dot{q}_i\dot{q}_k + \sum_{j,k} a_{kj}\dot{q}_j\dot{q}_k - T + V$$

$$= T + T - T + V$$

$$= T + V \qquad (A.25)$$

The extremely important result is that the Hamiltonian represents the total energy $(T + V)$ of the system. It is from this classical development that we begin most of the quantum-mechanical treatments.

Supplementary References

J. H. Goldstein, *Classical Mechanics* (Addison-Wesley, Reading, Mass., 1959).

L. D. Landau and E. M. Lifschitz, *Mechanics* (Addison-Wesley, Reading, Mass., 1958).

J. L. Synge and B. A. Griffith, *Principles of Mechanics* (McGraw-Hill, New York, 1959), 3rd ed.

J. M. Anderson, *Mathematics for Quantum Chemistry* (Benjamin, New York, 1966).

APPENDIX B

SOME FREQUENTLY USED QUANTUM MECHANICAL THEOREMS

We have made frequent use of several quantum-mechanical definitions and theorems. The more important of these have been reproduced briefly below.

1. Hermitian operators.

An operator $\tilde{A}$ is Hermitian if

$$\int \psi_i^* \tilde{A} \psi_j d\tau = \int \psi_j \tilde{A}^* \psi_i^* d\tau \tag{B.1}$$

Note that the left-hand integral might be considered to be the A_{ij} element of the matrix $\mathbf{A}$ in the ψ representation. The right-hand integral is then the A_{ji}^* element, and we see that the Hermitian operator definition of (B.1) is equivalent to the Hermitian matrix definition given in (C.25).

2. The eigenvalues of Hermitian operators are real.

Suppose

$$\tilde{R}\psi_i = R_i \psi_i \tag{B.2}$$

thus

$$\tilde{R}^* \psi_i^* = R_i^* \psi_i^* \tag{B.3}$$

Then from (B.2) and (B.3) we find

$$\int \psi_i^* \tilde{R} \psi_i d\tau = R_i \int \psi_i^* \psi_i d\tau$$

and

$$\int \psi_i \tilde{R}^* \psi_i^* d\tau = R_i^* \int \psi_i \psi_i^* d\tau \tag{B.4}$$

But if $\tilde{R}$ is Hermitian the left-hand sides of the equations of (B.4) must be equal, so

$$R_i = R_i^* \tag{B.5}$$

since the integrals on the right-hand sides of (B.4) are obviously identical. This theorem is equivalent to that of Appendix C which states that the diagonal elements of Hermitian matrices are real.

As a corollary to this theorem, we may prove that the diagonal elements of squared Hermitian matrices are greater than zero. If **A** is the matrix,

$$(\mathbf{A}^2)_{ii} = \sum_k A_{ik} A_{ki} \tag{B.6}$$

But since **A** is Hermitian,

$$(\mathbf{A}^2)_{ii} = \sum_k A_{ik} A_{ik}^*$$

$$= \sum_k |A_{ik}|^2 \tag{B.7}$$

and consequently the diagonal elements are necessarily positive.

3. Expansion theorem.

The expansion of a set of functions $\psi_1, \psi_2, \ldots, \psi_n$ in terms of a complete orthonormal set $\phi_1, \phi_2, \ldots, \phi_n$ may be written

$$\psi_i = \sum_j a_{ji} \phi_j \tag{B.8}$$

If we multiply both sides of (B.8) by ϕ_k^* and integrate over all space we obtain

$$\int \phi_k^* \psi_i d\tau = \sum_j a_{ji} \int \phi_k^* \phi_j d\tau \tag{B.9}$$

But the integral on the right-hand side is zero except when $k = j$ in which case it has the value unity. Thus the coefficients in the expansion (B.8) are given by

$$a_{ki} = \int \phi_k^* \psi_i d\tau \tag{B.10}$$

4. If two operators $\tilde{A}$ and $\tilde{B}$ commute, any eigenfunction of $\tilde{A}$ is simultaneously an eigenfunction of $\tilde{B}$.

To prove this theorem we define the eigenvalue and eigenfunction of the operator $\tilde{A}$

$$\tilde{A}\psi_i = A_i\psi_i \tag{B.11}$$

We wish to show that the ψ_i are also eigenfunctions of the operator $\tilde{B}$. We shall restrict our considerations to the case when the eigenvalues are non-degenerate, but the theorem may also be proved for a case when degeneracy is present.

Since $\tilde{A}$ and $\tilde{B}$ commute, we have

$$\tilde{A}\tilde{B}\psi_i = \tilde{B}\tilde{A}\psi_i \tag{B.12}$$

which by use of (B.11) becomes

$$\tilde{A}(\tilde{B}\psi_i) = A_i(\tilde{B}\psi_i) \tag{B.13}$$

Thus we see that $\tilde{B}\psi_i$ is an eigenfunction of $\tilde{A}$. Since by (B.11) the eigenfunctions are defined as the ψ_i, $\tilde{B}\psi_i$ can differ from ψ_i only by a constant factor for nondegenerate states. Thus

$$\tilde{B}\psi_i = B_i\psi_i \tag{B.14}$$

and we have proved the theorem.

The analogous theorem for commuting matrices may also be stated; namely, if two matrices commute, there exists a similarity transformation which simultaneously diagonalizes both matrices. Or, somewhat differently, if **A** and **B** commute and **A** is diagonal, then **B** is also diagonal.

These theorems for commuting operators or matrices are easily extendable to the case when there are more than two commuting operators or matrices. Thus if

$$[\tilde{A}, \tilde{B}] = \tilde{A}\tilde{B} - \tilde{B}\tilde{A} = 0, [\tilde{A}, \tilde{C}] = 0, \quad \text{and} \quad [\tilde{B}, \tilde{C}] = 0,$$

$$\tilde{A}\psi_i = A_i\psi_i$$

$$\tilde{B}\psi_i = B_i\psi_i$$

$$\tilde{C}\psi_i = C_i\psi_i$$

5. All operator equations are valid when rewritten as matrix equations.

For example, if

$$\tilde{A}\tilde{B} = \tilde{C} \tag{B.15}$$

which really means, of course,

$$\tilde{A}\tilde{B}\psi_i = \tilde{C}\psi_i$$

then

$$\mathbf{AB} = \mathbf{C} \tag{B.16}$$

must be true. Suppose the matrix elements of $\tilde{A}$ are obtained in the ψ representation. Then

$$A_{ki} = \int \psi_k^* \tilde{A} \psi_i d\tau$$

$$= \langle \psi_k \mid \tilde{A} \mid \psi_i \rangle \tag{B.17}$$

Written differently, the matrix elements are

$$\tilde{A}\psi_i = \sum_k A_{ik}\psi_k \tag{B.18}$$

Similarly the matrix elements of $\tilde{B}$ are given by

$$\tilde{B}\psi_j = \sum_i B_{ij}\psi_i \tag{B.19}$$

and the matrix elements of C are given by

$$\tilde{C}\psi_j = \sum_k C_{kj}\psi_k \tag{B.20}$$

According to Eq. (B.15), $\tilde{C} = \tilde{A}\tilde{B}$ so the left-hand side of Eq. (B.20) becomes

$$\tilde{A}\tilde{B}\psi_j = \tilde{A} \sum_i B_{ij}\psi_i$$

$$= \sum_k \sum_i A_{ki}B_{ij}\psi_k$$

$$= \sum_k \left[\sum_i A_{ki}B_{ij}\right]\psi_k \tag{B.21}$$

using also (B.18) and (B.19). Comparing the right-hand sides of Eqs. (B.20) and (B.21) we see that

$$C_{kj} = \sum_i A_{ki}B_{ij} \tag{B.22}$$

which is the general element of

$$\mathbf{C} = \mathbf{AB}$$

Thus the matrix equation is of identical form to the operator equation (B.15).

The theorem is valid for any form of the operator equation, as may be verified by extension of these arguments.

APPENDIX C

PROPERTIES OF MATRICES

For our purposes we define a matrix as an ordered array of numbers or elements in square or rectangular form. If A is a square matrix of order N, then the elements of A are labeled A_{ij} (or a_{ij}) where i is the row label and j the column label, and i and j run from $1, \ldots, N$. Sometimes it is convenient to use other numbers as indices, but the form is always the same. The same form applies to an $N \times M$ rectangular matrix, R, with elements R_{ij}, where $i = 1, \ldots, N$ and $j = 1, \ldots, M$. A column matrix,

$$
C = \begin{pmatrix} c_1 \\ c_2 \\ c_3 \\ \vdots \\ c_N \end{pmatrix}
\tag{C.1}
$$

needs only one index, as does a row matrix.

We list below some properties of matrices and some special square matrices.

1. Matrix multiplication.
Two square matrices are multiplied together to form a third matrix,

$$
A = BC
\tag{C.2}
$$

according to the rule

$$A_{ij} = \sum_k B_{ik} C_{kj} \tag{C.3}$$

That is, the ijth element of **A** is obtained by summing up the products of the elements of the ith *row* of **B** and the jth *column* of **C**. In short, the multiplication is of the form $ROW \times COLUMN$. The multiplication of several matrices follows the rule

$$\mathbf{A} = \mathbf{BCD} = (\mathbf{BC})\mathbf{D} = \mathbf{B}(\mathbf{CD}) \tag{C.4}$$

Two rectangular matrices may be multiplied according to (C.3) only if **B** is $N \times M$ and **C** is $M \times N$. The resulting matrix is square and has a dimension equal to the row dimension N of the matrix **B**. Note that a row matrix times a column matrix yields a matrix containing only one element.

    2.   Multiplication by a constant.

If c is a constant, then

$$\mathbf{B} = c\mathbf{A} \tag{C.5}$$

means

$$B_{ij} = cA_{ij} \tag{C.6}$$

    3.   Addition of matrices (square).

$$\mathbf{A} = \mathbf{B} + \mathbf{C} \tag{C.7}$$

means

$$A_{ij} = B_{ij} + C_{ij} \tag{C.8}$$

    4.   Noncommutative behavior.

In general, matrix multiplication is not commutative, that is

$$\mathbf{AB} \neq \mathbf{BA} \tag{C.9}$$

    5.   Unit and null matrices.

Corresponding to the 1 and 0 of ordinary algebra are the unit matrix

$$\mathbf{I} = \begin{pmatrix} 1 & 0 & 0 & 0 & \cdots \\ 0 & 1 & 0 & 0 & \\ 0 & 0 & 1 & 0 & \\ 0 & 0 & 0 & 1 & \\ \vdots & & & & \ddots \end{pmatrix} \tag{C.10}$$

and the null matrix

$$0 = \begin{pmatrix} 0 & 0 & 0 & \cdots \\ 0 & 0 & 0 & \\ 0 & 0 & 0 & \\ \cdot & & \cdot & \\ \cdot & & \cdot & \\ \cdot & & & \cdot \end{pmatrix} \tag{C.11}$$

Very often we simply write **0** as 0 for simplicity, and occasionally **E** is used in place of **I**.

6. Inverse matrix.

The matrix inverse to **A** is symbolized by $\mathbf{A}^{-1}$ and is defined by

$$\mathbf{AA}^{-1} = \mathbf{A}^{-1}\mathbf{A} = \mathbf{I} \tag{C.12}$$

Every nonsingular matrix has an inverse. A nonsingular matrix is one whose determinant is nonzero, that is

$$|\mathbf{A}| \neq 0 \tag{C.13}$$

See references for practical means of finding $\mathbf{A}^{-1}$ given **A**.

7. Diagonal matrix.

A diagonal matrix is one for which

$$A_{ij} = 0, \, i \neq j \tag{C.14}$$

while at least one of the diagonal elements A_{ii} are nonzero.

8. Complex conjugate matrix.

The complex conjugate of **A** is denoted by $\mathbf{A}^*$ and defined by

$$(\mathbf{A}^*)_{ij} = A_{ij}^* \tag{C.15}$$

9. Transpose matrix.

The transpose of **A** is denoted by $\mathbf{A}'$ and defined by

$$(\mathbf{A}')_{ij} = A_{ji} \tag{C.16}$$

Note that the transpose is obtained simply by interchanging rows and columns, or by reflecting the matrix across the principal diagonal. Note also that the transpose of a column (row) matrix is a row (column) matrix.

10. A special multiplication property.

It may be proved that

$$A^{-1}B^{-1}C^{-1} = (CBA)^{-1} \qquad (C.17)$$

$$A'B'C' = (CBA)' \qquad (C.18)$$

$$A^*B^*C^* = (ABC)^* \qquad (C.19)$$

11. Unitary matrix.

A matrix is said to be unitary if its inverse is equal to its complex conjugate transpose, that is, if U is unitary,

$$U^{-1} = U'^* \qquad (C.20)$$

or

$$(U^{-1})_{ij} = (U'^*)_{ij} \qquad (C.21)$$

The following results may then be proved:

$$\sum_j U_{ij}{}^* U_{ij} = 1$$

$$\sum_i U_{ij}{}^* U_{ij} = 1$$

$$\sum_j U_{ij}{}^* U_{kj} = 0 \qquad (C.22)$$

$$\sum_i U_{ij}{}^* U_{ik} = 0$$

If a unitary matrix is composed entirely of *real* elements, it is known as a *real orthogonal* matrix. For a real orthogonal matrix, R,

$$R^{-1} = R' \qquad (C.23)$$

12. Hermitian matrix.

A matrix is Hermitian if it is equal to its complex conjugate transpose,

$$H = H^{*\prime} \qquad (C.24)$$

which means

$$H_{ij} = H_{ji}{}^* \qquad (C.25)$$

One of the important properties of Hermitian matrices is that the diagonal elements must be real, as is seen by setting $i = j$ in Eq. (C.25). In quantum mechanics all matrices corresponding to observable quantities must be Hermitian.

13. Unitary and real orthogonal transformations.

If V is a real vector whose components are arranged as a column matrix, then a real orthogonal matrix R transforms V to the new vector W in the following way.

$$W = RV \qquad (C.26)$$

The reverse transformation is given by

$$\mathbf{R'W} = \mathbf{V} \tag{C.27}$$

If the vector $\mathbf{V}$ is not real, the transformation matrix $\mathbf{R}$ must be replaced by a unitary matrix. Rather than considering (C.26) to be a vector transformation, it may be considered as a coordinate transformation.

14. Similarity transformations.

A similarity transformation of the matrix $\mathbf{B}$ by the matrix $\mathbf{A}$ is written

$$\mathbf{C} = \mathbf{A^{-1}BA} \tag{C.28}$$

where $\mathbf{C}$ is the resulting matrix. We restrict $\mathbf{A}$ to be unitary or real orthogonal. If $\mathbf{A}$ is real orthogonal we may replace $\mathbf{A^{-1}}$ by $\mathbf{A'}$. Two important properties of the (unitary) similarity transformation are the following:

(a) The trace of a matrix is invariant to a similarity transformation,

$$\operatorname{Tr}\mathbf{C} = \operatorname{Tr}\mathbf{B} \tag{C.29}$$

where $\mathbf{B}$ and $\mathbf{C}$ are related by a similarity transformation as in Eq. (C.28) and the trace is defined by

$$\operatorname{Tr}\mathbf{C} = \sum_i C_{ii} \tag{C.30}$$

(b) Matrix equations are invariant to similarity transformations.
Thus if

$$\mathbf{D} = \mathbf{AB} + \mathbf{C} \tag{C.31}$$

then

$$\mathbf{X^{-1}DX} = (\mathbf{X^{-1}AX})(\mathbf{X^{-1}BX}) + \mathbf{X^{-1}CX} \tag{C.32}$$

is true also.

15. Matrix diagonalization.

If $\mathbf{L}$ is a Hermitian or unitary matrix it is always possible to find a similarity transformation which transforms $\mathbf{L}$ into a diagonal matrix, $\mathbf{\Lambda}$. That is,

$$\mathbf{A^{-1}LA} = \mathbf{\Lambda} \tag{C.33}$$

This process is known as matrix diagonalization, and the elements Λ_i are known as the eigenvalues of $\mathbf{L}$. In Sec. 2.2 it has been shown that the eigenvalues are the roots Λ of the secular equation

$$|\mathbf{L} - \Lambda\mathbf{I}| = 0 \tag{C.34}$$

16. Multiplication and addition of block-diagonal matrices.

Consider two matrices $\mathbf{A}$ and $\mathbf{B}$ which have identical block-diagonal structures as shown in Eqs. (C.35). These matrices have no nonzero elements except in the *square* blocks along the principal diagonal. These blocks form

$$
\mathbf{A} = \begin{pmatrix} \boxed{\mathbf{A}^{(1)}} & & \mathbf{0} \\ & \boxed{\mathbf{A}^{(2)}} & \\ \mathbf{0} & & \boxed{\mathbf{A}^{(3)}} \end{pmatrix}
\tag{C.35a}
$$

$$
\mathbf{B} = \begin{pmatrix} \boxed{\mathbf{B}^{(1)}} & & \mathbf{0} \\ & \boxed{\mathbf{B}^{(2)}} & \\ \mathbf{0} & & \boxed{\mathbf{B}^{(3)}} \end{pmatrix}
\tag{C.35b}
$$

matrices of lower dimension and have been labeled $\mathbf{A}^{(i)}$ and $\mathbf{B}^{(i)}$. By the definition of matrix addition it is clear that the sum of $\mathbf{A}$ and $\mathbf{B}$ must necessarily have zeros except in the positions not within the smaller blocks, that is, if

$$
\mathbf{D} = \mathbf{A} + \mathbf{B}
\tag{C.36}
$$

$\mathbf{D}$ will, in general, be in the same block-diagonal form. The elements of the blocks of $\mathbf{D}$ are given by

$$
\mathbf{D}^{(j)} = \mathbf{A}^{(j)} + \mathbf{B}^{(j)}
\tag{C.37}
$$

If the product of $\mathbf{A}$ and $\mathbf{B}$ is formed

$$
\mathbf{C} = \mathbf{AB}
\tag{C.38}
$$

the matrix $\mathbf{C}$ is also in block-diagonal form. This follows by close inspection of the product rule (C.3). Thus any element in $\mathbf{C}$ which does not fall within a block must be zero since the terms in the sum (C.3) may have only the forms $A_{ij}(0)$, $(0)(0)$, or $(0)B_{jk}$. The nonzero values of $\mathbf{C}$ are therefore obtained by multiplying the small submatrices,

$$
\mathbf{C}^{(j)} = \mathbf{A}^{(j)}\mathbf{B}^{(j)}
\tag{C.39}
$$

The result of these observations is that matrix equations involving block-diagonal matrices (of identical form) may be manipulated simply by working with the individual blocks.

17. A useful matrix multiplication identity.

If $\mathbf{a}$ is a 3-dimensional row matrix, $\mathbf{a}'$ is its transpose (a column matrix), and $\mathbf{B}$ is a 3-dimensional square matrix,

$$\mathbf{aBa}' = (a_x a_y a_z) \begin{pmatrix} B_{xx} & B_{xy} & B_{xz} \\ B_{yx} & B_{yy} & B_{yz} \\ B_{zx} & B_{zy} & B_{zz} \end{pmatrix} \begin{pmatrix} a_x \\ a_y \\ a_z \end{pmatrix}$$

$$= a_x^2 B_{xx} + a_y^2 B_{yy} + a_z^2 B_{zz} \qquad \text{(C.40)}$$
$$+ a_x a_y B_{xy} + a_y a_x B_{yx} + a_x a_z B_{xz}$$
$$+ a_z a_x B_{zx} + a_y a_z B_{yz} + a_z a_y B_{zy}$$

Very often $\mathbf{aBa}'$ is written $\mathbf{aBa}$, leaving for the reader the task of writing $\mathbf{a}$ in its correct row-column form. The form (C.40) is not uncommon in physical problems, the matrix $\mathbf{B}$ normally having the mathematical properties of a tensor. $\mathbf{a}$ has the properties of a vector $\mathbf{a}$, and we see that the result of the matrix multiplication is a *scalar*. If $\mathbf{B}$ is a symmetrical tensor, that is, $B_{ij} = B_{ji}$,

$$\mathbf{aBa} = a_x^2 B_{xx} + a_y^2 B_{yy} + a_z^2 B_{zz} + 2a_x a_y B_{xy} + 2a_x a_z B_{xz} + 2a_y a_z B_{yz} \quad \text{(C.41)}$$

Supplementary References
J. M. Anderson, *Mathematics for Quantum Chemistry* (Benjamin, New York, 1966).
H. Margenau and G. M. Murphy, *The Mathematics of Physics and Chemistry* (Van . Nostrand, Princeton, 1956).

APPENDIX D

SIGN CHANGE IN ANGULAR MOMENTUM COMMUTATION RELATIONS

In Sec. 4.2 the angular momentum commutation relations in molecule-fixed axes were written down with no proof. They may be derived directly from Eqs. (4.56) and the commutators between the direction cosines and the angular momenta as given by Eqs. (4.111). For example, $[\tilde{P}_x, \tilde{P}_y]$ can be evaluated by substituting Eq. (4.107a) and expanding.

$$[\tilde{P}_x, \tilde{P}_y] = \tilde{P}_x\tilde{P}_y - \tilde{P}_y\tilde{P}_x$$

$$= (\Phi_{xX}\tilde{P}_X + \Phi_{xY}\tilde{P}_Y + \Phi_{xZ}\tilde{P}_Z)(\Phi_{yX}\tilde{P}_X + \Phi_{yY}\tilde{P}_Y + \Phi_{yZ}\tilde{P}_Z) \quad \text{(D.1)}$$

$$- (\Phi_{yX}\tilde{P}_X + \Phi_{yY}\tilde{P}_Y + \Phi_{yZ}\tilde{P}_Z)(\Phi_{xX}\tilde{P}_X + \Phi_{xY}\tilde{P}_Y + \Phi_{xZ}\tilde{P}_Z)$$

Upon expanding and grouping terms we get

$$
\begin{aligned}
[\tilde{P}_x, \tilde{P}_y] = {} & \Phi_{xX}(\tilde{P}_X\Phi_{yY})\tilde{P}_Y - \Phi_{yX}(\tilde{P}_X\Phi_{xY})\tilde{P}_Y \\
& + \Phi_{xX}(\tilde{P}_X\Phi_{yZ})\tilde{P}_Z - \Phi_{yX}(\tilde{P}_X\Phi_{xZ})\tilde{P}_Z \\
& + \Phi_{xY}(\tilde{P}_Y\Phi_{yX})\tilde{P}_X - \Phi_{yY}(\tilde{P}_Y\Phi_{xX})\tilde{P}_X \\
& + \Phi_{xY}(\tilde{P}_Y\Phi_{yZ})\tilde{P}_Z - \Phi_{yY}(\tilde{P}_Y\Phi_{xZ})\tilde{P}_Z \\
& + \Phi_{xZ}(\tilde{P}_Z\Phi_{yX})\tilde{P}_X - \Phi_{yZ}(\tilde{P}_Z\Phi_{xX})\tilde{P}_X \\
& + \Phi_{xZ}(\tilde{P}_Z\Phi_{yY})\tilde{P}_Y - \Phi_{yZ}(\tilde{P}_Z\Phi_{xY})\tilde{P}_Y \quad \text{(D.2)} \\
& + \Phi_{xX}(\tilde{P}_X\Phi_{yX})\tilde{P}_X - \Phi_{yX}(\tilde{P}_X\Phi_{xX})\tilde{P}_X \\
& + \Phi_{xY}(\tilde{P}_Y\Phi_{yY})\tilde{P}_Y - \Phi_{yY}(\tilde{P}_Y\Phi_{xY})\tilde{P}_Y \\
& + \Phi_{xZ}(\tilde{P}_Z\Phi_{yZ})\tilde{P}_Z - \Phi_{yZ}(\tilde{P}_Z\Phi_{xZ})\tilde{P}_Z
\end{aligned}
$$

But because $[\tilde{P}_F, \Phi_{gF}] = 0$ from Eq. (4.111c) and all the direction cosines commute with each other, the last six terms go to zero. If now the terms in parentheses are substituted by means of Eqs. (4.111d), the Φ_{gF} can be factored from the $\tilde{P}_F$. For example, the first term becomes

$$\Phi_{zX}(\tilde{P}_X\Phi_{yY})\tilde{P}_Y = \Phi_{zX}\left(-\frac{\hbar}{i}\Phi_{yZ} + \Phi_{yY}\tilde{P}_X\right)\tilde{P}_Y$$

$$= -\frac{\hbar}{i}\Phi_{zX}\Phi_{yZ}\tilde{P}_Y + \Phi_{zX}\Phi_{yY}\tilde{P}_X\tilde{P}_Y \qquad (D.3)$$

After all these substitutions are made and the terms are regrouped we find

$$\begin{aligned}
[\tilde{P}_x, \tilde{P}_y] = &(\Phi_{xY}\Phi_{yZ} - \Phi_{xZ}\Phi_{yY})(\tilde{P}_Y\tilde{P}_Z - \tilde{P}_Z\tilde{P}_Y) \\
&- (\Phi_{zX}\Phi_{yZ} - \Phi_{xZ}\Phi_{yX})(\tilde{P}_Z\tilde{P}_X - \tilde{P}_X\tilde{P}_Z) \\
&+ (\Phi_{zX}\Phi_{yY} - \Phi_{xY}\Phi_{yX})(\tilde{P}_X\tilde{P}_Y - \tilde{P}_Y\tilde{P}_X) \\
&+ \frac{\hbar}{i}(-\Phi_{xY}\Phi_{yZ} - \Phi_{xZ}\Phi_{yY} - \Phi_{yY}\Phi_{xZ} + \Phi_{yZ}\Phi_{xY})\tilde{P}_X \quad (D.4) \\
&+ \frac{\hbar}{i}(-\Phi_{zX}\Phi_{yZ} + \Phi_{xZ}\Phi_{yX} + \Phi_{yX}\Phi_{xZ} - \Phi_{yZ}\Phi_{xX})\tilde{P}_Y \\
&+ \frac{\hbar}{i}(\Phi_{zX}\Phi_{yY} - \Phi_{xY}\Phi_{yX} - \Phi_{yX}\Phi_{xY} + \Phi_{yY}\Phi_{zX})\tilde{P}_Z
\end{aligned}$$

Significant simplification of Eq. (D.4) is possible now by using the fact that each element of a real orthogonal matrix (such as the direction cosine matrix) is equal to the cofactor of the element in the determinant of the matrix. This is most easily shown by recalling that for a real orthogonal matrix the determinant of the matrix has value unity. Therefore, for example

$$1 = \Phi_{zX}C_{zX} + \Phi_{yX}C_{yX} + \Phi_{zX}C_{zX}$$

where the C_{gF} are the cofactors of the various direction cosines. Also, the sum of the squares of the elements in any row has value unity. Thus,

$$1 = \Phi_{zX}{}^2 + \Phi_{zY}{}^2 + \Phi_{zZ}{}^2$$

By comparison, we see that

$$\Phi_{zX} = C_{zX}$$

$$\Phi_{zY} = C_{zY}$$

$$\Phi_{zZ} = C_{zZ}$$

Thus, for example,

$$(\Phi_{xY}\Phi_{yZ} - \Phi_{xZ}\Phi_{yY}) = \Phi_{zX} \tag{D.5}$$

Using equations of the form of (D.5), Eq. (D.4) becomes

$$[\tilde{P}_x, \tilde{P}_y] = \Phi_{zX}(\tilde{P}_Y\tilde{P}_Z - \tilde{P}_Z\tilde{P}_Y) + \Phi_{zY}(\tilde{P}_Z\tilde{P}_X - \tilde{P}_X\tilde{P}_Z)$$

$$+ \Phi_{zZ}(\tilde{P}_X\tilde{P}_Y - \tilde{P}_Y\tilde{P}_X) + 2\frac{\hbar}{i}\Phi_{zX}\tilde{P}_X \tag{D.6}$$

$$+ 2\frac{\hbar}{i}\Phi_{zY}\tilde{P}_Y + 2\frac{\hbar}{i}\Phi_{zZ}\tilde{P}_Z$$

By using the commutation relations (4.56), we get finally

$$[\tilde{P}_x, \tilde{P}_y] = \frac{\hbar}{i}(\Phi_{zX}\tilde{P}_X + \Phi_{zY}\tilde{P}_Y + \Phi_{zZ}\tilde{P}_Z) \tag{D.7}$$

which leads to the desired result by substituting Eq. (4.107a).

$$[\tilde{P}_x, \tilde{P}_y] = \frac{\hbar}{i}\tilde{P}_z \tag{D.8}$$

Supplementary Reference
J. H. Van Vleck, *Rev. Mod. Phys.* **23,** 213 (1951).

APPENDIX E

CHARACTER TABLES

In this appendix we list the character tables of most of the symmetry groups which are of importance in molecular spectroscopy. These include the finite crystallographic point groups, the groups involving C_5 rotations (which cannot exist in a crystal), and the two infinite groups needed for linear molecules.

The character tables are in the standard format. Except for the two infinite groups, one-dimensional representations are labeled by A or B, two-dimensional by E and three-dimensional by T. Each character table also shows the transformation properties of the coordinates x, y, and z; the infinitesimal rotations R_x, R_y, and R_z; and certain quadratic functions of x, y and z.

Some groups may be written as direct products of groups of lower order. Thus $\{C_{4h}\} = \{C_4\} \times \{i\}$ where the brackets indicate the collection of elements making up a group, and the direct product multiplication means that the elements of the C_{4h} group are obtained by forming all the products of the elements of $\{C_4\}$ with those of $\{i\}$. Since C_4 consists of the elements E, C_4, C_2, $C_4{}^3$ and i consists of E and i, C_{4h} contains EE, EC_4, EC_2, $EC_4{}^3$, iE, iC_4, iC_2, $iC_4{}^3$. Of course, $ER = R$ for any R so

$$EE = E$$

$$EC_4 = C_4$$

$$EC_2 = C_2$$

$$EC_4{}^3 = C_4{}^3$$

$$iE = i$$

By geometrical considerations it is also clear that

$$iC_4 = S_4{}^3$$

$$iC_2 = \sigma_h$$

$$iC_4{}^3 = S_4$$

These are the elements of the group C_{4h}. The characters (of the irreducible matrix representations) are similarly obtained by multiplying each of those of C_4 by each of those of i. This procedure yields the table shown for C_{4h}. We have saved some space by merely indicating how some of the character tables may be formed by the direct-product principle.

It is seen from the following tables that many groups have matrix representations with complex or imaginary characters. In such cases, as with the group C_4, the complex or imaginary characters occur as complex-conjugate pairs for a given class. Thus in the group C_4 the last two irreducible representations have characters of i and $i^* = -i$ for C_4 and $i^* = -i$ and i for $C_4{}^3$. Note that all the mathematical properties (see chapter 6) of the group and the character table require the four irreducible representations as written. Nevertheless, we always group the complex conjugate pairs of representations together and give them *one* doubly degenerate label only, such as E for the C_4 group. This is because in practically all physical applications the complex-conjugate pairs may be regarded as a doubly degenerate representation whose characters are the *sum* of the characters of the individual representations. For C_4, the characters of the E representation may be taken as

$$(E) = 1 + 1 = 2$$

$$(C_4) = i + {-i} = 0$$

$$(C_2) = -1 - 1 = -2$$

$$(C_4{}^3) = -i + i = 0$$

With these characters it is easily shown that the real variables x, y form a basis for the E representation. The single symbol for the complex-conjugate pairs is also the correct symbol for deducing the degeneracy of a state belonging to the complex-conjugate pair representations.

C_1	E
A	1

$C_{1h} = \sigma_h = C_s$	E	σ_h		
A'	1	1	x, y, R_z	$x^2, y^2,$ z^2, xy
A''	1	-1	z, R_x, R_y	yz, xz

$S_2 = i = C_i$	E	i		
A_g	1	1	R_x, R_y, R_z	x^2, y^2, z^2
A_u	1	-1	x, y, z	xy, xz, yz

C^2	E	C_2		
A	1	1	R_z, z	x^2, y^2, z^2, xy
B	1	-1	$\begin{cases} x, y \\ \\ R_x, R_y \end{cases}$	xz, yz

C_3	E	C_3	$C_3{}^2$			
A	1	1	1		R_z, z	$x^2 + y^2, z^2$
E	$\begin{cases} 1 \\ \\ 1 \end{cases}$	$\begin{matrix} \omega \\ \\ \omega^2 \end{matrix}$	$\begin{matrix} \omega^2 \\ \\ \omega \end{matrix}$	$(\omega = e^{2\pi i/3})$	$\begin{matrix} (x, y) \\ \\ (R_x, R_y) \end{matrix}$	$\begin{matrix} (xz, yz) \\ \\ (x^2 - y^2, xy) \end{matrix}$

C_4	E	C_2	C_4	$C_4{}^3$		
A	1	1	1	1	R_z, z	$x^2 + y^2, z^2$
B	1	1	-1	-1		$x^2 - y^2, xy$
E	$\begin{cases} 1 \\ \\ 1 \end{cases}$	$\begin{matrix} -1 \\ \\ -1 \end{matrix}$	$\begin{matrix} i \\ \\ -i \end{matrix}$	$\begin{matrix} -i \\ \\ i \end{matrix}$	$\begin{matrix} (x, y) \\ \\ (R_x, R_y) \end{matrix}$	(xz, yz)

C_5	E	C_5	$C_5{}^2$	$C_5{}^3$	$C_5{}^4$			
A	1	1	1	1	1		R_z, z	$x^2 + y^2, z^2$
E'	$\begin{cases} 1 \\ \\ 1 \end{cases}$	$\begin{matrix} \omega \\ \\ \omega^4 \end{matrix}$	$\begin{matrix} \omega^2 \\ \\ \omega^3 \end{matrix}$	$\begin{matrix} \omega^3 \\ \\ \omega^2 \end{matrix}$	$\begin{matrix} \omega^4 \\ \\ \omega \end{matrix}$	$(\omega = e^{2\pi i/5})$	$\begin{matrix} (x, y) \\ \\ (R_x, R_y) \end{matrix}$	(xz, yz)
E''	$\begin{cases} 1 \\ \\ 1 \end{cases}$	$\begin{matrix} \omega^2 \\ \\ \omega^3 \end{matrix}$	$\begin{matrix} \omega^4 \\ \\ \omega \end{matrix}$	$\begin{matrix} \omega \\ \\ \omega^4 \end{matrix}$	$\begin{matrix} \omega^3 \\ \\ \omega^2 \end{matrix}$			$\{(x^2 - y^2, xy)$

C_6	E	C_6	C_3	C_2	$C_3{}^2$	$C_6{}^5$			
A	1	1	1	1	1	1		R_z, z	$x^2 + y^2, z^2$
B	1	-1	1	-1	1	-1			
E' $\Big\{$	1	ω	ω^2	ω^3	ω^4	ω^5	$(\omega = e^{2\pi i/6})$	$\Big\{ \begin{matrix}(x, y)\\ \\ (R_x, R_y)\end{matrix}$	(xz, yz)
	1	ω^5	ω^4	ω^3	ω^2	ω			
E'' $\Big\{$	1	ω^2	ω^4	1	ω^2	ω^4			$\Big\{ (x^2 - y^2, xy)$
	1	ω^4	ω^2	1	ω^4	ω^2			

C_{2v}	E	C_2	$\sigma_v(xz)$	$\sigma_v'(yz)$		
A_1	1	1	1	1	z	x^2, y^2, z^2
A_2	1	1	-1	-1	R_z	xy
B_1	1	-1	1	-1	R_y, x	xz
B_2	1	-1	-1	1	R_x, y	yz

C_{3v}	E	$2C_3$	$3\sigma_v$		
A_1	1	1	1	z	$x^2 + y^2, z^2$
A_2	1	1	-1	R_z	
E	2	-1	0	$\begin{cases}(x, y)\\ \\ (R_x, R_y)\end{cases}$	$\begin{matrix}(x^2 - y^2, xy)\\ \\ (xz, yz)\end{matrix}$

C_{4v}	E	C_2	$2C_4$	$2\sigma_v$	$2\sigma_d$		
A_1	1	1	1	1	1	z	$x^2 + y^2, z^2$
A_2	1	1	1	-1	-1	R_z	
B_1	1	1	-1	1	-1		$x^2 - y^2$
B_2	1	1	-1	-1	1		xy
E	2	-2	0	0	0	$\begin{cases}(x, y)\\ \\ (R_x, R_y)\end{cases}$	(xz, yz)

C_{5v}	E	$2C_5$	$2C_5^2$	$5\sigma_v$			
A_1	1	1	1	1		z	$x^2 + y^2, z^2$
A_2	1	1	1	-1	$x = \dfrac{2\pi}{5}$	R_z	
E_1	2	$2\cos x$	$2\cos 2x$	0		$\begin{cases}(x, y) \\ (R_x, R_y)\end{cases}$	(xz, yz)
E_2	2	$2\cos 2x$	$2\cos 4x$	0			$(x^2 - y^2, xy)$

C_{6v}	E	C_2	$2C_3$	$2C_6$	$3\sigma_d$	$3\sigma_v$		
A_1	1	1	1	1	1	1	z	$x^2 + y^2, z^2$
A_2	1	1	1	1	-1	-1	R_z	
B_1	1	-1	1	-1	-1	1		
B_2	1	-1	1	-1	1	-1		
E_1	2	-2	-1	1	0	0	$(x, y), (R_x, R_y)$	(xz, yz)
E_2	2	2	-1	-1	0	0		$(x^2 - y^2, xy)$

C_{2h}	E	C_2	i	σ_h		
A_g	1	1	1	1	R_z	x^2, y^2, z^2, xy
B_g	1	-1	1	-1	R_x, R_y	xz, yz
A_u	1	1	-1	-1	z	
B_u	1	-1	-1	1	x, y	

C_{3h}	E	C_3	C_3^2	σ_h	S_3	S_3^5		
A'	1	1	1	1	1	1	R_z	$x^2 + y^2, z^2$
A''	1	1	1	-1	-1	-1	z	
E'	$\begin{cases}1 \\ 1\end{cases}$	$\begin{matrix}\omega \\ \omega^2\end{matrix}$	$\begin{matrix}\omega^2 \\ \omega\end{matrix}$	$\begin{matrix}1 \\ 1\end{matrix}$	$\begin{matrix}\omega \\ \omega^2\end{matrix}$	$\begin{matrix}\omega^2 \, (\omega = e^{2\pi i/3}) \\ \omega\end{matrix}$	(x, y)	$(x^2 - y^2, xy)$
E''	$\begin{cases}1 \\ 1\end{cases}$	$\begin{matrix}\omega \\ \omega^2\end{matrix}$	$\begin{matrix}\omega^2 \\ \omega\end{matrix}$	$\begin{matrix}-1 \\ -1\end{matrix}$	$\begin{matrix}-\omega \\ -\omega^2\end{matrix}$	$\begin{matrix}-\omega^2 \\ -\omega\end{matrix}$	(R_x, R_y)	(xz, yz)

C_{4h}	E	C_4	C_2	C_4^3	i	S_4^3	σ_h	S_4		
A_g	1	1	1	1	1	1	1	1	R_z	$x^2 + y^2, z^2$
B_g	1	-1	1	-1	1	-1	1	-1		$x^2 - y^2, xy$
E_g	1	i	-1	$-i$	1	i	-1	$-i$		
	1	$-i$	-1	i	1	$-i$	-1	i	(R_x, R_y)	(xz, yz)
A_u	1	1	1	1	-1	-1	-1	-1	z	
B_u	1	-1	1	-1	-1	1	-1	1		
E_u	1	i	-1	$-i$	-1	$-i$	1	i		
	1	$-i$	-1	i	-1	i	1	$-i$	(x, y)	

$$C_{5h} = C_5 \times \sigma_h$$

$$C_{6h} = C_6 \times i$$

S_4	E	C_2	S_4	S_4^3		
A	1	1	1	1	R_z	$x^2 + y^2, z^2$
B	1	1	-1	-1	z	
E	1	-1	i	$-i$	(x, y)	(xy, yz)
	1	-1	$-i$	i	(R_x, R_y)	$(x^2 - y^2, xy)$

$$S_6 = C_3 \times i$$

$V = D_2$	E	$C_2(z)$	$C_2(y)$	$C_2(z)$		
A	1	1	1	1		x^2, y^2, z^2
B_1	1	1	-1	-1	z, R_z	xy
B_2	1	-1	1	-1	y, R_y	xz
B_3	1	-1	-1	1	x, R_x	yz

D_3	E	$2C_3$	$3C_2$			
A_1	1	1	1			$x^2 + y^2, z^2$
A_2	1	1	-1	z, R_z		
E	2	-1	0	(x, y)	(R_x, R_y)	$(x^2 - y^2, xy)\,(xz, yz)$

D_4	E	$2C_4$	$C_2(=C_4^2)$	$2C_2'$	$2C_2''$		
A_1	1	1	1	1	1		$x^2 + y^2, z^2$
A_2	1	1	1	-1	-1	z, R_z	
B_1	1	-1	1	1	-1		$x^2 - y^2$
B_2	1	-1	1	-1	1		xy
E	2	0	-2	0	0	$(x, y), (R_x, R_y)$	(xz, yz)

D_5	E	$2C_5$	$2C_5^2$	$5C_2$		
A_1	1	1	1	1		$x^2 + y^2, z^2$
A_2	1	1	1	-1	R_z, z	
E_1	2	$2\cos x$	$2\cos 2x$	$0 \left(x = \dfrac{2\pi}{5} \right)$	$(x, y), (R_x, R_y)$	(xz, yz)
E_2	2	$2\cos 2x$	$2\cos 4x$	0		$(x^2 - y^2, xy)$

D_6	E	$2C_6$	$2C_3$	C_2	$3C_2'$	$3C_2''$		
A_1	1	1	1	1	1	1		$x^2 + y^2, z^2$
A_2	1	1	1	1	-1	-1	z, R_z	
B_1	1	-1	1	-1	1	-1		
B_2	1	-1	1	-1	-1	1		
E_1	2	1	-1	-2	0	0	$(x, y) (R_x, R_y)$	(xz, yz)
E_2	2	-1	-1	2	0	0		$(x^2 - y^2, xy)$

$V_d = D_{2d}$	E	C_2	$2S_4$	$2C_2'$	$2\sigma_d$		
A_1	1	1	1	1	1		$x^2 + y^2, z^2$
A_2	1	1	1	-1	-1	R_z	
B_1	1	1	-1	1	-1		$x^2 - y^2$
B_2	1	1	-1	-1	1	z	xy
E	2	-2	0	0	0	$(x, y), (R_x, R_y)$	(xz, yz)

D_{3d}	E	$2C_3$	$3C_2$	i	$2S_6$	$3\sigma_d$		
A_{1g}	1	1	1	1	1	1		$x^2 + y^2, z^2$
A_{2g}	1	1	-1	1	1	-1	R_z	
E_g	2	-1	0	2	-1	0	(R_x, R_y)	$(x^2 - y^2, xy), (xz, yz)$
A_{1u}	1	1	1	-1	-1	-1		
A_{2u}	1	1	-1	-1	-1	1	z	
E_u	2	-1	0	-2	1	0	(x, y)	

$$D_{2h} = D_2 \times i$$

D_{3h}	E	$2C_3$	$3C_2$	σ_h	$2S_3$	$3\sigma_v$		
A_1'	1	1	1	1	1	1		$x^2+y^2,\ z^2$
A_2'	1	1	-1	1	1	-1	R_z	
E'	2	-1	0	2	-1	0	(x,y)	$(x^2-y^2,\ xy)$
A_1''	1	1	1	-1	-1	-1		
A_2''	1	1	-1	-1	-1	1	z	
E''	2	-1	0	-2	1	0	(R_x, R_y)	(xz, yz)

D_{4h}	E	$2C_4$	C_2	$2C_2'$	$2C_2''$	i	$2S_4$	σ_h	$2\sigma_v$	$2\sigma_d$		
A_{1g}	1	1	1	1	1	1	1	1	1	1		$x^2+y^2,\ z^2$
A_{2g}	1	1	1	-1	-1	1	1	1	-1	-1	R_z	
B_{1g}	1	-1	1	1	-1	1	-1	1	1	-1		x^2-y^2
B_{2g}	1	-1	1	-1	1	1	-1	1	-1	1		xy
E_g	2	0	-2	0	0	2	0	-2	0	0	(R_x, R_y)	(xz, yz)
A_{1u}	1	1	1	1	1	-1	-1	-1	-1	-1		
A_{2u}	1	1	1	-1	-1	-1	-1	-1	1	1	z	
B_{1u}	1	-1	1	1	-1	-1	1	-1	-1	1		
B_{2u}	1	-1	1	-1	1	-1	1	-1	1	-1		
E_u	2	0	-2	0	0	-2	0	2	0	0	(x, y)	

$$D_{5h} = D_5 \times \sigma_h$$

D_{6h}	E	$2C_6$	$2C_3$	C_2	$3C_2'$	$3C_2''$	i	$2S_3$	$2S_6$	σ_h	$3\sigma_d$	$3\sigma_v$		
A_{1g}	1	1	1	1	1	1	1	1	1	1	1	1		$x^2+y^2,\ z^2$
A_{2g}	1	1	1	1	-1	-1	1	1	1	1	-1	-1	R_z	
B_{1g}	1	-1	1	-1	1	-1	1	-1	1	-1	1	-1		
B_{2g}	1	-1	1	-1	-1	1	1	-1	1	-1	-1	1		
E_{1g}	2	1	-1	-2	0	0	2	1	-1	-2	0	0	(R_x, R_y)	(xz, yz)
E_{2g}	2	-1	-1	2	0	0	2	-1	-1	2	0	0		(x^2-y^2, xy)
A_{1u}	1	1	1	1	1	1	-1	-1	-1	-1	-1	-1		
A_{2u}	1	1	1	1	-1	-1	-1	-1	-1	-1	1	1	z	
B_{1u}	1	-1	1	-1	1	-1	-1	1	-1	1	-1	1		
B_{2u}	1	-1	1	-1	-1	1	-1	1	-1	1	1	-1		
E_{1u}	2	1	-1	-2	0	0	-2	-1	1	2	0	0	(x, y)	
E_{2u}	2	-1	-1	2	0	0	-2	1	1	-2	0	0		

T	E	$3C_2$	$4C_3$	$4C_3'$		
A	1	1	1	1		
	1	1	ω	ω^2		
E						
	1	1	ω^2	ω	$(\omega = e^{2\pi i/3})$	(R_x, R_y, R_z)
T	3	-1	0	0		(x, y, z)

T_d	E	$8C_3$	$3C_2$	$6S_4$	$6\sigma_d$		
A_1	1	1	1	1	1		$x^2 + y^2, z^2$
A_2	1	1	1	-1	-1		
E	2	-1	2	0	0		$(2z^2 - x^2 - y^2, x^2 - y^2)$
T_1	3	0	-1	1	-1	(R_x, R_y, R_z)	
T_2	3	0	-1	-1	1	(x, y, z)	(xy, xz, yz)

$$T_h = T \times i$$

O	E	$8C_3$	$3C_2$	$6C_2$	$6C_4$		
A_1	1	1	1	1	1		
A_2	1	1	1	-1	-1		
E	2	-1	2	0	0		
T_1	3	0	-1	-1	$+1$	(R_x, R_y, R_z)	
T_2	3	0	-1	$+1$	-1	(x, y, z)	

O_h	E	$8C_3$	$6C_2$	$6C_4$	$3C_2(=C_4^2)$	i	$6S_4$	$8S_6$	$3\sigma_h$	$6\sigma_d$		
A_{1g}	1	1	1	1	1	1	1	1	1	1		$x^2+y^2+z^2$
A_{2g}	1	1	-1	-1	1	1	-1	1	1	-1		
E_g	2	-1	0	0	2	2	0	-1	2	0		$(2z^2-x^2-y^2,$ $x^2-y^2)$
T_{1g}	3	0	-1	1	-1	3	1	0	-1	-1	(R_x,R_y,R_z)	
T_{2g}	3	0	1	-1	-1	3	-1	0	-1	1		(xz, yz, xy)
A_{1u}	1	1	1	1	1	-1	-1	-1	-1	-1		
A_{2u}	1	1	-1	-1	1	-1	1	-1	-1	1		
E_u	2	-1	0	0	2	-2	0	1	-2	0		
T_{1u}	3	0	-1	1	-1	-3	-1	0	1	1	(x, y, z)	
T_{2u}	3	0	1	-1	-1	-3	1	0	1	-1		

$C_{\infty v}$	E	$2C_\phi$	$\infty\,\sigma_v$		
$A_1 = \Sigma^+$	1	1	1		$x^2 + y^2,\, z^2$
$A_2 = \Sigma^-$	1	1	-1	z	
$E_1 = \Pi$	2	$2\cos\phi$	0	$(x, y),\, (R_x, R_y)$	(xz, yz)
$E_2 = \Delta$	2	$2\cos 2\phi$	0		$(x^2 - y^2,\, xy)$
$\cdots$	$\cdots$	$\cdots$	$\cdots$		

$D_{\infty h}$	E	$2C_\phi$	$\infty\,\sigma_v$	i	$2S_\phi$	$\infty\,C_2$		
Σ_g^+	1	1	1	1	1	1		$x^2 + y^2,\, z^2$
Σ_u^+	1	1	1	-1	-1	-1	z	
Σ_g^-	1	1	-1	1	1	-1	R_z	
Σ_u^-	1	1	-1	-1	-1	1		
Π_g	2	$2\cos\phi$	0	2	$-2\cos\phi$	0	(R_x, R_y)	(xz, yz)
Π_u	2	$2\cos\phi$	0	-2	$2\cos\phi$	0	(x, y)	
Δ_g	2	$2\cos 2\phi$	0	2	$2\cos 2\phi$	0		$(x^2 - y^2,\, xy)$
Δ_u	2	$2\cos 2\phi$	0	-2	$-2\cos 2\phi$	0		

APPENDIX F

ATOMIC MASSES AND NATURAL ABUNDANCES OF THE STABLE ISOTOPES OF ELEMENTS WITH ATOMIC NUMBER ≦ 54[1]

Atomic number	Element	Mass number	Atomic mass	Abundance %
1	H	1	1.007825	99.99
		2	2.01410	0.01
2	He	3	3.01603	1×10^{-4}
		4	4.00260	~100
3	Li	6	6.01513	7.5
		7	7.01601	92.5
4	Be	9	9.01219	100
5	B	10	10.01294	~19
		11	11.00931	~81
6	C	12	12.00000	98.89
		13	13.00335	1.11
7	N	14	14.00307	99.6
		15	15.00011	0.4
8	O	16	15.99491	99.76
		17	16.99914	0.04
		18	17.99916	0.20
9	F	19	18.99840	100
10	Ne	20	19.99244	90.9
		21	20.99395	0.3
		22	21.99138	8.8

[1] *Handbook of Chemistry and Physics* (Chemical Rubber Co., Cleveland, 1969), 50th ed.; J. H. E. Mattauch, W. Thiele, and A. H. Wapstra, *Nucl. Phys.* **67,** 1 (1965); **67,** 32 (1965); **67,** 73 (1965).

APPENDIX F. (Continued)

Atomic number	Element	Mass number	Atomic mass	Abundance %
11	Na	23	22.98977	100
12	Mg	24	23.98504	78.6
		25	24.98584	10.1
		26	25.98259	11.3
13	Al	27	26.98153	100
14	Si	28	27.97693	92.3
		29	28.97649	4.7
		30	29.97376	3.0
15	P	31	30.97376	100
16	S	32	31.97207	95.0
		33	32.97146	0.8
		34	33.96786	4.2
17	Cl	35	34.96885	75.4
		37	36.96590	24.6
18	Ar	36	35.96755	0.34
		38	37.96272	0.06
		40	39.96238	99.6
19	K	39	38.96371	93.1
		40	39.96	0.01
		41	40.96184	6.9
20	Ca	40	39.96259	97.0
		42	41.95863	0.6
		43	42.95878	0.1
		44	43.95549	2.0
		46	45.9537	0.003
		48	47.9524	0.2
21	Sc	45	44.95592	100
22	Ti	46	45.95263	7.95
		47	46.9518	7.75
		48	47.94795	73.5
		49	48.94787	5.5
		50	49.9448	5.3
23	V	50	49.9472	0.2
		51	50.9440	99.8
24	Cr	50	49.9461	4.3
		52	51.9405	83.8
		53	52.9407	9.6
		54	53.9389	2.4
25	Mn	55	54.9381	100
26	Fe	54	53.9396	5.8
		56	55.9349	91.7
		57	56.9354	2.2
		58	57.9333	0.3
27	Co	59	58.9332	100

APPENDIX F. (*Continued*)

Atomic number	Element	Mass number	Atomic mass	Abundance %
28	Ni	58	57.9353	67.8
		60	59.9332	26.2
		61	60.9310	1.2
		62	61.9283	3.7
		64	63.9280	1.2
29	Cu	63	62.9298	69.1
		65	64.9278	30.9
30	Zn	64	63.9291	48.9
		66	65.9260	27.8
		67	66.9271	4.1
		68	67.9249	18.6
		70	69.9253	0.6
31	Ga	69	68.9257	60.2
		71	70.9249	39.8
32	Ge	70	69.9243	20.6
		72	71.9217	27.4
		73	72.9234	7.7
		74	73.9212	36.7
		76	75.9214	7.7
33	As	75	74.9216	100
34	Se	74	73.9225	0.9
		76	75.9192	9.0
		77	76.9199	7.6
		78	77.9173	23.5
		80	79.9165	49.8
		82	81.9167	9.2
35	Br	79	78.9183	50.5
		81	80.9163	49.5
36	Kr	78	77.9204	0.4
		80	79.9164	2.3
		82	81.9135	11.6
		83	82.9141	11.6
		84	83.9115	56.9
		86	85.9106	17.4
37	Rb	85	84.9117	72.2
		87	86.91	27.8
38	Sr	84	83.9134	0.6
		86	85.9094	9.9
		87	86.9089	7.0
		88	87.9056	82.6
39	Y	89	88.9054	100
40	Zr	90	89.9043	51.5
		91	90.9053	11.2
		92	91.9046	17.1
		94	93.9061	17.4
		96	95.9082	2.8

APPENDIX F. (*Continued*)

Atomic number	Element	Mass number	Atomic mass	Abundance %
41	Nb	93	92.9060	100
42	Mo	92	91.9063	15.9
		94	93.9047	9.1
		95	94.9046	15.7
		96	95.9046	16.5
		97	96.9058	9.4
		98	97.9055	23.8
		100	99.9076	9.6
43	Tc	99	unstable	
44	Ru	96	95.9076	5.7
		98	97.9055	2.2
		99	98.9061	12.8
		100	99.9030	12.7
		101	100.9041	17.0
		102	101.9037	31.3
		104	103.9055	18.3
45	Rh	103	102.9048	100
46	Pd	102	101.9049	0.8
		104	103.9036	9.3
		105	104.9046	22.6
		106	105.9032	27.2
		108	107.9030	26.8
		110	109.9045	13.5
47	Ag	107	106.9041	51.4
		109	108.9047	48.6
48	Cd	106	105.907	1.2
		108	107.9050	0.9
		110	109.9030	12.4
		111	110.9042	12.8
		112	111.9028	24.1
		113	112.9046	12.3
		114	113.9036	28.9
		116	115.905	7.6
49	In	113	112.9043	4.2
		115	114.9041	95.8
50	Sn	112	111.9040	0.9
		114	113.9030	0.6
		115	114.9035	0.3
		116	115.9021	14.2
		117	116.9031	7.6
		118	117.9018	24.0
		119	118.9034	8.6
		120	119.9021	33.0
		122	121.9034	4.7
		124	123.9052	6.0

APPENDIX F. *(Continued)*

Atomic number	Element	Mass number	Atomic mass	Abundance %
51	Sb	121	120.9038	57.2
		123	122.9041	42.8
52	Te	120	119.9045	0.1
		122	121.9030	2.5
		123	122.9042	0.9
		124	123.9028	4.6
		125	124.9044	7.0
		126	125.9032	18.7
		128	127.9047	31.8
		130	129.9067	34.5
53	I	127	126.9044	100
54	Xe	124	123.9061	0.1
		126	125.9042	0.1
		128	127.9035	1.9
		129	128.9048	26.4
		130	129.9035	4.1
		131	130.9051	21.2
		132	131.9042	26.9
		134	133.9054	10.4
		136	135.9072	8.9

APPENDIX G

PHYSICAL CONSTANTS AND CONVERSION FACTORS[1]

Physical Constants

Speed of light	$c = 2.997925 \times 10^{10}$ cm sec^{-1}
Electron charge	$e = 1.60210 \times 10^{-20}$ emu
	$= 4.80298 \times 10^{-10}$ esu
Avogadro's number	$N = 6.02252 \times 10^{23}$ mole^{-1}
Electron mass	$m_e = 9.10908 \times 10^{-28}$ g
Proton mass	$m_p = 1.67252 \times 10^{-24}$ g
Planck's constant	$h = 6.62559 \times 10^{-27}$ erg sec
Bohr radius	$a_0 = 0.52917 \times 10^{-8}$ cm
Bohr magneton	$\beta_e = 9.2732 \times 10^{-21}$ erg G^{-1}
Nuclear magneton	$\beta_n = 5.0505 \times 10^{-24}$ erg G^{-1}
Boltzmann constant	$k = 1.38054 \times 10^{-16}$ erg deg^{-1}
	$= 0.695029$ cm^{-1} deg^{-1}
Gas constant	$R = 1.9872$ cal deg^{-1} mole^{-1}
	$= 0.082053$ liter atm mole^{-1} deg^{-1}
	$= 8.31434 \times 10^7$ erg deg^{-1} mole^{-1}

---

[1] See E. R. Cohen and J. W. M. Dumond, *Rev. Mod. Phys.* **37,** 537 (1965).

Conversion Factors

$$1 \text{ eV} = 8065.73 \text{ cm}^{-1}$$
$$= 23061 \text{ cal/mole}$$
$$= 1.60210 \times 10^{-12} \text{ erg}$$

$$1 \text{ cm}^{-1} = 29979.25 \text{ MHz}$$
$$= 2.8591 \text{ cal/mole}$$

$$1 \text{ hartree} = 27.212 \text{ eV}$$
$$= 627.52 \text{ kcal/mole}$$

$$BI = 505375 \text{ MHz amu } Å^2$$

APPENDIX H

DERIVATION OF EQ. (8.128)

The derivation of Eq. (8.128) has been given by Wilson, Decius, and Cross.[1] We give it here for completeness. To begin with we need to show the relationship of the G matrix to the kinetic energy, namely

$$2T = \dot{S}'G^{-1}\dot{S} \qquad (H.1)$$

For simplicity we treat only the case where all quantities are real.

In chapter 3 we wrote the kinetic energy as

$$2T = \sum_i^{3N} \dot{q}_i^2 \qquad (H.2a)$$

Since the conjugate momentum is $p_j = \dot{q}_j$, this may also be written

$$2T = \sum_i p_i^2 \qquad (H.2b)$$

If we write the velocities $\dot{q}_i$ as elements of the column matrix $\dot{q}$, so that $\dot{q}'$ is a row matrix, we get

$$2T = \dot{q}'\dot{q} \qquad (H.3a)$$

In a similar fashion

$$2T = p'p \qquad (H.3b)$$

[1] E. B. Wilson, Jr., J. C. Decius, and P. C. Cross, *Molecular Vibrations* (McGraw-Hill, New York, 1955), Appendices VIII and IX.

The $3N$ mass-weighted coordinates are related to the $3N - 6$ internal coordinates by a linear transformation,

$$S_i = \sum_{j}^{3N} D_{ij}q_j \qquad i = 1\ldots 3N - 6 \tag{H.4}$$

or

$$\mathbf{S} = \mathbf{Dq} \tag{H.5}$$

where $\mathbf{D}$ is a rectangular matrix. Using Eq. (A.11) and the fact that $q_i = q_i(S_1\ldots S_{3N-6})$, we write

$$p_j = \frac{\partial T}{\partial \dot{q}_j} = \sum_{i}^{3N-6} \frac{\partial T}{\partial \dot{S}_i}\frac{\partial \dot{S}_i}{\partial \dot{q}_j} \tag{H.6}$$

Defining as in Eq. (A.11) the momentum conjugate to the coordinate S_i,

$$P_i = \frac{\partial T}{\partial \dot{S}_i} \tag{H.7}$$

and noting that

$$\frac{\partial \dot{S}_i}{\partial \dot{q}_j} = \frac{\partial S_i}{\partial q_j} = D_{ij} \tag{H.8}$$

by use of Eq. (H.4), we rewrite Eq. (H.6) as

$$p_j = \sum_{i}^{3N-6} P_i D_{ij} \qquad j = 1\ldots 3N \tag{H.9}$$

Introducing matrix notation this becomes

$$\mathbf{p'} = \mathbf{P'D} \tag{H.10a}$$

where $\mathbf{P'}$ is a row matrix of dimension $3N - 6$. The transpose of $\mathbf{p'}$ is then

$$\mathbf{p} = (\mathbf{P'D})' = \mathbf{D'P} \tag{H.10b}$$

using Eq. (C.18). Thus the kinetic energy may be written

$$2T = \mathbf{P'DD'P} \tag{H.11}$$

Now $\mathbf{D}$ is related to $\mathbf{G}$ as defined by Eq. (8.125). To find this relation we rewrite (H.4) as

$$S_i = \sum_{j} D_{ij}\sqrt{m_i}\,\xi_i \tag{H.12}$$

using the definition of mass-weighted coordinates [Eq. (3.70)]. But by Eq. (8.123a)

$$S_i = \sum_j B_{ij}\xi_i \tag{8.123a}$$

so

$$D_{ij} = m_j^{-1/2}B_{ij} \tag{H.13}$$

Thus

$$(\mathbf{DD'})_{ik} = \sum_j^{3N} D_{ij}(\mathbf{D'})_{jk} = \sum_j D_{ij}D_{kj}$$

$$= \sum_j m_j^{-1}B_{ij}B_{kj}$$

$$= G_{ik} \tag{H.14}$$

Equation (H.11) then takes the form

$$2T = \mathbf{P'GP} \tag{H.15a}$$

or

$$2T = \sum_{i,k}^{3N-6} P_iG_{ik}P_k \tag{H.15b}$$

According to Hamilton's equation (A.16),

$$\dot{S}_l = \frac{\partial \mathcal{K}}{\partial P_l} = \frac{\partial T}{\partial P_l} \tag{H.16}$$

which becomes by use of (H.15b)

$$\dot{S}_l = \sum_k G_{lk}P_k \tag{H.17a}$$

or

$$\dot{\mathbf{S}} = \mathbf{GP} \tag{H.17b}$$

If $\mathbf{G}^{-1}$ exists (i.e., $|\mathbf{G}| \neq 0$)

$$\mathbf{P} = \mathbf{G}^{-1}\dot{\mathbf{S}} \tag{H.18a}$$

and

$$\mathbf{P'} = (\mathbf{G}^{-1}\dot{\mathbf{S}})'$$

$$= \dot{\mathbf{S}}'(\mathbf{G}^{-1})' = \dot{\mathbf{S}}'\mathbf{G}^{-1} \tag{H.18b}$$

since $\mathbf{G}^{-1}$ (and $\mathbf{G}$) are symmetric. Substituting Eqs. (H.18) into Eq. (H.11) we obtain the desired Eq. (H.1).

Equations (H.1) and (8.126) provide expressions for the kinetic and potential energies in terms of internal coordinates and the $\mathbf{F}$ and $\mathbf{G}$ matrix elements. Note that Eq. (8.126) may be written in matrix form as

$$2V = \mathbf{S'FS} \tag{H.19}$$

The internal coordinates have been related to the normal coordinates by Eqs. (8.127). The restrictions on **L** are that it must cause the kinetic and potential energies to have the forms given by Eqs. (8.60). These latter equations may be written in matrix form as

$$2T = \dot{\mathbf{Q}}'\dot{\mathbf{Q}} = \dot{\mathbf{Q}}'\mathbf{E}\dot{\mathbf{Q}} \tag{H.20}$$

and

$$2V = \mathbf{Q}'\mathbf{\Lambda}\mathbf{Q} \tag{H.21}$$

where **E** is the unit matrix and **Λ** is a diagonal matrix whose elements are the $3N - 6$ eigenvalues, λ_i. In all of what follows we assume that the zero roots of **Λ** have been removed, and that **Q** contains only the $3N - 6$ (or 5) genuine vibrational normal modes.

If Eq. (8.127) and its transpose are substituted into (H.1) and (H.19), we obtain

$$2T = (\mathbf{L}\dot{\mathbf{Q}})'\mathbf{G}^{-1}(\mathbf{L}\dot{\mathbf{Q}})$$

$$= \dot{\mathbf{Q}}'\mathbf{L}'\mathbf{G}^{-1}\mathbf{L}\dot{\mathbf{Q}} \tag{H.22}$$

and

$$2V = \mathbf{Q}'\mathbf{L}'\mathbf{F}\mathbf{L}\mathbf{Q} \tag{H.23}$$

Comparing the last four equations we find

$$\mathbf{L}'\mathbf{G}^{-1}\mathbf{L} = \mathbf{E} \tag{H.24}$$

$$\mathbf{L}'\mathbf{F}\mathbf{L} = \mathbf{\Lambda} \tag{H.25}$$

Post-multiplying (H.24) by $\mathbf{L}^{-1}\mathbf{G}$ gives

$$\mathbf{L}' = \mathbf{E}\mathbf{L}^{-1}\mathbf{G} = \mathbf{L}^{-1}\mathbf{G} \tag{H.26}$$

which when substituted into (H.25) gives

$$\mathbf{L}^{-1}\mathbf{G}\mathbf{F}\mathbf{L} = \mathbf{\Lambda} \tag{8.128}$$

or

$$\mathbf{G}\mathbf{F}\mathbf{L} = \mathbf{L}\mathbf{\Lambda} \tag{H.27}$$

This matrix equation leads to the $3N - 6$ equations given by Eq. (8.129) and the secular equation for the roots, Eq. (8.130). Pre- and post-multiplication of (H.27) by $\mathbf{L}^{-1}$ gives

$$\mathbf{L}^{-1}\mathbf{G}\mathbf{F} = \mathbf{\Lambda}\mathbf{L}^{-1} \tag{H.28a}$$

whose transpose is

$$\mathbf{F}\mathbf{G}(\mathbf{L}^{-1})' = (\mathbf{L}^{-1})'\mathbf{\Lambda} \tag{H.28b}$$

Note that this latter step uses the fact that $\mathbf{F}$ and $\mathbf{G}$ are symmetrical matrices. Equation (H.28b) leads to the alternate forms given by Eqs. (8.131).

A word of caution is perhaps in order in closing this appendix. Equation (8.128) looks like a standard matrix diagonalization, but it is not. First, we point out that $\mathbf{GF}$ is not necessarily symmetric, although both $\mathbf{G}$ and $\mathbf{F}$ are symmetric. Consequently $\mathbf{GF}$ is not diagonalized by a (unitary) similarity transformation. Indeed, $\mathbf{L}$ is not a unitary matrix in general. Thus while Eq. (8.128) is formally correct, the eigenvalues of $\mathbf{GF}$ (or $\mathbf{FG}$) must be found by solving the secular equations.

INDEX

Absorption coefficient, 6–7, 104–107, 168–172, 182–184, 238–239
  See also Extinction coefficient
Angular momentum, 40, 65–78, 531–533
Anharmonic oscillator, 232–236, 240–243
Anisotropy, 418, 420, 427–431
Asymmetric rotor, 184–204
Asymmetry parameter, 185
Atomic masses, 545–549
Average value, 10
Axial symmetry, see Cylindrical symmetry

Backward-wave oscillator, 151
Badger's rule, 307
Balanced bridge, 396
Band origin, 247
Barriers to internal rotation, see Internal rotation
Basis functions, 17, 478–479
Biradicals, 411
Birge–Sponer plot, 461–463
Bohr frequency relation, 1
Bohr magneton, 19, 397
  nuclear, 320
Born–Oppenheimer approximation, 28–33
Bose–Einstein particles, 20

Center of mass, 155
Central force potential, 289
Centrifugal distortion, 162–164, 178–179, 193
Character, of matrix representations, 125
  tables, 127–131, 535–544
Chemical shift, 327–331, 379–383
Chopper, 227
Class, 120–122

Coalescence temperature, 354
Combination bands, 280–283
Commutation relations, 18, 54–55, 63, 65–69, 81
Configuration interaction, 480–482
Conjugate elements, 120–122
Conjugate momenta and coordinates, 515
Continuous absorption, 459
Correlation diagram, 185, 470–471, 490
Coulomb integral, 500
Coulombic interaction, 25–26
Coupling constant, electron spin–nuclear spin, 398, 437–442
  nuclear spin–spin, 332, 337–338, 340–341, 383–387
Crystal field, 421, 424–427, 507–512
Curie law, 393
Cylindrical symmetry, 212, 432, 434

De Broglie wavelength, 4
Degrees of freedom, 34, 264
Detectors, 151, 229
Diamagnetic screening, 323–324, 380
Dipole–dipole constants in EPR, 442–444
Direct product, multiplication, 132–135
  representation, 132–135
Direction cosines, 78–83, 194
Displacement coordinates, mass-weighted, 42, 247–248, 252
  non mass-weighted, 230, 266–268, 284–285
Dissociation energy, 460, 476
Doppler broadening, 101–102

Einstein's coefficients, 97
Electric dipole moment, 95, 165–166, 193–194, 204–209

Electric field gradient, 210, 212–215
Electric quadrupole moment of nucleus, 209, 320–321, 352
Electrical anharmonicity, 240
Electromagnet, 314, 395
Electromagnetic spectrum, 4, 5
Electron configuration, of diatomic molecules, 472–478
   of polyatomic molecules, 489–491
Electronic energies, 448
   of conjugated hydrocarbons, 498–506
   of diatomic molecules, 450, 464–485
   of polyatomic molecules, 485–498
   of transition metal complexes, 506–512
Electron paramagnetic resonance energy levels, 397–400, 404, 410–418, 427–435
Electron paramagnetic resonance spectrometer, 395–397
Equivalent nuclei, 324, 326–327, 333–336, 403
Exchange narrowing, 437–438
Exchange processes, 352–360, 437
Expectation value, 10
Extinction coefficient, 7, 104–107, 497–498, 506

Far infrared, 225
Fast exchange, 353
Fermi contact coupling, 384, 398, 437–438
Fermi–Dirac particles, 20
Fermi resonance, 257–258
FG matrix theory, 291–305, 553–557
First-order NMR spectra, 333–337, 377–379
Fluorescence, 3
Force constant, 43, 230–236, 288–289, 292, 305–311
   stretching and bending, 306–307
Franck–Condon principle, 458–461
Free radical, 394
Frequency standard, 153
Full-rotational group, 508
Fundamental vibrational frequency, 231, 249, 279–280

g-factor, 403, 432–433, 443
Globar, 228
Great orthogonality theorem, 125–127
Group absorption wavelengths, 494–498
Group multiplication, 116–120
Group vibrational frequencies, 305–311
Gyromagnetic ratio, 320

Hamiltonian, electron dipole–dipole, 410
   electron spin-nuclear spin, 398
   electron Zeeman, 397–398, 429
   electronic, 30, 448
   H_2, 481
   H_2^+, 464
   nuclear, 39
   nuclear dipole–dipole, 388
   nuclear spin–spin, 332, 360, 373
   nuclear Zeeman, 322, 360, 373
   rotational, 39–41, 158, 177, 186
   vibrational, 41–46, 230, 247, 255
Hamiltonian function, 516
Hamiltonian matrix, 11–17, 57
Harmonic oscillator, 42, 46, 53–65, 230–232, 242–257
Heisenberg uncertainty principle, 20, 93, 317
Heitler–London method, see Valence-bond method
Hermitian, matrix, 13, 526
   operator, 519–520
Hertz, 4
Hindered rotation, see Internal rotation
Hooke's law, 46, 54, 230
Hot bands, 239
Hund's rules, 422, 473
Hyperfine splitting, in electron paramagnetic resonance, 398–400
   in nuclear magnetic resonance, 331–344, 368–371, 377–378
   in rotational spectra, 209–212

Identity element, 116
Inertial tensor, 40, 155–158
Infrared activity, 260, 279–282
Infrared spectrometer, 227–229
Intensity, of electronic spectra, 457–461, 497–498, 505–506
   of EPR spectra, 400–402, 418
   of NMR spectra, 327, 334–337, 345
   of rotational spectra, 164–173, 181–184, 193–195
   of vibrational spectra, 237–245
Interaction force constant, 288
Internal coordinates, 269–273, 283–305
   equivalent, 269–273
Internal rotation, 215–220, 355
Internuclear distance, 160, 220–223, 390
Inverse element, 116
Inverse matrix, 525

Isotope effect, nuclear magnetic resonance, 339–340
  rotational spectra, 173–177, 220–223
  vibrational spectra, 234
Isotropic coupling constant, *see* Coupling constant

Klystron, 151, 395
Kronecker delta, 11

Lagrangian function, 514–515
Lambert–Beer relation, 6
Lifetime, 93, 97
Ligand field, *see* Crystal field
Line shapes, 100–103, 353–354, 435–437
  Gaussian, 101–103
  Lorentzian, 102–103, 168, 351

Magic-T, 396
Magnetic dipole moment, electron, 18–19, 397
  nuclear, 19, 320
Matrices, 523–529
Matrix elements, 16
  angular momentum, 69–78
  asymmetric rotor, 191
  direction cosine, 78–83
  group theory, 138–140
  harmonic oscillator, 53–65
  NMR Hamiltonian, 361–364
Matrix mechanics, 10–17
Microwave cavity, 396
Microwave spectrometer, 151–153
Minimal basis set, 479
Molecular orbitals, 464–472, 478–483, 492–506
  antibonding, 470
  bonding, 469
  FEMO, 498
  Hückel, 500
  LCAO, 479
  SCF, 479
Molecular structure, 7, 220–223
Molecule-fixed axes, 35–36, 68, 155–158, 165
Moments of inertia, 40, 155–158
  effective, 160
Morse curve, 235–236

Nanometer, 4
Near infrared, 225

Neighbor anisotropy screening, 382
Nernst glower, 228
No-crossing rule, 471–472
Normal coordinate, 247–254, 264–269, 283–305
Normal frequencies, *see* Fundamental vibrational frequency
Normal mode, 247–254
Nuclear magnetic double resonance, 387–388
Nuclear magnetic energy levels, 321–327, 360–379, 388–389
Nuclear magnetic resonance spectrometer, 314–318
Nuclear moments, 318–321

Orbital angular momentum, 453, 465–466, 473
Orientation effects in NMR, 388–400
Orthogonality theorem, 125–127
Oscillator strength, 98
Overtone bands, 240–243, 280–281

Paramagnetic screening, 323–324, 380
Paramagnetism, 393–394
Pauli exclusion principle, 454, 472
Pauli matrices, 77
Phase sensitive detection, 152, 316
Phosphorescence, 3
Photon, 4
Physical constants, 551–552
Planck radiation law, 228–229
Point group, 112–113
Potential energy curve, 59–60, 216, 458–461, 467
Predissociation, 461
Pressure broadening, 168
Principal axes, dipolar, 412
  inertial, 153–158
Principal-axis transformation, 15
Probe, 315
Products of inertia, 40, 155–158
Progression, 456

Quadrupole coupling constant, 210–215
Quenching, 427–428

Reduced mass, 41, 45, 160, 230–236
Redundant coordinates, 36, 37, 43–46, 270

Relative coordinates, 26–27
Relaxation mechanisms, 337, 347–348, 351–360, 435–437
Relaxation time, 347–352, 435–437
Representations of groups, 122–135
  direct product, 132–135
  irreducible, 124–127, 131–132
  reducible, 122–125, 131–132, 134–135
  totally symmetric, 127
Resolution, infrared, 226–227
  microwave, 149–150
  NMR, 313, 316–318
Resonance integral, 500
Resonance spectra, 3, 313–445
Rigid rotor, 39–41, 158, 177–178, 186–193
Ring currents, 383
Rotational constant, 158, 178, 186–193
Rotational energy, separation of, 33–46
Rotational energy levels, asymmetric rotor, 184–193
  linear molecule, 158–164
  symmetric rotor, 177–181
Rotational spectra selection rules, asymmetric rotor, 193–194, 196–200
  linear molecule, 164–173
  symmetric rotor, 181
Rotation-vibration spectra, 243–247, 261–263

Sample spinning, 316–318
Satellite lines, 172
Saturation, 350
Schrödinger's equation, 10–17
Screening constant, 323–324, 379–383
Selection rules, 97
  electron paramagnetic resonance, 400–402, 404–405
  electronic spectra, 450–455, 486–489, 496–497, 505
  nuclear magnetic resonance, 322, 344–347
  rotational spectra, 164–168, 181, 193–194, 196–200, 202–203, 211, 218
  vibrational spectra, 237–238, 240–241, 243, 258–261, 278–283
Separated-atom states, 468–470
Similarity transformation, 11, 13, 15, 155–157, 188–190, 527
Slater determinant, 472
Space degeneracy, 159, 202
Space-fixed axes, 24, 25, 34–35, 67

Spherical rotor, 177
Spin, intrinsic, 17–20, 318–320, 397
Spin angular momentum, electron, 18, 397, 473
  fictitious, 432
  matrix elements, 75–77
  nuclear, 19, 318
Spin decoupling, 388
Spin density, 440
Spin Hamiltonian, 431–432
Spin-lattice relaxation, 348–352, 435, 436
Spin-orbit coupling, 20, 421, 428–429
Spin polarization, 438–439
Spin selection rule, 454
Spin–spin interaction, 331–332, 360
Spin–spin relaxation, 351–352, 435–436
Spontaneous emission, 3, 97
Square wave modulation, 152
Stark, components, 150, 152, 208–209
  effect, 47, 202–203, 204–209
  modulation, 152
Stationary state, 2
Statistical weight, 182
Stimulated emission, 3, 459
Structure, *see* Molecular structure
Subgroup, 120
Substitution coordinates, 175–177, 220–223
Symmetric rotor, 177–184
Symmetry, of asymmetric rotor states, 196–198
  of electronic states, 452–453, 486–489, 507–510
  of F and G matrices, 291–299
  of normal modes, 263–273
  of vibrational energy states, 273–278
Symmetry coordinate, 291, 293–305
Symmetry elements, *see* Symmetry operators
Symmetry functions, 140–146, 272–273, 502–503
Symmetry operators, 111–116, 145–146

Term symbols for diatomics, 472–478
Time-dependent perturbation theory, 86–90
Trace of matrix, 527
Transformation operators, 113–116
Transition metal ions, EPR spectra of, 420–435
  in crystal field, 421–427

Transition moment, 92
  electric dipole, 96, 167, 181, 194, 237, 278, 450, 452
  magnetic dipole, 345, 400
Transition probability, 86–98
  electric dipole, 95–98
  electric quadrupole, 99
  magnetic dipole, 99
  in periodic perturbation, 90–95
Translational energy, 23–28
Transpose matrix, 525
Triplet-state radicals, 410–419

Unitary matrix, 11, 188, 526
United atom states, 466

Valence-bond method, 483–485
Valence-bond potential function, 288, 301, 304
Van Vleck–Weisskopf lineshape, 105–106

Vibrational band structure, 455–464
Vibrational energy, separation of, 33–46
Vibrational energy levels, diatomic molecule, 230–236
  polyatomic molecule, 254–258, 291–305
Vibration-rotation coupling, 38–40, 160–161, 178, 217
Vibronic states, 451–452

Wang transformation, 189
Wave mechanics, 10–17
Waveguide, 151
Wavemeter, 152
Wavenumber, 4

Zeeman effect, electron, 397–400
  nuclear, 321–327
Zero-field splitting, 402, 417–418
Zero-point vibrations, 220